自然资源与生态文明译丛

生态系统服务中的对地观测

［西］多明戈·阿尔卡拉斯-塞古拉
［阿根廷］卡洛斯·马塞洛·迪贝拉　编
［阿根廷］朱丽叶·维罗妮卡·斯特拉施诺伊

薛超　蒋捷　译

EARTH OBSERVATION OF ECOSYSTEM SERVICES

Domingo Alcaraz-Segura　Carlos Marcelo Di Bella
Julieta Veronica Straschnoy

商务印书馆　CRC Press
Taylor & Francis Group

EARTH OBSERVATION OF ECOSYSTEM SERVICES

1st Edition

9781138073920

edited by Alcaraz-Segura, Domingo; Bella, Carlos Marcelo Di; Straschnoy, Julieta Veronica

All Rights Reserved.

Authorized translation from the English language edition published by CRC Press, a member of the Taylor & Francis Group, LLC.

Copies of this book sold without a Taylor & Francis sticker on the cover are unauthorized and illegal.

中文版经泰勒·弗朗西斯出版集团授权。

根据泰勒·弗朗西斯出版集团旗下 CRC 出版社 2014 年平装本译出。

本书封面贴有 Taylor & Francis 公司防伪标签，无标签图书为非法且未经授权的。

"自然资源与生态文明"译丛
"自然资源保护和利用"丛书
总序

（一）

新时代呼唤新理论，新理论引领新实践。中国当前正在进行着人类历史上最为宏大而独特的理论和实践创新。创新，植根于中华优秀传统文化，植根于中国改革开放以来的建设实践，也借鉴与吸收了世界文明的一切有益成果。

问题是时代的口号，"时代是出卷人，我们是答卷人"。习近平新时代中国特色社会主义思想正是为解决时代问题而生，是回答时代之问的科学理论。以此为引领，亿万中国人民驰而不息，久久为功，秉持"绿水青山就是金山银山"理念，努力建设"人与自然和谐共生"的现代化，集聚力量建设天蓝、地绿、水清的美丽中国，为共建清洁美丽世界贡献中国智慧和中国力量。

伟大时代孕育伟大思想，伟大思想引领伟大实践。习近平新时代中国特色社会主义思想开辟了马克思主义新境界，开辟了中国特色社会主义新境界，开辟了治国理政的新境界，开辟了管党治党的新境界。这一思想对马克思主义哲学、政治经济学、科学社会主义各个领域都提出了许多标志性、引领性的新观点，实现了对中国特色社会主义建设规律认识的新跃升，也为新时代自然资源治理提供了新理念、新方法、新手段。

明者因时而变，知者随事而制。在国际形势风云变幻、国内经济转型升级的背景下，习近平总书记对关系新时代经济发展的一系列重大理论和实践问题进行深邃思考和科学判断，形成了习近平经济思想。这一思想统筹人与自然、经济与社会、经济基础与上层建筑，兼顾效率与公平、局部与全局、当前与长远，为当前复杂条件下破解发展难题提供智慧之钥，也促成了新时代经济发展举世瞩目的辉煌成就。

生态兴则文明兴——"生态文明建设是关系中华民族永续发展的根本大计"。在新时代生态文明建设伟大实践中，形成了习近平生态文明思想。习近平生态文明思想是对马克思主义自然观、中华优秀传统文化和我国生态文明实践的升华。马克思主义自然观中对人与自然辩证关系的诠释为习近平生态文明思想构筑了坚实的理论基础，中华优秀传统文化中的生态思想为习近平生态文明思想提供了丰厚的理论滋养，改革开放以来所积累的生态文明建设实践经验为习近平生态文明思想奠定了实践基础。

自然资源是高质量发展的物质基础、空间载体和能量来源，是发展之基、稳定之本、民生之要、财富之源，是人类文明演进的载体。在实践过程中，自然资源治理全力践行习近平经济思想和习近平生态文明思想。实践是理论的源泉，通过实践得出真知：发展经济不能对资源和生态环境竭泽而渔，生态环境保护也不是舍弃经济发展而缘木求鱼。只有统筹资源开发与生态保护，才能促进人与自然和谐发展。

是为自然资源部推出"自然资源与生态文明"译丛、"自然资源保护和利用"丛书两套丛书的初衷之一。坚心守志，持之以恒。期待由见之变知之，由知之变行之，通过积极学习而大胆借鉴，通过实践总结而理论提升，建构中国自主的自然资源知识和理论体系。

（二）

如何处理现代化过程中的经济发展与生态保护关系，是人类至今仍然面临的难题。自《寂静的春天》(蕾切尔·卡森，1962)、《增长的极限》(德内拉·梅多斯，1972)、《我们共同的未来》(布伦特兰报告，格罗·哈莱姆·布伦特兰，1987)这些经典著作发表以来，资源环境治理的一个焦点就是破解保护和发展的难题。从世界现代化思想史来看，如何处理现代化过程中的经济发展与生态保护关系，是人类至今仍然面临的难题。"自然资源与生态文明"译丛中的许多文献，运用技术逻辑、行政逻辑和法理逻辑，从自然科学和社会科学不同视角，提出了众多富有见解的理论、方法、模型，试图破解这个难题，但始终没有得出明确的结论性认识。

全球性问题的解决需要全球性的智慧，面对共同挑战，任何人任何国家都无法独善其身。2019年4月习近平总书记指出，"面对生态环境挑战，人类是一荣俱荣、一损俱损的命运共同体，没有哪个国家能独善其身。唯有携手合作，我们才能有效应对气候变化、海洋污染、生物保护等全球性环境问题，实现联合国2030年可持续发展目标"。共建人与自然生命共同体，掌握国际社会应对资源环境挑战的经验，加强国际绿色合作，推动"绿色发展"，助力"绿色复苏"。

文明交流互鉴是推动人类文明进步和世界和平发展的重要动力。数千年来，中华文明海纳百川、博采众长、兼容并包，坚持合理借鉴人类文明一切优秀成果，在交流借鉴中不断发展完善，因而充满生机活力。中国共产党人始终努力推动我国在与世界不同文明交流互鉴中共同进步。1964年2月，毛主席在中央音乐学院学生的一封信上批示说"古为今用，洋为中用"。1992年2月，邓小平同志在南方谈话中指出，"必须大胆吸收和借鉴人类社会创造的一切文明成果"。2014年5月，习近平总书记在召开外国专家座谈会上强调，"中国要永远做一个学习大国，不论发展到什么水平都虚心向世界各国人民学习"。

"察势者明，趋势者智"。分析演变机理，探究发展规律，把握全球自然资源治理的态势、形势与趋势，着眼好全球生态文明建设的大势，自觉以回答中国之问、世界之问、人民之问、时代之问为学术己任，以彰显中国之路、中国之治、中国之理为思想追求，在研究解决事关党和国家全局性、根本性、关键性的重大问题上拿出真本事、取得好成果。

是为自然资源部推出"自然资源与生态文明"译丛、"自然资源保护和利用"丛书两套丛书的初衷之二。文明如水，润物无声。期待学蜜蜂采百花，问遍百家成行家，从全球视角思考责任担当，汇聚全球经验，破解全球性世纪难题，建设美丽自然、永续资源、和合国土。

（三）

2018年3月，中共中央印发《深化党和国家机构改革方案》，组建自然资源部。自然资源部的组建是一场系统性、整体性、重构性变革，涉及面之广、难度之大、问题之多，前所未有。几年来，自然资源系统围绕"两统一"核心职责，不负重托，不辱使命，开创了自然资源治理的新局面。

自然资源部组建以来，按照党中央、国务院决策部署，坚持人与自然和谐共生，践行绿水青山就是金山银山理念，坚持节约优先、保护优先、自然恢复为主的方针，统筹山水林田湖草沙冰一体化保护和系统治理，深化生态文明体制改革，夯实工作基础，优化开发保护格局，提升资源利用效率，自然资源管理工作全面加强。一是，坚决贯彻生态文明体制改革要求，建立健全自然资源管理制度体系。二是，加强重大基础性工作，有力支撑自然资源管理。三是，加大自然资源保护力度，国家安全的资源基础不断夯实。四是，加快构建国土空间规划体系和用途管制制度，推进国土空间开发保护格局不断优化。五是，加大生态保护修复力度，构筑国家生态安全屏障。六是，强化自然资源节约集约利用，促进发展方式绿色转型。七是，持续推进自然资源法治建设，自然资源综合监管效能逐步提升。

当前正值自然资源综合管理与生态治理实践的关键期，面临着前所未有的知识挑战。一方面，自然资源自身是一个复杂的系统，山水林田湖草沙等不同资源要素和生态要素之间的相互联系、彼此转化以及边界条件十分复杂，生态共同体运行的基本规律还需探索。自然资源既具系统性、关联性、实践性和社会性等特征，又有自然财富、生态财富、社会财富、经济财富等属性，也有系统治理过程中涉及资源种类多、学科领域广、系统庞大等特点。需要遵循法理、学理、道理和哲理的逻辑去思考，需要斟酌如何运用好法律、经济、行政等政策路径去实现，需要统筹考虑如何采用战略部署、规划引领、政策制定、标准规范的政策工具去落实。另一方面，自然资源综合治理对象的复杂性、系统性特点，对科研服务支撑决策提出了理论前瞻性、技术融合性、知识交融性的诉求。例如，自然资源节约集约利用的学理创新是什么？动态监测生态系统稳定性状况的方法有哪些？如何评估生态保护修复中的功能次序？等等不一而足，一系列重要领域的学理、制度、技术方法仍待突破与创新。最后，当下自然资源治理实践对自然资源与环境经济学、自然资源法学、自然地理学、城乡规划学、生态学与生态经济学、生态修复学等学科提出了理论创新的要求。

中国自然资源治理体系现代化应立足国家改革发展大局，紧扣"战略、战役、战术"问题导向，"立时代潮头、通古今之变，贯通中西之间、融会文理之壑"，在"知其然知其所以然，知其所以然的所以然"的学习研讨中明晰学理，在"究其因，思其果，寻其路"的问题查摆中总结经验，在"知识与技术的更新中，自然科学与社会科学的交融中"汲取智慧，在国际理论进展与实践经验的互鉴中促进提高。

是为自然资源部推出"自然资源与生态文明"译丛、"自然资源保护和利用"丛书这两套丛书的初衷之三。知难知重，砥砺前行。要以中国为观照、以时代为观照，立足中国实际，从学理、哲理、道理的逻辑线索中寻找解决方案，不断推进自然资源知识创新、理论创新、方法创新。

（四）

文明互鉴始于译介，实践蕴育理论升华。自然资源部决定出版"自然资源与生态文明"译丛、"自然资源保护和利用"丛书系列著作，办公厅和综合司统筹组织实施，中国自然资源经济研究院、自然资源部咨询研究中心、清华大学、自然资源部海洋信息中心、自然资源部测绘发展研究中心、商务印书馆、《海洋世界》杂志等单位承担完成"自然资源与生态文明"译丛编译工作或提供支撑。自然资源调查监测司、自然资源确权登记局、自然资源所有者权益司、国土空间规划局、国土空间用途管制司、国土空间生态修复司、海洋战略规划与经济司、海域海岛管理司、海洋预警监测司等司局组织完成"自然资源保护和利用"丛书编撰工作。

第一套丛书"自然资源与生态文明"译丛以"创新性、前沿性、经典性、基础性、学科性、可读性"为原则，聚焦国外自然资源治理前沿和基础领域，从各司局、各事业单位以及系统内外院士、专家推荐的书目中遴选出十本，从不同维度呈现了当前全球自然资源治理前沿的经纬和纵横。

具体包括：《自然资源与环境：经济、法律、政治和制度》，《环境与自然资源经济学：当代方法》（第五版），《自然资源管理的重新构想：运用系统生态学范式》，《空间规划中的生态理性：可持续土地利用决策的概念和工具》，《城市化的自然：基于近代以来欧洲城市历史的反思》，《城市生态学：跨学科系统方法视角》，《矿产资源经济（第一卷）：背景和热点问题》，《海洋和海岸带资源管理：原则与实践》，《生态系统服务中的对地观测》，《负排放技术和可靠封存：研究议程》。

第二套丛书"自然资源保护和利用"丛书基于自然资源部组建以来开展生态文明建设和自然资源管理工作的实践成果，聚焦自然资源领域重大基础性问题和难点焦点问题，经过多次论证和选题，最终选定七本（此次先出版五本）。在各相关研究单位的支撑下，启动了丛书撰写工作。

具体包括：自然资源确权登记局组织撰写的《自然资源和不动产统一确权登记理论与实践》，自然资源所有者权益司组织撰写的《全民所有自然资源资产所有者权益管理》，自然资源调查监测司组织撰写的《自然资源调查监测实践与探索》，国土空间规划局组织撰写的《新时代"多规合一"国土空间规划理论与实践》，国土空间用途管制司组织撰写的《国土空间用途管制理论与实践》。

"自然资源与生态文明"译丛和"自然资源保护和利用"丛书的出版，正值生态文明建设进程中自然资源领域改革与发展的关键期、攻坚期、窗口期，愿为自然资源管理工作者提供有益参照，愿为构建中国特色的资源环境学科建设添砖加瓦，愿为有志于投身自然资源科学的研究者贡献一份有价值的学习素材。

百里不同风，千里不同俗。任何一种制度都有其存在和发展的土壤，照搬照抄他国制度行不通，很可能画虎不成反类犬。与此同时，我们探索自然资源治理实践的过程，也并非一帆风顺，有过积极的成效，也有过惨痛的教训。因此，吸收借鉴别人的制度经验，必须坚持立足本国、辩证结合，也要从我们的实践中汲取好的经验，总结失败的教训。我们推荐大家来读"自然资源与生态文明"译丛和"自然资源保护和利用"丛书中的书目，也希望与业内外专家同仁们一道，勤思考，多实践，提境界，在全面建设社会主义现代化国家新征程中，建立和完善具有中国特色、符合国际通行规则的自然资源治理理论体系。

在两套丛书编译撰写过程中，我们深感生态文明学科涉及之广泛，自然资源之于生态文明之重要，自然科学与社会科学关系之密切。正如习近平总书记所指出的，"一个没有发达的自然科学的国家不可能走在世界前列，一个没有繁荣的哲学社会科学的国家也不可能走在世界前列"。两套丛书涉及诸多专业领域，要求我们既要掌握自然资源专业领域本领，又要熟悉社会科学的基础知识。译丛翻译专业词汇多、疑难语句多、习俗俚语多，背景知识复杂，丛书撰写则涉及领域多、专业要求强、参与单位广，给编译和撰写工作带来不小的挑战，丛书成果难免出现错漏，谨供读者们参考交流。

编写组

前　　言

趋势监测对于更好地了解环境变化愈加重要。地球是一个生命系统，物理与生物过程间存在多种相互作用，并不断改变地球景观。除了这些自然过程之外，由于人类与大气、海洋与陆地流动之间存在各种各样的相互作用，人类活动在解释环境过程中发挥了非常重要的作用。此等相互作用存在两个互补的因素：人类如何影响生态过程及生态过程如何影响人类。地球是我们的家园，我们依靠自然资源找到食物和住所，甚至是我们的精神指导。几千年来，我们以不同方式使用这些资源，认为它们可无限使用。我们认为地球体积巨大、资源多样，足以无条件满足人类的需要。然而，我们现在意识到，人类的足迹几乎无处不在。我们逐渐开始将自然资源视为宝贵和有限的商品。我们的家园变得太小了，至少我们是这样认为。我们可以忽视目前的生活方式可能给我们带来潜在的灾难，或者我们可以限制自己的发展，但首先，我们应更好地理解这个问题：此等看法是否属实，以及预期最新形势的发展趋势。

卫星对地观测已成为了解许多自然过程的全球视角，是监测其趋势不可缺少的工具。尽管卫星影像的历史档案极少（25～30年前才发射首批可靠卫星），但它们提供了诸多关键信息，涉及热带森林砍伐、土地利用趋势、水质、作物产量、冰雪范围、海岸带过程、海洋、云和气溶胶分布及许多其他对描述全球系统至关重要的变量。

生态系统服务是指对人类社会产生重大影响的所有自然过程。近几十年来，经济评价一直将其作为重要考量要素。大自然为人类提供了广泛服务，从水源到木材到牧场，从狩猎到捕鱼，从生物多样性保护到积雪，从碳储存到土壤侵蚀保护，从软木到坚果到蘑菇，涉及方方面面。本书涵盖了其中许多方面，涉及的方法关联性极高，有助于更新资料从而更好地理解如何在监测生态系统服务的过程中使用卫星影像。本书可能是第一本涵盖这些主题的书籍，并对这个极具创新的研究领域进行了全面分析,不久将会得以推广。本书邀请了各领域的作者探讨了非常广泛的主题,对此深表钦佩与感谢。

本书将为生态学家、水文学家、生物学家、地理学家和许多其他环境科学家提供大量参考材料。学者们可进一步依靠不断增多的卫星影像,更好地了解和监测我们脆弱的环境。

美国阿尔卡拉大学地理学教授　楚维科·埃米利奥(Emilio Chuvieco)

目　录

作者简介 ·· xv
撰稿人 ·· xvii
审稿人 ·· xxxi

第一部分　导言

第一章　利用卫星传感器监测生态系统服务的全球展望 ······················· 3
　　　　D. 阿尔卡拉斯-塞古拉　C.M. 迪贝拉

第二部分　碳循环相关生态系统服务

第二章　利用遥感技术评估碳循环提供的生态系统服务 ······················· 19
　　　　J.M. 帕鲁埃洛　M. 瓦莱霍斯

第三章　碳吸收相关陆地生态系统服务的光合胁迫估算最新进展 ············ 41
　　　　M.F. 加布尔斯基　I. 菲莱利亚　J. 佩努埃拉斯

第四章　北方高纬度系统碳循环池和过程的对地观测 ···························· 64
　　　　H.E. 爱泼斯坦

第五章　监测牧草生产的生态系统服务 ··· 88
　　　　J.G.N. 伊里萨里　M. 奥斯特赫尔德　M. 奥雅扎巴尔　J.M. 帕鲁埃洛　M. 杜兰特

第六章　从光能利用效率模型估算碳收益服务中的缺失差距 ·················· 106
　　　　A.J. 卡斯特罗·马丁内斯　J.M. 帕鲁埃洛　D. 阿尔卡拉斯-塞古拉　J. 卡贝洛　M. 奥亚尔扎巴尔　E. 洛佩斯-卡里克

第七章　亚马孙热带森林生物质燃烧排放估算·················· 126
　　　　Y.E. 岛袋　G. 佩雷拉　F.S. 卡多佐　R. 斯托克勒　S.R. 弗雷塔斯　S.M.C. 库拉

第三部分　生物多样性相关生态系统服务

第八章　基于对地观测的物种多样性评价和监测·················· 153
　　　　N. 费尔南德斯
第九章　利用遥感对国家公园网络的生态系统服务功能多样性和碳保护战略进行评估·················· 179
　　　　J. 卡贝洛　P. 劳伦　A. 雷耶斯　D. 阿尔卡拉斯-塞古拉
第十章　利用对地观测数据在流域尺度上分析河岸植被对河流生态完整性的影响·················· 199
　　　　T. 托莫斯　K. 范洛伊　P. 科苏特　B. 维伦纽夫　Y. 苏雄

第四部分　水循环相关生态系统服务

第十一章　遥感水文生态系统服务评价·················· 227
　　　　C. 卡瓦略-桑托斯　B. 马科斯　J. 埃斯皮尼亚·马克斯　D. 阿尔卡拉斯-塞古拉　L. 海因　J. 洪拉多
第十二章　用于生态系统服务评估的水文模型的遥感数据同化·················· 256
　　　　J. 埃雷罗　A. 米拉尔斯　C. 阿吉拉尔　F.J. 博内特　M.J. 波罗
第十三章　基于卫星绿度异常和时间动力学的地下水生态系统依赖探测·················· 277
　　　　S. 孔特雷拉斯　D. 阿尔卡拉斯-塞古拉　B. 斯坎伦　E.G. 乔巴吉
第十四章　地表土壤水分遥感监测：在生态系统过程和尺度效应中的应用·················· 296
　　　　M.J. 保罗　M.P. 冈萨雷斯-杜戈　C. 阿吉拉尔　A. 安德鲁
第十五章　积雪作为山区生态系统服务的关键要素：有效监测计划的设计思路·················· 320
　　　　F.J. 博内特　A. 米拉尔斯　J. 埃雷罗

第五部分　地表能量平衡相关生态系统服务

第十六章　气候调节服务的表征与监测·················· 341
　　　　　　D. 阿尔卡拉斯-塞古拉　E. H. 博柏利　O. V. 穆勒　J. M. 帕雷洛

第十七章　与能量平衡相关的生态系统服务：以湿地反射能量为例 ········ 369
　　　　　　C. M. 迪贝拉　M. E. 贝格特

第十八章　能量平衡和蒸散：一种评价生态系统服务的遥感方法 ········ 390
　　　　　　V. A. 马尔切西尼　J. P. 古尔施曼　J. A. 索布里诺

第十九章　城市热岛效应······················· 407
　　　　　　J. A. 索布里诺　R. 奥尔特拉-卡里奥　G. 索里亚

第六部分　生态系统服务的其他维度

第二十章　生态系统服务评估的多维方法··············· 431
　　　　　　A. J. 卡斯特罗·马丁内斯　M. 加西亚-略伦特　B. 马丁-洛佩斯
　　　　　　I. 帕洛莫　I. 伊涅斯塔-阿兰迪亚

索引···································· 459

译后记··································· 474

彩插

作者简介

多明戈·阿尔卡拉斯-塞古拉：1978年出生于西班牙阿尔梅里亚。2000年在阿尔梅里亚大学(University of Almería)获得环境科学学士学位，2005年获得博士学位。曾在弗吉尼亚大学、布宜诺斯艾利斯大学、德克萨斯大学奥斯汀分校、马里兰大学和西班牙国家科学研究委员会
（多尼亚纳生物站）担任博士后职位。目前担任格拉纳达大学(University of Granada)（西班牙）教授和安达卢西亚全球变化评估与监测中心(Andalusian Center for the Assessment and Monitoring of Global Change)助理研究员。他教授植物学、地球植物学、全球变化、生物多样性保护和人类福祉及卫星影像时间序列分析课程。目前的研究领域为生物多样性的环境控制、土地覆盖和土地利用变化对生态系统功能与服务及水文气候的影响，以及全球变化对保护区影响监测和预警系统的开发。他的研究以实地调查、遥感技术、时间序列分析和地理信息系统为基础。

卡洛斯·马塞洛·迪贝拉：1969年出生于阿根廷布宜诺斯艾利斯。1994年以农学家身份毕业于阿根廷布宜诺斯艾利斯大学(University of Buenos Aires)农学系，2002年获得巴黎-格里农国立农业研究院博士学位。自1998年起，他在阿根廷气候与水研究所(暨阿根廷国家农业技术研究所担任研究员，2006年起在阿根廷国家
科学与技术研究理事会担任研究员。他还担任应用于自然资源和农业生产研究的遥感与地理信息系统方向(阿根廷布宜诺斯艾利斯大学农学院暨阿尔贝托·索利亚诺研究生院)的研究生导师。他的研究重点是遥感应用与发展,以及地理信息系统在自然资源和农业生态系统研究、管理与监测中的应用。

朱丽叶·维罗妮卡·斯特拉施诺伊:1974年出生于阿根廷布宜诺斯艾利斯。1997年以数学和物理教师身份毕业,于2002年获得环境管理学士学位,目前正在布宜诺斯艾利斯商业与社会科学大学(University of Business and Social Sciences)攻读环境研究硕士学位。自2003年起,她一直是阿根廷气候与水研究所(暨阿根廷国家农业技术研究所)研究人员。参与了农业生态系统长期观测领域不同国家和国际项目的开发。

撰 稿 人

C. 阿吉拉尔
西班牙
  科尔多瓦大学
  农产品管理国际卓越园区
  安达卢西亚地球系统研究所
  河流动力学和水文研究组

C. Aguilar
Spain
  University of Córdoba
  Agrifood Campus of International Excellence
  Andalusian Institute of Earth System Research
  Fluvial Dynamics and Hydrology Research Group

D. 阿尔卡拉斯-塞古拉
西班牙
  格拉纳达大学
  自然科学学院
  植物学系
西班牙
  阿尔梅里亚大学
  安达卢西亚全球变化评估和监测中心（CAESCG）

D. Alcaraz-Segura
Spain
  University of Granada
  Faculty of Sciences
  Botany Department
Spain
  University of Almería
  Andalusian Center for the Assessment and Monitoring of Global Change

A. 安德鲁
西班牙
  安达卢西亚农业和渔业研究和培训研究所

A. Andreu
Spain
  Andalusian Institute for Agricultural and Fisheries Research and Training

M. E. 贝格特
阿根廷

M. E. Beget
Argentina

国家农业技术研究所 National Institute of Agricultural Technology

自然资源研究中心气候与水研究所 Climate and Water Institute Research Center of Natural Resources

E. H. 博柏利   **E. H. Berbery**

美国   United States

   马里兰大学   University of Maryland

   地球系统科学跨学科中心   Earth System Science Interdisciplinary Center

   气候和卫星合作研究所   Cooperative Institute for Climate and Satellites

F. J. 博内特   **F. J. Bonet**

西班牙   Spain

西班牙格拉纳达大学   University of Granada

安达卢西亚地球系统研究所   Andalusian Institute for Earth System Research

陆地生态研究组   Terrestrial Ecology Research Group

J. 卡贝罗   **J. Cabello**

西班牙   Spain

   西班牙阿尔梅里亚大学   University of Almería

   安达卢西亚全球变化评估和监测中心   Andalusian Center for the Assessment and Monitoring of Global Change

   生物与地质系   Department of Biology and Geology

F. S. 卡多佐   **F. S. Cardozo**

巴西   Brazil

   国家空间研究所   National Institute for Space Research

   巴西圣何塞多斯坎波斯   São José dos Campos, Brazil

C. 卡瓦略-桑托斯   **C. Carvalho-Santos**

葡萄牙   Porto

波尔图大学	University of Porto
生物多样性与遗传资源科学研究中心	Department of Biology Faculty of Sciences and Research Centre in Biodiversity and Genetic Resources
荷兰	Netherlands
瓦赫宁根大学	Wageningen University
环境系统分析组	Environmental Systems Analysis Group

A. J. 卡斯特罗·马丁内斯   **A. J. Castro Martínez**

美国	United States
俄克拉何马大学	University of Oklahoma
俄克拉何马州生物调查局	Oklahoma Biological Survey
西班牙	Spain
阿尔梅里亚大学	University of Almería Almería
植物生物学与生态学系	Department of Plant Biology and Ecology
安达卢西亚全球变化评估和监测中心	Andalusian Center for the Assessment and Monitoring of Global Change

S. 孔特雷拉斯   **S. Contreras**

西班牙	Spain
西班牙科学研究理事会	Spanish Council for Scientific Research
塞古拉土壤学和应用生物学中心	Centre of Pedology and Applied Biology of Segura

S. M. C. 库拉   **S. M. C. Coura**

| 巴西 | Brazil |
| 国家空间研究所 | National Institute for Space Research |

C. M. 迪贝拉   **C. M. Di Bella**

| 阿根廷 | Argentina |
| 国家科学技术研究委员会 | National Scientific and Technical Research Council |

国家农业技术研究所 National Institute of Agricultural Technology

自然资源研究中心气候与水研究所 Climate and Water Institute Research Center of Natural Resources

M. 杜兰特

阿根廷

 国家农业技术研究所

M. Durante

Argentina

 National Institute of Agricultural Technology

H. E. 爱泼斯坦

美国

 弗吉尼亚大学

 环境科学系

H. E. Epstein

United States

 University of Virginia

 Department of Environmental Sciences

J. 埃斯皮尼亚·马克斯

葡萄牙

 波尔图大学

 自然科学学院

 环境与空间规划

 地质中心和地球科学系

J. Espinha Marques

Porto

 University of Porto

 Faculty of Sciences

 Environment and Spatial Planning

 Geology Centre and Department of Geosciences

N. 费尔南德斯

西班牙

 科学研究理事会

 多尼亚纳生物站

N. Fernández

Spain

 National Research Council

 Doñana Biological Station

I. 菲莱利亚

西班牙

 西班牙科学研究理事会

 生态研究和林业应用中心

I. Filella

Spain

 National Research Council

 Center for Ecological Research and Forestry Applications

撰 稿 人

S. R. 弗雷塔斯
巴西
    国家空间研究所

M. F. 加布尔斯基
阿根廷
    布宜诺斯艾利斯大学
    农业学院
    国家科学技术研究委员会

    农业植物生理生态学研究所

M. 加西亚-略伦特
西班牙
    马德里卡洛斯三世大学
    社会分析学系
    社会学与环境研究领域
西班牙
    西班牙马德里自治大学
    生态系
    社会生态系统实验室

M. P. 冈萨雷斯-杜戈
西班牙
    安达卢西亚农业和渔业研究和培训研究所

J. P. 格尔施曼

S. R. Freitas
Brazil
    National Institute for Space Research

M. F. Garbulsky
Argentina
    University of Buenos Aires
    School of Agriculture
    National Scientific and Technical Research Council
    Institute for Agricultural Plant Physiology and Ecology

M. García-Llorente
Spain
    Carlos Ⅲ University of Madrid
    Social Analysis Department
    Sociology and the Environment Research Area
Spain
    Autonomous University of Madrid
    Department of Ecology
    Social-Ecological Systems Laboratory

M. P. González-Dugo
Spain
    Andalusian Institute for Agricultural and Fisheries Research and Training

J. P. Guerschman

澳大利亚　　　　　　　　　　　　　　Australia
　　澳大利亚联邦科学和工业研究组织　　　　Commonwealth Scientific and Industrial Research Organisation Land and Water
　　土地和水部门

L. 海因　　　　　　　　　　　　　　**L. Hein**
荷兰　　　　　　　　　　　　　　　　Netherlands
　　瓦赫宁根大学　　　　　　　　　　　　Wageningen University
　　环境系统分析小组　　　　　　　　　　Environmental Systems Analysis Group

J. 埃雷罗　　　　　　　　　　　　　**J. Herrero**
西班牙　　　　　　　　　　　　　　　Spain
　　西班牙格拉纳达大学　　　　　　　　　University of Granada
　　安达卢西亚地球系统研究所　　　　　　Andalusian Institute of Earth System Research
　　河流动力学和水文研究组　　　　　　　Fluvial Dynamics and Hydrology Research Group

J. 洪拉多　　　　　　　　　　　　　**J. Honrado**
葡萄牙　　　　　　　　　　　　　　　Porto
　　波尔图大学　　　　　　　　　　　　University of Porto
　　生物多样性和遗传资源科学研究中心　　Faculty of Sciences and Research Centre in Biodiversity and Genetic Resources
　　生物系　　　　　　　　　　　　　　Department of Biology

I. 伊涅斯塔-阿兰迪亚　　　　　　　　**I. Iniesta-Arandia**
西班牙　　　　　　　　　　　　　　　Spain
　　马德里自治大学　　　　　　　　　　　Autonomous University of Madrid
　　生态系　　　　　　　　　　　　　　Department of Ecology
　　社会生态系统实验室　　　　　　　　　Social-Ecological Systems Laboratory

J. G. N. 伊里萨里　　　　　　　　　**J. G. N. Irisarri**

| 撰 稿 人 | xxiii |

阿根廷　　　　　　　　　　　　　　Argentina
　布宜诺斯艾利斯大学　　　　　　　　School of Agriculture University of Buenos Aires
　农业学院　　　　　　　　　　　　　Institute for Agricultural
　区域分析和遥感实验室　　　　　　　Regional Analysis Laboratory and Remote Sensing

　国家科学技术研究委员会　　　　　　National Scientific and Technical Research Council
　农业植物生理生态学研究所　　　　　Plant Physiology and Ecology

E. G. 乔巴吉　　　　　　　　　　　**E. G. Jobbágy**
阿根廷　　　　　　　　　　　　　　Argentina
　国家科学技术研究委员会　　　　　　National Scientific and Technical Research Council

　圣路易斯应用数学研究所　　　　　　San Luis Institute of Applied Mathematics
　环境研究小组　　　　　　　　　　　Environmental Research Group

P. 科苏特　　　　　　　　　　　　**P. Kosuth**
法国　　　　　　　　　　　　　　　France
　国家环境与农业科学技术研究所　　　National Research Institute of Science and Technology for Environment and Agriculture

　环境和土地管理对地观测和地理信息小组　Earth Observation and Geo-Information for Environment and Land Management Unit

E. 洛佩斯-卡里克　　　　　　　　　**E. López-Carrique**
西班牙　　　　　　　　　　　　　　Spain
　阿尔梅里亚大学　　　　　　　　　　University of Almería
　安达卢西亚全球变化评估和监测中心　Andalusian Center for the Assessment and Monitoring of Global Change

P. 洛伦索　　　　　　　　　　　　**P. Lourenço**
西班牙　　　　　　　　　　　　　　Spain

阿尔梅里亚大学 University of Almería
生物地质系 Department of Biology and Geology
安达卢西亚全球变化评估和监测中心 Andalusian Center for the Assessment and Monitoring of Global Change

V. A. 马尔切西尼 **V. A. Marchesini**
阿根廷 Argentina
  布宜诺斯艾利斯大学 University of Buenos Aires
  农业学院 School of Agriculture
  区域分析和遥感实验室 Regional Analysis Laboratory and Remote Sensing
  农业植物生理生态学研究所(IFEVA) Institute for Agricultural Plant Physiology and Ecology
  国家科学技术研究委员会 National Scientific and Technical Research Council
  阿根廷布宜诺斯艾利斯 Buenos Aires, Argentina
澳大利亚 Australia
  西澳大学 The University of Western Australia
  植物生物学学院 School of Plant Biology

B. 马科斯 **B. Marcos**
葡萄牙 Porto
  波尔图大学 University of Porto
  生物系 Department of Biology
  生物多样性和遗传资源科学研究中心 Faculty of Sciences and Research Centre in Biodiversity and Genetic Resources

B. 马丁-洛佩斯 **B. Martín-López**
西班牙 Spain
  马德里自治大学 Autonomous University of Madrid
  生态学系 Department of Ecology

撰 稿 人

社会生态系统实验室

Social-Ecological Systems Laboratory

A. 米拉尔斯
西班牙
　格拉纳达大学
　安达卢西亚地球系统研究所
　河流动力学和水文研究组

A. Millares
Spain
　University of Granada
　Andalusian Institute for Earth System Research
　Fluvial Dynamics and Hydrology Research Group

O. V. 穆勒
阿根廷
　国立利多拉大学
　工程与水资源学院
　国家科学技术研究委员会

O. V. Müller
Argentina
　National University of Litoral
　Faculty of Engineering and Water Resources
　National Scientific and Technical Research Council

M. 奥斯特赫尔德
阿根廷
　布宜诺斯艾利斯大学
　农业学院
　区域分析和遥感实验室

　农业植物生理生态学研究所

　国家科学技术研究委员会

M. Oesterheld
Argentina
　University of Buenos Aires
　School of Agriculture
　Regional Analysis Laboratory and Remote Sensing
　Institute for Agricultural Plant Physiology and Ecology
　National Scientific and Technical Research Council

R. 奥尔特拉-卡里奥
西班牙
　瓦伦西亚大学
　全球变化部门

R. Oltra-Carrió
Spain
　University of València
　Global Change Unit

影像处理实验室　　　　　　　　　　　Image Processing Laboratory

M. 奥雅扎巴尔　　　　　　　　　　**M. Oyarzabal**
阿根廷　　　　　　　　　　　　　　　Argentina
　布宜诺斯艾利斯大学　　　　　　　　　University of Buenos Aires
　农业学院　　　　　　　　　　　　　　School of Agriculture
　定量方法和信息系统系　　　　　　　　Department of Quantitative Methods and Information Systems
　区域分析和遥感实验室　　　　　　　　Regional Analysis Laboratory and Remote Sensing
　农业植物生理生态学研究所　　　　　　Institute for Agricultural Plant Physiology and Ecology
　国家科学技术研究委员会　　　　　　　National Scientific and Technical Research Council

I. 帕洛莫　　　　　　　　　　　　**I. Palomo**
西班牙　　　　　　　　　　　　　　　Spain
　马德里自治大学　　　　　　　　　　　Autonomous University of Madrid
　生态系　　　　　　　　　　　　　　　Department of Ecology
　社会生态系统实验室　　　　　　　　　Social-Ecological Systems Laboratory

J. M. 帕雷洛　　　　　　　　　　　**J. M. Paruelo**
阿根廷　　　　　　　　　　　　　　　Argentina
　布宜诺斯艾利斯大学　　　　　　　　　University of Buenos Aires
　农业学院　　　　　　　　　　　　　　School of Agriculture
　定量方法和信息系统系　　　　　　　　Department of Quantitative Methods and Information Systems
　区域分析和遥感实验室　　　　　　　　Regional Analysis Laboratory and Remote Sensing
　农业植物生理生态学研究所　　　　　　Institute for Agricultural Plant Physiology and Ecology

国家科学技术研究委员会	National Scientific and Technical Research Council

J. 佩努埃拉斯

西班牙

 西班牙科学研究理事会

 生态研究和林业应用中心

J. Peñuelas

Spain

 National Research Council

 Center for Ecological Research and Forestry Applications

G. 佩雷拉

巴西

 国家空间研究所

巴西

 圣若昂德尔雷联邦大学

G. Pereira

Brazil

 National Institute for Space Research

Brazil

 Federal University of São João del-Rei

M. J. 保罗

西班牙

 科尔多瓦大学

 农产品管理国际卓越园区

 安达卢西亚地球系统研究所

 河流动力学和水文研究组

M. J. Polo

Spain

 University of Córdoba

 Agrifood Campus of International Excellence

 Andalusian Institute of Earth System Research

 Fluvial Dynamics and Hydrology Research Group

A. 雷耶斯

西班牙

 阿尔梅里亚大学

 生物地质系

 安达卢西亚全球变化评估和监测中心

A. Reyes

Spain

 University of Almería

 Department of Biology and Geology

 Andalusian Center for the Assessment and Monitoring of Global Change

B. 斯坎伦   B. Scanlon

美国   United States

  德克萨斯大学奥斯汀分校   The University of Texas at Austin Austin, Texas

  杰克逊地球科学学院   Jackson School of Geosciences

  经济地质局   Bureau of Economic Geology

Y. E. 岛袋   Y. E. Shimabukuro

巴西   Brazil

  国家空间研究所   National Institute for Space Research

J. A. 索布里诺   J. A. Sobrino

西班牙   Spain

  瓦伦西亚大学   University of València

  全球变化部门   Global Change Unit

  影像处理实验室   Image Processing Laboratory

G. 索里亚   G. Sòria

西班牙   Spain

  瓦伦西亚大学   University of València

  全球变化部门   Global Change Unit

  影像处理实验室   Image Processing Laboratory

Y. 苏雄   Y. Souchon

法国   France

  国家环境与农业科学技术研究所   National Research Institute of Science and Technology for Environment and Agriculture

  水环境、生态与污染   Aquatic environments, ecology and pollution

  河流水文生态组   River Hydro-Ecology Unit

R. 斯托克勒   R. Stockler

| 巴西 | Brazil |
| 国家空间研究所 | National Institute for Space Research |

T. 托莫斯

法国

国家环境与农业科学技术研究所

水环境、生态与污染

法国国家水和水环境局

河流水文生态组

T. Tormos

France

National Research Institute of Science and Technology for Environment and Agriculture

Aquatic environments, ecology and pollution

French National Agency for Water and Aquatic Environments

River Hydro-Ecology Unit

M. 瓦列霍斯

阿根廷

布宜诺斯艾利斯大学农业学院

定量方法和信息系统系

区域分析和遥感实验室

国家科学技术研究委员会

农业植物生理生态学研究所

M. Vallejos

Argentina

School of Agriculture University of Buenos Aires

Department of Quantitative Methods and Information Systems

Regional Analysis Laboratory and Remote Sensing

National Scientific and Technical Research Council

Institute for Agricultural Plant Physiology and Ecology

K. 范洛伊

法国

国家环境与农业科学技术研究所

水环境、生态与污染

河流水文生态组

K. Van Looy

France

National Research Institute of Science and Technology for Environment and Agriculture

Aquatic environments, ecology and pollution

River Hydro-Ecology Unit

B. 维伦纽夫

法国

 国家环境与农业科学技术研究所

 水环境、生态与污染

 河流水文生态组

B. Villeneuve

France

 National Research Institute of Science and Technology for Environment and Agriculture

 Aquatic environments, ecology and pollution

 River Hydro-Ecology Unit

审 稿 人

我们衷心感谢为本书做出贡献的众多审稿人。他们的深思熟虑、建设性的评论大大改善了章节的内容和表述。我们也感谢为这项工作提供资金的众多大学、研究中心、公共机构和纳税人。

弗洛尔·阿尔瓦雷斯-塔博阿达	Flor Álvarez-Taboada
西班牙	Spain
莱昂大学	University of León
罗克萨娜·阿拉贡	Roxana Aragón
阿根廷	Argentina
图库曼国立大学	National University of Tucumán
奥尔加·巴伦	Olga Barron
澳大利亚	Australia
联邦科学和工业研究组织土地和水研究部门	CSIRO Land and Water
J. 赫苏斯·卡萨斯	J. Jesús Casas
西班牙	Spain
阿尔梅里亚大学	University of Almería
安东尼奥·卡斯特罗	Antonio Castro
美国	United States
俄克拉何马大学	University of Oklahoma

塞尔吉奥·布鲁诺·科斯塔　　　　　　Sérgio Bruno Costa
葡萄牙　　　　　　　　　　　　　　Portugal
　辛比安特公司　　　　　　　　　　　　Simbiente

皮达·克里斯蒂亚诺　　　　　　　　　Piedad Cristiano
阿根廷　　　　　　　　　　　　　　Argentina
　布宜诺斯艾利斯大学　　　　　　　　　University of Buenos Aires

米格尔·德利布斯　　　　　　　　　　Miguel Delibes,
西班牙　　　　　　　　　　　　　　Spain
　科学研究理事会　　　　　　　　　　　CSIC

科恩·德·里德　　　　　　　　　　　Koen De Ridder
比利时　　　　　　　　　　　　　　Belgium
　法兰德斯技术研究院　　　　　　　　　VITO

埃里贝托·迪亚斯-索利斯　　　　　　 Heriberto Díaz-Solís
墨西哥　　　　　　　　　　　　　　Mexico
　安东尼奥·纳罗农业自治大学　　　　　Antonio Narro Agrarian Autonomous University

马夏尔·杜盖　　　　　　　　　　　　Martial Duguay
意大利　　　　　　　　　　　　　　Italy
　EURAC 研究中心　　　　　　　　　　 EURAC

迈克尔·埃克　　　　　　　　　　　　Michael Ek
美国　　　　　　　　　　　　　　　United States
　国家海洋和大气管理局天气和气候预测中心　NOAA Center for Weather and Climate Prediction

马丁·加布尔斯基	**Martín Garbulsky**
阿根廷	Argentina
布宜诺斯艾利斯大学	University of Buenos Aires
莫妮卡·加西亚	**Monica García**
美国	United States
哥伦比亚大学	Columbia University
格雷戈里奥·加维尔-皮萨罗	**Gregorio Gavier-Pizarro**
阿根廷	Argentina
国家农业技术研究所	National Institute of Agricultural Technology
阿图尔·吉尔	**Artur Gil**
葡萄牙	Portugal
亚速尔群岛大学	University of the Açores
阿纳托利·吉特尔森	**Anatoly Gitelson**
美国	United States
内布拉斯加州大学	University of Nebraska
西尔瓦娜·戈伊兰	**Silvana Goirán**
阿根廷	Argentina
库约国立大学	National University of Cuyo
亚历山大·格拉夫	**Alexander Graf**
德国	Germany
于利希研究中心	Forschungszentrum Jülich
迭戈·古尔维奇	**Diego Gurvich**
阿根廷	Argentina

科尔多瓦国立大学 National University of Córdoba

迈克尔·海因 Michael Heinl
奥地利 Austria
  因斯布鲁克大学   University of Innsbruck

罗伯特·霍夫特 Robert Höft
加拿大 Canada
  生物多样性公约秘书处   Secretariat of the Convention on Biological Diversity

若昂·洪拉多 João Honrado
葡萄牙 Portugal
  波尔图大学   University of Porto

奈德·霍宁 Ned Horning
美国 United States
  自然历史博物馆   American Museum of Natural History

查尔斯·伊楚库 Charles Ichoku
美国 United States
  宇航局戈达德太空飞行中心   NASA Goddard Space Flight Center

伊藤明彦 Akihiko Ito
日本 Japan
  国家环境研究所   National Institute for Environmental Studies

伊娃·伊维茨 Eva Ivits
意大利 Italy
  环境与可持续研究所   Institute for Environment and Sustainability

贾根锁	Gensuo Jia
中国	China
科学院大学	Chinese Academy of Sciences
胡安·卡洛斯·希门尼斯-穆尼奥斯	Juan Carlos Jiménez-Muñoz
西班牙	Spain
瓦伦西亚大学	University of Valencia
埃里克·卡西施克	Eric Kasischke
美国	United States
马里兰大学	University of Maryland
威廉·劳恩罗斯	William Lauenroth
美国	United States
怀俄明州大学	University of Wyoming
费利西亚娜·利恰儿代洛	Feliciana Licciardello
意大利	Italy
卡塔尼亚大学	University of Catania
塞萨尔·阿古斯丁·洛佩斯·圣地亚哥	César Agustín López Santiago
西班牙	Spain
马德里自治大学	University Autonomous of Madrid
内斯托·奥斯卡·马塞拉	Néstor Oscar Maceira
阿根廷	National Institute of Agricultural Technology
国家农业技术研究所	Argentina
普里西拉·米诺蒂	Priscilla Minotti
阿根廷	Argentina

圣马丁大学	National University of San Martín
克劳迪娅·诺塔尼科拉	**Claudia Notarnicola**
意大利	Italy
EURAC 研究中心	EURAC
米伦·奥纳因迪亚	**Miren Onaindia**
西班牙	Spain
巴斯克大学	University of the Basque Country
佩德罗·佩尼亚·加西良	**Pedro Peña Garcillán**
墨西哥	Mexico
西北部生物研究中心	Biological Research Center of the Northwest
塞萨尔·佩雷斯-克鲁萨多	**César Pérez-Cruzado**
德国	Germany
哥廷根大学	University of Göttingen
加布里埃拉·波塞	**Gabriela Posse**
阿根廷	Argentina
国家农业技术研究所	National Institute of Agricultural Technology
塞尔吉·兰巴尔	**Serge Rambal**
加拿大	Canada
国家科学研究中心功能与进化生态学中心	CEFE CNRS
杜乔·罗基尼	**Duccio Rocchini**
意大利	Italy
埃德蒙·马赫基金会	Edmund Mach Foundation

尼尔达·桑切斯·马丁　　　　　　　　Nilda Sánchez Martín

西班牙　　　　　　　　　　　　　　Spain

　　萨拉曼卡大学 CIALE 研究所　　　　　　CIALE，University of Salamanca

费尔南多·桑托斯·马丁　　　　　　　Fernando Santos-Martín

西班牙　　　　　　　　　　　　　　Spain

　　马德里自治大学　　　　　　　　　　Autonomous University of Madrid

大卫·希兰　　　　　　　　　　　　David Sheeren

法国　　　　　　　　　　　　　　　France

　　图卢兹国家理工学院　　　　　　　　National Polytechnic Institute of Toulouse

鲍勃·苏　　　　　　　　　　　　　Bob Su

荷兰　　　　　　　　　　　　　　　Netherlands

　　特温特大学　　　　　　　　　　　　University of Twente

安克·特茨拉夫　　　　　　　　　　Anke Tetzlaff

意大利　　　　　　　　　　　　　　Italy

　　EURAC 研究中心　　　　　　　　　　EURAC

圣地亚哥·韦龙　　　　　　　　　　Santiago Verón

阿根廷　　　　　　　　　　　　　　Argentina

　　国家农业技术研究所　　　　　　　　National Institute of Agricultural Technology

唐纳德·杨　　　　　　　　　　　　Donald Young

美国　　　　　　　　　　　　　　　United States

　　弗吉尼亚联邦大学　　　　　　　　　Virginia Commonwealth University

朱莉·辛纳特　　　　　　　　Julie Zinnert

美国　　　　　　　　　　　　United States

　弗吉尼亚联邦大学　　　　　　Virginia Commonwealth University

第一部分

导　言

第一章 利用卫星传感器监测生态系统服务的全球展望

D. 阿尔卡拉斯-塞古拉 C. M. 迪贝拉

一、生态系统服务遥感概况

生态系统服务可以定义为"为人类提供利益(或偶尔造成损害)的生态系统的一项活动或功能"(Mace et al., 2012; Burkhard et al., 2012; Crossman et al., 2012)。从科斯坦萨等人(Costanza et al., 1997)和千年生态系统评估(MA 2005)的早期尝试，到生物多样性和生态系统服务政府间科学政策平台(Intergovernmental Platform on Biodiversity and Ecosystem Services, IPBES; IPBES, 2013)和全球生态系统服务变化监测系统(Tallis et al., 2012; 2013年对地观测生物多样性观测网)，人们一直在努力量化、评估、绘制、监测和分析维持人类福祉的各种生态系统服务的组成部分。

在对地观测小组的主持下，生物多样性和生态系统服务政府间科学政策平台与50余个国际组织和80个政府代表联合开发全球对地观测系统(Global Earth Observation System of Systems, GEOSS)。他们的目标是对多种服务进行广泛和标准的监测，使决策者和科学家能够探索、更好地理解和优先考虑社会生态环境和规模(从国家到全球层面)之间的权衡。对地观测组织生物多样性观测网络(Group on Earth Observations Biodiversity Observation Network, GEO-BON)确定了四种主要数据来源：国家统计、实地观测、遥感和数值模拟模型(Tallis et al., 2012)。其中，遥感技术提供了从地方到全球范围以及在整个时间内使用相同标准协议的潜力。这对于长期监测和跨区域权衡评估至关重要。基于卫星的对地观测可能是在大范围内以高时间、空间和光谱分辨率系统地检索全球信息的最经济可行的手段(Ayanu et al.，2012)。随着获得这些数据所需成本的降低，以及更高分辨率传感器处理更大数据集的计算能力的增加，将最大限度地发挥这一优势(Kreuter et al.，2001)。

近年来，生态系统服务遥感研究飞速发展。在 Scopus 数据库中搜索标题、摘要或关键词中包含"遥感""对地观测""生态系统服务"的文献时，我们发现总共 270 个文献(表 1-1)。搜索显示，这是一个新的研究领域(第一篇文章可追溯到 2001 年)，最近发展迅速(从 2011 年至今 50% 的文章是在过去两年半内发表)。值得注意的是，中国的研究在通过遥感工具对生态系统服务进行评估方面发挥了相对积极的作用，50% 的文章是由中国撰写。在其余出版物中，25% 对应于美国机构，其余的则来自其他各个国家。在前 15 个机构中(发表该领域 5 份或更多文件的机构)，只有 5 个为非中国机构(即布宜诺斯艾利斯大学、德国航空航天中心、弗吉尼亚理工学院和州立大学、纽约哥伦比亚大学)。吸引较多文章(5 篇或更多)的期刊包括《环境遥感》(影响因子 4.574)、《生态学报》《生态学杂志》、《生态学指标》(影响因子 2.695)、《应用生态学报》、《地理学报》、《农业生态系统与环境》(影响因子 3.004)等。

本文已对近期的一些文章进行汇编，并强调了遥感为生态系统服务制图提供的许多可能性(如 Naidoo et al.，2008；Feld et al.，2010；Feng et al.，2010；Ayanu et al.，2012；Crossman et al.，2012；Martínez-Harms and Balvanera，2012；Tallis et al.，2012)。

然而，仍然迫切需要制订一种标准化和一致的方法，或甚至一份蓝图，以便绘制一整套更完整的生态系统服务的存量和流量以及供需关系(Crossman et al.，2012；Martínez-Harms and Balvanera，2012；Palomo et al.，2012)。在过去的 12 年中，从生态系统服务供应的生物物理维度考虑，遥感信息主要用于估算生态系统服务的供给和调节功能，但由于其需求的内在社会经济特征，很少用于测算生态系统服务的文化服务功能。卫星影像为从多维角度评估生态系统服务提供了相关信息(见第二十章)，角度涵盖"供应侧"(自然供应什么?)和"需求侧"(人类需求什么?)。只有少数文章使用遥感工具作为评估生态系统服务需求侧的附带信息(如 Sutton and Costanza，2002；Scullion et al.，2011)。

表1-1 每年发表生态系统服务对地观测文章数量、排名第一的国家、作者、期刊

年份	数量	国家	数量	作者	数量	期刊	数量
2013	18	中国	134	Wang, K	5	生态学	17
2012	53	美国	66	Paruelo, J.	4	环境遥感	11
2011	61	德国	19	Kuenzer, C	4	生态学杂志	9
2010	43	英国	15	Pan, Y.	4	应用生态学杂志	6
2009	35	澳大利亚	11	Gu, X.	4	生态指标	6
2008	17	荷兰	8	Zhu, W	4	地理学报	6
2007	21	阿根廷	7	DeFries, R.	3	农业生态系与环境	5
2006	6	西班牙	7	Chong, J.	3	应用地理学	4
2005	5	意大利	7	He, H	3	国际应用地球观测和地球信息杂志	4
2004	2	印度	6	Li, J.	3	环境地球科学	4
2003	3	加拿大	6	Shi, P.	3	生态经济学	4
2002	3	芬兰	6	Zhang, C.	3	环境监测与评价	4
2001	3	瑞士	5	Gond, V.	2	国际遥感杂志	4
合计	270	法国	5	DECH, S.	2	遥感	4

注:2001年1月1日~2013年4月1日Scopus数据库中标题、摘要或关键词中包含"遥感""对地观测""生态系统服务"的文献。

截至目前,卫星影像在供应侧的最常见用途是制作土地利用/覆盖图。这些地图后来被用于模型中,用于模拟生态系统服务的提供及其在时间和空间上的变化(Kreuter et al.,2001;Konarska et al.,2002;Zhao et al.,2004;De-Yong et al.,2005;Wang et al.,2006,2009;Li et al.,2007,2011,2012;Hu et al.,2008;Du et al.,2009;Huang et al.,2009,2011;Liu et al.,2009,2012;Feng et al.,2010;McNally et al.,2011;Burkhard et al.,2012;Estoque and Murayama 2012;Hao et al.,2012;Bian and Lu,2013;Duan et al.,2013;Verburg et al.,2013;Zhao and Tong,2013)。前者旨在通过将生态系统服务价值与影像分类中存在的不同土地利用、覆盖类型进行关联,以绘制生态系统服务地图。然后,根据土地利用、覆盖类型的变化估算生态系统服务交付的变化。然而,土地覆盖类型存在空间差异,相同土地覆盖可能提供不同景观功能,因此可提供多种生态系统服务(Verburg et al.,2009)。例如,半干旱草原的放牧能力在北方和南方光照照射及干旱与潮湿年份间之间差异很大。同样,同一耕地类别(例如大豆种植园)可能具有非常不同的生态系统功能,并可能根据土地管理方式(如耕作与不耕作)提供不同的生态系统服务(Jayawickreme et al.,2011;Viglizzo et al.,2011)。这是费尔堡等人(Verburg et al.,2009)提出在生态系统服务评估中使用土地功能动态评估(连续评估)而非土地覆盖变化特征(分类评估)的原因之一。同样,尤利斯等人(Euliss et al.,2010)指出,有必要采取更全面和综合的方法来监测生态系统过程和相关服务。例如,施奈德等人(Schneider et al.,2012)从中分辨率成像光谱仪(Moderate Resolution Imaging Spectroradiometer,MODIS)卫星影像时间序列得出的与地表反照率和能量平衡动力学相关的生态系统服务的直接估算,与在土地覆盖图上运行模型(Ago-IBIS 动态全球植被模型)所得其他模拟生态系统服务相结合(Kucharik,2003),涵盖对碳和水循环及能量平衡的影响。一般来说,生态系统服务建模和制图中使用的所有模型都利用了与土地覆盖、气候变量和地形有关的空间明确信息。此外,它们还可能包含从遥感数据中得出的蒸散量(Evapotranspiration,ET)等其他生物物理变量。目前,分布最广的生态系统服务模型是 InVEST、ARIES 和 PolySCAPE(有关回顾和更多细节,见第二十章)。

卫星影像也可以直接被用于量化一个或一组特定生态系统服务,即无须通过对应特定土地覆盖类型从而进行建模。遥感文献中有大量概念和经验模型,将光谱信息与关键的生物物理变量和生态系统过程进行联系,如初级生产力、生物量、表面温度、反照率、蒸散量(ET)、土壤湿度、表面粗糙度、物种丰富度等(Verstraete et al.,2000;Jensen, 2007;Chuvieco, 2008)。

这些生物物理变量与生态系统功能及其相关生态系统服务紧密相关。因此,它们已被广泛应用于生态系统服务制图和评估(Jin et al.,2009;Krishnaswamy et al., 2009; Malmstrom et al., 2009; Newton et al., 2009; Tenhunen et al.,2009; Feng et al.,2010; Lane and D'Amico, 2010; Porfirio et al.,2010; Rocchini et al., 2010; Woodcock et al.,2010; McPherson et al., 2011; Turner et al.,2011; Caride et al.,2012; Frazier et al.,2012;Ivits et al., 2012; Martínez-Harms and Balvanera, 2012; Politi et al.,2012; Shi et al., 2012; Tallis et al.,2012; Volante et al.,2012; Forsius et al.,2013; Oki et al., 2013)。

二、章节概述

本书其余十九章共分为五部分:碳循环、生物多样性、水循环、地表能量平衡和生态系统服务五个维度。每章均回顾了将遥感数据与关键生态系统服务相关的生物物理变量和生态系统功能联系起来的概念和经验方法、技术和案例研究。本书并非旨在详尽地涵盖所有生态系统服务类别;相反,本书从全球视角审视了利用卫星数据估算关键生态系统服务的最相关方法。

(一) 碳循环相关生态系统服务

区域和全球初级生产可能是最常通过遥感技术研究的生态系统过程之一(关于估算的综述见伊藤2011年文章)。诸多研究使用了初级生产的直接估算值或替代值,以便从供应侧评估生态系统服务(如 Malmstrom et al.,2009; Paruelo et al.,2011; Turner et al.,2011; Caride et al., 2012 ; Volante et al., 2012)。第二章中,帕鲁埃洛和瓦莱霍斯(Paruelo and Vallejos)回顾了利用光学和有源传感器估算碳循环通量和存量的遥感理论基础和技术。其提出了生态系统服务提供与碳循环关键过程间的概念联系,并将其应用于可持续土地利用规划。由加布尔斯基(Garbulsky)、菲莱利亚(Filella)和佩努埃拉斯(Peñuelas)撰写的第三章和由卡斯特罗·马丁内斯(Castro)等人撰写的第六章评估了目前遥感在估算光使用效率方面的知识、进展和挑战,而光使用效率是对调节碳收益的服务进行量化和建模的难点所在。在第四章中,爱泼斯坦(Epstein)介绍了通过遥感技术在北方高纬度地区研究的碳循环的所有组成部分,从碳吸收和火灾排放到甲烷通量和土壤微生物呼吸。第五章中,伊里萨里(Irisarri)等人提出了一种牧草生产监测系统的遥感方法。最后,岛袋(Shimabukuro)等人在第七章讨论了使用遥感数据估算热带森林生物质燃烧排放的相关内容。

(二) 生物多样性相关生态系统服务

生物多样性在生态系统服务评估中可能发挥三种不同作用(Mace et al., 2012):生物多样性可作为支撑生态系统过程的调节器(如控制养分循环的初级生产者或土壤微生物),作为最终生态系统服务(例如野生作物、牲畜或渔业)或作为一种本身具有内在价值的商品(例如有魅力物种、旗舰物种或保护伞物种)。一些研究利用遥感工具进行生态系统服务评估和生物多样性保护(Naidoo et al., 2008; Krishnaswamy et al., 2009; Rocchini, 2009; DeFries et al., 2010; Feld et al., 2010; Jones et al., 2010; Rocchini et al., 2010; Cabello et al., 2012; Alcaraz-Segura et al., 2013)。第八章,费尔南德斯(Fernández)全面回顾了利用遥感数据估算生态系统服务评估中生物多样性所涵盖的各个方面的研究:调查与模拟物种分布、与关键生态系统过程的关系。第九章,卡贝洛(Cabello)等人就利用遥感帮助评估保护区网络在代表生物多样性和提供与碳循环有关的服务有效性方面提供了一个独创性实践。最后,第十章,托莫斯(Tormos)等人使用高空间分辨率影像和基于对象的影像分析,评估了河岸植被是否对河流生态系统的完整性和生态系统服务起到了增强作用。

(三) 水循环相关生态系统服务

生态系统提供诸多与水循环相关的供应和调节服务,即淡水供应与防洪。诸多遥感产品可用于估算和模拟从降雨到积雪、蒸散量(ET)、径流或地下水消耗等多个水循环组成部分。卡瓦略-桑托斯(Carvalho-Santos)等人在第十一章首先提出了水文服务的概念,以及其如何与水循环关键要素的联系。随后,作者全面回顾了最相关的卫星传感器和水文模型,以便对这些水循环要素进行评估与监测。埃雷罗(Herrero)等人在第十二章中集中讨论了遥感信息的同化对水文建模和水文生态系统服务功能评估所起到的重要作用。第十三章,孔特雷拉斯(Contreras)等人提出了一种确定和量化生态系统对除降水以外其他水输入(包括地下水和地表水输入)依赖性的遥感方法。最后,保罗(Polo)等人(第十四章)与博尼特(Bonet)、米拉尔(Millares)和埃雷罗(Herrero)(第十五章)分别侧重回顾了水文循环的两个关键要素及其应用——蒸散量(ET)和积雪。

(四) 地表能量平衡相关生态系统服务

生态系统影响气候的方式众多:生物地球化学过程、向大气排放或吸收温室气体和气溶胶及作为反照率、潜热和感热的函数控制地表能量平衡。在第十六章,阿尔卡拉斯-塞古拉(Alcaraz-Segura)等人解释了如何利用生态系统服务评估,估算气候调节的生物物理或能量平衡分量的方法。此外,这些学者们还介绍了一种原始的遥感方法,以便间接监测与气候模拟相关的生态系统关键生物物理特性。第十七章,迪贝拉(DiBella)和贝格特(Beget)讨论了陆地表面反照率的估算,这是生物地球物理反馈的主要组成部分之一。他们还评估了土地利用变化对地表反照率和光谱反照率的影响。第十八章,马尔切西尼(Marchesini)、古尔施曼(Guerschman)和索布里诺(Sobrino)讨论了蒸散量的潜热分量、用于通过遥感估算蒸散量的不同模型及用于预测潜热通量和显热通量之间分配的能量平衡方程。第十九章,索布里诺(Sobrino)等人讨论了地表温度,这是与能量平衡中显热通量相关的生物物理性质。他们利用温度与发射率分离算法评估城市热岛效应,并确定了遥感的最佳空间分辨率和卫星过顶时间,以便通过遥感平台对城市热岛现象进行监测。

(五) 生态系统服务的其他维度

如前文所述,制定一种标准化、跨学科的框架甚至一份蓝图以便绘制生态系统服务的存量、流量及供需关系是非常有必要的(Crossman et al., 2012; Martínez-Harms and Balvanera, 2012)。第二十章,卡斯特罗·马丁内斯(Castro)等人回顾了现有的从供应侧到需求侧的生态系统服务评估方法,综合考虑了生态系统服务的多维度性质(生物物理、社会文化和经济方面),并强调了遥感工具可能有助于推进服务评估的跨学科综合。

致谢

我们衷心感谢为本书作出贡献的诸多作者与审稿人，你们深思熟虑的评价与建设性的评论极大地完善了本书各章节的内容与呈现方式。我们也真诚地感谢为我们工作提供资金的众多大学、研究中心、公共机构和纳税人。

最后，我们感谢系列编辑埃米利奥·楚维科（Emilio Chuvieco）、艾玛·布里顿（Irma Britton）和劳里·施拉格斯（Laurie Schlags）在整个编辑过程中的耐心、慷慨帮助，以及他们对我们的信任与支持。

参 考 文 献

Alcaraz-Segura, D., J. Paruelo, H. Epstein, and J. Cabello. 2013. Environmental and human controls of ecosystem functional diversity in temperate South America. *Remote Sensing* 5:127–154.

Ayanu, Y. Z., C. Conrad, T. Nauss, M. Wegmann, and T. Koellner. 2012. Quantifying and mapping ecosystem services supplies and demands: A review of remote sensing applications. *Environmental Science and Technology* 46:8529–8541.

Bian, Z., and Q. Lu. 2013. Ecological effects analysis of land use change in coal mining area based on ecosystem service valuing: A case study in Jiawang. *Environmental Earth Sciences* 68:1619–1630.

Burkhard, B., F. Kroll, S. Nedkov, and F. Müller. 2012. Mapping ecosystem service supply, demand and budgets. *Ecological Indicators* 21:17–29.

Cabello, J., N. Fernández, D. Alcaraz-Segura, et al. 2012. The ecosystem functioning dimension in conservation: Insights from remote sensing. *Biodiversity and Conservation* 21:3287–3305.

Caride, C., G. Piñeiro, and J. M. Paruelo. 2012. How does agricultural management modify ecosystem services in the Argentine Pampas? The effects on soil C dynamics. *Agriculture Ecosystems and Environment* 154:23–33.

Chuvieco, E. 2008. *Earth observation of global change: The role of satellite remote sensing in monitoring the global environment.* Berlin: Springer-Verlag.

Costanza, R., R. d'Arge, R. de Groot, et al. 1997. The value of the world's ecosystem services and natural capital. *Nature* 387:253–260.

Crossman, N. D., B. Burkhard, and S. Nedkov. 2012. Quantifying and mapping ecosystem services. *International Journal of Biodiversity Science, Ecosystem Services and Management* 8:1–4.

DeFries, R., K. K. Karanth, and S. Pareeth. 2010. Interactions between protected areas and their surroundings in human-dominated tropical landscapes. *Biological Conservation* 143:2870–2880.

De-Yong, Y., P. Yao-Zhong, W. Yan-Yan, L. Xin, L. Jing, and L. Zhong-Hua. 2005. Valuation of ecosystem services for Huzhou City, Zhejiang Province from 2001 to 2003 by remote sensing data. *Journal of Forestry Research* 16:223–227.

Du, Z., Y. Shen, J. Wang, and W. Cheng. 2009. Land-use change and its ecological responses: A pilot study of typical agro-pastoral region in the Heihe River, northwest China. *Environmental Geology* 58:1549–1556.

Duan, J., Y. Li, and J. Huang. 2013. An assessment of conservation effects in Shilin Karst of South China Karst. *Environmental Earth Sciences* 68:821–832.

Estoque, R. C., and Y. Murayama. 2012. Examining the potential impact of land use/cover changes on the ecosystem services of Baguio City, the Philippines: A scenario-based analysis. *Applied Geography* 35:316–326.

Euliss, N. H., Jr., L. M. Smith, S. Liu, et al. 2010. The need for simultaneous evaluation of ecosystem services and land use change. *Environmental Science and Technology* 44:7761–7763.

Feld, C. K., J. P. Sousa, P. M. da Silva, and T. P. Dawson. 2010. Indicators for biodiversity and ecosystem services: Towards an improved framework for ecosystems assessment. *Biodiversity and Conservation* 19:2895–2919.

Feng, X., B. Fu, X. Yang, and Y. Lü. 2010. Remote sensing of ecosystem services: An opportunity for spatially explicit assessment. *Chinese Geographical Science* 20:522–535.

Forsius, M., S. Anttila, L. Arvola, et al. 2013. Impacts and adaptation options of climate change on ecosystem services in Finland: A model based study. *Current Opinion in Environmental Sustainability* 5:26–40.

Frazier, A. E., C. S. Renschler, and S. B. Miles. 2012. Evaluating post-disaster ecosystem resilience using MODIS GPP data. *International Journal of Applied Earth Observation and Geoinformation* 21:43–52.

GEO BON ES (Group on Earth Observations. Biodiversity Observation Network. Ecosystem Services Working Group). 2013. Available from: http://www.earthobservations.org/geobon_wgs.shtml (accessed April 4, 2013).

Hao, F., X. Lai, W. Ouyang, Y. Xu, X. Wei, and K. Song. 2012. Effects of land use changes on the ecosystem service values of a reclamation farm in northeast China. *Environmental Management* 50:888–899.

Hu, H., W. Liu, and M. Cao. 2008. Impact of land use and land cover changes on ecosystem services in Menglun, Xishuangbanna, southwest China. *Environmental Monitoring and Assessment* 146:147–156.

Huang, Q., D. D. Li, and H. B. Zhang. 2009. Effects of land use and land cover change on ecosystem service values in oasis region of northwest China. *Proceedings of SPIE—The International Society for Optical Engineering* 7384. doi:10.1117/12.834897.

Huang, X., Y. Chen, J. Ma, and X. Hao. 2011. Research of the sustainable development of Tarim River based on ecosystem service function. *Procedia Environmental Sciences* 10:239–246.

IPBES (Intergovernmental Science-Policy Platform on Biodiversity and Ecosystem Services). 2013. IPBES Draft Work Programme 2014–2018. Available from: http://www.ipbes.net (accessed April 4, 2013).

Ito, A. 2011. A historical meta-analysis of global terrestrial net primary productivity: Are estimates converging? *Global Change Biology* 17:3161–3175.

Ivits, E., M. Cherlet, G. Tóth, et al. 2012. Combining satellite derived phenology with climate data for climate change impact assessment. *Global and Planetary Change* 88–89:85–97.

Jayawickreme, D. H., C. S. Santoni, J. H. Kim, E. G. Jobbágy, and R. B. Jackson. 2011. Changes in hydrology and salinity accompanying a century of agricultural conversion in Argentina. *Ecological Applications* 21:2367–2379.

Jensen, J. R. 2007. *Remote sensing of the environment: An earth resource perspective.* Boston, MA: Pearson Prentice Hall.

Jin, Y., J. F. Huang, and D. L. Peng. 2009. A new quantitative model of ecological compensation based on ecosystem capital in Zhejiang Province, China. *Journal of Zhejiang University: Science B* 10:301–305.

Jones, K. B., E. T. Slonecker, M. S. Nash, et al. 2010. Riparian habitat changes across the continental United States (1972–2003) and potential implications for sustaining ecosystem services. *Landscape Ecology* 25:1261–1275.

Konarska, K. M., P. C. Sutton, and M. Castellon. 2002. Evaluating scale dependence of ecosystem service valuation: A comparison of NOAA-AVHRR and Landsat TM datasets. *Ecological Economics* 41:491–507.

Kreuter, U. P., H. G. Harris, M. D. Matlock, and R. E. Lacey. 2001. Change in ecosystem service values in the San Antonio area, Texas. *Ecological Economics* 39:333–346.

Krishnaswamy, J., K. S. Bawa, K. N. Ganeshaiah, and M. C. Kiran. 2009. Quantifying and mapping biodiversity and ecosystem services: Utility of a multi-season NDVI based Mahalanobis distance surrogate. *Remote Sensing of Environment* 113:857–867.

Kucharik, C. J. 2003. Evaluation of a process-based agro-ecosystem model (Agro-IBIS) across the U.S. corn belt: Simulations of the interannual variability in maize yield. *Earth Interactions* 7:1–33.

Lane, C. R., and E. D'Amico. 2010. Calculating the ecosystem service of water storage in isolated wetlands using LIDAR in north central Florida, USA. *Wetlands* 30:967–977.

Li, P., L. Jiang, Z. Feng, and X. Yu. 2012. Research progress on trade-offs and synergies of ecosystem services: An overview. *Shengtai Xuebao/Acta Ecologica Sinica* 32:5219–5229.

Li, R. Q., M. Dong, J. Y. Cui, L. L. Zhang, Q. G. Cui, and W. M. He. 2007. Quantification of the impact of land-use changes on ecosystem services: A case study in Pingbian County, China. *Environmental Monitoring and Assessment* 128:503–510.

Li, W. J., S. H. Zhang, and H. M. Wang. 2011. Ecosystem services evaluation based on geographic information system and remote sensing technology: A review. *Chinese Journal of Applied Ecology* 22:3358–3364.

Liu, J., J. Gao, and Y. Nie. 2009. Measurement and dynamic changes of ecosystem services value for the Tibetan Plateau based on remote sensing techniques. *Geography and Geo-Information Science* 3:022.

Liu, Y., J. Li, and H. Zhang. 2012. An ecosystem service valuation of land use change in Taiyuan City, China. *Ecological Modelling* 225:127–132.

MA (Millennium Ecosystem Assessment). 2005. *Ecosystems and human well-being. Synthesis report.* Washington, DC: Island Press.

Mace, G. M., K. Norris, and A. H. Fitter. 2012. Biodiversity and ecosystem services: A multilayered relationship. *Trends in Ecology and Evolution* 27:19–26.

Malmstrom, C. M., H. S. Butterfield, C. Barber, et al. 2009. Using remote sensing to evaluate the influence of grassland restoration activities on ecosystem forage provisioning services. *Restoration Ecology* 17:526–538.

Martínez-Harms, M. J., and P. Balvanera. 2012. Methods for mapping ecosystem service supply: A review. *International Journal of Biodiversity Science, Ecosystem Services and Management* 8:17–25.

McNally, C. G., E. Uchida, and A. J. Gold. 2011. The effect of a protected area on the tradeoffs between short-run and long-run benefits from mangrove ecosystems. *Proceedings of the National Academy of Sciences of the United States of America* 108:13945–13950.

McPherson, E. G., J. R. Simpson, Q. Xiao, and C. Wu. 2011. Million trees Los Angeles canopy cover and benefit assessment. *Landscape and Urban Planning* 99:40–50.

Naidoo, R., A. Balmford, and R. Costanza. 2008. Global mapping of ecosystem services and conservation priorities. *Proceedings of the National Academy of Sciences of the United States of America* 105:9495–9500.

Newton, A. C., R. A. Hill, C. Echeverría, et al. 2009. Remote sensing and the future of landscape ecology. *Progress in Physical Geography* 33:528–546.

Oki, T., E. M. Blyth, E. H. Berbery, and D. Alcaraz-Segura. 2013. Land cover and land use changes and their impacts on hydroclimate, ecosystems and society. In *Climate science for serving society: Research, modeling and prediction priorities,* eds. G. R. Asrar and J. W. Hurrel, 185–203. Dordrecht, The Netherlands: Springer Science and Business Media.

Palomo, I., B. Martín-López, M. Potschin, R. Haines-Young, and C. Montes. 2012. National parks, buffer zones and surrounding landscape: Mapping ecosystem services flows. *Ecosystem Services Journal* 4:104–116.

Paruelo, J., D. Alcaraz-Segura, and J. N. Volante. 2011. El seguimiento del nivel de provisión de los servicios ecosistémicos [Monitoring ecosystem services provision]. In *Valoración de servicios ecosistémicos: Conceptos, herramientas y aplicaciones para el ordenamiento territorial* [Ecosystem services valuation: concepts, tools and applications for land use planning], eds. P. Laterra, E. G. Jobbágy, and J. M. Paruelo, 141–160. Buenos Aires: Instituto Nacional de Tecnología Agropecuaria.

Politi, E., M. E. J. Cutler, and J. S. Rowan. 2012. Using the NOAA advanced very high resolution radiometer to characterise temporal and spatial trends in water temperature of large European lakes. *Remote Sensing of Environment* 126:1–11.

Porfirio, L. L., W. Steffen, D. J. Barrett, and S. L. Berry. 2010. The net ecosystem carbon exchange of human-modified environments in the Australian capital region. *Regional Environmental Change* 10:1–12.

Rocchini, D. 2009. Commentary on Krishnaswamy et al. Quantifying and mapping biodiversity and ecosystem services: Utility of a multi-season NDVI based Mahalanobis distance surrogate. *Remote Sensing of Environment* 113:904–906.

Rocchini, D., N. Balkenhol, G. A. Carter, et al. 2010. Remotely sensed spectral heterogeneity as a proxy of species diversity: Recent advances and open challenges.

Ecological Informatics 5:318–329.

Schneider, A., K. E. Logan, and C. J. Kucharik. 2012. Impacts of urbanization on ecosystem goods and services in the U.S. corn belt. *Ecosystems* 15:519–541.

Scullion, J., C. W. Thomas, K. A. Vogt, O. Pérez-Maqueo, and M. G. Logsdon. 2011. Evaluating the environmental impact of payments for ecosystem services in Coatepec (Mexico) using remote sensing and on-site interviews. *Environmental Conservation* 38:426–434.

Shi, Y., R. S. Wang, J. L. Huang, and W. R. Yang. 2012. An analysis of the spatial and temporal changes in Chinese terrestrial ecosystem service functions. *Chinese Science Bulletin* 57:2120–2131.

Sutton, P. C., and R. Costanza. 2002. Global estimates of market and non-market values derived from nighttime satellite imagery, land cover, and ecosystem service valuation. *Ecological Economics* 41:509–527.

Tallis, H., H. Mooney, S. Andelman, et al. 2012. A global system for monitoring ecosystem service change. *BioScience* 62:977–986.

Tenhunen, J., R. Geyer, S. Adiku, et al. 2009. Influences of changing land use and CO_2 concentration on ecosystem and landscape level carbon and water balances in mountainous terrain of the Stubai Valley, Austria. *Global and Planetary Change* 67:29–43.

Turner, D. P., W. D. Ritts, Z. Yang, et al. 2011. Decadal trends in net ecosystem production and net ecosystem carbon balance for a regional socioecological system. *Forest Ecology and Management* 262:1318–1325.

Verburg, P., S. van Asselen, E. van der Zanden, and E. Stehfest. 2013. The representation of landscapes in global scale assessments of environmental change. *Landscape Ecology* 28(6):1067–1080.

Verburg, P. H., J. van de Steeg, A. Veldkamp, and L. Willemen. 2009. From land cover change to land function dynamics: A major challenge to improve land characterization. *Journal of Environmental Management* 90:1327–1335.

Verstraete, M. M., M. Menenti, and J. Peltoniemi. 2000. *Observing land from space: Science, customers and technology*. Berlin: Springer-Verlag.

Viglizzo, E. F., F. C. Frank, and L. V. Carreño. 2011. Ecological and environmental footprint of 50 years of agricultural expansion in Argentina. *Global Change Biology* 17:959–973.

Volante, J. N., D. Alcaraz-Segura, M. J. Mosciaro, E. F. Viglizzo, and J. M. Paruelo. 2012. Ecosystem functional changes associated with land clearing in NW Argentina. *Agriculture, Ecosystems and Environment* 154:12–22.

Wang, C., P. D. Van Meer, M. Peng, W. Douven, R. Hessel, and C. Dang. 2009. Ecosystem services assessment of two watersheds of Lancang River in Yunnan, China with a decision tree approach. *Ambio* 38:47–54.

Wang, Z., B. Zhang, S. Zhang, et al. 2006. Changes of land use and of ecosystem service values in Sanjiang Plain, northeast China. *Environmental Monitoring and Assessment* 112:69–91.

Woodcock, B. A., J. Redhead, A. J. Vanbergen, et al. 2010. Impact of habitat type and landscape structure on biomass, species richness and functional diversity of ground beetles. *Agriculture, Ecosystems and Environment* 139:181–186.

Zhao, B., U. Kreuter, B. Li, Z. Ma, J. Chen, and N. Nakagoshi. 2004. An ecosystem service value assessment of land-use change on Chongming Island, China. *Land Use Policy* 21:139–148.

Zhao, X., and P. Tong. 2013. Ecosystem services valuation based on land use change in a typical waterfront town, Poyang Lake basin, China. *Applied Mechanics and Materials* 295:722–725.

第二部分

碳循环相关生态系统服务

第二章 利用遥感技术评估碳循环提供的生态系统服务

J. M. 帕鲁埃洛　M. 瓦莱霍斯

一、简介

为了实现社会、经济和环境一体化的可持续发展,制定政策时,必须明确考虑和评估人类活动对提供生态系统服务(ESs)的影响。土地利用规划的决策制定过程中,需要对生态系统服务、生态系统服务的供给率和相关人类活动的影响同时开展调查。生态系统服务评估采用的指标,通常无法正确反映整个地区的状况且(或)无法体现生态系统服务供给随时间的动态变化(Carpenter and Folke,2006)。

从这个角度上来看,遥感技术可实现生态系统服务的时空动态连续制图(Paruelo,2008)。

生态系统服务框架的优点之一在于其与生态系统功能及人类福祉存在直接联系。此外,《千年生态系统评估》(MA,2004)的定义明确将生态系统服务供给与生态系统的物质和能量交换(即碳收益的营养循环过程)联系起来。迪亚斯等人(Díaz et al.,2007)确定了多个生态系统服务和与之相关的生态系统过程,将生态系统服务供应水平的变化与相关功能变化,特别是植物功能多样性的变化相关联。麦克诺顿等人(McNaughton et al.,1989)提出碳收益是生态系统功能的一个综合体现,因为诸多其他过程均与碳通量密切相关。碳储量(活生物量或死生物量)也是对生态系统过程和干扰的综合描述。土壤有机碳的变化反映了干扰机制和土地利用的变化对土壤中碳输入和输出的影响。全球碳平衡是分析气候变化的一个关键问题,对大气二氧化碳浓度以及辐射平衡至关重要(Canadell et al.,2004)。碳循环的关键通量和存量可综合反映生态系统的状况及其供应生态系统服务的能力,因此定量估算这些参量至关重要(Cabello et al.,2012)。

在本章中,我们讨论了利用遥感技术给碳循环的生态系统服务评估带来的机遇,还介绍了碳循环的关键过程,利用光谱数据进行评估的理论基础及其与生态系统服务供应在概念上的联系。

二、碳循环的关键过程

生物群与大气间的碳交换动态与其他物质的能量流动和循环密切相关,是生态系统功能的一个综合表现。植物、动物和微生物的光合作用与呼吸间的平衡是碳动态的主要决定因素。光合与呼吸作用之差为净生态系统生产力(Net Ecosystem Production,NEP)或净生态系统碳交换(Net Ecosystem Exchange,NEE),是陆地生态系统的一个基本特征,因为其直接与碳汇有关。植被和土壤固碳的动态变化(假设无横向流动)可用两个不同的方程来描述:

$$\Delta C = \Delta AGB + \Delta BGB + \Delta L + \Delta S \quad \text{物质平衡方程} \quad \text{式}2-1$$

$$\Delta C = GPP - RA - RH - D \quad \text{过程方程} \quad \text{式}2-2$$

其中,ΔC 是植被和土壤带来的碳储量的变化;ΔAGB 是地上生物量的变化;ΔBGB 是地下生物量的变化;ΔL 是凋落物的变化;ΔS 是土壤碳的变化;GPP 是总初级生产力;RA 是自养呼吸;RH 是异养呼吸;D 是干扰引起的碳损失。虽然式 2-1 可以视为分配式,其中明确包括生物量,但式 2-2 反映了不同储层间的碳通量。

遥感数据是从区域到全球尺度上进行生物量估算的主要数据来源(Goetz et al.,2009)。然而,地上生物量(Above Ground Biomass,AGB)不能通过任何传感器从太空直接测量(Sun et al.,2011)。了解陆地碳过程需要整合多种类型和来源的信息,包括地面数据、生态模型和遥感数据。目前,有三种不同的遥感技术可用于估算生态系统的生物量:光学遥感、合成孔径雷达(Synthetic Aperture Radar,SAR)和激光雷达(Laser Imaging Detection and Ranging,LI-

DAR）。这些方法之间可高度互补。

三、连接碳动态和生态系统服务的概念框架

目前人们对生态系统服务用不同的方式进行了多种定义（Fisher et al.，2009）。《千年生态系统评估》（MA，2004）给出的定义为：生态系统服务是人们从生态系统中获得的好处。《千年生态系统评估》中的定义和其他相关的定义（Costanza et al.，1997；Daily，1997）考虑了生态系统之外的主观和文化因素，以确定生态系统服务供应水平特征化的好处。《千年生态系统评估》中的定义将生态系统服务分为供应生态系统服务、调节生态系统服务、文化生态系统服务和支持性生态系统服务（图2-1）。在《千年生态系统评估》的方案中，生态系统服务的提供、调节或支持水平不仅与生态系统功能的基本方面（如物质和能量的生态系统交换；Virginia and Wall，2001）有关，还与价值观、利益和需求的社会背景有关。

博伊德和班茨哈夫（Boyd and Banzhaf，2007）将生态系统服务称为直接消费或享用以创造人类福祉的生态成分，而不考虑主观和文化背景。在此基础上，费希尔等人（Fisher et al.，2009）将生态系统服务定义为（主动或被动）用于创造人类福祉的生态系统的各个方面。

（一）连接中间和最终生态系统服务的生产函数

费希尔等人（Fisher et al.，2009）提出了生态系统服务的分类方案，其中生态系统服务功能和结构被认为是"中间"服务，其反过来决定了"最终"服务（图2-1）。

一些中间服务（如初级生产力或物种组成）可能决定最终服务供应的水平（如牧草生产或碳汇，见第五章）。生态系统功能和结构（中间服务）与最终服务间的联系由"生产功能"定义（图2-2(a)）。对于具有市场价值的最终生态系统服务（如谷物产量），这类生产函数得到了很好的定义，其中，产量由诸多生物物理因素（水和养分的可用性、温度等）和管理因素（播种日期、栽培技术等）而定。由中间生态系统服务（如净初级生产力、植被结构或土壤特征）定义最终生态系统服务的生产函数，已被确定为将生态系统理念纳入决策过程的重要步

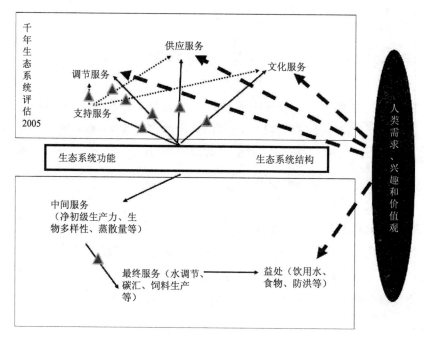

图 2-1

注:《千年生态系统评估》(2005 年)采用的由费希尔等人(Fisher et al.,2009)开发的与生态系统服务分类方案相关的主要概念。黑色箭头表示不同类型的生态系统服务与生态系统的结构和功能间的关系。该关系用生产函数(三角形)来定义。点虚线表示生态系统服务种类间的关系,虚线代表了人类的需求、兴趣和价值观对生态系统服务的定义和对两种分类方案的益处的影响。(摘自 Volante, J. N., et al., 2012。)

骤(Latera et al.,2011)。

(二) 连接干扰或压力因素与生态系统服务供应的影响函数

为增加其他生态系统服务的供应,人类活动大大减少了一些最终生态系统服务的供应。生态系统服务间的取舍导致了一些生态系统服务(如粮食产量)的供应水平提高,而其他生态系统服务(如土壤保护、水量调节、碳封存或碳汇等)的供应水平降低(de Groot et al.,2010)。理解最终生态系统服务(即粮食生产和饮用水质量)间的平衡问题较为困难,但却是土地利用规划的一项关键任务

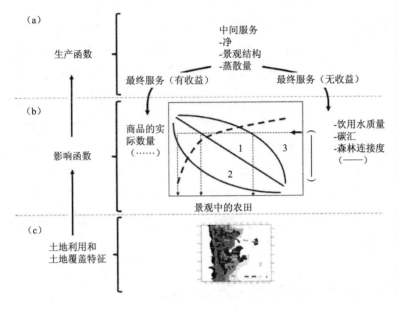

图 2-2①

注：(a)生产函数、(b)影响函数和(c)土地利用和土地覆盖特征之间的关联图。生产函数连接中间服务（生态过程）和最终服务（有或没有货币价值）。影响函数将服务生产水平的变化与和人类活动相关的压力或干扰因素联系起来。虚线表示商品生产实际数量的变化，实线表示不同类型的最终生态系统服务供应水平变化。箭头表示假设的社会决定可容忍的生态系统服务供给减少的水平。

(Viglizzo et al.,2012)。最终生态系统服务供应的变化由结构和功能的变化调节，例如生物多样性丧失和碳、水通量的变化（中间服务）(Guerschman et al.,2003；Guerschman and Paruelo,2005；Jackson et al.,2005；Nosetto et al.,2005；Fisher et al.,2009)。为了定义解释此类变化的"影响函数"，必须确定主要的干扰因素和压力因素，并量化其影响——例如生态系统服务水平（如碳汇）如何随着特定的压力或干扰（如森林被砍伐的区域）而变化。

给定与人类活动相关的压力因素或干扰因素的大小的变化（即景观中耕地面积的增加或农业活动的加剧），最终生态系统服务也将发生变化（图 2-2(b)）。总之，有些将会增加（即具有市场价值的商品的实际数量），而有些将会

---

① 此书中插附地图系原文插附地图。——译者注

减少(即水质、生物多样性保护、大气循环,价值不可估量)。了解压力因素大小和生态系统服务间的函数关系对于确定社会能够容忍的变化程度至关重要。社会愿意容忍的生态系统服务供应的减少程度(为简化分析,将社会视为以统一方式表达的实体)由图2-2(b)右轴水平箭头表示;景观的转变程度将根据生态系统服务的功能关系而不同。若关系为线性关系(曲线1,图2-2(b)),转换为耕地的土地面积应不同于曲线2(应较低)或曲线3(应较高)描述的关系。制定影响函数是将生态系统服务概念纳入土地利用规划或其他决策过程的关键步骤(Paruelo et al.,2011)。遥感技术在量化产生最终服务(如碳收益、蒸散量和反照率)的生态系统过程(图2-2(a))、描述人类干预的强度(土地覆盖和土地利用变化的空间和时间动态;图2-2(b)),以及土地利用和土地覆盖特征(图2-2(c))方面都具有重要的作用。

四、评估碳相关生态系统服务的尺度问题

定义开展分析工作的空间与时间尺度是生态研究的关键步骤(O'Neill et al.,1986;Peterson et al.,1998)。生态系统服务具有一个相关联的空间尺度,但该尺度可能与支持这些服务的生态系统过程的尺度不同。例如,微生物活动对于解毒残留物或调节甲烷或一氧化二氮排放的能力,其背后的机制涉及在亚细胞尺度中进行的复杂代谢途径。虽然生理机制在微观尺度进行,但生理过程的净结果(就调节大气气体的浓度或解毒而言)在较粗的尺度上具有相关性。就调节大气气体而言,这个尺度是全球性的。除了由提供服务或处于特定配置(如下游)的生态系统提供的益处之外,还有其他益处(Fisher et al.,2009)。

侵蚀控制体现了需要如何从空间角度来评估生态系统服务的供应:尽管特定地块的植被类型和数量、坡度和土壤质地是表征侵蚀风险的关键因素,但景观背景(相对位置、干扰状况、相邻地块的特征)也至关重要。进一步,例如,生态演替、养分再分配、径流或局部灭绝等都是景观尺度与环境相关的生态过程,这些过程与生态系统服务的供应直接相关。所有这些例子都突出了景观尺度在评价生态系统服务中的重要性。虽然景观维度存在差异,但通常景观的延伸范围在10公顷~105公顷内变化,并且这些限制与分水岭的限制相关联。对地观测方

案必须考虑生态系统服务的空间分辨率等尺度问题。

五、应当监测的中间服务类别

就操作方面来看,待评估的生态系统方面(中间服务)必须可靠,易于在不同的尺度上进行测量或估算,并且应与最终服务有逻辑地连接起来。碳循环的某些方面由于其整合生态系统碳动态的能力而特别适合于监测。对地上生物量(AGB)及其随时间的变化进行估算可显著减少质量平衡方程中的任何不确定性(Le Toan et al.,2004)。然而,对中高生物量森林的碳储量进行直接估算仍然是遥感面临的一个主要挑战。

我们特别强调监测的两个方面:净初级生产力(Net Primary Production,NPP)和生物量存量。这两个属性整合了生态系统的一些其他功能和结构方面(McNaughton et al.,1989)。

布雷肯里奇等人(Breckenridge et al.,1995)列举了在选择指标时应考虑的几个标准:
- 适用于不同地区的普适性和简便性;
- 与关键生态系统过程的相关性;
- 时间和空间变异性;
- 记录自动化的可能性;
- 关系成本效益;
- 对变化的响应/敏感性;
- 取样的环境影响;
- 对观测方案的经验和概念的支持。

他们得出结论,卫星平台提供的频谱数据特别适合满足这些标准。光谱数据不仅能够体现景观的结构方面(即土地覆盖类型在空间和时间上的分布),还能够体现生态系统的功能方面(即碳收益动态变化、蒸散量、干扰状况)(Wessman,1992;Kerr and Ostrovsky,2003;Pettorelli et al.,2005;Paruelo,2008;Cabello et al.,2012)。

本章中,我们将会回顾生态系统的功能(净初级生产力)和结构(生物量)属性,这些属性可在碳循环过程中融入中间服务,并体现式2-1和式2-2中的关键术语。这些属性可通过运用成熟的技术和仿真模型来根据遥感数据进行估算。明确的实施方案使得这些属性可以集成到景观尺度的监测计划中。遥感技术还使得调节碳存量和流量的干扰机制(洪水和火灾)的关键方面得以描述出来(Di Bella et al.,2008)。我们讨论了监测生物量燃烧过程中释放的能量(和碳),特别是评估火灾辐射强度(式2-1的地上生物量变化的分量和式2-2中的因子D)的可能性。

(一) 净初级生产力估算

生物量收获测量法对于进行大尺度的NPP时空动态分析的贡献有限(Singh et al.,1975;Lauenroth et al.,1986)。卫星影像为监测不同植被类型的净初级生产力提供了有价值的数据(Prince,1991;Running et al.,2000),这些数据具有覆盖面积大、时间分辨率高和空间分辨率适中的特点。一些记录红光和近红外段地物电磁波谱反射率的光学传感器和平台已得到广泛使用(即Landsat MSS、TM和ETM+、MODIS、Vegetation和AVHRR-NOAA)。

波谱指数,特别是归一化差异植被指数(Normalized Difference Vegetation Index,NDVI)和增强植被指数(Enhanced Vegetation Index,EVI),与绿色植被的光合有效辐射吸收比例(fraction of the Absorbed Photosynthetically Active Radiation,fAPAR)呈密切正相关关系(Sellers et al.,1992;Huete et al.,2002;Di Bella et al.,2004)。因此,可将从气象站获得的光合有效辐射吸收比例和入射的光合有效辐射(Photosynthetically Active Radiation,PAR)相乘,以此来估算吸收的光合有效辐射(Absorbed Photosynthetically Active Radiation,APAR)。净初级生产力可根据Monteith模型估算:

$$NPP = fAPAR \times PAR \times RUE \qquad 式2-3$$

其中,RUE是辐射利用效率,以每兆焦干物质克数为单位(Monteith,1972)。根据以530纳米和570纳米为中心的两个波段计算的遥感波谱指数,即光化学反射率指数(Photochemical Reflectance Index,PRI;Garbulsky et al.,2008)(见第三章),可估算出辐射利用效率。

蒙蒂斯(Monteith)模型已用于从1平方千米到64平方千米的多个空间分辨率下估算净初级生产力(Running et al.,2004)。皮涅罗等人(Piñeiro et al.,

2006)、巴埃萨等人(Baeza et al.,2010)和伊里萨里等人(Irisarri et al.,2012)使用蒙蒂思模型估算了南美洲草原上大片区域的地上净初级生产力(Aboveground Net Primary Production,ANPP)。瓦萨洛等人(Vasallo et al.,2013)使用蒙蒂思模型来比较当地草原和用造林区替代草原带来的碳收益。

佩脱雷利等人(Pettorelli et al.,2005)研究表明,利用光谱指数评估的季节碳收益和年际碳收益提供了对于生态系统功能的综合描述。增强植被指数和归一化差异植被指数的季节曲线的两个特征尤其有效,即年积分和季节变化(Paruelo and Laurenenth,1998)。沃兰特等人(Volante et al.,2012)分析了在南美查科地区土地清理对这两个碳相关的中间生态系统服务的影响。虽然出于农业和牧业目的对土地进行清理,对于年度地上净初级生产力总量的影响相对较小,但一旦森林被砍伐,地块的季节性较被取代的自然植被的季节性更明显。此等季节性的增加与一年中部分时间段内(休耕)光合能力减弱有关。这期间内光合作用的减少可能会对一些例如侵蚀控制和水调节(由于土壤裸露面积更大)的生态系统服务和生物多样性(由于栖息地丧失或栖息地质量下降及休耕期间初级消费者可获得的绿色生物量减少)产生直接影响。帕鲁埃洛等人(Paruelo et al.,2004)研究表明,同样以查科地区为例,在不同的尺度上,随着农田在景观中所占的比例增加,总碳收益(以年度 NDVI 积分来描述)下降(图 2-3)。阿根廷的潘帕斯草原(Guerschman et al.,2003)和美国大平原(Paruelo et al.,2001)也呈现了类似的模式。对于南美洲的温带草原和林地,农业用地的扩张可能会导致净初级生产力下降,取决于当地景观的原始植被覆盖和年均降水量(Paruelo et al.,2001b)的净初级生产力。与较干燥的地区相比,降水量较高地区的净初级生产力下降更明显(图 2-4(a))。从碳封存到水调节,许多最终生态系统服务与总碳收益直接相关。

碳收益的季节性(全年变化)随景观中农业用地比例的增加而增加。然而,碳汇的削减程度则取决于种植制度;双季作物(小麦、大豆)带来的碳汇减少低于单季作物(图 2-4(b))。因此,净初级生产力的季节性等中间生态系统服务的变化将决定最终生态系统服务的变化,比如侵蚀控制(由于不同季节植物覆盖的变化)和气候控制(由于不同季节的叶面积指数、潜热量或反照率大小的变化)等。

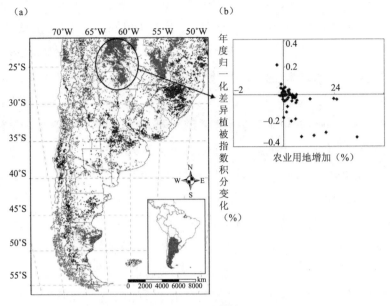

图 2-3 （见彩插）

注：(a)里的阿根廷地图显示了 1981~2000 年间植被吸收的光合有效辐射与时间之间关系的斜率。红色像素点和蓝色像素点分别表示负斜率和正斜率。吸收的光合有效辐射是根据来自 AVHRR/NOAA 卫星 PAL 系列的归一化差异植被指数计算得出的。(b)显示了阿根廷西北部森林覆盖各县（萨尔塔、查科、朱瑞和图库曼）的耕地面积的年际变化与归一化差异植被指数年际变化之间的关系。（改编自 Paruelo, J. M., et al., 2004）。

放牧被认为是生态系统中的一个主要干扰和(或)压力因素。放牧对草地、灌木地和热带稀树草原的结构和功能的影响引起了学术界的热议和争论(McNaughton, 1979; Milchunas and Lauenroth, 1993; Oesterheld et al., 1999; Chase et al., 2000)。净初级生产力可能对长期放牧压力有复杂的反应，这取决于资源可用性和长期放牧历史(Milchunas and Lauenroth, 1993; Oesterheld et al., 1999)。阿吉亚尔等人(Aguiar et al., 1996)提出了将巴塔哥尼亚大草原净初级生产力的影响函数作为历史放牧压力的函数。此外，本文使用厄斯特黑尔德等人(Oesterheld et al., 1992)提出的模型，给出了来自中间服务(草地地上净初级生产力；图2-4(c))的一种最终服务(国内草食动物生物量)的生产函数。

保护区和未退化的草地或灌木地对降水变化的敏感性低于放牧现象严重的地区(Paruelo et al., 2005; Verón et al., 2011)。在这种情况下，中间生态系统服

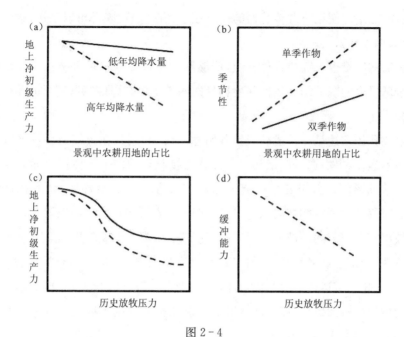

图 2-4

注：(a)，(b)是假设的关于地上净初级生产力、ANPP 季节性（中间服务）和景观中农耕用地的占比（干扰因素）的影响函数。不同的线条对应不同的气候条件——高、低年平均降水量（Mean Annual Precipitation, MAP）或管理（单季作物或双季作物）。(c)，(d)是假设的关于地上净初级生产力和生态系统在功能层面缓冲气候波动的能力（中间因素）的影响函数，作为关于天然牧场的历史放牧压力（干扰因素）的函数。(c)还显示了放牧对家养食草动物生物量的影响（最终服务）。（基于 Aguiar, M. R., et al., 1996；Guerschman, J. P., et al., 2003；Paruelo, J. M., et al., 2005，以及 Verón, S. R., et al., 2011. 等的分析）

务是生态系统的缓冲能力，即就碳收益而言功能属性相对于环境波动的相对变率（图 2-4(d)）。

土地清理增加了碳收益的年际差异程度。这表明，与耕地相比，天然植被对气候波动的缓冲能力较低(Volante et al., 2012)。

（二）地上生物量(AGB)的估算

众所周知，要从局部尺度(Fang et al., 2006)到大陆尺度(Houghton et al., 2001)对生物量做出真实的估算困难重重。人们为此作出了大量努力，比如利用实地和遥感技术估算生物量（主要在森林范围里），但很少进行影响函数的开发。生物量的估算一直是基于有源传感器，即这些传感器发出信号，然后测量反射回传感器的能量。使用有源传感器的优点在于其白天、夜晚都可以工作,微波可穿过雾霾、烟雾和云层。

传输的能量也可穿透森林林冠,能够测量林冠高度和垂直结构。一直用于测量全球尺度的森林结构属性和生物量的两类主要有源传感器是合成孔径雷达和激光雷达。

合成孔径雷达是一种机载或星载雷达系统,利用其在天线与目标区域间的相对向前运动,提供通过记录和组合传感器接收到的多个个体信号而生成的高分辨率遥感影像。由于其穿透能力和对植被含水量的敏感性,合成孔径雷达(SAR)对森林空间结构非常敏感(Le Toan et al.,2004)。后向散射是输出雷达信号被重定向回天线的部分。后向散射受表面参数(目标的粗糙度、几何形状和介电特性)和雷达观测参数(发射电磁波的频率、偏振率和入射角)的影响。众所周知,合成孔径雷达用于感测树冠体积(特别是在较长波长下)并提供具有后向散射能量值的影像数据,该后向散射能量值在很大程度上取决于叶、枝、茎等树冠结构元素的大小和方向。

信号的频率(f)描述了信号与森林结构的相互作用以及波的穿透能力。波长越长,对植被垂直结构的敏感性越强,在林冠里的穿透性更好。雷达数据是使用 X、C、L 和 P 波段获取的。较短波长(X 和 C 波段,分别为 2.5 厘米和 7.5 厘米)对较小的林冠元素(如叶、小分枝)敏感,较长波长(L 和 P 波段,分别为 23.5 厘米和 70 厘米)则对大分枝和树干敏感。极化(p)是电磁波中电场的方向,是信号与反射物相互作用的主要因素。大多微波传感器发射和接收水平(H)极化或垂直(V)极化的信号。测量发射和接收电磁波的极化使得地上生物量(AGB)的测量灵敏度进一步提升(Goetz et al.,2009)。干涉测量法计算由星载合成孔径雷达在两个不同的时间获取的两幅影像间的相位差造成的干涉影像,由此得到的干涉图是地面和合成孔径雷达仪器间距离变化的等高线图(Feigl,1998)。卡西希克等人(Kasischke et al.,1997)提出,使用具有交叉极化(HV 或 VH)信道的低频(P 和 L 波段)雷达系统可以实现对生物量的最佳估算。

利用合成孔径雷达估算生物量最简单的方法是利用回归分析将后向散射系数与田间生物量测量值联系起来。这种方法已经在不同地区进行了试验,在针叶林中取得了良好的结果(Dobson et al.,1992;Le Toan,1992)。间接方法也被用来估算地上生物量,包括由此得出的关于森林结构的估算值(例如树高或林冠高度),以通过干涉测量法推断森林生物量。此外,偏振法和干涉法的合成孔径雷达数据已用于森林生物量的估算(Dobson et al.,1992,1995;Ranson and Sun,1994;Kasischke et al.,1995)和林冠高度的估算(Treuhaft et al.,1996,2004;Kobayashi et al.,2000;Kellndorfer et al.,2004;Walker et al.,2007)。

这些应用需要用于培训和验证目的的地面采样数据(Sun et al.,2011)。利用合成孔径雷达估算生物量的主要问题在于饱和水平。在不同类型的森林(温带、北部和热带)内使用合成孔径雷达的实验研究表明,饱和分别发生在C、L和P波段的每公顷30吨、50吨和150~200吨(Le Toan et al.,2004)。这些饱和度值都很接近,取决于试验条件和森林特征。使用长波长或将极化分集与干涉测量技术相结合(偏振干涉测量)的先进的机载合成孔径雷达系统已被证明在估算森林生物量方面能力显著增强(Le Toan,2002)。雷达影像干涉测量(interferometric SAR,InSAR)也被用于改进对于地上生物量的估算(Walker et al.,2007),其中异速方程被用于建立结构模式(例如树高)和森林其他性质(例如生物量)间的定量关系。

激光雷达是一种相对较新的有源遥感技术,由于其能够以高精度确定三维(3D)测量结果,因此特别适合用于再现林分的三维结构。激光雷达仪器使用了发射脉冲的激光扫描仪,记录发射光脉冲与接收光脉冲间的延迟时间,以计算高程值。每个数据点均记录有精确的水平位置、垂直高程和其他属性值。记录多个收回的信号,并为每个点分配一个分类,以识别景观特征。激光雷达也会获取反射能量的强度并且对其进行分析,以提供关于地形特征的额外信息。激光雷达度量指标是从3D点云(3D坐标系中的一组顶点)创建的统计表征数,并且通常在根据激光雷达的数据预测森林参量时使用。目前已经有很多类激光雷达系统应用于获取越来越广的植被特征和生物量估算。基于激光雷达的地上生物量估算可通过点云和光栅化数据实现。为存储3D点云数据和点属性,必须以标准化的二进制格式管理和处理3D数据的点集合。基于激光雷达原始点云的三维单树建模过程由王等人(Wang et al.,2007)首次公开。当处理用于生成区域范围的地上生物量(AGB)地图的大数据集时,点云数据处理是一个对计算效率要求很高的任务(Jochem et al., 2011)。使用光栅化数据需要将3D点云聚集成单元,这意味着林冠表面由单值函数表示。该过程伴随着3D结构的不可逆损失,但使得处理更省时并且存储体积明显减小。激光雷达系统被分类为离散返回记录系统或全波形记录系统,并且可进一步分为轮廓(仅沿着传感器正下方的窄线记录)或扫描系统(在传感器两侧的宽条带上记录)(Lefsky et al., 2002; Lim et al., 2003)。

在森林里,全波形系统记录用于分析的整个波形,而离散返回系统记录表示已截取特征的点云(Wulder et al., 2012)。

激光雷达传感器已用于将地块水平估计值扩展到更大的空间和生态尺度(Lefsky et al., 2002; Zhao et al., 2009)。使用激光雷达数据估算碳储量的异速生长方程通常具有区域性,模型标定费力,地面调查数据获取代价高昂(Lefsky et al., 1999; Nelson et al., 2012)。然而,阿斯纳等人(Asner et al., 2012)使用了单一通用激光雷达模型来预测实测样地的地上碳密度,发展了热带树种通用的生物量异速生长方程。在这种方法中,作者将森林结构特性简化为平均林冠剖面高度(亦称MCH),即林冠体积剖面的垂直中心(与简单的林冠顶部高度相反),并详细说明了其与碳密度和茎面积的关系。

(三) 野火释放的碳与能量

火是陆地生态系统中能量和碳释放的一个重要途径。吉利奥等人(Giglio et al., 2010)报告称,全球每年有330万~430万平方千米的地域发生过火灾。火主要以颗粒物质和温室气体的形式释放碳,包括二氧化碳和甲烷(van der Werf et al., 2010),并通过改变水文循环、引起土壤侵蚀和大气辐射强迫等严重影响生态系统服务(Lohmann and Feichter, 1997; DeFries et al., 2002; Hoffmann et al., 2002, 2003; Mouillot and Field, 2005; van der Werf et al., 2008)。

火灾辐射功率产品(Fire Radiative Power product, FRP)定量描述了特定大火的辐射热输出(单位:兆瓦)。FRP与已发生的火灾消耗的生物量有关(另见第七章)。现已证明(在小规模的实验大火中)每单位时间释放的辐射热能量的值(FRP)与燃料消耗速率有关(Wooster et al., 2005),它代表燃烧过程的直接输出。多年对火焰辐射功率的积分提供了对火焰辐射能量(Fire Radiative Energy,FRE)的估算。就野火而言,火焰辐射能量应与燃烧的总生物量成比例(Verón et al.,2012)。

为了得到FRP产品,要从识别火灾像元开始。火灾热异常(Fire Thermal Anomaly,FTA)算法测试了红外波谱部分辐亮度的升高。该算法包括用于区分火灾与其他可能在该光谱带中引起类似响应的现象(即镜面反射和云边缘)(Roberts et al.,2005)的额外测试,它主要对在3.9~11.0微米间的亮温数据及其差异有效。火灾检测阈值基于调整紧邻的无火灾背景像素的检测结果的相关测试。一旦检测到火灾像素,就使用从中红外(Middle-Infrared Radur,MIR,3.9微米)通道和在无火灾情况下在相同位置观察到的背景辐射亮度来估算火焰辐射功率(Giglio et al.,2003)。

火焰辐射功率数据可从Terra和Aqua平台搭载的MODIS(中分辨率成像光谱辐射计)传感器第5版数据集生成的MOD和MYD14CMG火灾产品(Giglio et al.,2006)获得。该数据集将时间分辨率为1~2天、空间分辨率为1平方千米的数据转换为月尺度0.5°×0.5°网格。

根据对火灾的能源产生和空间分布进行的全球分析,韦龙等人(Verón et al.,2012)研究表明,在2003~2010年间,全球火灾消耗的能量约为8 300±592PJ·yr^{-1},相当于2008年全球电力消耗的36%~44%,比57个国家全国电力消耗量的100%还多。森林/林地、耕地、灌木地和草地分别贡献了全球火灾释放能量的53%、19%、16%和3.5%。

六、结束语

遥感技术为估算碳平衡的两个关键方面提供了机会:净初级生产力和生物量。根据费希尔等人(Fisher et al.,2009)确定的生态系统服务的定义,这两个变量代表中间服务,并涵盖生态系统结构和功能的诸多基本方面。此外,它们与从牧草、木材生产到碳固定的重要最终服务存在明确的联系。将这两方面与其他中间服务结合,有助于确定气候调节、土壤侵蚀控制或水供应等其他中间服务。

卫星观测是采用相同观测方法实现大范围、近实时指标估算的重要一步。

这些特征在用来监测生态系统服务供应水平变化的计划中体现出明显优势。遥感技术不仅能估算与中间生态系统服务相关的碳，还能估算与能量平衡相关的过程，如潜热通量和反照率。

致谢

本项目由 UBACYT、FONCYT 和 CONICET 资助，同时得到了美洲国家间全球变化研究所（IAI）CRN-3095 的援助，并得到了美国国家科学基金会（Grant GEO-1128040）的支持。

参 考 文 献

Aguiar, M. R., J. M. Paruelo, C. E. Sala, and W. K. Lauenroth. 1996. Ecosystem responses to changes in plant functional type composition: An example from the Patagonian steppe. *Journal of Vegetation Science* 7:381–390.

Asner, G. P., J. Mascaro, H. C. Muller-Landau, et al. 2012. A universal airborne LiDAR approach for tropical forest carbon mapping. *Oecologia* 168:1147–1160.

Baeza, S., F. Lezama, G. Piñeiro, A. Altesor, and J. M. Paruelo. 2010. Spatial variability of aboveground net primary production in Uruguayan grasslands: A remote sensing approach. *Applied Vegetation Science* 13:72–85.

Boyd, J., and S. Banzhaf. 2007. What are ecosystem services? The need for standardized environmental accounting units. *Ecological Economics* 63:616–626.

Breckenridge, R. P., W. G. Kepner, and D. A. Mouat. 1995. A process for selecting indicators for monitoring conditions of rangeland health. *Environmental Monitoring and Assessment* 36:45–60.

Cabello, J., D. Alcaraz-Segura, R. Ferrero, A. J. Castro, and E. Liras. 2012. The role of vegetation and lithology in the spatial and inter-annual response of EVI to climate in drylands of Southeastern Spain. *Journal of Arid Environments* 79:76–83.

Canadell, J., P. Ciais, P. Cox, and M. Heimann. 2004. Quantifying, understanding and managing the carbon cycle in the next decades. *Climatic Change* 67:147–160.

Carpenter, S. R., and C. Folke. 2006. Ecology for transformation. *TRENDS in Ecology and Evolution* 21:309–315.

Chase, J. M., M. A. Leibold, A. L. Downing, and J. B. Shurin. 2000. The effects of productivity, herbivory, and plant species turnover in grassland food webs. *Ecology* 81:2485–2497.

Costanza, R., R. d'Arge, R. de Groot, et al. 1997. The value of the world's ecosystem services and natural capital. *Nature* 387:253–260.

Daily, G. C. (ed.). 1997. *Nature's services: Societal dependence on natural ecosystems.* Washington, DC: Island Press.

DeFries, R. S., R. A. Houghton, M. C. Hansen, C. B. Field, D. Skole, and J. Townshend. 2002. Carbon emissions from tropical deforestation and regrowth based on satellite observations for the 1980s and 1990s. *Proceedings of the National Academy of Sciences* 99:14256–14261.

de Groot, R. S., R. Alkemade, L. Braat, L. Hein, and L. Willeman. 2010. Challenges in integrating the concept of ecosystem services and values in landscape planning, management and decision making. *Ecological Complexity* 7:260–272.

Díaz, S., S. Lavorel, F. S. Chapin, P. A. Tecco, D. E Gurvich, and K. Grigulis. 2007. Functional diversity at the crossroads between ecosystem functioning and environmental filters. In *Terrestrial ecosystems in a changing world*, eds. J. Canadell, L. F. Pitelka, and D. Pataki, 81–91. New York: Springer-Verlag.

Di Bella, C. M., J. M. Paruelo, J. E. Becerra, C. Bacour, and F. Baret. 2004. Experimental and simulated evidences of the effect of senescent biomass on the estimation of fPAR from NDVI measurements on grass canopies. *International Journal of Remote Sensing* 25:5415–5427.

Di Bella, C. M., G. Posse, M. E. Beget, M. A. Fischer, N. Mari, and S. Veron. 2008. La teledetección como herramienta para la prevención, seguimiento y evaluación de incendios e inundaciones [Remote sensing as a tool for the prevention, monitoring and evaluation of fires and floods]. *Ecosistemas* 17:39–52.

Dobson, M. C., F. T. Ulaby, T. Le Toan, A. Beaudoin, E. S. Kasischke, and N. Christensen. 1992. Dependence of radar backscatter on coniferous forest biomass. *IEEE Transactions on Geoscience and Remote Sensing* 30:412–415.

Dobson, M. C., F. T. Ulaby, L. E. Pierce, et al. 1995. Estimation of forest biophysical characteristics in Northern Michigan with SIR-C/X-SAR. *IEEE Transactions on Geoscience and Remote Sensing* 33:877–895.

Fang, J., S. Brown, Y. Tang, G. J. Nabuurus, X. Wang, and S. Haihua. 2006. Overestimated biomass carbon pools of the northern mid- and high latitude forests. *Climatic Change* 74:355–368.

Feigl, K. L. 1998. RADAR interferometry and its application to changes in the earth surface. *Reviews of Geophysics* 36:441–500.

Fisher, B., K. R. Turner, and P. Morling. 2009. Defining and classifying ecosystem services for decision making. *Ecological Economics* 68:643–653.

Garbulsky, M. F., J. Peñuelas, J. M. Ourcival, and I. Filella. 2008. Estimación de la eficiencia del uso de la radiación en bosques mediterráneos a partir de datos MODIS. Uso del Índice de Reflectancia Fotoquímica (PRI) [Radiation use efficiency estimation in Mediterranean forests using MODIS Photochemical Reflectance Index (PRI)]. *Ecosistemas* 17:89–97.

Giglio, L., J. Descloitres, C. O. Justice, and Y. J. Kaufman. 2003. An enhanced contextual fire detection algorithm for MODIS. *Remote Sensing of Environment* 87:273–282.

Giglio, L., J. T. Randerson, G. R. van der Werf, et al. 2010. Assessing variability and long-term trends in burned area by merging multiple satellite fire products. *Biogeosciences* 7:1171–1186.

Giglio, L., G. R. van der Werf, J. T. Randerson, G. J. Collatz, and P. Kasibhatla. 2006. Global estimation of burned area using MODIS active fire observations. *Atmospheric Chemistry and Physics* 6:957–974.

Goetz, S. J., A. Baccini, N. T. Laporte, et al. 2009. Mapping and monitoring carbon stocks with satellite observations: A comparison of methods. *Carbon Balance and Management* 4:2.

Guerschman, J. P., and J. M. Paruelo. 2005. Agricultural impacts on ecosystem functioning in temperate areas of North and South America. *Global and Planetary Change* 47:170–180.

Guerschman, J. P., J. M. Paruelo, C. M. Di Bella, M. C. Giallorenzi, and F. Pacín. 2003. Land classification in the Argentine Pampas using multitemporal landsat TM data. *International Journal of Remote Sensing* 17:3381–3402.

Hoffmann, W. A., W. Schroeder, and R. B. Jackson. 2002. Positive feedbacks of fire, climate, and vegetation change and the conversion of tropical savannas. *Geophysical Research Letters* 29:1–9.

Hoffmann, W. A., W. Schroeder, and R. B. Jackson. 2003. Regional feedbacks among climate, fire, and tropical deforestation. *Journal of Geophysical Research: Atmospheres* 108:1–11.

Houghton, R. A., K. T. Lawrence, J. L. Hackler, and S. Brown. 2001. The spatial distribution of forest biomass in the Brazilian Amazon: A comparison of estimates. *Global Change Biology* 7:731–746.

Huete, A., K. Didan, T. Miura, E. P. Rodriguez, X. Gao, and L. G. Ferreira. 2002. Overview of the radiometric and biophysical performance of the MODIS vegetation indices. *Remote Sensing of Environment* 83:195–213.

Irisarri, G., M. Oesterheld, J. M. Paruelo, and M. Texeira. 2012. Patterns and controls of above-ground net primary production in meadows of Patagonia. A remote sensing approach. *Journal of Vegetation Science* 23:114–126.

Jackson, R. B., E. G. Jobbágy, R. Avissar, et al. 2005. Trading water for carbon with biological carbon sequestration. *Science* 310:1944–1947.

Jochem, A., M. Hollaus, M. Rutzinger, K. Schadauer, and B. Maier. 2011. Estimation of aboveground biomass using airborne LiDAR data. *Sensors* 11:278–295.

Kasischke, E. S., N. L. Christensen, and L. L. Bourgeauchavez. 1995. Correlating radar backscatter with components of biomass in loblolly-pine forests. *IEEE Transactions on Geoscience and Remote Sensing* 33:643–659.

Kasischke, E. S., J. M. Melack, and M. C. Dobson. 1997. The use of imaging radars for ecological applications—A review. *Remote Sensing of Environment* 59:141–156.

Kellndorfer, J., W. Walker, L. Pierce, et al. 2004. Vegetation height estimation from Shuttle Radar Topography Mission and National Elevation Datasets. *Remote Sensing of Environment* 93:339–358.

Kerr, J. T., and M. Ostrovsky. 2003. From space to species: Ecological applications for remote sensing. *TRENDS in Ecology and Evolution* 18:299–305.

Kobayashi, Y., K. Sarabandi, L. Pierce, and M. C. Dobson. 2000. An evaluation of the JPL TOPSAR for extracting tree heights. *IEEE Transactions on Geoscience and Remote Sensing* 38:2446–2454.

Laterra, P., J. M. Paruelo, and E. G. Jobbagy (eds.). 2011. *Valoración de servicios ecosistémicos: Conceptos, herramientas y aplicaciones para el ordenamiento territorial* [Appraisal of ecosystem services: concepts, tool and applications for territorial organization]. Buenos Aires: Ediciones INTA.

Lauenroth, W. K., H. W. Hunt, D. M. Swift, and J. S. Singh. 1986. Estimating aboveground net primary productivity in grasslands: A simulation approach. *Ecological*

Modeling 33:297–314.

Lefsky, M. A., W. B. Cohen, G. G. Parker, and D. J. Harding. 2002. Lidar remote sensing for ecosystem studies. *BioScience* 1:19–30.

Lefsky, M. A., D. Harding, W. B. Cohen, and G. G. Parker. 1999. Surface Lidar remote sensing of the basal area and biomass in deciduous forests of eastern Maryland, USA. *Remote Sensing of Environment* 67:83–98.

Le Toan, T. 1992. Relating forest biomass to SAR data. *IEEE Transactions on Geoscience and Remote Sensing* 30:403–411.

Le Toan, T. 2002. BIOMASCA: Biomass Monitoring Mission for Carbon Assessment. A proposal in response to the ESA Second Call for Earth Explorer Opportunity Missions. Available from: http://www.cesbio.ups-tlse.fr/data_all/pdf/biomasca.pdf (accessed July, 2013).

Le Toan, T., S. Quegan, I. A. N. Woodward, M. Lomas, N. Delbart, and G. Picard. 2004. Relating radar remote sensing of biomass to modeling of forest carbon budgets. *Climatic Change* 67:379–402.

Lim, K., P. Treitz, M. A. Wulder, B. St-Onge, and M. Flood. 2003. Lidar remote sensing of forest structure. *Progress in Physical Geography* 27:88–106.

Lohmann, U., and J. Feichter. 1997. Impact of sulfate aerosols on albedo and lifetime of clouds: A sensitivity study with the ECHAM GCM. *Journal of Geophysical Research* 102:685–700.

MA (Millennium Ecosystem Assessment). 2004. *Ecosystems and human well-being: Our human planet*. Washington, DC: Island Press.

MA (Millennium Ecosystem Assessment). 2005. *Ecosystems and human well-being: General synthesis*. Washington, DC: Island Press and World Resources Institute.

McNaughton, S. J. 1979. Grazing as an optimization process: Grass–ungulate relationships in the Serengeti. *The American Naturalist* 113:691–703.

McNaughton, S. J., M. Oesterheld, D. A. Frank, and K. J. Williams. 1989. Ecosystem-level patterns of primary productivity and herbivory in terrestrial habitats. *Nature* 341:142–144.

Milchunas, D. G., and K. W. Lauenroth. 1993. Quantitative effects of grazing on vegetation and soils over a global range of environments. *Ecological Monographs* 63:327–366.

Monteith, J. 1972. Solar radiation and productivity in tropical ecosystems. *Journal of Applied Ecology* 9:747–766.

Mouillot, F., and C. B. Field. 2005. Fire history and the global carbon budget: A $1° \times 1°$ fire history reconstruction for the 20th century. *Global Change Biology* 11:398–420.

Nelson, R., T. Gobakken, E. Næsset, et al. 2012. Lidar sampling—Using an airborne profiler to estimate forest biomass in Hedmark County, Norway. *Remote Sensing of Environment* 123:563–578.

Nosetto, M. D., E. G. Jobbágy, and J. M. Paruelo. 2005. Land use change and water losses: The case of grassland afforestation across a soil textural gradient in central Argentina. *Global Change Biology* 11:1101–1117.

Oesterheld, M., J. Loreti, M. Semmartin, and J. M. Paruelo. 1999. Grazing, fire, and climate effects on primary productivity of grasslands and savannas. In *Ecosystems of disturbed ground*, ed. L. R. Walker, 287–306. Amsterdam: Elsevier.

Oesterheld, M., O. E. Sala, and S. J. McNaughton. 1992. Effect of animal husbandry on herbivore-carrying capacity at a regional scale. *Nature* 356:234–236.

O'Neill, R. V., D. L. de Angelis, J. B. Waide, and T. F. Allen. 1986. *A hierarchical concept of ecosystems*. Princeton, NJ: Princeton University Press.

Paruelo, J. M. 2008. La caracterización funcional de ecosistemas mediante sensores remotos. *Ecosistemas* 17:3.

Paruelo, J. M., I. C. Burke, and W. K. Lauenroth. 2001a. Land use impact on ecosystem functioning in eastern Colorado, USA-*Global Change Biology* 7:631–639.

Paruelo, J. M., M. F. Garbulsky, J. P. Guerschman, and E. G. Jobbágy. 2004. Two decades of NDVI in South America: Identifying the imprint of global changes. *International Journal of Remote Sensing* 25:2793–2806.

Paruelo, J. M., E. G. Jobbagy, and O. E. Sala. 2001b. Current distribution of ecosystem functional types in temperate South America. *Ecosystems* 4:683–698.

Paruelo, J. M., and W. K. Lauenroth. 1998. Interannual variability of NDVI and their relationship to climate for North American shrublands and grasslands. *Journal of Biogeography* 25:721–733.

Paruelo, J. M., G. Piñeiro, C. Oyonarte, D. Alcaraz, J. Cabello, and P. Escribano. 2005. Temporal and spatial patterns of ecosystem functioning in protected arid areas of southeastern Spain. *Applied Vegetation Science* 8:93–102.

Paruelo, J. M., S. R. Verón, J. N. Volante, et al. 2011. Elementos conceptuales y metodológicos para la Evaluación de Impactos Ambientales Acumulativos (EIAAc) en bosques subtropicales. El caso del Este de Salta, Argentina [Conceptual and Methodological Elements for Cumulative Environmental Effects Assessment (CEEA) in Subtropical Forests. The Case of Eastern Salta, Argentina]. *Ecología Austral* 21:163–178.

Peterson, G. D., C. R. Allen, and C. S. Holling. 1998. Ecological resilience, biodiversity and scale. *Ecosystems* 1:6–18.

Pettorelli, N., J. O. Vik, A. Mysterud, J. M. Gaillard, C. J. Tucker, and N. C. Stenseth. 2005. Using the satellite-derived Normalized Difference Vegetation Index (NDVI) to assess ecological effects of environmental change. *TRENDS in Ecology and Evolution* 20:503–510.

Piñeiro, G., M. Oesterheld, and J. M. Paruelo. 2006. Seasonal variation in aboveground production and radiation use efficiency of temperate rangelands estimated through remote sensing. *Ecosystems* 9:357–357.

Prince, S. D. 1991. A model of regional primary production for use with coarse resolution satellite data. *International Journal of Remote Sensing* 12:1313–1330.

Ranson, K. J., and G. Sun. 1994. Mapping biomass for a northern forest ecosystem using multifrequency SAR data. *IEEE Transactions on Geoscience and Remote Sensing* 32:388–396.

Roberts, G., M. J. Wooster, G. L. W. Perry, et al. 2005. Retrieval of biomass combustion rates and totals from fire radiative power observations: Application to southern Africa using geostationary SEVIRI imagery. *Journal of Geophysical Research* 110:1–19.

Running, S., R. R. Nemani, F. A. Heinsch, M. Zhao, M. Reeves, and H. Hashimoto. 2004. A continuous satellite-derived measure of global terrestrial primary production. *BioScience* 54:547–560.

Running, S. W., P. E. Thornton, R. R. Nemani, and J. M. Glassy. 2000. Global terrestrial gross and net primary productivity from the earth observing system. In *Methods in ecosystem science*, eds. O. Sala, R. Jackson, and H. Mooney, 44–57. New York: Springer-Verlag.

Sellers, P. J., J. A. Berry, G. J. Collatz, C. B. Field, and F. G. Hall. 1992. Canopy reflectance, photosynthesis, and transpiration. A reanalysis using improved leaf models and a new canopy integration scheme. *Remote Sensing of Environment* 42:187–216.

Singh, J. S., W. K. Lauenroth, and R. K. Sernhorst. 1975. Review and assessment of various techniques for estimating net aerial primary production in grasslands from harvest data. *Botanical Review* 41:181–232.

Sun, G., K. J. Ranson, Z. Guo, Z. Zhang, P. Montesano, and D. Kimes. 2011. Forest biomass mapping from Lidar and radar synergies. *Remote Sensing of Environment* 115:2906–2916.

Treuhaft, R. N., B. E. Law, and G. P. Asner. 2004. Forest attributes from radar interferometric structure and its fusion with optical remote sensing. *BioScience* 54:561–571.

Treuhaft, R. N., S. N. Madsen, M. Moghaddam, and J. J. van Zyl. 1996. Vegetation characteristics and underlying topography from interferometric radar. *Radio Science* 31:1449–1485.

van der Werf, G. R., J. T. Randerson, L. Giglio, et al. 2010. Global fire emissions and the contribution of deforestation, savanna, forest, agricultural, and peat fires (1997–2009). *Atmospheric Chemistry and Physics* 10:11707–11735.

van der Werf, G. R., J. T. Randerson, L. Giglio, N. Gobron, and A. J. Dolman. 2008. Climate controls on the variability of fires in the tropics and subtropics. *Global Biogeochemical Cycle* 22:28–36.

Vasallo, M. M., H. D. Dieguez, M. F. Garbulsky, E. G. Jobbágy, and J. M. Paruelo. 2013. Grassland afforestation impact on primary productivity: A remote sensing approach. *Applied Vegetation Science.* 16(3): 390–403.

Verón, S. R., E. G. Jobbágy, C. M. Di Bella, et al. 2012. Assessing the potential of wildfires as a sustainable bioenergy opportunity. *Global Change Biology Bioenergy* 4:634–641.

Verón, S. R., J. M. Paruelo, and M. Oesterheld. 2011. Grazing-induced losses of biodiversity affect the transpiration of an arid ecosystem. *Oecologia* 165:501–510.

Viglizzo, E. F., J. M. Paruelo, P. Laterra, and E. G. Jobbágy. 2012. Ecosystem service evaluation to support land-use policy. *Agriculture, Ecosystems & Environment* 154:78–84.

Virginia, R. A., and D. H. Wall. 2001. Principles of ecosystem function. In *Encyclopedia of biodiversity*, ed. S. A. Levin, 345–352. San Diego, CA: Academic Press.

Volante, J. N., D. Alcaraz-Segura, M. J. Mosciaro, E. F. Viglizzo, and J. M. Paruelo. 2012. Ecosystem functional changes associated with land clearing in NW Argentina. *Agriculture, Ecosystems & Environment* 154:12–22.

Walker, W. S., J. M. Kellndorfer, and L. E. Pierce. 2007. Quality assessment of SRTM C- and X-band interferometric data: Implications for the retrieval of vegetation canopy height. *Remote Sensing of Environment* 109:482–499.

Wang, Y., H. Weinacker, and B. Koch. 2007. Development of a procedure for vertical structure analysis and 3D single tree extraction within forest based on Lidar point cloud. *Proceedings of the ISPRS Workshop Laser Scanning 2007 and SilviLaser 2007* 36(3/W52):419–423.

Wessman, C. A. 1992. Spatial scales and global change: Bridging the gaps from plots to GCM grids cells. *Annual Review of Ecology and Systematics* 23:175–200.

Wooster, M. J., G. Roberts, G. L. W. Perry, and Y. J. Kaufman. 2005. Retrieval of biomass combustion rates and totals from fire radiative power observations: FRP derivation and calibration relationships between biomass consumption and fire radiative energy release. *Journal of Geophysical Research* 110:1–24.

Wulder, M. A., J. C. White, R. F. Nelson, et al. 2012. Lidar sampling for large-area forest characterization: A review. *Remote Sensing of Environment* 121:196–209.

Zhao, K., S. Popescu, and R. Nelson. 2009. Lidar remote sensing of forest biomass: A scale-invariant estimation approach using airborne lasers. *Remote Sensing of Environment* 113:182–196.

第三章 碳吸收相关陆地生态系统服务的光合胁迫估算最新进展

M. F. 加布尔斯基 I. 菲莱利亚 J. 佩努埃拉斯

一、简介

对陆地植被碳吸收量的估算仍是评价生态系统服务功能的主要挑战。一般而言,这些服务是生态系统对人类福祉的直接和间接贡献,换句话说,是由生态系统提供的产品与服务。由于光合作用是调节90%碳和水通量的关键过程(Joiner et al., 2011),因此它是生态系统所提供的诸多服务(例如气候调节、碳汇、碳储存、食品或草原畜牧业)的主要驱动力之一(Costanza et al., 1997; de Groot et al., 2002; Naidoo et al., 2008)。生态系统服务的完整性是人类福祉的基础。我们需要了解这些服务与生态系统过程间的关联。估算在当地、地区和全球范围内的这些关联的规模,对于长期维持这些关联来说至关重要(Haines-Young and Potschin, 2010)。

光合组织的分布随着空间、时间尺度以及土地利用方式的不同发生变化(如Paruelo et al., 2004)。传统的遥感技术可以评估绿色植物生物量,从而评估植物的光合能力。然而,检测光合能力的实际实现程度,是一个更具挑战性的目标,它需要估算光合碳通量或总初级生产力(Gross Primary Productivity, GPP),而总初级生产力是光合碳吸收的生态系统水平的体现。由于光合作用的周期性环境和生理限制,植物光合组织将吸收的光转化为有机化合物的效率(即陆地植被的光能利用效率, Light Use Effciency, LUE)也随时间和空间而变化(Runyon et al., 1994; Gamon et al., 1995; Garbulsky et al., 2010)。将功能类型进行对比（Gamon et al.，1997；Huemmrich et al.，2010），干旱、极端温度（Landsberg and Waring，1997；Sims et al.，2006）及营养水平（Gamon et al.，1997；Ollinger et al.，2008）是导致这种变化的因素。

在各种形式下，初级生产力、植被吸收的辐射量、植物将辐射转化为生物量的效率之间的简单关系（Monteith，1977），已经成为从树冠到全球范围内评估光合作用和初级生产力的许多方法的基础（Field et al.，1995；Running et al.，2004）。蒙蒂斯（Monteith，1977）最初提出这种关系，用于估算净初级生产力（NPP）。然而，生产效率模型（Production Effciency Models，PEMs）是基于GPP 的 LUE 形成的理论，该理论认为光合碳吸收与冠层吸收的辐射之间存在相对恒定的关系。估算碳吸收量的不同方法都是基于这种 LUE 模型。其中许多方法假定设了一个恒定效率（Myneni et al.，1995），或从文献中引用的生物群落的恒定值导出这一术语（Ruimy et al.，1994）。

另一种方法是利用气象变量，如蒸汽压差（Vapor Pressure Defcit，VPD）和温度，作为光合胁迫的替代指标，来降低生物群落的最大效率（Running et al.，2004）。由于蒸汽压差和温度本身并不总是效率降低的良好指标（Garbulsky et al.，2010），基于气象的方法并不总能充分地解释效率变化。因此，为了更好地评估衍生的生态系统服务，还需要其他方法来确定光合胁迫或 LUE，来估算碳吸收量。因此，直接远程估算 LUE 非常重要，可能有多种应用，比如用于估算生产力（二氧化碳固定），或在叶面积减少之前，检测环境胁迫对植被碳吸收的影响（Garbulsky et al.，2008）。

由于最近科学界和社会对碳循环变化的全球影响提出了担忧，我们远程估算 LUE 的能力有了很大的提高（Grace et al.，2007）。远程估算 LUE 的方法与我们如何从反射光、热量或叶绿素荧光中收集光合效率的信号有关（图 3-1）。虽然从这些路径中远程估算的理论似乎足够清晰，并且估算是在不同的尺度上进行，但随着新平台和传感器的出现，仍有复杂的问题尚待解决。

与碳吸收量和初级生产力的远程估算有关的主要挑战，在于阐明如何将信号从叶片放大到整个生态系统，以及远程传感器如何评估从叶片到生态系统的功能特征。尤其是估算实际光合性能或叶片光合作用胁迫，并将其扩大到生态系统水平这一挑战。本章中，我们回顾并综合了利用当前和未来数据远程估算

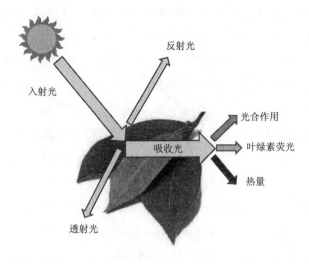

图 3-1（见彩插）

注：到达叶片的入射光和被树冠截留后吸收光的补充路径。叶绿素吸收的光可用于驱动光化学，该过程可使用高达 80% 的吸收辐射。此外，它可以以荧光（吸收辐射的 0.5%~2%）或热量（18%~98%）的方式而损失。光合作用、叶绿素荧光和热损失是密切相关的、并且是直接竞争的。因此，一个过程的加速将不可避免地降低其他过程的速率。

LUE 的最新方法。我们提出了估算 LUE 的最新方法及现有遥感技术和数据的最新进展。

二、远程估算陆地植被光合胁迫的替代方法

（一）叶片色素

叶片色素变化是诊断一系列植物生理特性与过程的关键工具（Peñuelas and Filella，1998；Blackburn，2007）。那些与叶片色素含量和循环相关的、试图估算碳吸收效率的不同方法，都是建立在与叶片级别生化过程关系的基础上。

1. 叶绿素含量

该方法是基于对植被叶绿素含量（Chl）的远程估算，从而对总初级生产力（GPP）进行估算。在一年生作物中，由于冠层叶绿素含量的长期或中期变化与作物物候学、冠层胁迫和植被光合能力有关（如 Ustin et al.，1998；Zarco-Tejada *et al.*,2002),这些变化也可能与总初级生产力相关。在冠层级别上,叶绿素含量可能是与预测生产力最相关的群落属性(Dawson *et al.*,2003)。叶绿素含量不是光能利用效率的替代指标,但它可用于估算总光合能力。这种方法并不完全依赖于光能利用效率的估算,而是假设叶绿素含量等于光合有效辐射吸收比率与光能利用效率的乘积。原则上,它不依赖于广泛使用的归一化差异植被指数(即由冠层的红光和红外反射率得出的光谱指数)与光合有效辐射之间的关系。本章中提出的光能利用效率模型可表达为总初级生产力 GPP＝VI×PAR,其中 PAR 为光合有效辐射,VI 是叶绿素含量的光谱指数代替指标。

前文提到的植被指数(Vegetation Indices,VIs)是基于两个光谱通道的反射率(ρ):近红外(near-infrared,NIR)和绿或者红边通道的反射率。一些植被指数(VIs),如 MERIS 陆地叶绿素指数(MERIS Terrestrial Chlorophyll Index,MTCI,MTCI＝[ρNIR－ρred edge]/[ρred edge－ρred])和叶绿素指数(绿光叶绿素指数 CIgreen＝ρNIR/ρgreen－1;红边叶绿素指数 CIred edge＝ρNIR/ρred edge－1)已被专门提出来估算总叶绿素含量(Gitelson *et al.*,2003,2005;Dash and Curran,2004)。在一年生作物(如玉米)中,总叶绿素含量与植被指数(VIs)间的关系表明,一些植被指数(VIs)可解释 87% 以上的叶绿素含量变化(Peng *et al.*,2011)。玉米和大豆中绿光叶绿素指数或红边叶绿素指数,与总叶绿素含量的关系的测定系数均超过 0.92(Gitelson *et al.*,2005)。因此,这些与叶绿素和叶面积指数(Leaf Area Index,LAI)相关的植被指数可作为叶绿素在总初级生产力 GPP＝VI×PAR 模型中的替代指标,特别是针对那些,水分或营养胁迫导致总叶绿素损失、从而引起碳吸收量快速降低的一年生草本植物。在生态系统层面,在涵盖北美植被组成广泛变化的 15 个涡度协方差塔中,对于农田和落叶林来说,MERIS 陆地叶绿素指数和塔式总初级生产力间的相关程度比中分辨率成像光谱仪(MODIS)的总初级生产力或增强植被指数和塔式总初级生产力间的相关程度更强,而对于草原来说则相对较小。然而,常绿林(Harris and Dash,2010)或泥炭地(Harris and Dash,2011)并非如此。这些分析表明,来自 MERIS 传感器的数据可作为 MODIS 的替代数据用于估算碳通量。塔式总初级生产力与两种植被指数(增强植被指数和 MERIS 陆地叶绿素指数)之间的相关性仅对于落叶植被而言是相似的,这表明生理驱动的光谱指数,如 MERIS 陆地叶绿素指数，也可以在基于卫星的碳通量建模工作中补充现有的基于结构的指数。因此，总初级生产力和任何叶绿素指数间的关系可能高度依赖于不同植被功能类型的参数和强度（图3-2）。

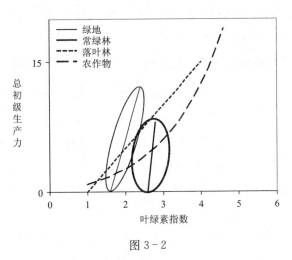

图3-2

注：不同物候和总初级生产力动态的生态系统中午间总初级生产力与叶绿素指数的关系示意图。农作物和落叶林的关系得到了文献的充分支持，而椭圆代表的数据强度较低、分散性较高，即草原和常绿林的关系缺乏文献支持。

基于总叶绿素含量和附带光合有效辐射的模型的最终评估表明，该模型可用于准确估算灌溉及雨养玉米和大豆的总初级生产力（Peng and Gitelson，2012）。由于叶片结构和冠层结构的差异，总初级生产力估算算法对于玉米和大豆存在物种特异性，尤其是采用基于近红外和红绿光反射率的植被指数时。然而，采用对作物种类差异最不敏感的MERIS光谱带的红边叶绿素指数、MERIS陆地叶绿素指数和红边归一化植被指数，就可能在玉米和大豆中实现估算总初级生产力的统一算法。红边叶绿素指数和红边归一化植被指数（红边波段约720纳米）对玉米和大豆来说不存在物种差异，且在估算玉米和大豆综合总初级生产力时准确性极高。

由于光能利用效率与叶绿素含量之间存在相当稳定的关系，这种技术可以准确地估算玉米和大豆作物、甚至其他一年生作物在雨养和灌溉条件下的午间总初级生产力。相比之下，几乎没有证据表明这种方法在常绿林或其他适应干旱的树冠等生态系统类型中有用。

进一步的研究采用绿光叶绿素指数（主要用于估算冠层叶绿素含量（Gitelson et al.，2005），利用无云 MODIS 影像（500 米）和玉米的通量测量（Wu et al.，2012），作为午间光能利用效率的估算值。这一关系随后被成功应用于估算针叶林和草地的午间光能利用效率。

2. 叶片色素循环和光化学反射率指数

植物色素循环的遥感研究已被验证是用于检测光能利用效率的空间及时间变化的有效途径。这种用于估算光能利用效率的遥感方法，以叶黄素循环的去环氧化状态为基础，而这种状态与叶片的散热有关（Demmig-Adams and Adams，1996）。这是一个激发叶绿素的衰变过程，它与光合电子传输相竞争，并与之互补（Niyogi，1999）。高光谱遥感已被用于开发技术和分析方法，以便在一系列空间尺度上无损地重复量化色素。在推导与植物色素相关的高光谱反射率数据的各种特征和转换之间的预测关系方面所获最新进展表明，在生态生理学、环境、农业和林业科学中，这种方法的应用范围不断扩大（Blackburn，2007；Nichol et al.，2012）。

在 20 世纪 90 年代，利用地面或低平台的近距遥感，在叶片和近冠层水平上进行了一系列研究，能够基于同时发生的叶黄素色素变化来评估这一效率参数（光能利用效率）（Gamon et al.，1990，1992，1997；Peñuelas et al.，1994，1995，1997，1998；Filella et al.，1996；Gamon and Surfus，1999）。因为 531 纳米波段的反射率在功能上与叶黄素循环的去环氧化状态是相关的（Gamon et al.，1990，1992；Peñuelas et al.，1995），光化学反射率指数（通常计算式为[R531－R570]/[R531＋R570]，其中 R 表示反射率，数字表示波段中心的波长纳米），是一种利用窄波段反射率远程评估光合效率的方法（Gamon et al.，1992；Peñuelas et al.，1995）。

利用光化学反射率指数度量绿光反射率峰（550 纳米；图 3-3）两侧的相对反射率，它还比较了光谱中蓝光（叶绿素和类胡萝卜素吸收）区域的反射率与红光（仅叶绿素吸收）区域的反射率（Peñuelas et al.，2011）。因此，它可作为相对叶绿素/类胡萝卜素水平的指标，通常称为本体色素比率。据报道，在较长的时间范围内（几周至几个月），由于叶片发育、老化或慢性胁迫，导致本体色素含量和比率的变化,与叶黄素色素环氧化一同在光化学反射率指数信号中发挥着重要作用(Peñuelas et al.,1997;Gamon et al.,2001;Stylinski et al.,2002)。因此,光化学反射率指数(PRI)也经常与在不同物种、年龄和条件下的叶片中的类胡萝卜素/叶绿素比率相关(Stylinski et al.,2002;Filella et al.,2009)。

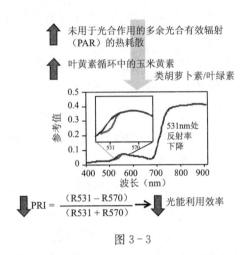

图 3-3

注:利用光化学反射率指数遥感陆地植被总光能利用效率的色素循环方法。叶黄素循环的遥感为估算在区域范围内,未用于光合作用的多余辐射从叶片到初级生产力的耗散提供了替代指标。箭头和反射光谱的深灰色表示辐射过剩、光能利用效率较低的条件,而531nm附近的灰色光谱表示光能利用效率较高时的条件。

因此,在光合活性与叶绿素、类胡萝卜素比例变化相关的情况下,光化学反射率指数可能提供一个相对光合作用速率的有效方法。季节性变化的色素水平也会极大地影响光化学反射率指数。这种季节变化可能有助于解释光化学反射率指数预测光能利用效率时的良好表现,因为叶绿素/类胡萝卜素比率与叶黄素色素水平共同变化。这里描述的所有关系表明,这种方法为显著改善全球和区域陆地生态系统的二氧化碳吸收监测提供了巨大的可能性。

尽管这些波长选择的机理在叶片尺度上已被充分探讨(Gamon et al.,1993),但在冠层或更大尺度上的理论支持较少。其中使用了多种交替波段,这些波段通常基于统计相关性(Gamon et al.,1992;Inoue et al.,2008)或由仪器限制确定(Garbulsky et al.,2008)。关于哪种光化学反射率指数(PRI)波长能够最好地估算光能利用效率,文献中缺乏明确的共识,阻碍了交叉研究比较。因此，目前还不完全清楚，在叶片尺度(531 纳米和 570 纳米)测量该特征的最佳波长是否必然是逐步扩大的尺度下的最佳波长，而在更大尺度下，多重散射和其他混合效应可能改变叶黄素循环特征的光谱响应，就像色素吸收峰会因其化学和散射介质而变化一样。因此，可能需要更多的工作来确定机载或星载平台的理想光化学反射率指数算法；而合适的机载和航天仪器的有限可用性以及高成本，已经对这些研究产生了阻碍。

总之，这些对叶黄素循环的去环氧化状态和类胡萝卜素、叶绿素比率的反应，确保了光化学反射率指数与光合效率的尺度在不同的条件、物种和功能类型中的变化。现有证据表明，光化学反射率指数是一个可靠的生态生理变量估算指标，在广泛的物种、植物功能类型和时间尺度上，与叶片和冠层水平的光合效率密切相关(Garbulsky et al.，2011)。

自 2000 年以来，MODIS 传感器提供的全球数据的可用性一直是在生态系统水平测试光化学反射率指数效能的主要工具。卫星搭载的 MODIS 传感器提供的 530 纳米(526~536 纳米)波段被用作不同植被类型生态系统规模的光能利用效率指标，取得了显著成功(Rahman et al.，2004；Drolet et al.，2005，2008；Garbulsky et al.，2008a，2008b；Goerner et al.，2009)。目前，具备高光谱分辨率的空间遥感仪器很少(请注意，Hyperion 和 CHRIS/PROBA 是例外，这些示范性仪器可获得性有限)。但现在可从一系列新型直升机和飞行器(Malenovsky et al.，2009)和计划中的新卫星数据中收集这些类型数据。

除了在轨的传感器外，还有其他各种各样的传感器(表 3-1)，它们将基于色素生理学，从太空提供更好的数据来估算光能利用效率。新影像光谱仪的推出(如美国宇航局的 HyspIRI 或德国航空航天中心牵头的 E 纳米 AP 项目)，将允许在高空间分辨率下计算光化学反射率指数，从而为植物色素循环的遥感研究提供巨大的新潜力。

3. 缺点、注意事项和警示说明

有一些不同的问题仍然阻碍着光化学反射率指数在生态系统和生物圈范围内的普遍使用，以及作为光能利用效率估算器在全球和操作上的应用(Grace et al.，2007)。文献综述显示，在光能利用效率-光化学反射率指数关系中，可能存

表3-1 现有和未来的通过植物色素估算光合效率的卫星工具

卫星	传感器	发射/服务期	索引	数据的主要特征
地球观测1号（Earth Observing-1，EO-1）	Hyperion	2000年11月	http://eo1.gsfc.nasa.gov	下降的极地轨道，穿越赤道时间为10:03。具备220个光谱通道（波长范围400~2 500纳米）的高分辨率高光谱成像仪。分辨率为30米，运行周期为16天
PROBA（卫星自主计划）	CHRIS（Compact High Resolution Imaging Spectrometer，小型高分辨率成像光谱仪）	2001年10月22日	https://earth.esa.int/web/guest/missions/esa-operating-Eo-missions/propa	太阳同步轨道，运行周期为7天。根据设置模式，最低点设置的空间分辨率为18或34米。沿着轨道对最多五个角度的光化学反射率指数进行窄波段光谱观测
TERRA-AQUA	中分辨率成像光谱仪（Moderate Resolution Imaging Spectroradiometer，MODIS）	Terra:2000年1月，Aqua:2002年5月	http://modis.gsfc.nasa.gov	两颗卫星上的同一传感器。运行周期每天两次（上午和下午），具备16个光谱带，波长范围400~1 000纳米，空间分辨率为250米，500米和1 000米，带宽为10纳米
先进地球观测卫星II（ADvanced Earth Observing Satellite II，ADEOS-2）	全球成像器（GLobal Imager，GLI）	2002年12月~2003年10月	http://sharaku.eorc.jaxa.jp/ADEOS2	6个250米分辨率频道和30个1千米分辨率频道，运行周期为4天
环境测绘和分析程序（Environmental Mapping and Analysis Program，EnMAP）	高光谱成像仪（HyperSpectral Imager，HSI）	计划2015年	HTTP://WWW.ENMAP.ORG	太阳同步轨道，运行周期为4天，空间分辨率为30米×30米，94个光谱带，波长范围420~1 000纳米之间
高光谱红外成像仪（HyspIRI）	VSWIR	目前处于研究阶段	http://hyspiri.jpl.nasa.gov	尺度为10纳米的连续光谱带，波长范围380~2 500纳米，最低点的空间分辨率为60米，运行周期19天

注：当前发射情况已有变化。

在多种生物化学、生态和物理混杂因素,这些因素在多个聚集层面上起作用(Garbulsky et al.,2011)。在叶片层面,包括光呼吸、PSI 循环电子运输和硝酸盐还原在内的生化过程可与二氧化碳的固定竞争光合电子运输产生的还原剂(Niyogi,1999)。这可导致 PSII(二号光合体系)效率(光化学反射率指数)和二氧化碳同化发生分歧。甚至还有其他色素循环,例如热带树木的叶黄素环氧化物循环中所包含的色素循环(Matsubara et al.,2008;Esteban et al.,2009),也可能在光化学反射率指数信号中产生干扰。尽管存在这些潜在的并发问题,但总体光合系统似乎通常被充分调节,以维持 PSII 过程和二氧化碳固定间的一致关系(Gamon et al.,1997;Stylinski et al.,2002)。另一方面,如前所述,由于色素比率与光能利用效率间没有密切关系,不断变化的色素比率就成为混淆变量。

在冠层层面,存在的问题则与冠层的结构差异、卫星数据的不同背景效应(如土壤颜色、湿度、阴影或其他非绿色景观成分的存在)、光照和视角变化导致的不同反射信号等方面有关,以及冠层和林分结构的其他物理效应(如 LAI 变化),包括叶片移动、太阳和视角、土壤背景和阴影等(Filella et al.,2004;Sims et al.,2006;Hilker et al.,2010),这些都会显著影响光化学反射率指数信号(Barton and North,2001;Gamon et al.,1995)。不同研究表明,光化学反射率指数反射率可能受到以太阳为目标的传感器的几何形状和支架结构的影响(Asner,1998;Barton and North,2001;Drolet et al.,2005;Hall et al.,2008;Hilker et al.,2008)。最新进展(来自拥有 34 米空间分辨率的 PROBA 卫星上的 CHRIS 传感器的多角度卫星观测分析)表明,光化学反射率指数取决于常绿或混交林地低水平光能利用效率的冠层阴影比例。光化学反射率指数和冠层阴影比例之间的关系斜率与总光能利用效率之间存在负对数关系(Hilker et al.,2011)。相比之下,最近的其他分析显示,与增强植被指数相比,在三个森林地点的 10 年间,MODIS、光化学反射率指数和归一化植被指数不受视角的影响(Sims et al.,2011),这一结果可能与其他研究相矛盾。在任何情况下,阴影比例都可能是影响数据空间分辨率的一个变量。

光化学反射率指数、光能利用效率和二氧化碳吸收之间关系的一致性,在生态系统的不同研究中得到了越来越多的发现(Garbulsky et al.,2011;Hilker et al.,2011;Peñuelas et al.,2011),这表明在生态系统类型之间影响生态系统碳通量的生化、生理和结构成分有很大程度的功能趋同(Field,1991)。新兴的生态系统特性可允许通过简单的光谱方法来探索其复杂的光合行为,例如通过光化学反射率指数来度量植物色素循环。理解这种趋同的基础(并找到控制这些反应的生态生理学原理)仍是当前研究的主要目标。同时,光化学反射率指数对二氧化碳吸收的实用经验性远程测量更为重要,特别是近最低点卫星观测(Goerner et al.,2010)和多角度大气校正(Lyapustin and Wang,2009;Hilker et al.,2010),可成为全球持续性监测总初级生产力的优秀手段,这对于跟踪气候变化下的碳封存至关重要。

需要使用统一的协议来生成可比较的数据,并最终对光化学反射率指数-光能利用效率关系进行可能的一般校准(Peñuelas et al.,2011)。还需要进一步的研究来厘清光化学反射率指数信号的几个驱动因素,并解决潜在的混淆因素,以改进使用高光谱或窄波段遥感对诸多不同生物群落二氧化碳通量的评估。

(二) 叶绿素荧光

从太空观测叶绿素荧光来估算光能利用效率和碳吸收,无疑是近年来取得巨大进步的一个知识领域。空间对地植被荧光的遥感是非常有意义的,因为它可以提供包括光能利用效率和总初级生产力在内的植被功能状态的全球信息。对陆地植被发出的太阳诱导荧光的全球检索,开始为光合作用效率提供一个前所未有的衡量标准。

1. 荧光的基本原理、来源和特征

荧光测量技术的改进使荧光法成为基础和应用植物生理学研究的重要工具(Krause and Weis,1991)。荧光信号来源于光合组织的核心,在光合组织中,吸收的光合有效辐射被转换成化学能。荧光反映了天线中捕获的激发物在多条路径之间的竞争情况。当光化学作用以最大效率发生时,激发物主要传递到光反应。当光化学反应停止时,由于荧光和非辐射耗散路径间的竞争,激发物损失,后者将能量转化为热量。由于荧光产量与开放反应中心的比例成反比,它为研究光合过程提供了有用途径。同化叶片中叶绿素吸收的一小部分光可以被光合系统 II 的叶绿素分子重新发射为荧光,从而为反射的太阳辐射增加了微弱信号。由于这些过程发生在竞争中,通过测量叶绿素荧光,我们还可知道光化学和

散热的效率,这与二氧化碳同化有关(Baker,2008)。

传统上,叶绿素荧光测量是使用外部光源在叶片层面上进行的。荧光光谱峰的波长比吸收光长;因此,荧光产量可通过将叶片暴露于限定波长的光下,并在光关闭时测量在较长波长下重新发射的光量来量化。这种技术可以量化与碳固定的光相有关的不同参数。大量实验和理论研究表明,叶绿素荧光是实际光合作用的代表物,因此与光能利用效率和二氧化碳吸收直接相关(Seaton and Walker,1990);由于发射荧光与植物建立的适应/保护机制竞争,它也可以作为植物生命力和植物胁迫的指标。因此,测量荧光可以提供关于光合作用性能变量的缺失信息(Maxwell and Johnson,2000)。

这种实验室测量方法远未达到估算生态系统规模中荧光所需的要求;因此,需要新的方法从太空来测量叶绿素荧光。由于在环境条件下对光照和水分胁迫等干扰反应迅速,太阳诱导叶绿素荧光(F)可以为评估植被的实际功能状态提供早期和直接的方法;因此,太阳诱导叶绿素荧光可在叶绿素含量或 LAI 显著降低之前检测到胁迫状况。绿色植被的荧光由蓝绿色荧光(最大值在 440～520 纳米)、红色和远红色叶绿素荧光(最大值在 690～740 纳米)组成。为了监测植被的光合作用,需要分析叶片组织光合活性部分中的红色和远红色叶绿素荧光。最大值在 685～740 纳米的两个宽峰的大小可能与光合效率有关。由于植被和大气反射的太阳辐射强度可能是大气顶部的太阳诱导叶绿素荧光强度的 100～150 倍,因此从遥感被动测量中估算太阳诱导叶绿素荧光强度的主要挑战是,将太阳诱导叶绿素荧光强度信号与太阳辐射分离(Meroni et al.,2009)。

2. 最新进展

尽管首次尝试在没有人工激发源的情况下,对陆地植被的叶绿素进行量化可追溯到 20 世纪 70 年代,但叶绿素荧光的遥感仍处于发展阶段。尽管如此,研究仍在不断推进。太阳诱导叶绿素荧光强度信号可在太阳和大气光谱的窄吸收线(约 2～3 纳米)中被动地检测到,在这些吸收线中,辐照度会显著降低(即弗劳恩霍夫线)。利用可见光和近红外光谱中的三条主要弗劳恩霍夫线可估算太阳诱导叶绿素荧光强度:由于太阳大气中的氢(H)吸收,以 656.4 纳米为中心的 Hα 波段;地球大气中两个碲氧(O_2)吸收波段,即以 687.0 纳米为中心的 O_2-B

和760.4纳米的O_2-A。结合弗劳恩霍夫线和O_2线,就有可能测量所有的主要荧光带。两个O_2带(A和B)和$H\alpha$波段被认为是最有用的(Meroni et al.,2009)。

高光谱分辨率下的辐射测量利用弗劳恩霍夫线将太阳诱导叶绿素荧光强度从反射通量中分离。从太阳弗劳恩霍夫线中获取太阳诱导叶绿素荧光强度是基于对太阳诱导叶绿素荧光强度及以后引起的弗劳恩霍夫线填充的评估。由于弗劳恩霍夫线的分数深度不受大气散射影响,也不受不含碲化物吸收特征的狭窄光谱窗口的影响,因此所需的大气建模比用大气波段更简单。

虽然已有很多荧光测量方法,特别是最近的通过地面和机载仪器测量荧光的方法(Meroni et al.,2009),但从卫星获得的信息很少。利用ENVISAT中分辨率成像光谱仪(MEdium Resolution Imaging Spectrometer,MERIS)提供的O_2-A吸收带数据,首次对太阳诱导叶绿素荧光强度进行了基于太空的估算(Guanter et al.,2007)。MERIS测出的荧光与由紧凑型机载光谱成像仪(Compact Airborne Spectrographic Imager,CASI-1500)传感器获得的数据和基于地面的估算具有很好的相关性($R^2=0.85$)。最新的研究(Joiner et al.,2011;Frankenberg et al.,2011)提出了在全球范围内碳观测热近红外传感器-傅里叶变换光谱仪(Thermal and Near-infrared Sensor for carbon Observation-Fourier Transform Spectrometer,TANSO-FTS/GOSAT)的高光谱分辨率数据的第一个结果。这两项研究采用了不同的方法。乔伊纳等人(Joiner et al.,2011)使用的方法利用一条强弗劳恩霍夫线(KI在770.1纳米处)。它采用了TANSO-FTS的真实太阳辐照度测量,以避免仪器线形函数(ILSF)的显式建模。进而,在弗兰肯伯格等人(Frankenberg et al.,2011)提出的方法中将这种单线方法扩展到以755纳米和770纳米为中心的两个更宽的光谱窗口。这种方法利用了包含多条弗劳恩霍夫线的更宽的光谱窗口,预计对仪器噪声的敏感度会低于基于一条单线的估算方法,而两个单独窗口执行的估算提供了更独立的测量,将会增强最终太阳诱导叶绿素荧光强度结果的信噪比。目前,来自22个月系列的数据显示,与太阳诱导叶绿素荧光强度水平和空间模式相比,太阳诱导叶绿素荧光强度基于物理的估算方法是准确的。然而,有必要将生物群落相关性从太阳诱导叶绿素荧光强度扩展到总初级生产力(Guanter et al.,2012)。

3. 叶绿素荧光的未来

有几个项目正在开发卫星平台和传感器，以便远程测定叶绿素荧光，估算陆地植被的碳循环(表3-2)。由于植物荧光可转化为光合活性的指标，通过使用荧光数据，我们可以更好地理解植物吸收了多少碳，以及它们在碳循环和水循环中的作用。除了本文发表时在轨的传感器外，还有许多与植被荧光测量相关的卫星项目。从区域到全球范围内，详细的荧光数据的有效性是向前迈出的一大步，将代表着生产效率模型评估陆地植被碳吸收的时空变化方面能力的巨大提升。欧洲航天局(European Space Agency)牵头了一个更具野心的重大卫星项目，计划于2018年进入轨道。它被称为荧光探索者(FLuorescence EXplorer，FLEX)，其目标是观察光合作用，以便更好地了解碳循环，提供全球植被荧光地图。主要仪器为荧光成像光谱仪(Fluorescence Imaging Spectrometer，FIS)，它覆盖了光谱带为20纳米的O_2-A(760纳米)和O_2-B(677纳米)吸收线。最低点的地面空间分辨率为300米，运行周期为7天。

另一个重要的计划卫星是美国宇航局的轨道碳观测卫星二号(Orbiting Carbon Observatory-2，OCO-2)，它将专门用于从太空研究大气中的二氧化碳。三个高分辨率光栅光谱仪(Day et al.，2011)——光谱带位于O_2A波段(757~772纳米)、弱二氧化碳波段(1 590~1 621纳米)和强二氧化碳波段(2 041~2 082纳米)的每个光谱仪都将与气象观测和地面二氧化碳测量相结合，以1.29千米×2.25千米的空间分辨率，持续两年在区域范围内对二氧化碳源和汇进行月度观测。

地球同步碳过程测绘仪(Geostationary Carbon Process Mapper，GCPM)拥有三个计划中的地球同步平台。该项目旨在通过高时间分辨率测量与气候变化和人类活动相关的关键大气微量气体和过程示踪剂。这种理解来自于从地球同步轨道(Geostationary Orbit，GEO)上以相对较高的空间分辨率(约4千米×4千米)收集的二氧化碳、甲烷、一氧化碳和太阳诱导叶绿素荧光强度的连续地图，每天收集可高达10次。这些测量将捕捉碳循环在昼夜、天气、季节和年际等时间尺度上的时空变化。高分辨率测绘和高测量频率的结合实现了准连续监

表3-2 太空叶绿素荧光遥感评估工具

人造卫星	传感器	发射、在机	索引	数据的主要特征
ENVISAT	中分辨率成像光谱仪	2002年3月~2012年4月	http://wdc.dlr.de/sensors/meris	空间分辨率：260米×300米，O_2-A波段附近的两个通道，中心位于753.8纳米和760.6纳米，带宽分别为7.5纳米和3.75纳米
	大气图扫描成像吸收光谱仪（SCanning Imaging Absorption spectroMeter for Atmospheric CHartography, SCIMACHY）	2002年3月~2012年4月	http://www.sciamachy.org wdc.dlr.de/sensors/SCI-AMachy	波长范围为0.2~0.5纳米，240~1 700纳米，2 000~2 400纳米，边缘垂直3千米×132千米，最低点水平32千米×215千米
GOAST（Greenhouse gases Observing SATellite，温室气体观测卫星）	碳观测用热和近红外传感器-傅里叶变换光谱仪	2009年1月23日	http://www.gosat.nies.go.jp	太阳同步轨道，穿越赤道时间13：00，运行周期3天
轨道碳观测卫星-2（Orbiting Carbon Observatory-2，OCO-2）	带有三个经典光栅光谱仪的单一仪器	2014年7月	http://oco.jpl.nasa.gov	太阳同步轨道，穿越赤道时间为中午；三个测量角度
Landsat-8	荧光探测器荧光成像光谱仪	2018年或之后，目前正在开发中	http://esamultimedia.esa.int/docs/SP1313-4_FLEX.pdf	下行的太阳同步轨道，赤道穿越时间10：00
地球同步碳过程测绘仪	地球同步傅里叶变换光谱仪（Geostationary Fourier Transform Spectrometer, GeoFTS）	项目阶段	Key et al.，2012	与地球同步：每天10次，空间分辨率约4千米×4千米

注：当前发射情况已有变化。

测,有效地消除了源/汇反演模型中大气传输的不确定性。二氧化碳/甲烷/一氧化碳/太阳诱导叶绿素荧光强度的测量还可以提供必要的信息,以区分自然和人为因素对大气碳浓度的贡献。

4. 关于从太空中测算叶绿素荧光的最终评论

目前,在从太空测算叶绿素荧光估算植被碳吸收方面,有许多新的令人鼓舞的成果和正在进行的项目。利用弗劳恩霍夫线反演方法从太空对叶绿素荧光进行测算是可行的,该方法简单、快速、稳定,目前已得到地面实际数据的验证。荧光显示的信息与反射率数据无关(如 fAPAR)。来自 GOSAT 和 OCO-2 的叶绿素荧光测算,结合其全球大气二氧化碳测量,将为全球碳额度限制提供一个植被和大气视角的特殊组合,约束了我们对未来大气二氧化碳盈余的模型预测。最重要的是,这种方法在很大程度上不受大气散射的影响,甚至能够通过薄云感知太阳诱导叶绿素荧光强度。

虽然这种方法看起来很有前途,但现有数据也存在不同问题。在北方夏季,由于草原和农田中产生的高估和北方针叶林中产生的低估,太阳诱导叶绿素荧光强度和总初级生产力间的相关性不那么强。因此,需要进一步的研究来厘清光照、冠层结构、卫星数据的时空分辨率等干扰因素对太阳诱导荧光与光合作用之间关系的影响。

(三) 结语

虽然涡动协方差塔代表了当前生态系统碳通量总初级生产力估算的标准,但如果我们的目标是研发可靠的生态系统碳通量远程采样方法,我们必须学会用新的遥感产品对这些估算进行适当校准。

这仍是一个重大挑战,因为通量塔通过时间采样,而遥感影像通过空间采样(Rahman et al.,2001)。为了进行校准,我们应该在与通量塔足迹测量相同的时空尺度上,通过运用遥感飞行器和卫星测量,来混合这些采样领域,而这一点很少有人做到。因此,需要从不同的生物群落协调获取通量和光学数据。此外,应扩大通量塔的标准化地面光学采样计划(Gamon et al.,2006)。我们必须对不同生态系统或植被类型的光能利用效率替代指标进行适当校准,然后才能运用遥感技术从通量塔位置进行时间和空间上的推断。通量塔在不同植被类型中的实际广泛分布(本方法的长期目标)以及在不同空间、时间和光谱分辨率下遥感工具可用性的提升,是进一步研究的主要优势和原因。

本章概述的不同方法强调了叶绿素荧光的远程估算如何在不久的将来为全球光能利用效率提供准确估算。尽管针对叶绿素荧光估算的研究工作超过了所有其他方法,但叶绿素荧光估算对生态系统的依赖性似乎比其他方法更小。考虑到总初级生产力估算的目标,每种遥感方法的时间反应是光能利用效率估算中需要考虑的一个问题。尽管叶绿素荧光对调节光合作用的环境条件变化的反应时间很短(几毫秒),但在草本植物中,叶绿素含量面对胁迫而发生变化的过程可能长达三天(Houborg et al.,2011)。相比之下,计划中用于捕获叶绿素荧光的地球同步卫星平台星座(表3-2)将提供特别的日常时间分辨率数据库。

尽管重要且专门设计的任务正在或将会提供关于光能利用效率的数据,但值得注意的是,三个新卫星任务——SUOMI国家极地轨道伙伴关系(National Polar-orbiting Partnership,NPP)、陆地卫星8号和哨兵——并不会提供关于光能利用效率时空变化的太多信息。就SUOMI国家极地轨道伙伴关系而言,由于MODIS任务的连续性,其设计不是为了提供计算光化学反射率指数的波段。在任何情况下,从卫星数据估算陆地植被光能利用效率都有一个广阔而有希望的途径。

在本章中,我们介绍并分析了与陆地生态系统碳吸收相关的服务量化方面的最新进展。作为决定诸多生态系统服务的主要能量输入,总初级生产力的级联效应保证了这些通量在时间和空间上的准确估算。遥感技术为这些估算提供了新的方法。

致谢

本研究得到了布宜诺斯艾利斯大学UBACyT 01/F362项目、西班牙政府CGC2010-17172项目和CSD2008-00040项目及加泰罗尼亚政府SGR 2009-458项目的支持。

参 考 文 献

Asner, G. P. 1998. Biophysical and biochemical sources of variability in canopy reflectance. *Remote Sensing of Environment* 64:234–253.

Baker, N. R. 2008. Chlorophyll fluorescence: A probe of photosynthesis in vivo. *Annual Review of Plant Biology* 59:89–113.

Barton, C. V. M., and P. R. J. North. 2001. Remote sensing of canopy light use efficiency using the photochemical reflectance index—Model and sensitivity analysis. *Remote Sensing of Environment* 78:264–273.

Blackburn, G. A. 2007. Hyperspectral remote sensing of plant pigments. *Journal of Experimental Botany* 58:855–867.

Costanza, R., R. d'Arge, R. de Groot, et al. 1997. The value of the world's ecosystem services and natural capital. *Nature* 387:253–260.

Dash, J., and P. J. Curran. 2004. The MERIS terrestrial chlorophyll index. *International Journal of Remote Sensing* 25:5003–5013.

Dawson, T. P., P. R. J. North, S. E. Plummer, and P. J. Curran. 2003. Forest ecosystem chlorophyll content: Implications for remotely sensed estimates of net primary productivity. *International Journal of Remote Sensing* 24:611–617.

Day, J. O., C. W. O'Dell, R. Pollock, et al. 2011. Preflight spectral calibration of the Orbiting Carbon Observatory. *IEEE Transactions on Geoscience and Remote Sensing* 49:2793–2801.

de Groot, R. S., M. A. Wilson, and R. M. J. Boumans. 2002. A typology for the classification, description and valuation of ecosystem functions, goods and services. *Ecological Economics* 41:393–408.

Demmig-Adams, B. B., and W. W. Adams. 1996. The role of xanthophyll cycle carotenoids in the protection of photosynthesis. *Trends in Plant Science* 1:21–26.

Drolet, G. G., K. F. Huemmrich, F. G. Hall, et al. 2005. A MODIS-derived photochemical reflectance index to detect inter-annual variations in the photosynthetic light-use efficiency of a boreal deciduous forest. *Remote Sensing of Environment* 98:212–224.

Drolet, G. G., E. M. Middleton, K. F. Huemmrich, et al. 2008. Regional mapping of gross light-use efficiency using MODIS spectral indices. *Remote Sensing of Environment* 112:3064–3078.

Esteban, R., J. M. Olano, J. Castresana, et al. 2009. Distribution and evolutionary trends of photoprotective isoprenoids (*xanthophylls* and *tocopherols*) within the plant kingdom. *Physiologia Plantarum* 135:379–389.

Field, C. B. 1991. Ecological scaling of carbon gain to stress and resource availability. In *Response of plants to multiple stresses*, eds. H. A. Mooney, W. E. Winner, and E. J. Pell, 35–65. San Diego, CA: Academic Press.

Field, C. B., J. T. Randerson, and C. M. Malmstrom. 1995. Global net primary production: Combining ecology and remote sensing. *Remote Sensing of Environment* 51:74–88.

Filella, I., T. Amaro, J. L. Araus, and J. Peñuelas. 1996. Relationship between photosynthetic radiation-use efficiency of barley canopies and the photochemical reflectance index (PRI). *Physiologia Plantarum* 96:211–216.

Filella, I., J. Peñuelas, L. Llorens, and M. Estiarte. 2004. Reflectance assessment of seasonal and annual changes in biomass and CO_2 uptake of a Mediterranean shrubland submitted to experimental warming and drought. *Remote Sensing of Environment* 90:308–318.

Filella, I., A. Porcar-Castell, S. Munné-Bosch, J. Bäck, M. F. Garbulsky, and J. Peñuelas. 2009. PRI assessment of long-term changes in carotenoids/chlorophyll ratio and short-term changes in de-epoxidation state of the xanthophyll cycle. *International Journal of Remote Sensing* 30:4443–4455.

Frankenberg, C., A. Butz, and G. C. Toon. 2011a. Disentangling chlorophyll fluorescence from atmospheric scattering effects in O_2 A-band spectra of reflected sunlight. *Geophysical Research Letters* 38:L03801.

Frankenberg, C., J. B. Fisher, J. Worden, et al. 2011b. New global observations of the terrestrial carbon cycle from GOSAT: Patterns of plant fluorescence with gross primary productivity. *Geophysical Research Letters* 38:L17706.

Gamon, J. A., C. B. Field, W. Bilger, O. Björkman, A. Fredeen, and J. Peñuelas. 1990. Remote sensing of the xanthophyll cycle and chlorophyll fluorescence in sunflower leaves and canopies. *Oecologia* 85:1–7.

Gamon, J. A., C. B. Field, A. L. Fredeen, and S. Thayer. 2001. Assessing photosynthetic downregulation in sunflower stands with an optically-based model. *Photosynthesis Research* 67:113–125.

Gamon, J. A., C. B. Field, M. Goulden, et al. 1995. Relationships between NDVI, canopy structure, and photosynthetic activity in three Californian vegetation types. *Ecological Applications* 5:28–41.

Gamon, J. A., I. Filella, and J. Peñuelas. 1993. The dynamic 531-nanometer delta reflectance signal: A survey of twenty angiosperm species. In *Photosynthetic responses to the environment,* eds. H. Yamamoto and C. Smith, 172–177. Rockville, MD: American Society of Plant Physiologists.

Gamon, J. A., J. Peñuelas, and C. B. Field. 1992. A narrow-waveband spectral index that tracks diurnal changes in photosynthetic efficiency. *Remote Sensing of Environment* 41:35–44.

Gamon, J. A., A. F. Rahman, J. L. Dungan, M. Schildhauer, and K. F. Huemmrich. 2006. Spectral Network (SpecNet). What is it and why do we need it? *Remote Sensing of Environment* 103:227–235.

Gamon, J. A., L. Serrano, and J. S. Surfus. 1997. The photochemical reflectance index: An optical indicator of photosynthetic radiation use efficiency across species, functional types, and nutrient levels. *Oecologia* 112:492–501.

Gamon, J. A., and J. S. Surfus. 1999. Assessing leaf pigment content and activity with a reflectometer. *New Phytologist* 143:105–117.

Garbulsky, M. F., J. Peñuelas, J. A. Gamon, Y. Inoue, and I. Filella. 2011. The photochemical reflectance index (PRI) and the remote sensing of leaf, canopy and ecosystem radiation use efficiencies: A review and meta-analysis. *Remote Sensing of Environment* 115:281–297.

Garbulsky, M. F., J. Peñuelas, J. M. Ourcival, and I. Filella. 2008a. Estimación de la eficiencia del uso de la radiación en bosques mediterráneos a partir de datos MODIS. Uso del Índice de Reflectancia Fotoquímica (PRI) [Radiation use efficiency estimation in Mediterranean forests using MODIS Photochemical

Reflectance Index (PRI)]. *Ecosistemas* 17:89–97.

Garbulsky, M. F., J. Peñuelas, D. Papale, and I. Filella. 2008b. Remote estimation of carbon dioxide uptake of a Mediterranean forest. *Global Change Biology* 14:2860–2867.

Garbulsky, M. F., J. Peñuelas, D. Papale, et al. 2010. Patterns and controls of the variability of radiation use efficiency and primary productivity across terrestrial ecosystems. *Global Ecology and Biogeography* 19:253–267.

Gitelson, A. A., S. B. Verma, A. Viña, et al. 2003. Novel technique for remote estimation of CO2 flux in maize. *Geophysical Research Letter* 30:1486.

Gitelson, A. A., A. Viña, V. Ciganda, D. C. Rundquist, and T. J. Arkebauer. 2005. Remote estimation of canopy chlorophyll content in crops. *Geophysical Research Letter* 32:L08403.

Goerner, A., M. Reichstein, and S. Rambal. 2009. Tracking seasonal drought effects on ecosystem light use efficiency with satellite-based PRI in a Mediterranean forest. *Remote Sensing of Environment* 113:1101–1111.

Goerner, A., M. Reichstein, E. Tomelleri, et al. 2010. Remote sensing of ecosystem light use efficiency with MODIS-based PRI—The DOs and DON'Ts. *Biogeosciences Discusssion* 7:6935–6969.

Grace, J., C. Nichol, M. Disney, P. Lewis, T. Quaife, and P. Bowyer. 2007. Can we measure terrestrial photosynthesis from space directly, using spectral reflectance and fluorescence? *Global Change Biology* 13:1484–1497.

Guanter, L., L. Alonso, L. Gómez-Chova, J. Amorós, J. Vila, and J. Moreno. 2007. Estimation of solar-induced vegetation fluorescence from space measurements. *Geophysical Research Letters* 34:L08401.

Guanter, L., C. Frankenberg, A. Dudhia, et al. 2012. Retrieval and global assessment of terrestrial chlorophyll fluorescence from GOSAT space measurements. *Remote Sensing of Environment* 121:236–251.

Haines-Young, R., and M. Potschin. 2010. The links between biodiversity, ecosystem services and human well-being. In *Ecosystem ecology: A new synthesis*, eds. D. G. Raffaelli and C. L. J. Frid, 110–139. Cambridge, UK: Cambridge University Press.

Hall, F. G., T. Hilker, N. C. Coops, et al. 2008. Multi-angle remote sensing of forest light use efficiency by observing PRI variation with canopy shadow fraction. *Remote Sensing of Environment* 112:3201–3211.

Harris, A., and J. Dash. 2010. The potential of the MERIS Terrestrial Chlorophyll Index for carbon flux estimation. *Remote Sensing of Environment* 114:1856–1862.

Harris, A., and J. Dash. 2011. A new approach for estimating northern peatland gross primary productivity using a satellite-sensor-derived chlorophyll index. *Journal of Geophysical Research-Biogeosciences* 116:G04002.

Hilker, T., N. C. Coops, F. G. Hall, et al. 2008. Separating physiologically and directionally induced changes in PRI using BRDF models. *Remote Sensing of Environment* 112:2777–2788.

Hilker, T., N. C. Coops, F. G. Hall, et al. 2011. Inferring terrestrial photosynthetic light use efficiency of temperate ecosystems from space. *Journal of Geophysical Research* 116:G03014.

Hilker, T., F. G. Hall, N. C. Coops, et al. 2010. Remote sensing of photosynthetic light-use efficiency across two forested biomes: Spatial scaling. *Remote Sensing of Environment* 114:2863–2874.

Houborg, R., M. C. Anderson, C. S. T. Daughtry, W. P. Kustas, and M. Rodell. 2011. Using leaf chlorophyll to parameterize light-use-efficiency within a thermal-based carbon, water and energy exchange model. *Remote Sensing of Environment* 115:1694–1705.

Huemmrich, K. F., J. A. Gamon, C. E. Tweedie, et al. 2010. Remote sensing of tundra gross ecosystem productivity and light use efficiency under varying temperature and moisture conditions. *Remote Sensing of Environment* 114:481–489.

Inoue, Y., J. Peñuelas, A. Miyata, and M. Mano. 2008. Normalized difference spectral indices for estimating photosynthetic efficiency and capacity at a canopy scale derived from hyperspectral and CO_2 flux measurements in rice. *Remote Sensing of Environment* 112:156–172.

Joiner, J., Y. Yoshida, A. P. Vasilkov, Y. Yoshida, L. A. Corp, and E. M. Middleton. 2011. First observations of global and seasonal terrestrial chlorophyll fluorescence from space. *Biogeosciences* 8:637–651.

Key, R., S. Sander, A. Eldering, et al. 2012. The Geostationary Carbon Process Mapper. *IEEE Aerospace Conference Proceedings*. Big Sky, MT. 3–10 March 2012. Article no. 6187029.

Krause, G., and E. Weis. 1991. Chlorophyll fluorescence and photosynthesis: The basics. *Annual Review of Plant Biology* 42:313–349.

Landsberg, J. J., and R. H. Waring. 1997. A generalised model of forest productivity using simplified concepts of radiation-use efficiency, carbon balance and partitioning. *Forest Ecology and Management* 95:209–228.

Lyapustin, A., and Y. Wang. 2009. The time series technique for aerosol retrievals over land from MODIS. In *Satellite aerosol remote sensing over land*, eds. A. Kokhanovky and G. De Leeuw, 69–99. Heidelberg, Berlin: Springer.

Malenovsky, Z., K. B. Mishra, F. Zemek, U. Rascher, and L. Nedbal. 2009. Scientific and technical challenges in remote sensing of plant canopy reflectance and fluorescence. *Journal of Experimental Botany* 60:2987–3004.

Matsubara, S., G. H. Krause, M. Seltmann, et al. 2008. Lutein epoxide cycle, light harvesting and photoprotection in species of the tropical tree genus Inga. *Plant, Cell and Environment* 31:548–561.

Maxwell, K., and G. N. Johnson. 2000. Chlorophyll fluorescence—A practical guide. *Journal of Experimental Botany* 51:659–668.

Meroni, M., M. Rossini, L. Guanter, et al. 2009. Remote sensing of solar-induced chlorophyll fluorescence: Review of methods and applications. *Remote Sensing of Environment* 113:2037–2051.

Monteith, J. L. 1977. Climate and the efficiency of crop production in Britain. *Philosophical Transactions of the Royal Society—Biological Sciences* 281:277–294.

Myneni, R. B., S. O. Los, and G. Asrar. 1995. Potential gross primary productivity of terrestrial vegetation from 1982–1990. *Geophysical Research Letters* 22:2617–2620.

Naidoo, R., A. Balmford, R. Costanza, et al. 2008. Global mapping of ecosystem services and conservation priorities. *Proceedings of the National Academy of Sciences of the United States of America* 105:9495–9500.

Nichol, C. J., R. Pieruschka, K. Takayama, et al. 2012. Canopy conundrums: Building on the Biosphere 2 experience to scale measurements of inner and outer canopy photoprotection from the leaf to the landscape. *Functional Plant Biology* 39:1–24.

Niyogi, K. K. 1999. Photoprotection revisited: Genetic and molecular approaches. *Annual Review Plant Physiology and Plant Molecular Biology* 50:333–359.

Ollinger, S. V., A. D. Richardson, M. E. Martin, et al. 2008. Canopy nitrogen, carbon assimilation, and albedo in temperate and boreal forests: Functional relations and potential climate feedbacks. *Proceedings of the National Academy of Sciences of the United States of America* 105:19335–19340.

Paruelo, J. M., M. F. Garbulsky, J. P. Guerschman, and E. G. Jobbagy. 2004. Two decades of Normalized Difference Vegetation Index changes in South America: Identifying the imprint of global change. *International Journal of Remote Sensing* 25:2793–2806.

Peng, Y., and A. Gitelson. 2012. Remote estimation of gross primary productivity in soybean and maize based on total crop chlorophyll content. *Remote Sensing of Environment* 117:440–448.

Peng, Y., A. Gitelson, G. Keydan, D. C. Rundquist, and W. Moses. 2011. Remote estimation of gross primary production in maize and support for a new paradigm based on total crop chlorophyll content. *Remote Sensing of Environment* 115:978–989.

Peñuelas, J., and I. Filella. 1998. Visible and near-infrared reflectance techniques for diagnosing plant physiological status. *Trends in Plant Science* 3:151–156.

Peñuelas, J., I. Filella, and J. A. Gamon. 1995. Assessment of photosynthetic radiation-use efficiency with spectral reflectance. *New Phytologist* 131:291–296.

Peñuelas, J., I. Filella, J. A. Gamon, and C. Field. 1997. Assessing photosynthetic radiation-use efficiency of emergent aquatic vegetation from spectral reflectance. *Aquatic Botany* 58:307–315.

Peñuelas, J., I. Filella, J. Llusia, D. Siscart, and J. Piñol. 1998. Comparative field study of spring and summer leaf gas exchange and photobiology of the Mediterranean trees *Quercus ilex* and *Phillyrea latifolia*. *Journal of Experimental Botany* 49:229–238.

Peñuelas, J., J. A. Gamon, A. L. Fredeen, J. Merino, and C. B. Field. 1994. Reflectance indexes associated with physiological changes in nitrogen-limited and water-limited sunflower leaves. *Remote Sensing of Environment* 48:135–146.

Peñuelas, J., M. F. Garbulsky, and I. Filella. 2011. Photochemical reflectance index (PRI) and remote sensing of plant CO_2 uptake. *New Phytologist* 191:596–599.

Rahman, A. F., V. D. Cordova, J. A. Gamon, H. P. Schmid, and D. A. Sims. 2004. Potential of MODIS ocean bands for estimating CO_2 flux from terrestrial vegetation: A novel approach. *Geophysical Research Letters* 31:L10503.

Rahman, A. F., J. A. Gamon, D. A. Fuentes, D. A. Roberts, and D. Prentiss. 2001. Modeling spatially distributed ecosystem flux of boreal forest using hyperspectral indices from AVIRIS imagery. *Journal of Geophysical Research—Atmospheres* 106:33579–33591.

Ruimy, A., B. Saugier, and G. Dedieu. 1994. Methodology for the estimation of terrestrial net primary production from remotely sensed data. *Journal of Geophysical Research* 99:5263–5283.

Running, S. W., R. R. Nemani, F. A. Heinsch, M. Zhao, M. Reeves, and H. Hashimoto. 2004. A continuous satellite-derived measure of global terrestrial primary production. *BioScience* 54:547–560.

Runyon, J., R. H. Waring, S. N. Goward, and J. M. Welles. 1994. Environmental limits on net primary production and light-use efficiency across the Oregon transect. *Ecological Applications* 4:226–237.

Seaton, G. G. R., and D. A. Walker. 1990. Chlorophyll fluorescence as a measure of photosynthetic carbon assimilation. *Proceedings of the Royal Society of London* 242:29–35.

Sims, D. A., H. Luo, S. Hastings, W. C. Oechel, A. F. Rahman, and J. A. Gamon. 2006. Parallel adjustments in vegetation greenness and ecosystem CO_2 exchange in response to drought in a Southern California chaparral ecosystem. *Remote Sensing of Environment* 103:289–303.

Sims, D. A., A. F. Rahman, E. F. Vermote, and Z. Jiang. 2011. Seasonal and inter-annual variation in view angle effects on MODIS vegetation indices at three forest sites. *Remote Sensing of Environment* 115:3112–3120.

Stylinski, C. D., J. A. Gamon, and W. C. Oechel. 2002. Seasonal patterns of reflectance indices, carotenoid pigments and photosynthesis of evergreen chaparral species. *Oecologia* 131:366–374.

Ustin, S. L., D. A. Roberts, J. Pinzón, et al. 1998. Estimating canopy water content of chaparral shrubs using optical methods. *Remote Sensing of Environment* 65:280–291.

Wu, C., Z. Niu, and S. Gao. 2012. The potential of the satellite derived green chlorophyll index for estimating midday light use efficiency in maize, coniferous forest and grassland. *Ecological Indicators* 14:66–73.

Zarco-Tejada, P. J., J. R. Miller, G. H. Mohammed, T. L. Noland, and P. H. Sampson. 2002. Vegetation stress detection through chlorophyll a + b estimation and fluorescence effects on hyperspectral imagery. *Journal of Environmental Quality* 31:1433–1441.

第四章 北方高纬度系统碳循环池和过程的对地观测

H. E. 爱泼斯坦

一、北方高纬度碳循环过程遥感概论

北方陆地系统的生态系统服务包括所有三类主要服务：供应、调节和文化（Haynes-Young and Potschin，2010）。本章重点讨论的生态系统服务主要是供应（如生物量生产）和调节（如碳汇和储存）。北方高纬度地区的一项关键服务是其储存碳的能力。

在历史上，北极和亚北极的生态系统一直扮演着碳汇的角色，储存大量死亡的有机物，并保护其免于快速分解。北方生态系统对死亡有机物的保护主要是由于相对寒冷的土壤、浸渍土壤和永久冻土的存在，它们降低了分解速率。除了土壤有机质分解率低和碳释放率低之外，碳汇的另一个组成部分是植物通过光合作用吸收二氧化碳，并将这种碳结合到植物组织中。北方生态系统的历史碳汇表明，植物吸收二氧化碳的速率大于通过呼吸过程（包括分解）释放二氧化碳的速率。新的估算表明，高纬度多年冻土区 3 米深的土壤碳量是大气二氧化碳中碳量的两倍（Schuur *et al.*，2008；Tarnocai *et al.*，2009），北极最近的变暖无疑将改变该地区的陆地-大气碳交换（如 Euskirchen *et al.*，2009；Lee *et al.*，2012；Natali *et al.*，2012；Trucco *et al.*，2012）。

自 20 世纪 90 年代中叶以来，研究人员通过在高纬度地区使用遥感技术探究碳循环过程。由于遥感主要有助于观察地表特性，因此它主要用于通过植被指数来检查碳预算中的二氧化碳吸收分量，而非量化分解和北方土壤的二氧化碳流出速率。不过也有使用遥感信息来估算地下过程的例子，详细讨论参见相应章节。关于地面成分，遥感已被用于监测高纬度的植被，从早期可获得的卫星信息开始，研究人员利用此等信息推断植被对二氧化碳的吸收，以总初级生产力或净初级生产力表示（见第二章）。

一些最早发表的研究利用美国国家海洋和大气管理局（NOAA）地球轨道卫星上的先进甚高分辨率辐射计（Advanced Very High Resolution Radiometer，AVHRR）的记录，来检验北方高纬度地区归一化植被指数（NDVI）的空间模式和时间趋势（Myneni et al., 1997；Tucker et al., 2001；Zhou et al., 2001；Slayback et al., 2003）。NDVI 是植被光合作用活动的一个指标，由红色和近红外（NIR）波长的地表反射率计算得出；由于叶绿素的存在以及近红外光谱中植物细胞结构的散射而具有的反射性，植被通常对红色波长有吸收作用。前述研究表明，从 20 世纪 80 年代初到 90 年代末，北纬 35 度以北地区的 NDVI 有所增加，这意味着至少地面上的 NPP 有所增加。周等人（Zhou et al., 2001）指出，在这 20 年间，欧亚大陆的 NDVI 增长程度大于北美（北纬 40 度～70 度）；而斯莱贝克等人（Slayback et al., 2003）指出，自 1992 年以来，北美的 NDVI 增长程度大于欧亚大陆（北纬 45 度～75 度）。

进一步的研究开始扩展卫星记录，并集中于大尺度北方高纬度地区 NDVI 动态的一些空间变化。一个重要的区别开始凸显：北极苔原和北方森林生物群落通常朝着相反的方向发展，苔原表现出几乎无处不在的"绿化"（NDVI 增加），而北方森林则表现出"褐化"（NDVI 减少）（Jia et al., 2003；Bunn et al., 2005，2007；Goetz et al., 2005；Bunn and Goetz, 2006；Verbyla, 2008；Bhatt et al., 2010；Beck and Goetz, 2011；Baird et al., 2012），尽管文献中对这些趋势仍有一些讨论（如 Alcaraz-Segura et al., 2010）。为此，我们对高纬度地区初级生产力遥感的讨论将分为北极苔原和北方森林部分。因此，本章以下两节讨论了北极苔原植被生物量和初级生产力的遥感估算及北方森林植被生物量和初级生产力的遥感估算，这两节同时涉及供给（生物量积累）与调节（碳汇）。接下来的三部分涉及陆地-大气碳交换的遥感、土壤碳过程和碳存量及火灾产生的碳排放；这些主要是与大气二氧化碳和甲烷以及土壤碳储存有关的调节性生态系统服务。最后一节概述了如何利用地球观测来评估北方高纬度环境的碳相关生态系统服务。

二、北极苔原植被生物量和初级生产力的遥感估算

一些最早的研究,利用阿拉斯加北坡的重复照片研究,表明北极苔原的生物量和初级生产力存在潜在变化(Sturm et al.,2001;Tape et al.,2006)。这些研究发现,自20世纪中期以来,在以前无植被或被低矮的苔原植被主导的山坡和河岸低地,高灌木和低灌木(桤木、柳树、桦树)不断扩张。罗帕斯和布德罗(Ropars and Boudreau,2012)最近利用航空照片(1957)和WorldView-1卫星影像记录了低北极魁北克省北方林线的沙质梯田和山顶桦树灌木的增加。特伦布莱等人(Tremblay et al.,2012)还使用了重复的航拍相片(1964~2003年)显示加拿大东努纳维克的一个低北极地区的桦树灌木和东部落叶松覆盖率急剧增加。贾等人(Jia et al.,2003)利用8千米×8千米空间分辨率的AVHRR数据发现,1981~2001年间,阿拉斯加北坡NDVI峰值增加了16.9%。

韦尔比拉(Verbyla,2008)随后将阿拉斯加的AVHRR记录扩展到2003年,并证实了最大NDVI增加的发现,其中变化最大的是阿拉斯加沿海平原。

过去十年的研究利用AVHRR数据研究了北美大陆(阿拉斯加和加拿大)的NDVI动态,观察到冻土带NDVI自1981年以来持续增加(Stow et al.,2004;Bunn et al.,2005,2007;Goetz et al.,2005;Bunn and Goetz,2006;Jia et al.,2009)。贾等人(Jia et al.,2009)发现,1982~2006年期间,加拿大境内所有五个北极苔原生物气候亚区(Walker et al.,2005,从北到南的A~E亚区)均发生了苔原绿化。其中一些研究还注意到NDVI季节性动态变化,一些地区植被生长较早(Goetz et al.,2005;Jia et al.,2009),早季NDVI值更大(Verbyla,2008),NDVI峰值更早(特别是在北极高地区)(Jia et al.,2009)。一些研究在更高的空间分辨率(1千米AVHRR和30米Landsat卫星)下观察到,北美北极苔原的生产力(NDVI)持续增加(Neigh et al.,2008;Olthof et al.,2008;Pouliot et al.,2009),表明苔原植被的绿化趋势在景观尺度上可能相当均匀。普略特等人(Pouliot et al.,2009)发现,1985~2006年,22%的加拿大陆地表面有明显的NDVI正值趋势,其中大部分在加拿大北方。奥尔索夫等人(Olthof et *al.*，2008)指出，与地衣主导的景观相比，维管植物对加拿大苔原的绿化信号贡献更大。弗雷泽等人(Fraser *et al.*，2011)利用1984～2009年的Landsat卫星数据，发现加拿大北方四个国家公园的NDVI增加了6.1%～25.5%。

一些研究为前面提到的对高纬度地区环极NVDI的早期研究进一步丰富了信息(更大的时间范围和更精细的空间细节)。在全球范围内，北极苔原的生产力增长趋势基本上是连续的(Bunn and Goetz，2006；Bunn *et al.*，2007；Bhatt *et al.*，2010)。巴特等人(Bhatt *et al.*，2010)最明确地研究了北极圈苔原生产力动态的空间模式，特雷西尼科夫(Treshnikov，1985)将北极划分为海洋子区域，并分析了NDVI峰值和季节综合NDVI。他们发现，近岸苔原几乎普遍变绿，然而，白令海峡和西楚科奇地区的绿化有所下降。与早期显示北美和欧亚大陆之间NDVI趋势差异结果不一的研究相比，这些后来的研究表明，北美苔原植被的绿化程度似乎高于欧亚大陆(Bunn *et al.*，2007；Bhatt *et al.*，2010；Goetz *et al.*，2011)。巴特等人(Bhatt *et al.*，2010)发现，1982～2008年，北美苔原的最大NDVI增加了9%，但欧亚苔原的最大NDVI只增加了2%。贝克和戈茨(Beck and Goetz，2011)利用1982～2008年的AVHRR和中分辨率成像光谱仪数据发现，自20世纪90年代中期以来，阿拉斯加北坡的夏季NDVI(基本上不考虑灌木覆盖)增加幅度较大。

除了戴和塔克(Dye and Tucker，2003)对欧亚大陆北纬50度和71度的NDVI季节性、积雪覆盖和温度的一项研究及戈茨等人(Goetz *et al.*，2011)的信息合成之外，专门针对欧亚苔原的研究很少。

上述研究大多以NDVI作为地上植被的度量标准，并假设NDVI与地上初级生产力或地上生物量或两者均有关系。一些研究已建立了北极苔原NDVI和地上生物量之间的关系(Hope *et al.*，1993；Shippert *et al.*，1995；Walker *et al.*，2003；2012；Jia *et al.*，2006)，雷诺兹等人(Raynolds *et al.*，2012)发表了苔原植被的最大NDVI和地上生物量间的稳健关系。因此，可将NDVI与碳相关池和过程建立联系。爱泼斯坦等人(Epstein *et al.*，2012)利用雷诺兹等人(Raynolds *et al.*，2012)发表的NDVI-生物量对数关系，将北极苔原观测到的NDVI变化转化为地上生物量(或碳)变化。他们发现，在1982～2010年间，北极苔原的地上总生物量增加了0.40皮克Pg(图4-1)。若我们假设地下总生物量的增加相等,并且植物组织中的碳物质占比为50%,那么活体植物碳库在这段时间内增加了0.40皮克C。在这一点上,将该碳差额转化为自1982年以来的实际固碳增量是很有挑战性的,因为固碳将是以下因素的函数:(1)在所观察的时间段内该生物量增加的速率;(2)各种植物组织的周转速率;(3)衰老的植物材料的命运。爱泼斯坦等人(Epstein et al.,2012)得出的一个显著结论是,北极苔原带地上生物量平均增加了19.8%;然而,与更北方的苔原带亚带(2%~7%)相比,这一增加大部分发生在更南部的苔原带亚带(20%~26%)。

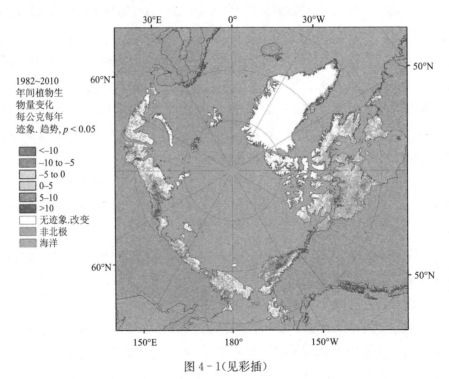

图4-1(见彩插)

注:1982~2010年间北极苔原带地上植物量($gm^{-2}y^{-1}$)的变化,来自AVHRR-NDVI和实地测量的生物量之间的关系。(摘自Epstein, H. E., et al.)

遥感产品还与模拟模型结合,用以评估北极苔原的总初级生产力和植被净初级生产力在各种空间尺度上的现状和未来趋势(如Williams et al.,2001; Turner et al.,2005; Kimball et al.,2007)。威廉姆斯等人(Williams et al., 2001)利用AVHRR NDVI产品和土壤-植物-大气模型,预测了阿拉斯加北方库帕鲁克河流域总初级生产力的季节动态（每日预测）。金伯尔等人（Kimball et al.，2007）将 AVHRR-NDVI 与生产效率模型、BIOME-BGC 和陆地生态系统模型（Terrestrial Ecosystem Model，TEM）结合起来，得出北极苔原的总初级生产力和植被净初级生产力。利用 AVHRR PEM，他们估算，1982～2000 年，苔原植被净初级生产力每 10 年增加 7.6%（21.1 克碳/平方米），这与其他苔原绿化观测结果一致。在阿拉斯加北方土壤温度和湿度受到控制的地点，胡姆里奇等人（Huemmrich et al.，2010）进行的一项研究支持使用遥感驱动的效率模型来评估苔原对气候变化的反应。最近，塔格森等人（Tagesson et al.，2012）为格陵兰岛东北方扎肯伯格的高北极湿润苔原地点建立了高分辨率卫星 NDVI 和光合有效辐射吸收比之间的关系。然后他们利用这一关系，加上一个简单的光利用效率模型，估算了 1992～2008 年的总初级生产力，发现旺季总初级生产力（及气温）在这段时间内有所增加。斯蒂奇等人（Sitch et al.，2007）回顾了遥感和基于过程的模拟模型在评估北极苔原带碳循环中的应用，对预测的综合分析也表明，在整个 21 世纪，北极苔原将是一个小的碳汇（$10\sim40\ gCm^{-2}y^{-1}$）。

三、北方森林植被生物量和初级生产力的遥感估算

关于利用遥感估算北方森林生态系统的生物量和初级生产力，有大量文献可供参考，并且已有对这些文献的评论（如 Gamon et al.，2004；Boyd and Danson 2005；Lutz et al.，2008），因此，并无必要在这里进行全面讨论（也是一个相当大的挑战）。

因此，重点是在相对较粗的空间尺度上进行的研究，及那些关于随着时间推移生物量的产生和碳汇的研究。戈茨和普林斯（Goetz and Prince，1996）对明尼苏达州东北方常绿、落叶混交林进行了一项相对较早的研究。他们使用来自 Landsat 卫星影像的植被绿色指数来估算冠层每年截获的光合有效辐射（IPAR），并建立 IPAR 与年度地上植被净初级生产力（Annual Aboveground NPP, AANPP）数据间的关系。通过对光利用效率的估算，他们得出了森林的落叶部分（363 毫克碳/平方千米）和常绿（125 毫克碳/平方千米）部分的 AANPP，并计算了 2 280 平方千米研究区域地上植被的年度净碳吸收量（256

毫克碳/平方千米)。

甘蒙等人(Gamon et al.,2004)总结了在北方生态系统大气研究(Boreal Ecosystem Atmosphere Study,BOREAS)期间利用遥感评估北方植被生物量和产量方面取得的诸多进展。霍尔等人(Hall et al.,1996)使用红色和近红外波段的 Landsat TM 反射率数据和辐射传输模型来估算以湿地黑云杉为主的南部 BOREAS 区域内的地上生物量密度。戈茨等人(Goetz et al.,1999)使用生产效率模型(GLO-PEM;Prince and Goward,1995),由 1 平方千米的 AVHRR 反射率数据和来自地球静止观测环境卫星(Geostationary Operational Environmental Satellite,GOES)的光合有效辐射驱动,估算加拿大中部整个 BOREAS 区域的碳循环变量,包括地上生物量、总初级生产力和植被净初级生产力。他们发现,该地区的平均总初级生产力为 604 克碳/每平方米每年,平均植被净初级生产力指数为 235 克碳/每平方米每年,这些数值因土地覆盖类型不同而有很大差异。使用基于过程的模型和 AVHRR 产品的其他估算得出了非常相似的结果(平均植被净初级生产力指数为 217 克碳/每平方米每年)(Liu et al.,1999)。BOREAS 项目中还使用了主动光学传感器激光雷达(光检测和测距)来可靠地估算北方黑云杉林分的地上生物量(Lefsky et al.,2002)。此外,乐博等人(Leboeuf et al.,2007)使用从高分辨率 QuickBird 影像获得的树冠阴影部分来绘制加拿大东部北方黑云杉林的地上森林生物量图。

利用遥感技术评估欧亚北方森林植被生物量和初级生产力的研究较少。克兰基纳等人(Krankina et al.,2005)利用 Landsat 卫星影像估算了圣彼得堡地区活森林生物量中的总碳量和碳汇速率(0.36 毫克碳/每公顷每年),这些估算与 1990 年代初的森林清查数据相比是相当的。福希斯等人(Fuchs et al.,2009)利用 QuickBird 和 ASTER 影像建立了西伯利亚西北方的森林苔原集水区地上碳总量和波谱指数的关系。他们发现地上碳量的空间异质性较高,范围为 4.3 吨~28.8 吨每公顷不等。

郑等人(Zheng et al.,2004)使用 1 平方千米的 AVHRR 数据来估算芬兰和瑞典北方森林的植被净初级生产力指数,他们还发现了该区域的空间异质性存在数量级的差异。

在环极尺度上,AVHRR-NDVI 数据已被用于估算北方森林木本生物量中的碳(Myneni et al.,2001;Dong et al., 2003)。迈内尼等人(Myneni et al., 2001)在生长季节 NDVI 综合指数和木本生物量的森林清查估算量之间建立了强而饱和的相关关系;利用这些关系,他们计算了北半球森林的木质碳库和木质碳库的变化(1981~1999 年)。他们发现,在俄罗斯北方森林中,木质碳增加了 0.3 吨碳/每公顷每年,而在加拿大北方森林中,木质碳损失了 0.1ton C ha^{-1} y^{-1}。事实上,在 20 世纪 90 年代,BOREAS 区域被发现是大气碳的来源(Gamon et al.,2004);然而,这很可能是由于这段时间程度很高的火灾干扰(Steyaert et al., 1997;Li et al.,2000;Chen et al.,2003)。将遥感与过程模型相结合也是在粗略的空间尺度下估算北方森林中的碳循环过程(如植被净初级生产力指数)的有效方法(Hall et al.,2006;Kimball et al.,2007;Zhang et al., 2007;Smith et al.,2008;Tagesson et al.,2009)。

关于最近北方森林地上生物量的时序变化趋势,1982 年以来的 AVHRR 记录已被广泛用于确定这些动态(Bunn et al.,2005,2007;Goetz et al.,2005; Bunn and Goetz,2006;Verbyla 2008;Beck and Goetz,2011;Baird et al., 2012)。戈茨等人(Goetz et al.,2005)发现,在 1982~2003 年间,无火灾干扰的北美北方森林区域 NDVI 有所下降。邦恩和戈茨(Bunn and Goetz,2006)将这一分析在空间上扩展到环北极地区,发现"褐化"主要发生在森林密集的地区,而 NDVI 在生长季晚期(7~8 月)普遍下降。考虑到整个生长季节,北方森林 NDVI 趋势基本上是正负混合的。邦恩等人(Bunn et al.,2007)计算了稀疏落叶松林(落叶)的主要绿化趋势,与密度较大的常绿林的褐化形成对比。韦尔比拉(Verbyla,2008)的一项研究证实,无论是否考虑火灾,阿拉斯加内陆北方森林都存在强烈的褐化趋势,帕伦特和韦尔比拉(Parent and Verbyla,2010)还通过使用更高分辨率的 Landsat 卫星数据发现,阿拉斯加北方森林的 NDVI 正在下降。相比之下,阿尔卡拉斯-塞古拉等人(Alcaraz-Segura et al.,2010)使用加拿大遥感中心的 1 千米分辨率 AVHRR 数据集发现,加拿大北方地区不到 1% 的未火烧迹地森林像元显示在 1984~2006 年 NDVI 有所下降。贝克和戈茨(Beck and Goetz,2011)获得了迄今为止最长记录的 AVHRR 和 MODIS 数据(1982~2008 年),数据显示北美和欧亚大陆北方森林 NDVI 持续下降,同样是在密度更大、常绿为主的林分中。金伯尔等人(Kimball et al.,2007)使用基于

AVHRR 数据驱动的 PEM 模型，模拟出 1982～2000 年阿拉斯加北方和加拿大西北方的植被净初级生产力增长了 9.1%，但张等人（Zhang et al.，2008）也使用了由 AVHRR 和 MODIS 数据驱动的 PEM，却模拟出北美北方的森林植被净初级生产力从 20 世纪 90 年代末～2005 年有所下降。

四、北极苔原和北方森林碳陆地大气交换的遥感估算

前几节中，我们回顾了利用遥感技术评估北方高纬度生态系统植被地上部分的碳组分（生物量）和碳吸收（总初级生产力、植被净初级生产力）；然而，遥感技术也被用于研究包括地下部分在内的碳相关过程。净生态系统生产力（NEP，生态定义的净碳汇）和净生态系统交换（NEE，大气定义的净碳汇）在计算中包括生态系统呼吸——包括根呼吸和地下异养呼吸，使得从遥感角度进行评估更具挑战性。

这些过程都是调节大气功能的生态系统服务的组成部分，在这种情景下，与温室气体、二氧化碳和甲烷的浓度有关。

对于北极苔原，少数研究尝试使用遥感方法来估算或推断陆地-大气二氧化碳交换量。一些初步研究是在阿拉斯加北坡进行的（Stow et al.，1998；McMichael et al.，1999）。作为在遥感数据和北极苔原二氧化碳交换之间建立关系的第一步，斯托等人（Stow et al.，1998）将两种不同空间分辨率的 NDVI 数据与沿阿拉斯加库帕鲁克河流域断面飞行并进行采样的飞机测得的二氧化碳通量联系起来。麦米克等人（McMichael et al.，1999）在阿拉斯加北方的两个地点以更精细的空间分辨率，发现了野外测量的（手持式光谱辐射计）NDVI 和实测总光合作用和生态系统呼吸通量测量值之间的关系。邵佳德等人（Soegaard et al.，2000）利用生长旺季的 Landsat 卫星数据，在格陵兰岛东北，针对北极高生态系统中的三种不同植被类型，将现场测量的二氧化碳通量外推到景观尺度。在这个系统中，6 月景观尺度的碳损失被 7～8 月的碳汇所抵消。

最近，金伯尔等人（Kimball et al.，2009）在其陆地碳通量（TCF）模型中综合了 MODIS（土地覆盖、叶面积指数和植被净初级生产力指数）和先进的微波扫描辐射计地球观测系统（Advanced Microwave Scanning Radiometer Earth Observing System,AMSR-E,地表湿度和温度)的数据,以估算北美北极和北方地区的生态系统净交换(及呼吸和土壤碳储量),他们将这些估计值与涡动协方差二氧化碳通量测量值进行了比较。同样,洛兰蒂等人(Loranty et al.,2011)使用MODIS 叶面积指数产品作为简单模型(只有三个输入变量)的输入数据,用于在景观尺度上模拟北极苔原生态系统净交换。最后,罗查和谢弗(Rocha and Shaver,2011)使用 MODIS 增强植被指数对生态系统净交换模型进行参数化,并从阿拉斯加北方阿纳克图武克河火灾痕迹上的通量塔足迹中推断信息。他们发现,他们的 MODIS 双波段 EVI 解释了在观察到的火烧严重程度梯度上生态系统净交换 86%的变化。

就北方森林而言,使用遥感评估生态系统净生产或生态系统净交换的研究同样稀少,而且他们所使用的方法与北极苔原的方法非常相似。例如,使用遥感数据推断从涡度相关通量塔收集的二氧化碳交换信息,像梅罗尼等人(Meroni et al.,2002)利用 Landsat 卫星在西伯利亚中部所做的那样。另一种常见的方法是使用遥感数据作为模拟模型输入(或导出模拟模型输入),如阿米罗等人(Amiro et al.,2003)利用加拿大西部的 AVHRR 数据所做的那样。川田等人(Kushida et al.,2004)使用手持光谱测量数据和 Landsat ETM 影像结合起来来估算阿拉斯加内陆北方森林的生态系统净生产。利用野外实测的下垫面和云杉针叶的光谱数据,建立了反射率与叶面积指数和林下苔藓植物间的关系。

将这些关系与叶面积指数植被净初级生产力指数关系和土壤呼吸观测相结合,估算生态系统净生产力,是利用 Landsat 卫星数据进行空间外推得出的。如前所述,金伯尔等人(Kimball et al.,2009)使用 MODIS 和 AMSR-E 来估算具有现有通量塔的苔原和北方森林遗址的生态系统净交换。德雷泽特和奎根(Drezet and Quegan,2007)使用合成孔径雷达数据来估算英国的森林生物量,进而计算森林年龄结构,然后计算生态系统净交换,提供的碳吸收值显著高于先前国家清查数据估算的碳吸收量。最后,理查德森等人(Richardson et al.,2010)利用通量数据和 MODIS 数据作为物候指标,评估了物候对北方和温带森林 21 个 FLUXNET 站点生态系统生产力(生态系统净生产力和生态系统总光合作用)的影响。他们发现常绿针叶林的生产力对物候变化的敏感性低于落叶阔叶林。

五、北极苔原和北方森林土壤碳过程（土壤呼吸、分解、甲烷产生）和存量的遥感估算

用遥感方法评估几乎完全发生在地下的过程更具挑战性，如土壤呼吸、有机物分解和甲烷产生。土壤呼吸来自根呼吸和其他土壤动植物呼吸的二氧化碳通量（其中很大一部分是异养微生物）。因此，土壤呼吸包括与植被地上器官明确相关的成分（根），这使得遥感至少是一个合理的选择。另一方面，分解（在一定程度上是土壤呼吸的一部分）主要由地下控制驱动，间接由地上过程驱动，如通过根部的凋落物和光合产物渗出，因此可能更难用遥感评估。这里已提到少数使用遥感方法估算北极苔原和北方森林的土壤呼吸或分解的研究，因为这些变量通常与评估净生态系统碳通量结合使用。

这里最常见的方法是使用微波遥感，如来自 AMSR-E、合成孔径雷达、扫描多通道微波辐射计（SMMR）和特殊传感器微波/成像仪（Special Sensor Microwave/Imager, SSM/I）的数据，来估算北方和北极地区的土壤温度、土壤湿度、土壤冻融状态和地表湿度（对土壤呼吸和分解的主要控制条件）（如 Way et al., 1997; Smith et al., 2004; Jones et al., 2007; Kimball et al., 2009），这反过来又被用于模拟这些生态系统过程。

甲烷的产生和从土壤到大气的通量也在很大程度上受地下条件的控制，与地上过程间接相关。利用遥感技术估算北方高纬度地区甲烷通量的诸多技术与用于估算土壤呼吸和分解的技术类似：使用遥感工具来评估陆地表面和土壤条件，然后再利用这些工具来估算或模拟碳相关过程。对于来自植被景观（如湿地）的甲烷通量，评估地表水和土壤湿度是最常见的方法；然而，高纬度地区甲烷的大量通量是由永久湖泊、解冻湖泊和排水湖泊盆地产生，因此这些也经常使用遥感工具进行评估（McGuire et al., 2009）。

巴奇等人（Bartsch et al., 2007, 2008）利用欧洲环境卫星（European Environmental Satellite, ENVISAT）的先进合成孔径雷达（Advanced Synthetic Aperture Radar, ASAR）影像及美国航天局快速散射计卫星的数据，以 150 米的空

间分辨率绘制了俄罗斯泰米尔半岛不同湿地类型和永久水体的分布图。特别是对于开放水域,利用苔原湖泊甲烷排放的一些现场测量数据来推断泰米尔半岛的值。沃尔特等人(Walter et al.,2008)也使用来自雷达卫星 1 号 RADARSAT-1 的合成孔径雷达数据来估算北极湖泊的甲烷排放量,但采用的方法分辨率更高。他们使用雷达后向散射器来估算湖泊冰中不同类型气泡群的分布,其中甲烷沸腾可占总甲烷通量的 95% 以上(Casper et al.,2000)。结合对气泡群和甲烷沸腾的现场测量,作者提出了最终将雷达后向散射与整个湖泊甲烷通量联系起来的方程。最后,正如二氧化碳通量研究通常所做的那样,布比尔等人(Bubier et al.,2005)使用遥感数据(本例中为 Landsat 卫星),利用北方生态系统大气研究(BOREAS)期间收集的数据,将甲烷通量的现场测量外推到景观尺度。诸多研究已使用遥感工具,如 Landsat 卫星(如 Stow et al.,2004;Frohn et al.,2005;Hinkel et al.,2005;Plug et al.,2008;Morgenstern et al.,2011;Wang et al.,2012)评估了热融湖和干涸的热融湖盆底的分布。然而,目前的文献仅暗示使用遥感技术来量化来自这些地表特征的甲烷通量。

很少有研究能够达到使用遥感方法评估高纬度生态系统地下碳储量的程度(见下一节关于火灾干扰)。最近,胡格里乌斯(Hugelius,2012)使用基于 Landsat 卫星数据的土地覆盖分类对俄罗斯欧洲乌沙河流域北方的土壤有机碳进行了空间升尺度研究。乌尔里希等人(Ulrich et al.,2009)将土壤性质(包括土壤有机碳浓度)与西伯利亚勒拿河三角洲的野外光谱数据联系起来。土壤数据随后被放大到土地覆盖单元,用 Landsat ETM+ 和 CORONA 影像进行分类。我们再次看到了将模拟建模与遥感输入相结合以估算碳循环属性和过程的方法(如 Kimball et al.,2009),在本例中是针对土壤有机碳。

六、火灾碳排放的遥感估算

火灾是高纬度系统中的主要干扰类型,对土壤碳储存和大气二氧化碳调节都有重要影响,而遥感是评估这些影响的关键工具(见第七章)。遥感经常被用于探测北方森林中的火灾(如 Li et al.,2000;Soja et al.,2004),卫星遥感的数据是促进阿拉斯加和加拿大大部分地区火烧区地图开发的关键资源（如Epp and Lanoville,1996；Fraser et al.,2004；Soja et al.,2004；Giglio et al.,2009；Loboda et al.,2011）。对俄罗斯大部分地区来说，关于火烧迹地的大量信息已经不那么可靠。然而,苏希宁等人(Sukhinin et al.,2004)使用AVHRR开发了首个全面的俄罗斯东部火烧迹地产品,为随后MODIS的产品提供了西伯利亚北方最前线火灾的更全面的印象(Loboda and Csizsar,2007；Loboda et al.,2007,2012；Giglio et al.,2010)。除了从遥感数据获得的几个其他与火灾相关的变量外,被烧毁的区域还可用于评估区域、大陆和全球范围内火灾期间的碳消耗(如 Sukhinin et al., 2004；Kasischke et al.,2005；de Groot et al.,2007；Frolking et al.,2009；Stinson et al.,2011；Turetsky et al.,2011)。火灾严重程度是一个关键变量,遥感数据已成功地用于若干研究中,绘制火灾严重程度图,以估算北方森林火灾期间的碳消耗(如Michalek et al.,2000；Conrad et al.,2002；Isaev et al.,2002)。事实证明,发展遥感方法来持续绘制不同区域和不同环境条件下的火灾严重程度图具有挑战性(如 Allen and Sorbel,2008；French et al.,2008)。然而,最近一些研究人员已做出一致的努力来改进对火烧迹地严重程度的估算,例如巴雷特等人(Barrett et al.,2010,2011)对阿拉斯加内陆云杉森林的估算。基于热红外遥感数据的火灾活动的时间动态是估算火灾期间碳消耗的另一个重要信息(如Loboda and Csizsar 2007；Kasischke and Hoy,2012)。

在火灾中损失的碳当然不限于地上的植物材料,这些研究中有许多使用遥感来估算火灾后森林下垫面和土壤有机碳的损失。其中一项研究使用高分辨率机载高光谱影像来评估阿拉斯加内陆火灾后森林下垫面物质的损失(Lewis et al.,2011);结果表明,火灾引起的森林下垫面消耗与绿色苔藓的覆盖密切相关,这是根据影像估算的。卡西希克和霍伊(Kasischke and Hoy,2012)使用Landsat卫星数据(植被覆盖和火烧迹地严重程度)和MODIS(火灾日期)数据来估算阿拉斯加内陆火灾所消耗的碳；其结果包括地表有机层碳的损失。其他研究使用遥感影像,如Landsat卫星来绘制北美北方森林火灾期间土壤有机层的消耗图(French et al.,2008；Barrett et al.,2010)。

使用遥感评估高纬度地区火灾的碳排放也不限于北方森林（Palacios-

Orueta et al., 2005),最近的一些研究使用遥感数据估算了苔原火灾的碳损失(Mack et al., 2011; Rocha and Shaver, 2011; Rocha et al., 2012)。

七、总结与结论

过去几十年,遥感以各种不同的方法被广泛用于评估北极苔原和北方森林等北方高纬度生态系统中碳循环相关生态系统服务的供应和调节(图4-2)。这些生态系统服务包括:(1)植被的碳积累(总初级生产力、植被净初级生产力)和北极苔原和北方森林的植被生物量的存储(生物量供应、碳固定/储存和大气二氧化碳调节);(2)陆地和大气之间的碳净交换(生态系统净生产、生态系统净交换)、有机物的分解和向二氧化碳的转化(如土壤呼吸、分解),及湖泊和湿地产生的甲烷(碳固定/储存,及大气二氧化碳和甲烷调节);(3)土壤中碳的储存(土壤形成、碳固定/储存和大气二氧化碳调节)。随着涉及的碳库和碳过程从完全受下垫面影响(如总初级生产力、地上生物量)转移到很大程度上受地下影响(如甲烷生产、土壤碳),这些生态系统服务越来越难以用遥感方法进行估算。此外,沿着这个梯度,文献也越来越少。因此,我们看到诸多研究将植被指数(如NDVI)作为北方高纬度地区植物吸收碳的替代指标,并对北极苔原的"绿化"和卫星记录长度上伴随的北方森林的"褐化"有了合理解释。然而,利用遥感技术研究净生态系统碳交换(NEP)、土壤呼吸、分解和甲烷通量的碳循环过程的研究很少。这些研究通常是使用遥感观测的植被属性(如叶面积指数)作为模型输入数据,来模拟碳循环过程,或者是利用遥感地表分类对相关过程的实测数据进行空间外推的研究。同样的情况也适用于少数利用遥感估算土壤碳储量的研究。由于高纬度生态系统中有大量的碳储存,这些碳的命运是预测未来陆地-大气的一个重要不确定性来源。现有的了解碳循环的遥感方法有望继续得到改进,需要开发新的遥感方法来更直接地评估北方高纬度系统中碳相关池和过程。

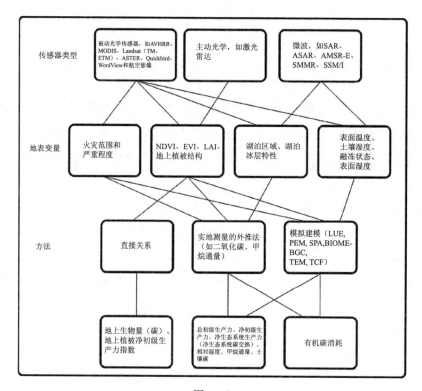

图 4-2

注：利用遥感技术评估北方高纬度生态系统碳循环变量的概念图。

致谢

本研究由美国国家科学基金会 ARC-0531166 和 ARC-0902152，及美国国家航空航天局 NNG6GE00A 和 NNX09AK56G 资助。杰丽卡·弗雷泽制作了本研究所含图表。

参 考 文 献

Alcaraz-Segura, D., E. Chuvieco, H. E. Epstein, E. Kasischke, and A. Trishchenko. 2010. The remotely sensed greening versus browning of the North American boreal forest. *Global Change Biology* 16:760–770.

Allen, J. L., and B. Sorbel. 2008. Assessing the differenced normalized burn ratio's ability to map burn severity in the boreal forest and tundra ecosystems of Alaska's national parks. *International Journal of Wildland Fire* 17:463–475.

Amiro, B. D., J. I. MacPherson, R. L. Desjardins, J. M. Chen, and J. Liu. 2003. Post-fire carbon dioxide fluxes in western Canadian boreal forest: Evidence from towers, aircraft and remote sensing. *Agricultural and Forest Meteorology* 115:91–107.

Baird, R. A., D. Verbyla, and T. N. Hollingsworth. 2012. Browning of the landscape of interior Alaska based on 1986–2009 Landsat sensor NDVI. *Canadian Journal of Forest Research* 42:1371–1382.

Barrett, K., E. S. Kasischke, A. D. McGuire, M. R. Turetsky, and E. S. Kane. 2010. Modeling fire severity in black spruce stands in the Alaskan boreal forest using spectral and non-spectral geospatial data. *Remote Sensing of the Environment* 114:1494–1503.

Barrett, K., A. D. McGuire, E. E. Hoy, and E. S. Kasischke. 2011. Potential shifts in dominant forest cover in interior Alaska driven by variations in fire severity. *Ecological Applications* 21:2380–2396.

Bartsch, A., R. A. Kidd, C. Pathe, K. Scipal, and W. Wagner. 2007. Satellite radar imagery for monitoring inland wetlands in boreal and sub-arctic environments. *Aquatic Conservation—Marine and Freshwater Ecosystems* 17:305–317.

Bartsch, A., C. Pathe, K. Scipal, and W. Wagner. 2008. Detection of permanent open water surfaces in central Siberia with ENVISAT ASAR wide swath data with species emphasis on the estimation of methane fluxes from tundra wetlands. *Hydrology Research* 39:89–100.

Beck, P. S. A., and S. J. Goetz. 2011. Satellite observations of high northern latitude vegetation productivity changes between 1982 and 2008: Ecological variability and regional differences. *Environmental Research Letters* 6:045501.

Beck, P. S. A., and S. J. Goetz. 2012. Corrigendum: Satellite observations of high northern latitude vegetation productivity changes between 1982 and 2008: Ecological variability and regional differences. *Environmental Research Letters* 7:029501.

Bhatt, U. S., D. A. Walker, M. K. Raynolds, et al. 2010. Circumpolar Arctic tundra vegetation change is linked to sea ice decline. *Earth Interactions* 14:8.

Boyd, D. S., and F. M. Danson. 2005. Satellite remote sensing of forest resources: Three decades of research development. *Progress in Physical Geography* 29:1–26.

Bubier, J., T. Moore, K. Savage, and P. Crill. 2005. A comparison of methane flux in a boreal landscape between a dry and a wet year. *Global Biogeochemical Cycles* 19:GB1023.

Bunn, A. G., and S. J. Goetz. 2006. Trends in satellite-observed circumpolar photosynthetic activity from 1982 to 2003: The influence of seasonality, cover type, and vegetation density. *Earth Interactions* 10:12.

Bunn, A. G., S. J. Goetz, and G. J. Fiske. 2005. Observed and predicted responses of plant growth to climate across Canada. *Geophysical Research Letters* 32:L16710.

Bunn, A. G., S. J. Goetz, J. S. Kimball, and K. Zhang. 2007. Northern high-latitude ecosystems respond to climate change. *EOS* 88:333–335.

Casper, P., S. C. Maberly, G. H. Hall, and B. J. Finlay. 2000. Fluxes of methane and carbon dioxide from a small productive lake to the atmosphere. *Biogeochemistry* 49:1–19.

Chen, J. M., W. M. Ju, J. Cihlar, et al. 2003. Spatial distribution of carbon sources and sinks in Canada's forests. *Tellus Series B—Chemical and Physical Meteorology* 55:622–641.

Conrad, S. G., A. I. Sukhinin, B. J. Stocks, D. R. Cahoon, E. P. Davidenko, and G. A. Ivanova. 2002. Determining effects of area burned and fire severity on carbon cycling and emissions in Siberia. *Climatic Change* 55:197–211.

de Groot, W. J., R. Landry, W. A. Kurz, et al. 2007. Estimating direct carbon emissions from Canadian wildland fires. *International Journal of Wildland Fire* 16:593–606.

Dong, J. R., R. K. Kaufmann, R. B. Myneni, et al. 2003. Remote sensing estimates of boreal and temperate forest woody biomass: Carbon pools, sources, and sinks. *Remote Sensing of the Environment* 84:393–410.

Drezet, P. M. L., and S. Quegan. 2007. Satellite-based radar mapping of British forest age and Net Ecosystem Exchange using ERS tandem coherence. *Forest Ecology and Management* 238:65–80.

Dye, D. G., and C. J. Tucker. 2003. Seasonality and trends of snow-cover, vegetation index, and temperature in northern Eurasia. *Geophysical Research Letters* 30:1405.

Epp, H., and R. Lanoville. 1996. Satellite data and geographic information systems for fire and resource management in the Canadian Arctic. *Geocarto International* 11:97–103.

Epstein, H. E., M. K. Raynolds, D. A. Walker, U. S. Bhatt, C. J. Tucker, and J. E. Pinzon. 2012. Dynamics of aboveground phytomass of the circumpolar Arctic tundra during the past three decades. *Environmental Research Letters* 7:015506.

Euskirchen, E. S., A. D. McGuire, F. S. Chapin, S. Yi, and C. C. Thompson. 2009. Changes in vegetation in northern Alaska under scenarios of climate change, 2003–2010: Implications for climate feedbacks. *Ecological Applications* 19:1022–1043.

Fraser, R. H., R. J. Hall, R. Landry, et al. 2004. Validation and calibration of Canada-wide coarse-resolution satellite burned-area maps. *Photogrammetric Engineering and Remote Sensing* 70:451–459.

Fraser, R. H., I. Olthof, M. Carriére, A. Deschamps, and D. Pouliot. 2011. Detecting long-term changes to vegetation in northern Canada using the Landsat satellite image archive. *Environmental Research Letters* 6:045502.

French, N. H. F., E. S. Kasischke, R. J. Hall, et al. 2008. Using Landsat data to assess fire and burn severity in the North American boreal forest region: An overview and summary of results. *International Journal of Wildland Fire* 17:443–462.

Frohn, R. C., K. M. Hinkel, and W. R. Eisner. 2005. Satellite remote sensing classification of thaw lakes and drained thaw lake basins on the North Slope of Alaska. *Remote Sensing of the Environment* 97:116–126.

Frolking, S., M. W. Palace, D. B. Clark, J. Q. Chambers, H. H. Shugart, and G. C. Hurtt. 2009. Forest disturbance and recovery: A general review in the context of spaceborne remote sensing of impacts on aboveground biomass and canopy structure. *Journal of Geophysical Research—Biogeosciences* 114:G00E02.

Fuchs, H., P. Magdon, C. Kleinn, and H. Flessa. 2009. Estimating aboveground carbon in a catchment of the Siberian forest tundra: Combining satellite imagery and field inventory. *Remote Sensing of the Environment* 113:518–531.

Gamon, J. A., K. F. Huemmrich, D. R. Peddle, et al. 2004. Remote sensing in BOREAS: Lessons learned. *Remote Sensing of the Environment* 89:139–162.

Giglio, L., T. Loboda, D. P. Roy, B. Quayle, and C. O. Justice. 2009. An active-fire based burned area mapping algorithm for the MODIS sensor. *Remote Sensing of the Environment* 113:408–420.

Giglio, L., J. T. Randerson, G. van der Werf, et al. 2010. Assessing variability and long-term trends in burned area by merging multiple satellite fire products. *Biogeosciences* 7:1171–1186.

Goetz, S. J., A. G. Bunn, G. J. Fiske, and R. A. Houghton. 2005. Satellite-observed photosynthetic trends across boreal North America associated with climate and fire disturbance. *Proceedings of the National Academy of Sciences of the United States of America* 102:13521–13525.

Goetz, S. J., H. E. Epstein, U. S. Bhatt, et al. 2011. Vegetation productivity and disturbance changes across arctic northern Eurasia: Satellite observations and simulation modeling. In *Eurasian Arctic land cover and land use in a changing climate*, eds. G. Gutman, P. Groisman, and A. Reissell, 9–36. Berlin: Springer-Verlag.

Goetz, S. J., and S. D. Prince. 1996. Remote sensing of net primary production in boreal forest stands. *Agricultural and Forest Meteorology* 78:149–179.

Goetz, S. J., S. D. Prince, S. N. Goward, M. M. Thawley, J. Small, and A. Johnston. 1999. Mapping net primary production and related biophysical variables with remote sensing: Applications to the BOREAS region. *Journal of Geophysical Research—Atmospheres* 104:22719–27734.

Hall, F. G., D. R. Peddle, and E. F. Ledrew. 1996. Remotes sensing of biophysical variables in boreal forest stands of *Picea mariana*. *International Journal of Remote Sensing* 17:3077–3081.

Hall, R. J., R. S. Skakum, E. J. Arsenault, and B. S. Case. 2006. Modeling forest stand structure attributes using Landsat ETM+ data: Application to mapping of aboveground biomass and stand volume. *Forest Ecology and Management* 225:378–390.

Haynes-Young, R., and M. Potschin. 2010. The links between biodiversity, ecosystem services and human well-being. In *Ecosystem ecology: A new synthesis*, eds. D. G. Rafaelli and C. L. J. Frid, 110–139. Cambridge, UK: Cambridge University Press.

Hinkel, K. M., R. C. Frohn, F. E. Nelson, W. R. Eisner, and R. A. Beck. 2005. Morphometric and spatial analysis of thaw lakes and drained thaw lake basins in the western Arctic Coastal Plain, Alaska. *Permafrost and Periglacial Processes* 16:327–341.

Hope, A. S., J. S. Kimball, and D. A. Stow. 1993. The relationship between tussock tundra spectral reflectance properties and biomass and vegetation composition. *International Journal of Remote Sensing* 14:1861–1874.

Huemmrich, K. F., G. Kinoshita, J. A. Gamon, S. Houston, H. Kwon, and W. C. Oechel. 2010. Tundra carbon balance under varying temperature and moisture regimes. *Journal of Geophysical Research—Biogeosciences* 115:G00I02.

Hugelius, G. 2012. Spatial upscaling using thematic maps: An analysis of uncertainties in permafrost soil carbon estimate. *Global Biogeochemical Cycles* 26:GB2026.

Isaev, A. S., G. N. Korovin, S. A. Bartalev, et al. 2002. Using remote sensing for assessment of forest wildfire carbon emissions. *Climatic Change* 55:231–255.

Jia, G. S., H. E. Epstein, and D. A. Walker. 2003. Greening of Arctic Alaska, 1981–2001. *Geophysical Research Letters* 30:2067.

Jia, G. S., H. E. Epstein, and D. A. Walker. 2006. Spatial heterogeneity of tundra vegetation response to recent temperature changes. *Global Change Biology* 12:42–55.

Jia, G. S., H. E. Epstein, and D. A. Walker. 2009. Vegetation greening in the Canadian arctic related to decadal warming. *Journal of Environmental Monitoring* 11:2231–2238.

Jones, L. A., J. S. Kimball, K. C. McDonald, S. T. K. Chan, E. G. Njoku, and W. C. Oechel. 2007. Satellite microwave remote sensing of boreal and arctic soil temperatures from AMSR-E. *IEEE Transactions on Geoscience and Remote Sensing* 45:2004–2018.

Kasischke, E. S., and E. E. Hoy. 2012. Controls on carbon consumption during Alaskan wildland fires. *Global Change Biology* 18:685–699.

Kasischke, E. S., E. Hyer, P. Novelli, et al. 2005. Influences of boreal fire emissions on Northern Hemisphere atmospheric carbon and carbon monoxide. *Global Biogeochemical Cycles* 19:GB1012.

Kimball, J. S., L. A. Jones, K. Zhang, F. A. Heinsch, K. C. McDonald, and W. C. Oechel. 2009. A satellite approach to estimate land-atmosphere CO_2 exchange for boreal and arctic biomes using MODIS and AMSR-E. *IEEE Transactions on Geoscience and Remote Sensing* 47:569–587.

Kimball, J. S., M. Zhao, A. D. McGuire, et al. 2007. Recent climate-drive increases in vegetation productivity for the western Arctic: Evidence of an acceleration of the northern terrestrial carbon cycle. *Earth Interactions* 11:4.

Krankina, O. N., R. A. Houghton, M. E. Harmon, et al. 2005. Effects of climate, disturbance, and species on forest biomass across Russia. *Canadian Journal of Forest Research* 35:2281–2293.

Kushida, K., Y. Kim, N. Tanaka, and M. Fukida. 2004. Remote sensing of net ecosystem productivity based on component spectrum and soil respiration observation in a boreal forest, interior Alaska. *Journal of Geophysical Research—Atmospheres* 109:D06108.

Leboeuf, A., A. Beaudoin, R. A. Fournier, L. Guindon, J. E. Luther, and M. C. Lambert. 2007. A shadow fraction method for mapping biomass of northern boreal black spruce forests using QuickBird imagery. *Remote Sensing of the Environment* 110:488–500.

Lee, H., E. A. G. Schuur, K. S. Inglett, M. Lavoie, and J. P. Chanton. 2012. The rate of permafrost carbon release under aerobic and anaerobic conditions and its potential effects on climate. *Global Change Biology* 18:515–527.

Lefsky, M. A., W. B. Cohen, D. J. Harding, G. G. Parker, S. A. Acker, and S. T. Gower. 2002. Lidar remote sensing of above-ground biomass in three biomes. *Global Ecology and Biogeography* 11:393–399.

Lewis, S. A., A. T. Hudak, R. D. Ottmar, et al. 2011. Using hyperspectral imagery to estimate forest floor consumption from wildfire in boreal forests of Alaksa, USA. *International Journal of Wildland Fire* 2:255–271.

Li, Z., S. Nadon, and J. Cihlar. 2000. Satellite-based detection of Canadian boreal forest fires: Development and application of the algorithm. *International Journal of Remote Sensing* 21:3057–3069.

Liu, J., J. M. Chen, J. Cihlar, and W. Chen. 1999. Net primary productivity distribution in the BOREAS region from a process model using satellite and surface data. *Journal of Geophysical Research—Atmospheres* 104:27735–27754.

Loboda, T., K. J. O'Neal, and I. Csiszar. 2007. Regionally adaptable dNBR-based algorithm for burned area mapping from MODIS data. *Remote Sensing of the Environment* 109:429–442.

Loboda, T. V., and I. A. Csiszar. 2007. Reconstruction of fire spread within wildland fire events in Northern Eurasia from the MODIS active fire product. *Global and Planetary Change* 56:258–273.

Loboda, T. V., E. E. Hoy, L. Giglio, and E. S. Kasischke. 2011. Mapping burned area in Alaska using MODIS data: A data limitations-driven modification to the regional burned area algorithm. *International Journal of Wildland Fire* 20:487–496.

Loboda, T. V., Z. Zhang, K. J. O'Neal, et al. 2012. Reconstructing disturbance history using satellite-based assessment of the distribution of land cover in the Russian Far East. *Remote Sensing of the Environment* 118:241–248.

Loranty, M. M., S. J. Goetz, E. B. Rastetter, et al. 2011. Scaling an instantaneous model of tundra NEE to the Arctic landscape. *Ecosystems* 14:76–93.

Lutz, D. A., R. A. Washington-Allen, and H. H. Shugart. 2008. Remote sensing of boreal forest biophysical and inventory parameters: A review. *Canadian Journal of Remote Sensing* 34:S286–S313.

Mack, M. C., M. S. Bret-Harte, T. N. Hollingsworth, et al. 2011. Carbon loss from an unprecedented Arctic tundra wildfire. *Nature* 475:489–492.

McGuire, A. D., L. G. Anderson, T. R. Christenson, et al. 2009. Sensitivity of the carbon cycling in the Arctic to climate change. *Ecological Monographs* 79:523–555.

McMichael, C. E., A. S. Hope, D. A. Stow, J. B. Fleming, G. Vourlitis, and W. Oechel. 1999. Estimating CO_2 exchange at two sites in Arctic tundra ecosystems during the growing season using a spectral vegetation index. *International Journal of Remote Sensing* 20:683–698.

Meroni, M., D. Mollicone, L. Belelli, et al. 2002. Carbon and water exchanges of regenerating forests in central Siberia. *Forest Ecology and Management* 169:115–122.

Michalek, J. L., N. H. F. French, E. S. Kasischke, R. D. Johnson, and J. E. Colwell. 2000. Using Landsat TM data to estimate carbon release from burned biomass in an Alaskan spruce complex. *International Journal of Remote Sensing* 21:323–338.

Morgenstern, A., G. Grosse, F. Gunther, I. Federova, and L. Schirmeister. 2011. Spatial analyses of thermokarst lakes and basins in Yedoma landscapes of the Lena Delta. *Cryosphere* 5:849–867.

Myneni, R. B., J. Dong, C. J. Tucker, et al. 2001. A large carbon sink in the woody biomass of Northern forests. *Proceedings of the National Academy of Sciences of the United States of America* 98:14784–14789.

Myneni, R. B., C. D. Keeling, C. J. Tucker, G. Asrar, and R. R. Nemani. 1997. Increased plant growth in the northern high latitudes from 1981 to 1991. *Nature* 386:698–702.

Natali, S. M., E. A. G. Schuur, and R. L. Rubin. 2012. Increased plan productivity in Alaska as a result of experimental warming of soil and permafrost. *Journal of*

Ecology 100:488–498.

Neigh, C. S. R., C. J. Tucker, and J. R. G. Townshend. 2008. North American vegetation dynamics observed with multi-resolution satellite data. *Remote Sensing of the Environment* 112:1749–1772.

Olthof, I., D. Pouliot, R. Latifovic, and W. Chen. 2008. Recent (1986–2006) vegetation-specific NDVI trends in northern Canada from satellite data. *Arctic* 61:381–394.

Palacios-Orueta, A., E. Chuvieco, A. Parra, and C. Carmona-Moreno. 2005. Biomass burning emissions: A review of models using remote-sensing data. *Environmental Monitoring and Assessment* 104:189–209.

Parent, M. B., and D. Verbyla. 2010. The browning of Alaska's boreal forest. *Remote Sensing* 2:2729–2747.

Plug, L. J., C. Walls, and B. M. Scott. 2008. Tundra lake changes from 1978 to 2001 on the Tuktoyaktuk Peninsula, western Canadian Arctic. *Geophysical Research Letters* 35:L03502.

Pouliot, D., R. Latifovic, and I. Olthof. 2009. Trends in vegetation NDVI from 1 km AVHRR data over Canada for the period 1985–2006. *International Journal of Remote Sensing* 30:149–168.

Prince, S. D., and S. N. Goward. 1995. Global primary production: A remote sensing approach. *Journal of Biogeography* 22:815–835.

Raynolds, M. K., D. A. Walker, H. E. Epstein, J. E. Pinzon, and C. J. Tucker. 2012. A new estimate of tundra-biome phytomass from trans-Arctic field data and AVHRR NDVI. *Remote Sensing Letters* 5:403–411.

Richardson, A. D., T. A. Black, P. Ciais, et al. 2010. Influence of spring and autumn phenological transitions on forest ecosystem productivity. *Philosophical Transactions of the Royal Society B—Biological Sciences* 365:3227–3246.

Rocha, A. V., M. M. Loranty, P. E. Higuera, et al. 2012. The footprint of Alaskan tundra fires during the past half-century: Implications for surface properties and radiative forcing. *Environmental Research Letters* 7:044039.

Rocha, A. V., and G. S. Shaver. 2011. Burn severity influences postfire CO_2 exchange in arctic tundra. *Ecological Applications* 21:477–489.

Ropars, P., and S. Boudreau. 2012. Shrub expansion at the forest-tundra ecotone: Spatial heterogeneity linked to local topography. *Environmental Research Letters* 7:015502.

Schuur, E. A. G., J. Bockheim, J. G. Canadell, et al. 2008. Vulnerability of permafrost carbon to climate change: Implications for the global carbon cycle. *BioScience* 58:701–714.

Shippert, M. M., D. A. Walker, N. A. Auerbach, and B. E. Lewis. 1995. Biomass and leaf-area index maps derived from SPOT images for Toolik Lake and Imnavait Creek areas, Alaska. *Polar Record* 31:147–154.

Sitch, S., A. D. McGuire, J. Kimball, et al. 2007. Assessing the carbon balance of circumpolar Arctic tundra using remote sensing and process modeling. *Ecological Applications* 17:213–234.

Slayback, D. A., J. E. Pinzon, S. O. Los, and C. J. Tucker. 2003. Northern hemisphere photosynthetic trends 1982–99. *Global Change Biology* 9:1–15.

Smith, B., W. Knorr, J. L. Widlowski, B. Pinty, and N. Gobron. 2008. Combining remote sensing data with process modeling to monitor boreal conifer forest carbon balances. *Forest Ecology and Management* 255:3985–3994.

Smith, N. V., S. S. Saatchi, and J. T. Randerson. 2004. Trends in high northern latitude soil freeze and thaw cycles from 1988 to 2002. *Journal of Geophysical Research—Atmospheres* 109:D12101.

Soegaard, H., C. Nordstroem, T. Friborg, B. U. Hansen, T. R. Christensen, and C. Bay. 2000. Trace gas exchange in a high-Arctic valley 3. Integrating and scaling CO_2 fluxes from canopy to landscape using flux data, footprint modeling, and remote sensing. *Global Biogeochemical Cycles* 14:725–744.

Soja, A. J., A. I. Sukhinin, D. R. Cahoon, H. H. Shugart, and P. W. Stackhouse. 2004. AVHRR-derived fire frequency, distribution and area burned in Siberia. *International Journal of Remote Sensing* 25:1939–1960.

Steyaert, L. T., F. G. Hall, and T. R. Loveland. 1997. Land cover mapping, fire regeneration, and scaling studies in the Canadian boreal forest with 1 km AVHRR and Landsat TM data. *Journal of Geophysical Research—Atmospheres* 102:29581–29598.

Stinson, G., W. A. Kurz, C. E. Smyth, et al. 2011. An inventory-based analysis of Canada's managed forest carbon dynamics, 1990 to 2008. *Global Change Biology* 17:2227–2244.

Stow, D., A. Hope, W. Boynton, S. Phinn, D. Walker, and N. Auerbach. 1998. Satellite-derived vegetation index and cover type maps for estimating carbon dioxide flux for arctic tundra regions. *Geomorphology* 21:313–327.

Stow, D. A., A. Hope, D. McGuire, et al. 2004. Remote sensing of vegetation and land-cover change in Arctic Tundra Ecosystems. *Remote Sensing of the Environment* 89:281–308.

Sturm, M., C. Racine, and K. Tape. 2001. Climate change—Increasing shrub abundance in the Arctic. *Nature* 411:546–547.

Sukhinin, A. I., N. H. F. French, E. S. Kasischke, et al. 2004. AVHRR-based mapping of fires in Russia: New products for fire management and carbon cycle studies. *Remote Sensing of the Environment* 93:546–564.

Tagesson, T., M. Mastepanov, M. P. Tamstorf, et al. 2012. High-resolution satellite data reveal an increase in peak growing season gross primary production in a high-Arctic wet tundra ecosystem 1992–2008. *International Journal of Applied Earth Observation and Geoinformation* 18:407–416.

Tagesson, T., B. Smith, A. Lofgren, A. Rammig, L. Eklundh, and A. Lindroth. 2009. Estimating net primary production of Swedish forest landscapes by combining mechanistic modeling and remote sensing. *Ambio* 38:316–324.

Tape, K., M. Sturm, and C. Racine. 2006. The evidence for shrub expansion in Northern Alaska and the Pan-Arctic. *Global Change Biology* 12:686–702.

Tarnocai, C., J. G. Canadell, E. A. G. Schuur, P. Kuhry, G. Mazhitova, and S. Zimov. 2009. Soil organic carbon pools in the northern circumpolar permafrost region. *Global Biogeochemical Cycles* 23:GB2023.

Tremblay, B., E. Lévesque, and S. Boudreau. 2012. Recent expansion of erect shrubs in Low Arctic: Evidence from Eastern Nunavik. *Environmental Research Letters* 7:035501.

Treshnikov, A. F. 1985. *Atlas of the Arctic* (in Russian). Moscow: Administrator of Geodesy and Cartography of the Soviet Ministry.

Trucco, C., E. A. G. Schuur, S. M. Natali, E. F. Belshe, R. Bracho, and J. Vogel. 2012. Seven-year trends of CO_2 exchange in a tundra ecosystem affected by long-term permafrost thaw. *Journal of Geophysical Research—Biogeosciences* 117:G02031.

Tucker, W. B., J. W. Weatherly, D. T. Eppler, L. D. Farmer, and D. L. Bently. 2001.

Evidence for rapid thinning of sea ice in the western Arctic Ocean at the end of the 1980s. *Geophysical Research Letters* 28:2851–2854.

Turetsky, M. R., E. S. Kane, J. W. Harden, et al. 2011. Recent acceleration of biomass burning and carbon losses in Alaskan forests and peatlands. *Nature Geosciences* 4:27–31.

Turner, D. P., W. D. Ritts, W. B. Cohen, et al. 2005. Site-level evaluation of satellite-based global terrestrial gross primary production and net primary production monitoring. *Global Change Biology* 11:666–684.

Ulrich, M., G. Grosse, S. Chabrillat, and L. Schirmeister. 2009. Spectral characterization of periglacial surfaces and geomorphological units in the Arctic Lena Delta using field spectrometry and remote sensing. *Remote Sensing of the Environment* 113:1220–1235.

Verbyla, D. 2008. The greening and browning of Alaska based on 1982–2003 satellite data. *Global Ecology and Biogeography* 17:547–555.

Walker, D. A., H. E. Epstein, G. J. Jia, et al. 2003. Phytomass, LAI, and NDVI in northern Alaska: Relationships to summer warmth, soil pH, plant functional types, and extrapolation to the circumpolar Arctic. *Journal of Geophysical Research—Atmospheres* 108:8169.

Walker, D. A., H. E. Epstein, M. K. Raynolds, et al. 2012. Environment, vegetation and greenness (NDVI) along the North America and Eurasia Arctic transects. *Environmental Research Letters* 7:015504.

Walker, D. A., M. K. Raynolds, F. J. A. Daniels, et al. 2005. The circumpolar Arctic vegetation map. *Journal of Vegetation Science* 16:267–282.

Walter, K. M., M. Engram, C. R. Duguay, M. O. Jeffries, and F. S. Chapin. 2008. The potential use of synthetic aperture radar for estimating methane ebullition from Arctic lake. *Journal of the American Water Resources Association* 44:305–315.

Wang, J. D., Y. W. Sheng, K. M. Hinkel, and E. A. Lyons. 2012. Drained thaw lake basin recovery on the western Arctic Coastal Plain of Alaska using high-resolution digital elevation models and remote sensing imagery. *Remote Sensing of the Environment* 119:325–336.

Way, J., R. Zimmerman, E. Rignot, K. McDonald, and R. Oren. 1997. Winter and spring thaw as observed with imaging radar at BOREAS. *Journal of Geophysical Research—Atmospheres* 102:29673–29684.

Williams, M., E. B. Rastetter, G. R. Shaver, J. E. Hobbie, E. Carpino, and B. L. Kwiatkowski. 2001. Primary production of an arctic watershed: An uncertainty analysis. *Ecological Applications* 11:1800–1816.

Zhang, K., J. S. Kimball, E. H. Hogg, et al. 2008. Satellite-based detection of recent climate-driven changes in northern high-latitude vegetation productivity. *Journal of Geophysical Research—Biogeosciences* 113:G03033.

Zhang, K., J. S. Kimball, M. S. Zhao, W. C. Oechel, J. Cassano, and S. W. Running. 2007. Sensitivity of pan-Arctic terrestrial net primary productivity simulations to daily surface meteorology from NCEP-NCAR and ERA-40 reanalyses. *Journal of Geophysical Research—Biogeosciences* 112:G01011.

Zheng, D., S. Prince, and T. Hame. 2004. Estimating net primary production of boreal forests in Finland and Sweden from field data and remote sensing. *Journal of Vegetation Science* 15:161–170.

Zhou, L. M., C. J. Tucker, R. K. Kaufmann, D. Slayback, N. V. Shabanov, and R. B. Myneni. 2001. Variations in northern vegetation activity inferred from satellite data of vegetation index during 1981 to 1999. *Journal of Geophysical Research—Atmospheres* 106:20069–20083.

第五章 监测牧草生产的生态系统服务

J. G. N. 伊里萨里　M. 奥斯特赫尔德
M. 奥雅扎巴尔　J. M. 帕鲁埃洛　M. 杜兰特

一、简介

在世界范围内,牛达到市场重量所需的80%的能量来自牧场草地和天然草地(Wheeler et al.,1981;Oltjen and Beckett,1996)。

管理这些牧草资源需要了解它们的产量。牧草产量,亦称牧草生长率,是地上净初级生产力的一部分,是关键的生态系统变量。在个体层面上,ANPP代表光合作用和呼吸作用之间的差异(Chapin et al.,2002);而在景观层面上,ANPP代表单位面积和时间的生物量生产率;在生态系统层面上,ANPP代表初级消费者可用能源的生产率。

ANPP是生态系统功能的核心,决定着许多生态系统服务的投入水平(Costanza et al.,1997)。例如,牧场草地和天然草地的地上净初级生产力信息的可用性,对于确定适当的载畜率和管理牧草的过剩或短缺至关重要。考虑到地上净初级生产力在牧场管理中具有高度相关性,但关于其空间和时间变化的可用数据很少。这类信息的缺乏导致很难估算实地的地上净初级生产力,也很难将为数不多的可用数据外推到其他空间或时间的情况中。基于遥感数据的地上净初级生产力模型(Paruelo et al.,1997,2000;Running et al.,2000;Piñeiro et al.,2006)可以解决这种信息缺乏的问题,因为它们覆盖的区域大、周期长。

地上净初级生产力在空间和时间上有不同尺度的变化。贯穿全部资源渐变,地上净初级生产力随着降雨量的增加而持续增加(Lauenroth,1979;Sala et al.，1988)。在景观中,地上净初级生产力在地形变化(Knapp et al.，1993；Buono et al. 2010；Irisarri et al.，2012)、土壤差异(Aragón and Oesterheld，2008)或干扰频率(Knapp et al.，1993；Oesterheld et al.，1999)方面存在很大差异。干草原、草原和热带稀树草原的地上净初级生产力的年际变化很大,给牧场管理带来极大的不确定性(Lauenroth and Sala，1992；Knapp and Smith，2001；Bai et al.，2008)。地上净初级生产力的季节性变化也很大,限制了牧场管理。例如,在拉普拉塔草原的温带地区,春季地上净初级生产力是冬季的10倍(Sala et al.，1981；Oesterheld and León，1987；Altesor et al.，2005；Semmartin et al.，2007)。

本章目标:(1)介绍两种估算地上净初级生产力的方法,一种是通过生物量收获的经典方法,另一种是基于遥感数据的最新实践;(2)介绍一种牧草生产监测系统,该系统根据第二种方法定期观察阿根廷和乌拉圭的牧场。在本章中,我们首先描述了通过连续的生物量收获来估算地上净初级生产力所涉及的操作方法与困难;其次,我们描述了一种基于遥感方法的操作方式;再次,对光能利用率(Radiation Use Effciency,RUE)的估算进行了较为详细的讨论;最后,我们介绍了目前在阿根廷和乌拉圭运行的牧草生产监测系统。

二、基于连续生物量收获估算地上净初级生产力

地上净初级生产力是由入射的光合有效辐射转化为生物量的结果。产生的生物量可由食草动物消耗,也可遵循碎屑流的连续步骤:静止的死生物量、凋落物和有机物,可能在这些步骤中被分解(图5-1)。

通过削减生物量来估算地上净初级生产力是最常用的方法,同时也是最具争议的方法(Singh et al.，1975；McNaughton et al.，1996；Sala and Austin，2000；Scurlock et al.，2002；Knapp et al.，2007)。其目的是获取一定时期内的累积生物量,这是植被生长的结果。使用这种方法至少涉及两个困难。第一个困难是,根据对某些流量大小的初始假设,相同的数据可能会导致地上净初级生产力的不同估算(Scurlock et al.，2002)。例如,可将地上净初级生产力视为单个活生物量峰值,或者,它也可包括诸如衰老或分解的其他流量(Sala and

Austin，2000）。根据这些假设，可能会生成截然不同的地上净初级生产力估算值。例如，基于相同的数据，草原的地上净初级生产力在 200～400 克/每平方米每年和 350～1 000 克/每平方米每年之间变化，这取决于所做的初始假设（Scurlock et al.，2002）。

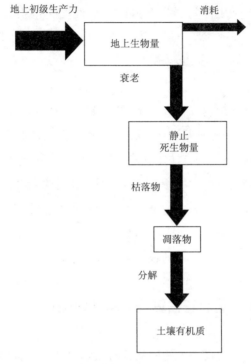

图 5-1

注：牧场草地和天然草地中与生物量产生和转化相关的流程与过程各分段部分的示意图。箭头和方框的大小并不代表流量和状态变量的大小。在该模型中不考虑通过呼吸作用或其他过程（如浸出或光降解）造成的生物量损失（Austin and Vivanco，2006）。

第二个困难是，许多天然草地和牧场草地的生长季节并不短，这与单峰变化生物量等同于地上净初级生产力的假设相一致（Semmartin et al.，2007）。因此，地上净初级生产力必须在至少两次连续收获的基础上进行量化，一次是在研究阶段开始时，另一次是在研究阶段结束时（Sala and Austin，2000）。这种情况必然涉及对图 5-1 中的不同方框进行量化。在两个连续日期之间，方框大小的变化将表示对箭头(通量)大小的估算。

在某些情况下,使用连续生物量收获法可能难以获取地上净初级生产力。当消耗和衰老为零时,两个连续日期之间的生物量差异明显代表地上净初级生产力。若在非常剧烈的落叶期后抑制消耗量,并且收获之间的间隔时间很短,足以将衰老保持在最低水平,就会发生上述这种情况。因此,当试图通过连续的生物量收获来估算地上净初级生产力时,必须排除放牧(Sala and Austin, 2000)因素。如上所述,还必须区分每个物种的活生物量、现存死生物量以及凋落物(Oesterheld and León, 1987; Sala and Austin, 2000)。这些分段部分是估算图5-1中描述的一些流量的基础。然而,量化每段单独部分的生物量需要更大的采样工作,而且总有一些流量无法完全估算(衰老或分解)。

随着越来越多的部分通量需要量化,采样挑战难度也逐渐增加。地上净初级生产力表示其他变量的组合,至少是两个连续日期的生物量值。因此,假设没有协方差,地上净初级生产力方差是两个连续日期的生物量方差之和。如果死生物量或凋落物包括在内,则它们的差异也必须包括在内。此外,在大多数采样程序的空间分辨率(小于1平方米)下,放牧草地的生物量方差通常较高。因此,生物量估算的精度通常较低(Knapp et al., 2007)。

连续收获法所代表的操作方法与困难可以通过类比估算流入水箱的水流量来解释,在图表中牛可以从水箱中自由饮水(图5-2(a))。在这个类比中,流入的流量代表地上净初级生产力,在某一时刻的水量代表生物量。如果水箱没有任何流出流量,则可通过两个连续周期之间的容积差来估算流入流量。然而,对流入流量的估算面临着来自三个不同来源的问题。

首先,水箱通过衰老和分解的"孔"流失水分;其次,必须将牛排除在外,以消除因饮水而造成的水分流失,这不可避免地导致现实性的流失(图5-2(a));再次,要实现高精度的体积测量是非常困难的,这种采样问题可以看作是水处于持续搅动状态,只有几秒钟的时间进行测量。上述问题的主要后果是,在空间和时间上对牧场区域的地上净初级生产力估算很少见,这种估算往往不够准确(Singh et al., 1975, Scurlock et al., 2002)。

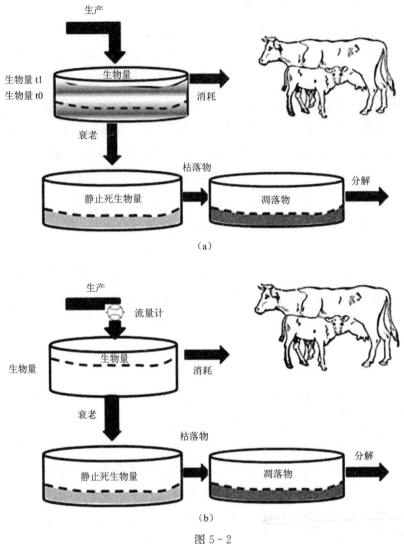

图 5-2

注:该图解释通过类比流入水箱的水量来估算地上净初级生产力遇到的问题,设定图中牛可以自由饮用水箱的水。(a)对于连续收获,进水按两个时刻的体积差来测量;(b)利用遥感技术,通过在进水管道上安装流量计来估算进水量。水位代表给定时间内的生物量。第一个水箱中的进水量代表地上净初级生产力。每个水箱代表一种特定的生物量状态:绿色生物量(主要输入的流量为地上净初级生产)、静止死生物量(输入的流量为衰老)和凋落物(输入流量为枯落物)。与连续收获方法相关的困难是水通过消耗、衰老和分解的"孔"泄漏。因此,牛必须被排除在外,即排除随之而来的水的流失。季节和空间的变化类似于水流的运动,并且增加了测量水量的难度。与遥感方法相关的困难是流量计的准确性(即遥感影像的时间和空间分辨率)以及需要将流量计读数校准为水流量(即要知道光能利用率)。

三、通过遥感估算地上净初级生产力

要了解基于遥感的方法，有必要了解蒙特斯模型（Monteith，1972）。该模型表明，地上净初级生产力是三种成分的产物，即光合有效辐射、光合有效辐射吸收比率和辐射利用率：

$$\text{ANPP}(g \cdot m^{-2} \cdot y^{-1}) = \text{PAR}(MJ \cdot m^{-2} \cdot y^{-1}) \times f\,\text{PAR} \times \text{RUE}(g \cdot MJ^{-1})$$

式 5-1

模型表明，随着光合辐射被吸收并转化为新的生物组织时，会产生新的生物量（地上净初级生产力）。在地上净初级生产力的特殊情况下，需要考虑到辐射利用率是指将所吸收的能量转化为新的地上生物量，其中包括地下生物量和地上生物量之间的资源分配。

通过这种方法进行地上净初级生产力估算需要解决三个问题。首先，有必要了解光合入射辐射。这相对简单，因为平均日摄入量可以从多种来源和许多地点获取。此外，光合入射辐射的年际变化较小。因此，一年中某个月份的值可以代替其平均值。如果需要更精确的精度，有一些模型可以根据气象数据、温度和降水来估算入射辐射（Thornton and Running，1999）。

其次，有必要知道光合活性组织吸收的入射辐射的比率。幸运的是，遥感可以提供此类信息（详见 Piñeiro et al.，2006）。传感器能够估算红色和近红外光谱反射率，也就可以估算绿色植被吸收的入射光合辐射的比率（Sellers et al.，1992）。

再欠，需要了解光能利用率。这是最难也是最不为人知的因素。光能利用率变化与光合途径（C_3-C_4）和与生命形式相关的光合作用和呼吸作用之间的比率，以及非生物因素有关，如水、养分供给或温度等。

它可以通过校准或遥感资料进行估算（详情请参见 Garbulsky et al.，2010）。幸运的是，通常与牧场管理相关的尺度上，无论在空间上还是在时间上，光能利用率是最小变量（Chapin et al.，2002）。

通过遥感估算地上净初级生产力的操作方法与困难可通过用相同的水箱模拟类比来解释（图 5-2）。在这种情况下，无须知道水位（生物量）。使用遥感方法类似于在进水管道上设置流量计(图5-2(b))。在这种情况下,牛可以自由饮水,因为流量是事先测量出来的。消耗对地上净初级生产力有影响,但这正是用遥感方法对光合有效辐射吸收比率估算得出的结果。通过遥感方法估算地上净初级生产力存在哪些困难?第一个困难是流量计具有最大分辨率,该分辨率与用于估算光合有效辐射吸收比率的像素大小有关。最大空间分辨率取决于像素大小,因此无法检测到一个像素内的差异。在被管理单元小于一个像素的情况下,这可能会带来一些不便。例如,一些中分辨率成像光谱仪产品的像素范围(5.3公顷)对于奶业系统的养殖围栏来说可能太大,但对于专门从事牛犊养殖的广阔牧场来说却是足够的。

第二个困难是估算光能利用率。应对这一限制有两种基本策略。第一种策略是完全不利用光能利用率,只利用关于吸收辐射的信息。在许多情况下,可以比较所评估的牧场单位或能率,因为可以假设其具有相似的光能利用率。例如,具有相似物种组成和拥有景观位置的牧场应该具有相似的光能利用率,可能存在的差异应该小于其他的变化来源。

在这些情况下,蒙特斯模型的前两个组成部分光合有效辐射和光合有效辐射吸收比率,代表了实际可吸收性光合有效辐射(兆焦/(公顷·天)),在许多情况下,这足以进行某些比较。第二个策略是估算光能利用率,如下一节内容所述。

四、基于地上净初级生产力和可吸收性光合有效辐射进行的光能利用率估算

光能利用率因植被类型和环境条件而异。它显示了生物群落之间的主要差异,从沙漠的 $0.27g·MJ^{-1}$ 到热带森林的 $0.71 g·MJ^{-1}$(Ruimy and Saugier,1994;Field et al.,1995)。半干旱草地的光能利用率年变化范围为 $0.27\sim0.35 g·MJ^{-1}$(Nouvellon et al.,2000)。在不同的季节之间,光能利用率变化甚至更大。例如,在潮湿的草原上,光能利用率在 $0.2\sim1.2g·MJ^{-1}$ 之间变化(Piñeiro et al.,2006)。光能利用率在不同物种组成和地形位置的牧场上也有所不同。例如,在以苜蓿为主的高地种植的牧草比以麦草为主的低地种植的牧草具有更

高的光能利用率(Grigera et al.,2007)。

在草原地区,通过间作技术改变物种组成也会影响光能利用率(Baeza et al.,2011)。正如许多补偿性生长案例所报道的那样,放牧改变了光能利用率。如上所述,在试图通过吸收的光合有效辐射估算地上净初级生产力时,都应考虑到所有这些变化的来源。

光能利用率可以通过三种方法进行估算(另见第六章)。一种是从蒙特斯(1972)公式推导出来的(式5-1),需要对地上净初级生产力和可吸收性光合有效辐射进行独立估算(Turner et al.,2003;Bradford et al.,2005;Piñeiro et al.,2006;Grigera et al.,2007;Irisarri et al.,2012)。另外两种方法都是基于气象数据或光谱指数,如光化学辐射指数(Garbulsky et al.,2011)(见第三章)。在本章中,我们将重点介绍第一种方法。

以下示例表示利用上述第一种方法对光能利用率进行估算。使用以下三个独立的信息来源:地上净初级生产力的连续生物量收获、遥感对光合有效辐射吸收比率的估算以及气象站的光合有效辐射值。一旦估算了光能利用率,并校准了模型(式5-1),就可以通过遥感独立估算几个围场的牧草产量(Piñeiro et al.,2006;Grigera et al.,2007)。该示例对应于阿根廷潮湿的潘帕斯地区西南端的高地栽培牧场草地(Grigera et al.,2007)——由冷季型草(高羊茅、鸭茅和多花黑麦草)和豆类植物(紫花苜蓿、红车轴草和白车轴草)混合组成。本例中使用的数据来源于先前的研究(Grigera et al.,2007)。与该项研究相比,我们将逐步解释地上净初级生产力与蒙特斯模型各组成部分之间的关系。

(一)地上净初级生产力估算

从2000年10月~2003年9月,在四个不同的围场进行了生物量收获。收获设计试图复制该地区实行的轮牧制度。它包含两个月休眠期后的生物量收获。在四个围场内,每个地方都使用了8个1平方米的笼子以限制放牧。在每个休眠期开始时,将每个笼子内的地上生物量修剪在4厘米高。在休眠期结束时,将积累的生物量收获到相同的高度。地上净初级生产力估算为每个休眠期间收获的生物量。

(二)可吸收性光合有效辐射估算

通过结合MODIS以及气象数据,对每个围场和休眠期吸收的光合有效辐射进行估算。对于每个围场,选择完全位于高地地区的 MODIS 像素(5.3 公顷)。因此,消除了造林、建筑或道路所影响的像素。采用归一化差异植被指数估算 fPAR 比率,特别是 MODIS 产品 NDVI。

MOD13Q1 的系列 5(https://lpdaac.usgs.gov/products/modis_products_table/mod13q1,植被指数产品),这个 MODIS 产品由 16 天的合成数据组成,包括用于缺失 NDVI 值的质量层。缺失的值用连续值插值计算得出,只有 1.5% 的 NDVI 值被插值。NDVI 计算为每个休眠期和围场的加权平均值。例如,为了估算 1 月 11 日~3 月 14 日休眠期间的 NDVI,使用了五个 NDVI 值。考虑到所述期间的时间重叠,对每个 NDVI 值进行加权。然后,取四个围场的 NDVI 值的平均值。

在每个休眠期内,绿色植被吸收的光合辐射的比例作为 NDVI 的非线性函数进行估算(光合有效辐射吸收比率,式 5-1)。波特等人(Potter et al.,1993)提出了将 NDVI 值转换为光合有效辐射吸收比率的经验模型。格里格拉等人(Grigera et al.,2007)针对研究区域校准了该模型。他们将用于农业裸露土壤区域的 NDVI 值指定为 0 光合有效辐射吸收比率。高覆盖栽培牧草和麦田的 NDVI 值(LAI>3)是光合有效辐射吸收比率的最大值,即 95%。最后,对每个围场取像素级的光合有效辐射吸收比率的平均值。光合有效辐射由气象站测量,该气象站距离最近的围场有 5 千米,距离最远的围场有 50 千米(37°24′S,61°26′W,200 m a.s.l.)。每个休眠期的光合有效辐射估算为全部入射辐射的总和乘以 0.48(McCree,1972)。每个围场可吸收性光合有效辐射估算为光合有效辐射吸收比率和光合有效辐射之间的乘积(式 5-1)。

NDVI 和地上净初级生产力之间存在一定程度的同步性(图 5-3(a))。NDVI 解释了 55% 的地上净初级生产力的变异性($R^2=0.55, p<0.001, n=18$)。与 NDVI 相比,光合有效辐射吸收比率增加了地上净初级生产力变异性的解释比例($R^2=0.58, p<0.001, n=18$)。但某些差异仍然存在,例如,在 2001 年 2 月~3 月之间(图 5-3(b))。光合有效辐射和地上净初级生产力的季节动态略有相似(图 5-3(c)),表明地上净初级生产力的变化仅与部分光合有效辐射相关($R^2=0.25; p<0.03; n=18$)。地上净初级生产力和可吸收性光合有效辐射的季节变化非常相似(图 5-3(d))(Grigera et al.,2007)。这表明,可吸收性光

合有效辐射与NDVI、光合有效辐射吸收比率或光合有效辐射相比,更能描述地上净初级生产力的季节变化(图5-3)。从观测模式中得到的主要结论是,可吸收性光合有效辐射解释了地上净初级生产力大部分的季节变化($R^2=0.87$;$p<0.0001$;$n=18$)。相比之下,皮涅罗等人(Piñeiro et al.,2006)在不同地区工作时发现,NDVI对地上净初级生产力季节变化的描述与可吸收性光合有效辐射相似。这与NDVI和光合有效辐射之间呈强正相关($r=0.56$;$n=21$)关系,而这种强正相关关系在我们的牧场草地实例中并不存在($r=0$;$n=18$)。

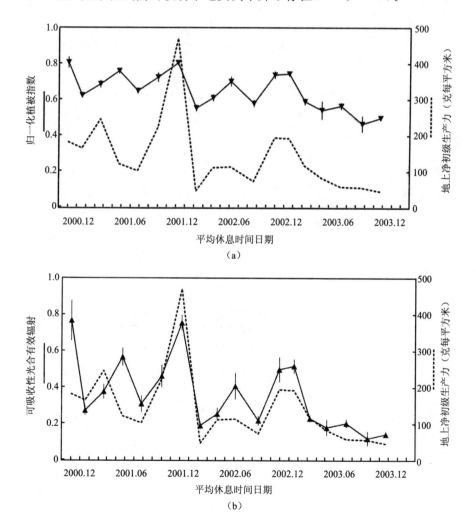

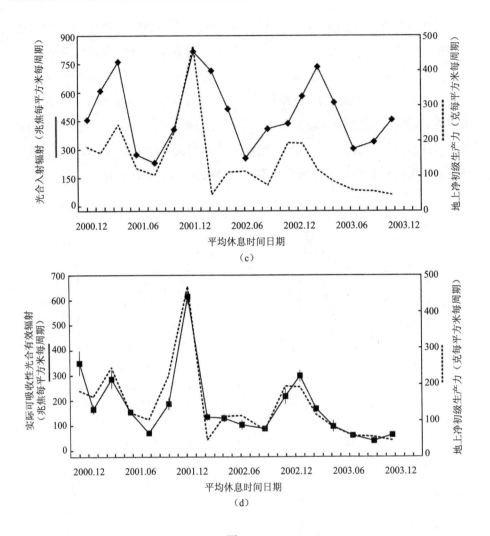

图 5-3

注:2000 年 10 月~2003 年 9 月布宜诺斯艾利斯省西南部高地牧场草场地上净初级生产力及其他组成部分的季节动态。(a)来自 MODIS 和地上净初级生产力的归一化差异植被指数;(b)绿色植被可吸收性光合有效辐射和季节性地上净初级生产力的比例。通过经验模型,根据 NDVI 值估算可吸收性光合有效辐射。(c)利用附近的气象站测量光合有效辐射。每一个值表示每个休眠期累积的光合有效辐射值。(d)高地牧场草地吸收的光合辐射(改编自 Grigera, G., et al., 2007)。折线图表示标准误差;当不存在时,误差小于符号。

(三) 光能利用率估算

光能利用率的估算有两种方法。第一种方法是考虑了每个日期的地上净初级生产力和可吸收性光合有效辐射之间的比率。

该方法得出了每个时期的光合有效辐射值,该值中也含有地上净初级生产力和可吸收性光合有效辐射值(图 5-3(d))。用该方法估算的辐射利用率在 0.3 克/兆焦(夏季)和 1.3 克/兆焦(冬季)之间变化,平均值为 0.82 克/兆焦。第一种方法的优点是可捕捉到光能利用率的季节和年际变化。

估算光能利用率的第二种方法是通过回归模型,其中地上净初级生产力是预测变量(Y 轴),可吸收性光合有效辐射是观测变量(X 轴)。如果模型的 Y 轴截距与 0 无差异,则光能利用率等于回归模型的斜率,且该斜率应与地上净初级生产力和可吸收性光合有效辐射之间的平均值相似。如果 Y 轴截距大于 0,则斜率将低估于光能利用率,因为当可吸收性光合有效辐射等于 0 时,回归模型预测地上净初级生产力值为正。若 Y 轴截距小于 0,则斜率将高估于光能利用率,因为当可吸收性光合有效辐射为正时,回归模型预测地上净初级生产力等于 0。在这两种情况下,可吸收性光合有效辐射值较低情况下的光能利用率的估计误差都大于可吸收性光合有效辐射值较高情况下的估计误差(Verón et al., 2005)。正如所预期的那样,就牧场草地而言,地上净初级生产力与可吸收性光合有效辐射呈正相关。它解释了 87% 的地上净初级生产力变化($R^2=0.87$; n=18)。模型的斜率等于 0.7 克/兆焦($p<0.0001$),Y 轴截距等于 22.3 克每平方米每周期,与 0 无差异($p=0.15$)。在这种情况下,回归模型的斜率表示光能利用率(Grigera et al., 2007)。

五、基于遥感的牧草监测系统

遥感监测牧场范围的 fPAR 在空间(像素尺寸过大)和时间(生长季节内的 fPAR 值稀疏)方面都受到限制。NASA 的 Terra 和 Aqua 卫星上的 MODIS 传感器将围场监测(约 5 公顷)的良好空间分辨率与日常数据相结合。这类遥感信息为阿根廷和乌拉圭的牧场地区开发 ANPP 监测系统提供了机会,该系统基于上述 Monteith 模型(Paruelo et al., 2000; Piñeiro et al., 2006; Grigera et al.,

2007)。

该监测系统基于三个信息来源。第一个来源是地理信息系统(Geographic Information Systems,GIS),其中包括牧场及其围场。第二个来源是数据库,其中包含每个围场的土地利用信息、2000年以来的年度基数、面积以及代表该围场的可变像素数的信息。第三个来源是PAR、NDVI值以及不同类型的牧场草地和天然草地区域和地理区域的RUE校准模型的数据库(Grigera et al.,2007; Irisarri et al.,2012)。将这三个信息来源结合起来估算每个围场的ANPP或APAR(对于没有校准模型的牧场草地和天然草地区域)。为此目的开发的特定软件将这三个信息来源进行组合并存储,同时生成输出数据。

一旦获得前一个月的气象和卫星信息,ANPP的估算值将在每个月的第15日生成。结果通过电子邮件发送给用户,并在网站上托管(http://larfile.agro.uba.ar/lab-sw/sw/gui/Inicial.page)。在这个网站上,每个特定的用户都有一个关键字来访问本人牧场中的数据。但也有关于不同地区内各种牧场草地和天然草地的ANPP平均值的公开数据。

六、遥感在畜牧系统中的其他用途

至少还有另外两个与ANPP相关的变量代表着未来通过遥感技术对其进行估算的挑战:与牧草种类相对应的ANPP比例以及这些牧草种类的营养价值。如上所述,在牧场规模上,由于消除了与fPAR估算相关的限制,因此通过遥感实现了季节性ANPP的估算。

几个例子表明,考虑到特定季节,可以通过遥感进行牧草种类识别(Schmidt and Skidmore,2001,2003;Mutanga et al.,2003)和营养价值量化(Curran,1989;Asner,1998;Curran et al.,2001;Hansen and Schjoerring,2003; Mutanga et al.,2004;Ferwerda et al.,2005;Beeri et al.,2007;Knyazikhin et al.,2013)。但只有两项研究涉及遥感能力,即区分牧草种类(Irisarri et al.,2009)或量化其营养价值与季节变化的关系(Pullanagari et al.,2013)。在生长季节,对区分牧草种类或量化牧草营养价值敏感的光谱范围发生了变化,这表明需要进行特别校准。捕捉这些变化所需的光谱分辨率比大多数正在运行的卫星传感器的分辨率都要高。

七、结论

缺乏关于 ANPP 时间和空间变化的信息,是因为难以通过连续的生物量收获来估算变化。基于遥感的 ANPP 估算必然涉及对每个 Monteith 模型分量的估算。其中 PAR 和 fPAR,这两项很容易被估算,因为气象站可提供 PAR 数据,遥感信息可估算 fPAR。

第三个组成部分光能利用率的估算是最难的,因为它需要通过连续生物量收获,根据地上净初级生产力实地估算进行校准。在本章中,我们介绍了 PAR 和 fPAR、APAR、连同特定校准的案例,为决策者使用提供了 ANPP 的时间和空间变化的估算值。

致谢

该项目由 ANPCYT、布宜诺斯艾利斯大学、CREA 和 IPCVA 资助。冈萨洛·伊里萨里通过博士后奖学金获得了科尼塞特大学的资助。

参 考 文 献

Altesor, A., M. Oesterheld, E. Leoni, F. Lezama, and C. Rodríguez. 2005. Effect of grazing on community structure and productivity of a Uruguayan grassland. *Plant Ecology* 179:83–91.

Aragón, R., and M. Oesterheld. 2008. Linking vegetation heterogeneity and functional attributes of temperate grasslands through remote sensing. *Applied Vegetation Science* 11:115–128.

Asner, G. P. 1998. Biophysical and biochemical sources of variability in canopy reflectance. *Remote Sensing of Environment* 64:234–253.

Austin, A. T., and L. Vivanco. 2006. Plant litter decomposition in a semi-arid ecosystem controlled by photodegradation. *Nature* 442:555–558.

Baeza, S., J. M. Paruelo, and W. Ayala. 2011. Eficiencia en el uso de la radiación y productividad primaria en recursos forrajeros del este del Uruguay [Radiation use efficiency and primary production of forage resources from Eastern Uruguay]. *Agrociencia* 15:48–59.

Bai, Y., J. Wu, Q. Xing, et al. 2008. Primary production and rain use efficiency across a precipitation gradient on the Mongolia Plateau. *Ecology* 89:2140–2153.

Beeri, O., R. Phillips, J. Hendrickson, A. B. Frank, and S. Kronberg. 2007. Estimating forage quantity and quality using aerial hyperspectral imagery for northern mixed-grass prairie. *Remote Sensing of Environment* 110:216–225.

Bradford, J. B., J. A. Hicke, and W. K. Lauenroth. 2005. The relative importance of light-use efficiency modifications from environmental conditions and cultivation for estimation of large-scale net primary productivity. *Remote Sensing of Environment* 96:246–255.

Buono, G., M. Oesterheld, V. Nakamatsu, and J. M. Paruelo. 2010. Spatial and temporal variation of primary production of Patagonian wet meadows. *Journal of Arid Environments* 74:1257–1261.

Chapin, F. S., P. A. Matson, and H. A. Mooney. 2002. *Principles of terrestrial ecosystem ecology*. New York: Springer.

Costanza, R., R. d'Arge, R. de Groot, et al. 1997. The value of the world's ecosystem services and natural capital. *Nature* 387:253–260.

Curran, P. J. 1989. Remote sensing of foliar chemistry. *Remote Sensing of Environment* 30:271–278.

Curran, P. J., J. L. Dungan, and D. L. Peterson. 2001. Estimating the foliar biochemical concentration of leaves with reflectance spectrometry: Testing the Kokaly and Clark methodologies. *Remote Sensing of Environment* 76:349–359.

Ferwerda, J. G., A. K. Skidmore, and O. Mutanga. 2005. Nitrogen detection with hyperspectral normalized ratio indices across multiple plant species. *International Journal of Remote Sensing* 26:4083–4095.

Field, C. B., J. T. Randerson, and C. M. Malmström. 1995. Global net primary production: Combining ecology and remote sensing. *Remote Sensing of Environment* 51:74–88.

Garbulsky, M. F., J. Peñuelas, J. Gamon, Y. Inoue, and I. Filella. 2011. The photochemical reflectance index (PRI) and the remote sensing of leaf, canopy and ecosystem radiation use efficiencies: A review and meta-analysis. *Remote Sensing of Environment* 115:281–297.

Garbulsky, M. F., J. Peñuelas, D. Papale, et al. 2010. Patterns and controls of the variability of radiation use efficiency and primary productivity across terrestrial ecosystems. *Global Ecology and Biogeography* 19:253–267.

Grigera, G., M. Oesterheld, and F. Pacín. 2007. Monitoring forage production for farmers' decision making. *Agricultural Systems* 94:637–648.

Hansen, P. M., and J. K. Schjoerring. 2003. Reflectance measurement of canopy biomass and nitrogen status in wheat crops using normalized difference vegetation indices and partial least squares regression. *Remote Sensing of Environment* 86:542–553.

Irisarri, J. G. N., M. Oesterheld, J. M. Paruelo, and M. A. Texeira. 2012. Patterns and controls of above-ground net primary production in meadows of Patagonia. A remote sensing approach. *Journal of Vegetation Science* 23:114–126.

Irisarri, J. G. N., M. Oesterheld, S. R. Verón, and J. M. Paruelo. 2009. Grass species differentiation through canopy hyperspectral reflectance. *International Journal of Remote Sensing* 30:5959–5975.

Knapp, A. K., J. M. Briggs, D. L. Childers, and O. E. Sala. 2007. Estimating aboveground net primary production in grassland and herbaceous dominated ecosystems. In *Principles and standards for measuring net primary production*, eds. T. J. Fahey and A. K. Knapp, 27–48. New York: Oxford University Press.

Knapp, A. K., J. T. Fahenstock, S. P. Hamburg, L. B. Statland, T. R. Seastedt, and D. S. Schimel. 1993. Landscape patterns in soil–plant water relations and primary production in tallgrass prairie. *Ecology* 74:549–560.

Knapp, A. K., and M. D. Smith. 2001. Variation among biomes in temporal dynamics of aboveground primary production. *Science* 291:481–484.

Knyazikhin, Y., M. A. Schull, P. Stenberg, et al. 2013. Hyperspectral remote sensing of foliar nitrogen content. *Proceedings of the National Academy of Sciences of the United States of America* 110:E185–E192.

Lauenroth, W. 1979. Grassland primary production: North American grasslands in perspective. In *Perspectives in grassland ecology*, ed. N. French, 3–24. New York: Springer-Verlag.

Lauenroth, W. K., and O. E. Sala. 1992. Long-term forage production of North American shortgrass steppe. *Ecological Applications* 2:397–403.

McCree, K. J. 1972. Test of current definitions of photosynthetically active radiation against leaf photosynthesis data. *Agricultural Meteorology* 10:442–453.

McNaughton, S. J., D. G. Milchunas, and D. A. Frank. 1996. How can net primary productivity be measured in grazing ecosystems? *Ecology* 77:974–977.

Monteith, J. L. 1972. Solar radiation and productivity in tropical ecosystems. *Journal of Applied Ecology* 9:747–766.

Mutanga, O., A. K. Skidmore, and H. H. T. Prins. 2004. Predicting in situ pasture quality in the Kruger National Park, South Africa, using continuum-removed absorption features. *Remote Sensing of Environment* 89:393–408.

Mutanga, O., A. K. Skidmore, and S. van Wieren. 2003. Discriminating tropical grass (*Cenchrus ciliaris*) canopies grown under different treatments using spectroradiometry. *ISPRS Journal of Photogrammetry and Remote Sensing* 57:263–272.

Nouvellon, Y., D. L. Seen, S. Rambal, et al. 2000. Time course of radiation use efficiency in a shortgrass ecosystem: Consequences for remotely sensed estimation of primary production. *Remote Sensing of Environment* 71:43–55.

Oesterheld, M., C. Di Bella, and K. Herdiles. 1998. Relation between NOAA-AVHRR satellite data and stocking rate of rangelands. *Ecological Applications* 8:207–212.

Oesterheld, M., and R. J. León. 1987. El envejecimiento de las pasturas implantadas: Su efecto sobre la Productividad Primaria [Sown pastures aging: its effect on primary production]. *Turrialba* 37:29–35.

Oesterheld, M., J. Loreti, M. Semmartin, and J. Paruelo. 1999. Grazing, fire, and climate effects on primary productivity of grasslands and savannas. In *Ecosystems of disturbed ground*, ed. L. Walker, 287–306. Amsterdam: Elsevier.

Oltjen, J. W., and J. L. Beckett. 1996. Role of ruminant livestock in sustainable agricultural systems. *Journal of Animal Science* 74:1406–1409.

Paruelo, J. M., H. E. Epstein, W. K. Lauenroth, and I. C. Burke. 1997. ANPP estimates from NDVI for the central grassland region of the United States. *Ecology* 78:953–958.

Paruelo, J. M., M. Oesterheld, C. M. Di Bella, et al. 2000. Estimation of primary production of subhumid rangelands from remote sensing data. *Applied Vegetations Science* 3:189–195.

Piñeiro, G., M. Oesterheld, and J. M. Paruelo. 2006. Seasonal variation in aboveground production and radiation-use efficiency of temperate rangelands estimated through remote sensing. *Ecosystems* 9:357–373.

Potter, C. S., J. T. Randerson, C. B. Field, et al. 1993. Terrestrial ecosystem production: A process model based on global satellite and surface data. *Global Biogeochemical Cycles* 7:811–841.

Pullanagari, R. R., I. J. Yule, M. P. Tuohy, M. J. Hedley, R. A. Dynes, and W. M. King. 2013. Proximal sensing of the seasonal variability of pasture nutritive value using multispectral radiometry. *Grass and Forage Science* 68:110–119.

Ruimy, A., and B. Saugier. 1994. Methodology for the estimation of terrestrial net primary production from remotely sensed data. *Journal of Geophysical Research* 99:5263–5283.

Running, S. W., P. E. Thornton, R. R. Nemani, and J. Glassy. 2000. Global terrestrial gross and net primary productivity from the Earth observing system. In *Methods in ecosystem science*, eds. O. E. Sala, R. B. Jackson, H. A. Mooney, and R. W. Howarth, 44–57. New York: Springer-Verlag.

Sala, O. E., and A. T. Austin. 2000. Methods of estimating aboveground primary production. In *Methods in ecosystem science*, eds. O. E. Sala, R. B. Jackson, H. A. Mooney, and R. W. Howarth, 31–43. New York: Springer-Verlag.

Sala, O. E., V. A. Deregibus, T. Schlichter, and H. Alippe. 1981. Productivity dynamics of a native temperate grassland in Argentina. *Journal of Range Management* 34:48–51.

Sala, O. E., W. J. Parton, L. A. Joyce, and W. K. Lauenroth. 1988. Primary production of the central grassland region of the United States. *Ecology* 69:40–45.

Schmidt, K. S., and A. K. Skidmore. 2001. Exploring spectral discrimination of grass species in African rangelands. *International Journal of Remote Sensing* 22:3421–3434.

Schmidt, K. S., and A. K. Skidmore. 2003. Spectral discrimination of vegetation types in a coastal wetland. *Remote Sensing of Environment* 85:92–108.

Scurlock, J. M. O., K. Johnson, and R. J. Olson. 2002. Estimating net primary productivity from grasslands biomass dynamics measurements. *Global Change Biology* 8:736–753.

Sellers, P., J. A. Berry, G. J. Collatz, C. B. Field, and F. G. Hall. 1992. Canopy reflectance, photosynthesis, and transpiration. III. A reanalysis using improved leaf models and a new canopy integration scheme. *Remote Sensing of Environment* 42:187–216.

Semmartin, M. A., M. Oyarzabal, J. Loreti, and M. N. Oesterheld. 2007. Controls of primary productivity and nutrient cycling in a temperate grassland with year-round production. *Austral Ecology* 32:416–428.

Singh, J. S., W. K. Lauenroth, and R. K. Steinhorst. 1975. Review and assessment of various techniques for estimating net aerial primary production in grasslands from harvest data. *The Botanical Review* 41:231–237.

Thornton, P. E., and S. W. Running. 1999. An improved algorithm for estimating incident daily solar radiation from measurements of temperature, humidity, and precipitation. *Agricultural and Forest Meteorology* 93:211–228.

Turner, D. P., S. Urbanski, D. Bremer, et al. 2003. A cross-biomes comparison of daily light use efficiency for gross primary production. *Global Change Biology* 9:383–395.

Verón, S. R., M. Oesterheld, and J. M. Paruelo. 2005. Production as a function of resource availability: Slopes and efficiencies are different. *Journal of Vegetation Science* 16:351–354.

Wheeler, R. D., G. L. Kramer, K. B. Young, and E. Ospina. 1981. *The world livestock product, feedstuff, and food grain system.* Morrilton, AR: Winrock International Livestock Research and Training Center.

第六章 从光能利用效率模型估算碳收益服务中的缺失差距

A.J. 卡斯特罗・马丁内斯　J.M. 帕鲁埃洛　D. 阿尔卡拉斯-塞古拉　J. 卡贝洛　M. 奥亚尔扎巴尔　E. 洛佩斯-卡里克

一、简介

科学界正被敦促投入更多的时间和经济资源,用以改进目前对全球和区域碳估算的方式(Scurlock et al., 1999)。人们常把碳收益当作一种中间服务(Fisher et al., 2009),或者是对提供和调节服务的支持(MA, 2005)。此外,净初级生产力是生态系统碳收益的估算值,通常把他当作是生态系统功能最综合的描述(McNaughton et al., 1989)。NPP估算值由生物量收获、通量塔测量、遥感和模型模拟得到(Ruimy et al., 1995; Sala et al., 2000; Still et al., 2004)。生物量收获成本高昂,并且不可避免地存在错误和方法问题。除此之外,此方法在空间和时间覆盖范围受到限制。考虑到植被吸收太阳辐射的比例与光谱植被指数间的线性关系(Sellers et al., 1992),蒙蒂斯模型(Monteith, 1972)提供了根据遥感数据估算碳收益季节变化的可能性(Potter, 1993)。蒙特斯模型指出,植被覆盖的碳收益(式6-1)是入射光合有效辐射(PAR)、光合有效辐射吸收比率(fPAR)以及光能利用效率(LUE; Still et al., 2004)的函数。使用蒙特斯模型估算的通量包括净初级生产力、总初级生产力以及净生态系统交换(NEE)(Ruimy et al., 1999;式6-2和式6-3)。

$$NPP = PAR \times fPAR \times LUE \qquad 式6-1$$
$$GPP = PAR \times fPAR \times LUE \qquad 式6-2$$

$$NEE = PAR \times fPAR \times LUE \quad \text{式6-3}$$

式中,PAR 可以使用辐射计直接测量;fPAR 可利用光谱指数,例如归一化植被指数(NDVI;Asrar et al.,1984)或者增强植被指数(EVI)来估算。fPAR 与光谱指数间的关系会因土地覆盖类型而异,但一些学者也提出了不同的经验关系:(1)线性(Choudhury,1987);(2)非线性(Potter,1993;Sellers et al.,1994);(3)两者的综合(Los et al.,2000)。

LUE 的最大值与在最佳条件下,叶片水平的光合效率或者量子产量值相当(Gower et al.,1999)。然而,低温、水分和营养压力会降低光能利用效率的值(Field et al.,1995;Gamon et al.,1995)。菲尔德等学者(Field et al.,1995)的报告给出了从沙漠的 0.27gC/MJ 到热带森林的 0.70gC/MJ 的可吸收性光合有效辐射光能利用效率的差异。

光能利用效率最初是在物种级别定义的,主要是针对作物物种(Andrade et al.,1993;Kiniry et al.,1998)。使用蒙特斯模型作为遥感估算地上净初级生产力的概念框架(见第五章)需要在生态系统级别定义光能利用效率(Ruimy et al.,1999;Sala et al.,2000;Fensholt et al.,2006)。通常,将大约 1gC/MJ 的可吸收性光合有效辐射的单个固定值用于大范围的时空情况(Maselli et al.,2009)。一些学者指出,光能利用效率在空间(Field et al.,1995;Paruelo et al.,2004;Tong et al.,2008;Garbulsky et al.,2010)和时间(Nouvellon et al.,2000;Piñeiro et al.,2006)上各不相同;同时,使用单一值可能会导致区域(Hilker et al.,2008)和全球(Turner et al.,2002,2003,2005;Tong et al.,2008)对碳收益的估算出现重大误差。

光能利用效率变化的时空模式受诸多因素影响。如物种组成、植物结构和生理学,包括叶形和 RUBISCO 酶(核酮糖-1,5-二磷酸羧化酶加氧酶)含量(Zhao et al.,2007),以及环境因素(即水分胁迫、二氧化碳浓度、温度)均会在生态系统级别改变光能利用效率的值。测量光能利用效率并不是一项简单的任务。在不同的结构水平(例如,从个体到生态系统)估算光能利用效率的值,可以使用单个个体的叶级估算或涡度协方差塔来推导生态系统水平的光能利用效率值(Garbulsky et al.,2010)。

光能利用效率是蒙特斯模型中较不确定的一个参数,因为无法通过直接测量获得,其值取决于对总初级生产力、净初级生产力、净生态系统碳交换量以及吸收辐射的估算(Gower et al.,1999;Ruimy et al.,1999)。本章回顾了前人给出的光能利用效率估算方法,以及从个体到生态系统的不同结构水平下,时间间隔对光能利用效率估算的影响。我们尝试回答以下几个问题:(1)光能利用效率是如何估算的?(2)光能利用效率在土地覆盖类型和生态系统结构水平的影响下有何不同?(3)根据估算的时间间隔,光能利用效率估算的变化量是多少?

二、材料与方法

我们查阅了1972～2007年,包含"光能利用效率"和"辐射利用效率"术语的125篇文章,其中只有101篇文章中提供了定量光能利用效率数据(见本章附录)。

该综述涵盖了65种不同的期刊,主要涉及生态学(占总研究的72%)和遥感领域(占总研究的22%)。根据已发表的研究,我们开发了一个数据库,其中包括结构水平、通量估算、光能利用效率估算以及研究地的地理坐标(表6-1)。我们分别从三个结构水平上对光能利用效率值进行了阐述:(1)"个体"是指基于单一物种的个体估算光能利用效率的区域尺度研究;(2)"单一物种主导的生态系统"是指当研究集中在一个优势物种地块时的情况(例如,占地面积为100平方米的涡度协方差通量塔中农业生态系统的净生态系统碳交换量 NEE);(3)"多物种主导的生态系统"。假设50%的干生物质对应于碳,光能利用效率值转化为最常见的单位系统:每兆焦耳吸收的入射光合有效辐射所固定的碳克数(g C/MJ)。为了分析光能利用效率数据的可变性,我们将每个数据分配到阿奇博尔德(Archibold,1995)土地覆盖类型分类的一个类别中。我们计算了光能利用效率的平均值、最大值和最小值,以及每个结构水平和土地覆盖类型的偏差。克鲁斯卡尔-沃利斯(Kruskal-Wallis)检验用于检测不同结构水平和土地覆盖类型的光能利用效率估算值($n=185$)的显著差异。

表6-1 1972~2007年光能利用效率数据汇总样本

土地覆盖类型	研究次数	地点	光能利用效率单位	碳通量模型(用生物群落内总研究的百分比表示)
PHM	1	阿拉斯加	g C/MJ APAR	(100%)其他*
CF	49	达勒姆,加拿大,威斯康星州,瑞典	g C/MJ APAR mol C/mol photons Moles CO_2/mol PAR	(8%) GPP = APAR × LUE (16%) NEE = APAR × LUE (32%) NPP = APAR × LUE (44%)其他*
TW	12	加拿大、欧洲、欧盟	g C/MJ APAR mol C/mol photons Moles CO_2/mol PAR	(8%) ANPP = APAR × LUE (29%) NPP = APAR × LUE (29%)其他*
Cr	38	爱尔兰、欧盟、中国、意大利、澳大利亚、英国、南非、印度	g C/MJ kg (CO_2/ha·h)/ (J/m^2·sg)	(30%) NPP = APAR × LUE (70%)其他*
TFE	65	欧洲、欧盟、日本、中国、新西兰	g C/MJ APAR Mol CO_2/ mol APAR mmol CO_2/mmol photons	(8%) GPP = APAR × LUE (32%) NEE = APAR × LUE (27%) NPP = APAR × LUE (33%)其他*
ME	12	西班牙、意大利、印度、欧盟	g C/MJ APAR Mol C/mol APAR	(20%) NEE = APAR × LUE (70%) NPP = APAR × LUE (10%)其他*
TG	16	加拿大、欧盟、阿根廷	g C/MJ APAR g DM/MJ	(13%) NEE = APAR × LUE (40%) NPP = APAR × LUE (47%)其他*
TpF	4	巴拿马,哥伦比亚,欧洲经济区	g C/MJ APAR kg (CO_2/ha·h)/ (J/m^2·sg)	(67%) GPP = APAR × LUE (33%)NEE=APAR×LUE
TpS	9	塞内加尔,阿根廷	g C/MJ APAR Mol C/mol APAR	(100%)其他*

续表

土地覆盖类型	研究次数	地点	光能利用效率单位	碳通量模型(用生物群落内总研究的百分比表示)
AR	8	撒哈拉,澳大利亚南部,马里,墨西哥	g C/MJ APAR gr DM/MJ	(17%) GPP = APAR × LUE (50%) NPP = APAR × LUE (33%)其他 *

注:其他 * 表示(a)当研究没有指定光能利用效率估算的碳通量模型时。(b)对蒙特斯(Monteith)模型进行修改,例如,NASA-CASA 模型(即卡内基-埃姆斯-斯坦福方法(Carnegie-Ames-Stanford))模拟了从区域到全球尺度的净初级生产力和土壤异养呼吸,再如利用 TURC 估算大陆总初级生产力和净初级生产力。再或者基于包含其他生理参数通过蒙特斯方法导出的模型。(c)恒定光能利用效率值。阿奇博尔德(Archibold)土地覆盖类型分类:PHM=极地高山冻原,CF=针叶林,TW=陆地湿地,Cr=作物,TFE=温带森林生态系统,ME=地中海生态系统,TG=温带草原,TpF=热带森林,TpS=热带稀树草原,AR=干旱区。

三、结果

(一) 估算方法与光能利用效率单位

光能利用效率值的估算主要采用了两种方法。在第一种方法(占总研究的 82%)中,光能利用效率值是在区域尺度上使用蒙特斯方程并基于先前对碳通量(即净初级生产力或总初级生产力)的估算得到的。这里,20%的碳通量估算值是使用涡度协方差技术观测大气和植被间的二氧化碳通量得出的(Ruimy et al.,1995;Zhao et al.,2007);在 44%的总数据中,光合有效辐射吸收比率数据是由遥感卫星衍生得到的归一化差异植被指数的线性函数计算得到的。其余的研究使用了其他研究报告的光合有效辐射吸收比率数据或是直接采用了测量冠层的数据。在大多数研究中,光合有效辐射是通过辐射计计算得到的。

在第二种方法(占总研究的 18%)中,光能利用效率是基于与其他变量(如叶面积指数或光化学反射指数)的相关模型估算得到的(Gu et al.,2002;Filella et al.,2004;Grace et al.,2007)(见第三章)。在这里,光能利用效率也可以通过在地块规模上收获的生物量与全年或生长季的可吸收性光合有效辐射之间的比率得出。

大多数文献中(占总研究的77%)都提供了光能利用效率的定量估算方法。

从这些研究中,我们得到185个光能利用效率值,其最初以四种不同的单位表示,包括克碳/兆焦耳 可吸收性光合有效辐射(C/MJ)(65%)、摩尔 二氧化碳/摩尔 可吸收性光合有效辐射(mol CO_2/mol)(14%)、克 干物质/兆焦耳 可吸收性光合有效辐射(g of dry matter/MJ)(9%)以及 摩尔 碳/摩尔 每分钟吸收的光子(mol C/mol)(3%)。在转换成gC/MJ 可吸收性光合有效辐射后,光能利用效率的平均值为0.99gC/MJ 可吸收性光合有效辐射(样本标准差SD=1.09),绝对最大值为8.2gC/MJ APAR,绝对最小值为0.05gC/MJ APAR。

(二) 跨结构水平和土地覆盖类型的光能利用效率估算

研究的次数因结构水平和土地覆盖类型而异(图6-1)。由图可知,以多物种主导的生态系统水平最为普遍(62%),其次是以单一物种主导的生态系统(19%)和个体(19%)。在查阅的文献中,温带森林、针叶林和农田是最具代表性的土地覆盖类型,而代表性最低的是极地高山冻原、热带森林和干旱地区(图6-1)。其中,在各结构水平研究中,农田和温带森林生态系统的土地覆盖类型具有独特性。

我们发现,个体水平(1.7C/MJ APAR;样本标准差 = 1.6)的平均光能利用效率值显著高于多物种主导(0.8 C/MJ APAR;样本标准差 = 0.9)和单一物种主导(0.7 C/MJ APAR;样本标准差= 0.4)的生态系统水平(图6-2)。

从图6-3可以看出,按土地覆盖类型划分的平均光能利用效率值存在显著差异。光能利用效率平均值在农田2.20gC/MJ APAR(样本标准差=1.67)和陆地湿地0.55gC/MJ APAR(样本标准差=0.23)之间变化。最大光能利用效率值出现在作物(8.20gC/MJ APAR)和温带草原(5.20gC/MJ APAR),而最小光能利用效率值出现在热带稀树草原(0.05gC/MJ APAR)和温带草原(0.06gC/MJ APAR)(图6-3)。

就个体水平而言,光能利用效率在热带稀树草原、热带森林和作物之间存在显著差异(图6-4)。在单一物种主导的生态系统水平下,针叶林和作物的光能利用效率值存在显著差异(图6-4)。在多物种主导的生态系统水平下,作物与

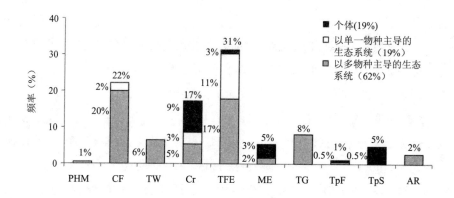

图 6-1 土地覆盖类型

注:在个体、单一物种主导的生态系统和多物种主导的生态系统水平上的土地覆盖类型的文章频率。总光能利用效率值=185。阿奇博尔德土地覆盖类型分类:PHM=极地高山冻原,CF=针叶林,TW=陆地湿地,Cr=作物,TFE=温带森林生态系统,ME=地中海生态系统,TG=温带草原,TpF=热带森林,TpS=热带稀树草原,AR=干旱区。

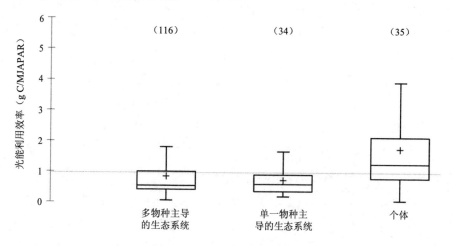

图 6-2 多物种主导的生态系统、单一物种主导的生态系统
以及个体水平的光能利用效率值箱线图

注:该图解释了光能利用效率值的最小值、第一四分位数(25%)、中位数、平均值以及第三四分位数(75%)。平均值用+表示,黑线对应于中位数。个体水平的最大光能利用效率值为 8.2 g C/MJ APAR,单一物种主导的生态系统水平的最大光能利用效率值为 2g C/MJ APAR,多物种主导的生态系统水平的最大光能利用效率值为 5.7g C/MJ APAR。水平虚线表示光能利用效率值的总平均值。每个箱线图的总光能利用效率值在括号中显示。

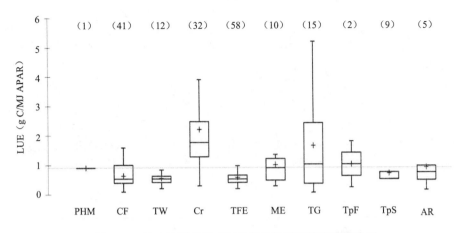

图6-3　按土地覆盖类型划分的光能利用效率值箱线图

注：该图解释了光能利用效率值的最小值、第一四分位数（25%）、中位数、平均值以及第三四分位数（75%）。平均值用+表示，黑线对应于中位数。总光能利用效率值=185。阿奇博尔德土地覆盖类型分类：PHM=极地高山冻原，CF=针叶林，TW=陆地湿地，Cr=作物，TFE=温带森林生态系统，ME=地中海生态系统，TG=温带草原，TpF=热带森林，TpS=热带稀树草原，AR=干旱区。

针叶林、陆地湿地、热带森林以及地中海生态系统显著不同。温带草原和热带森林生态系统与针叶林、陆地湿地以及温带森林生态系统有显著差异（图6-4）。

虽然文献中并没有对所有结构水平的光能利用效率值进行估算，但是我们观察到其在土地覆盖类型之间存在显著差异。

针叶林在单一物种主导和多物种主导的生态系统水平间存在显著差异。在温带森林生态系统中，我们观察到个体与单一物种主导的生态系统以及多物种主导的生态系统水平之间存在显著差异（图6-4）。在地中海生态系统中，个体和多物种主导的生态系统水平存在差异。在不同的结构水平中，作物是唯一没有显著差异的土地覆盖类型。热带森林在个体和多物种主导的生态系统水平都无显著差异。

（三）光能利用效率估算的时间间隔

由于估算的时间间隔（即在一天、一个季度或一年内采样）不同，因而光能利用效率值也显著不同。我们获取了所有结构水平的年度和季度的估算值，而每日光能利用效率的估算值仅在多物种主导的生态系统水平的文献中找到。在单一物种主导的生态系统水平上，年度和季度光能利用效率估算值之间的测量时间间隔存在显著差异。Kruskal-Wallis检验在个体水平上未发现年度和季度估

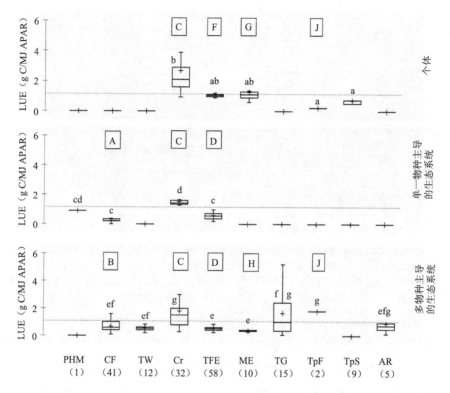

图 6-4 基于结构水平和土地覆盖类型的光能利用效率(LUE)值箱线图

注:该图解释了光能利用效率值的最小值、最大值、第一四分位数(25%)、中位数、平均值以及第三四分位数(75%)。平均值用+表示,黑线对应于中位数。字母表示显著不同的组(Kruskal Wallis 检验,$p<0.05$)。小写字母表示每个结构水平内,土地覆盖类型之间显著不同的组。大写字母表示不同结构水平,每种土地覆盖类型之间显著不同的组。总光能利用效率值=185。阿奇博尔德土地覆盖类型分类:PHM=极地高山冻原,CF=针叶林,TW=陆地湿地,Cr=作物,TFE=温带森林生态系统,ME=地中海生态系统,TG=温带草原,TpF=热带森林,TpS=热带稀树草原,AR=干旱区。

算值之间的显著差异;同时,也未发现在多物种主导的生态系统水平上,年度、季度与每日估算值之间的显著差异。

然而我们发现,在以单一物种主导的生态系统水平上,年度和季度估算值之间存在着显著差异。如图 6-5 所示,估算的时间间隔和结构水平之间的比较(图 6-5)揭示了在年度和季度观测中获得的光能利用效率值之间存在显著差异。在多物种主导和单一物种主导的生态系统水平上,年度和季度估算没有显著差异,但是我们发现在个体水平上获得的结果确实存在显著差异。

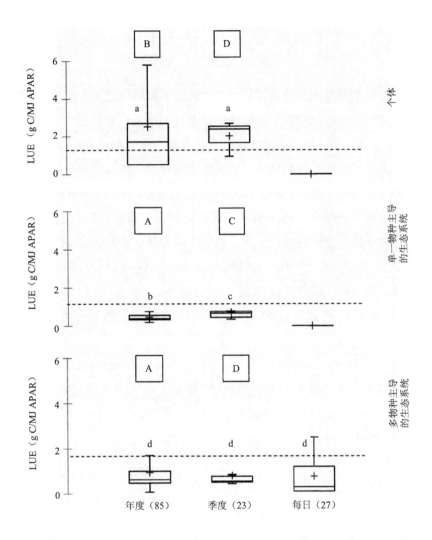

图 6-5　每个结构水平与估算时期的平均光能利用效率值随时间变化的箱线图

注：该图解释了光能利用效率值的最小值、最大值、第一四分位数(25%)、中位数、平均值以及第三四分位数(75%)。平均值用＋表示，黑线对应于中位数。不同的字母表示显著差异(Kruskal Wallis 检验，p<0.05)。小写字母表示每个结构水平内，估算时间间隔之间存在显著差异的组。大写字母表示每种估算时间间隔，结构水平之间存在显著差异的组。总光能利用效率值=185。阿奇博尔德土地覆盖类型分类：PHM＝极地高山冻原，CF＝针叶林，TW＝陆地湿地，Cr＝作物，TFE＝温带森林生态系统，ME＝地中海生态系统，TG＝温带草原，TpF＝热带森林，TpS＝热带稀树草原，AR＝干旱区。

四、结论

蒙特斯模型是基于光能利用效率和由遥感估算得到的光合有效辐射吸收比率的模型,众所周知,它构成了绘制陆地碳循环图的最广泛使用的方法(Jenkins et al.,2007;Pereira et al.,2007)。在不同方法中发现的内在时空变异性可以解释本研究中发现的光能利用效率估算值的变异性。估算的时间间隔和结构水平是这种变化的两个明显来源。在这种情况下,区域尺度的净初级生产力年度估算值不应用于个体水平的光能利用效率估算,而应用于短期(如数天)测量。光能利用效率估算值的可变性与环境和生理因素息息相关(如叶形、核糖二磷酸羧化酶含量、温度和湿度)(Ito and Oikawa,2007;Tong et al.,2008),如果将这些值外推到全球或区域尺度,那么光能利用效率估算值的可变性可能会导致较大误差(Nouvellon et al.,2000;Piñeiro et al.,2006)。结果表明,在不同土地覆盖类型的以个体和多物种主导的生态系统水平上,高时间变化的光能利用效率估算值(见 Grace et al.,2007;Cook et al.,2008;Hilker et al.,2008)并不能解释区域和全球净初级生产力估算值,因此通常采用恒定的光能利用效率值(Drolet et al.,2008;Maselli et al.,2009)。

致谢

作者非常感谢安达卢西亚政府环境部的工作人员提供获得必要帮助所需的设施。他们还感谢热尔瓦西奥·皮涅罗、多洛雷斯·阿罗塞纳、卡洛斯·迪贝拉、皮达·克里斯蒂亚诺和两位匿名审稿人有价值的评论。资金支持来自于 ERDF(FEDER)、安达卢西亚地区政府(Junta de Andalucía GLOCHARID & SEGALERT 项目,P09-RNM-5048)和科学与创新部(CGL2010-22314 项目)。安达卢西亚全球变化评估和监测中心(CAESCG)和俄克拉荷马大学的俄克拉荷马州生物调查局(OBS)也为 A. J. C. 提供了支持。

附录（1972—2007 年的文献综述）

Aalto, T., P. Ciais, A. Chevillard, and C. Moulin. 2004. Optimal determination of the parameters controlling biospheric CO_2 fluxes over Europe using eddy covariance fluxes and satellite NDVI measurements. *Tellus B* 56:93–104.

Ahl, D. E., S. T. Gower, D. S. Mackay, S. N. Burrows, J. M. Norman, and G. R. Diak. 2004. Heterogeneity of light use efficiency in a northern Wisconsin forest: Implications for modeling net primary production with remote sensing. *Remote Sensing of Environment* 93:168–178.

Ahl, D. E., S. T. Gower, D. S. Mackay, S. N. Burrows, J. M. Norman, and G. R. Diak. 2005. The effects of aggregated land cover data on estimating NPP in northern Wisconsin. *Remote Sensing of Environment* 97:1–14.

Anderson, M. C., W. P. Kustas, and J. M. Norman. 2007. Upscaling flux observations from local to continental scales using thermal remote sensing. *Agronomy Journal* 99:240–254.

Asner, G. P., K. M. Carlson, and R. E. Martin. 2005. Substrate age and precipitation effects on Hawaiian forest canopies from spaceborne imaging spectroscopy. *Remote Sensing of Environment* 98:457–467.

Asrar, G., M. Fuchs, E. T. Kanemasu, and J. L. Hatfield. 1984. Estimating absorbed photosynthetic radiation and leaf-area index from spectral reflectance in wheat. *Agronomy Journal* 76:300–306.

Baldocchi, D. D. 2003. Assessing the eddy covariance technique for evaluating carbon dioxide exchange rates of ecosystems: Past, present and future. *Global Change Biology* 9:479–492.

Black, T. A., D. Gaumont-Guay, R. S. Jassal, et al. 2005. Measurement of carbon dioxide exchange between the boreal forest and the atmosphere. In *Carbon balance of forest biomes*, eds. H. Griffiths and P. G. Jarvis, 151–185. Oxfordshire, UK: BIOS Scientific Publishers.

Boschetti, L., P. A. Brivio, H. D. Eva, J. Gallego, A. Baraldi, and J. M. Gregoire. 2006. A sampling method for the retrospective validation of global burned area products. *IEEE—Transactions on Geoscience and Remote Sensing* 44:1765–1773.

Bradford, J. B., J. A. Hickec, and W. K. Lauenroth. The relative importance of light-use efficiency modifications from environmental conditions and cultivation for estimation of large-scale net primary productivity. *Remote Sensing of Environment* 96:246–255.

Cannell, M. G. R., R. Milne, L. J. Sheppard, and M. H. Unsworth. 1987. Radiation interception and productivity of willow. *Journal of Applied Ecology* 24:261–278.

Christensen, S., and J. Goudriaan. 1993. Deriving light interception and biomass from spectral reflectance ratio. *Remote Sensing of the Environment* 43:87–95.

D'Antuono, L. F., and F. Rossini. 2006. Yield potential and ecophysiological traits of the Altamurano linseed (*Linum usitatissimum* L.), a landrace of southern Italy. *Genetic Resources and Crop Evolution* 53:65–75.

Drolet, G. G., K. F. Huemmrich, F. G. Hall, et al. 2005. A MODIS-derived photochemical reflectance index to detect inter-annual variations in the photosynthetic light-use efficiency of a boreal deciduous forest. *Remote Sensing of Environment* 98:212–224.

Dungan, R. J., and D. Whitehead. 2006. Modelling environmental limits to light use efficiency for a canopy of two broad-leaved tree species with contrasting leaf habit. *New Zealand Journal of Ecology* 30:251–259.

Fang, S., X. Xizeng, X. Xiang, and L. Zhengcai. 2005. Poplar in wetland agroforestry: A case study of ecological benefits, site productivity, and economics. *Wetlands Ecology and Management* 13:93–104.

Fensholt, R., I. Sandholt, and M. S. Rasmussen. 2004. Evaluation of MODIS LAI, fAPAR and the relation between fAPAR and NDVI in a semi-arid environment using in situ measurements. *Remote Sensing of Environment* 91:490–507.

Fensholt, R., I. Sandholt, M. S. Rasmussen, S. Stisen, and A. Diouf. 2006. Evaluation of satellite based primary production modeling in the semi-arid Sahel. *Remote Sensing of Environment* 105:173–188.

Fernández, M. E., J. E. Gyenge, and T. M. Schlichter. 2006. Growth of the grass *Festuca pallescens* in silvopastoral systems in a semi-arid environment, Part 1: Positive balance between competition and facilitation. *Agroforestry Systems* 66:259–269.

Field, C. B., J. T. Randerson, and C. M. Malmstrom. 1995. Global net primary production: Combining ecology and remote sensing. *Remote Sensing of Environment* 51:74–88.

Fleisher, D. H., D. J. Timlin, and V. R. Reddy. 2006. Temperature influence on potato leaf and branch distribution and on canopy photosynthetic rate. *Agronomy Journal* 98:1442–1452.

Fuentes, D., J. A. Gamon, Y. Cheng, et al. 2006. Mapping carbon and water flux in a chaparral ecosystem using vegetation indices derived from AVIRIS. *Remote Sensing of Environment* 103:312–323.

Gamon, J. A., K. Kitajima, S. S. Mulkey, L. Serrano, and S. J. Wright. 2005. Diverse optical and photosynthetic properties in a neotropical dry forest during the dry season: Implications for remote estimation of photosynthesis. *Biotropica* 37:547–560.

Gebremichael, M., and A. P. Barros. 2006. Evaluation of MODIS gross primary productivity (GPP) in tropical monsoon regions. *Remote Sensing of Environment* 100:150–166.

Goetz, S. J., and S. D. Prince. 1999. Modelling terrestrial carbon exchange and storage: Evidence and implications of functional convergence in light-use efficiency. *Advances in Ecological Research* 28:57–92.

Goetz, S. J., S. D. Prince, S. N. Goward, M. M. Thawle, J. Small, and A. Johnston. 1999. Mapping net primary production and related biophysical variables with remote sensing: Application to the Boreas region. *Journal of Geophysical Research—Atmospheres* 104:27719–27734.

Goulden, M. L., J. W. Munge, S. M. Fan, B. C. Daube, and S. C. Wofsy. 1996. Exchange of carbon dioxide by a deciduous forest: Response to interanual climate variability. *Science* 271:1576–1578.

Goward, S. N., and D. G. Dye. 1987. Evaluating North American net primary productivity with satellite observations. *Advances in Space Research* 7:165–174.

Goward, S. N., and K. F. Huemmrich. 1992. Vegetation canopy PAR absorbance and the normalized difference vegetation index: An assessment using the SAIL model. *Remote Sensing of the Environment* 39:119–140.

Guo, J., and C. M. Trotter. 2004. Estimating photosynthetic light-use efficiency using the photochemical reflectance index: Variations among species. *Functional Plant Biology* 31:255–265.

Hill, M., A. A. Held, R. Leuning, et al. 2006. MODIS spectral signals at a flux tower site: Relationships with high-resolution data, and CO_2 flux and light use efficiency measurements. *Remote Sensing of Environment* 103:351–368.

Inoue, Y., and J. Peñuelas. 2006. Relationships between light use efficiency and photochemical reflectance index in soybean leaves as affected by soil water content. *International Journal of Remote Sensing* 27:5109–5114.

Ito, A., and T. Oikawa. 2004. Global mapping of terrestrial primary productivity and light-use efficiency with a process-based model. In *Global environmental change in the ocean and on land*, eds. M. Shiyomi, H. Kawahata, H. Koizumi, A. Tsuda, and Y. Awaya, 343–358. Tokyo: Terrapub.

Kato, T., Y. Tang, S. Gu, et al. 2006. Temperature and biomass influences on interannual changes in CO_2 exchange in an alpine meadow on the Qinghai-Tibetan Plateau. *Global Change Biology* 12:1285–1298.

Kinirya, J. R., C. E. Simpsonb, A. M. Schubertc, and J. D. Reed. Peanut leaf area index, light interception, radiation use efficiency, and harvest index at three sites in Texas. *Field Crops Research* 91:297–306.

Krishnan, P., T. A. Black, N. J. Grant, et al. 2006. Carbon dioxide and water vapour exchange in a boreal aspen forest during and following severe drought. *Agricultural and Forest Meteorology* 139:208–223.

Lagergren, F. 2005. Net primary production and light use efficiency in a mixed coniferous forest in Sweden. *Plant, Cell & Environment* 28:412–423.

Leuning, R., H. A. Cleugh, S. J. Zegelin, and D. Hughes. 2005. Carbon and water fluxes over a temperate *Eucalyptus* forest and a tropical wet/dry savanna in Australia: Measurements and comparison with MODIS remote sensing estimates. *Agricultural and Forest Meteorology* 129:151–173.

Li, S. G., J. Asanuma, W. Eugster, et al. 2005. Net ecosystem carbon dioxide exchange over grazed steppe in central Mongolia. *Global Change Biology* 11:1941–1955.

Li, S. G., W. Eugster, J. Asanuma, et al. 2006. Energy partitioning and its biophysical controls above a grazing steppe in central Mongolia. *Agricultural and Forest Meteorology* 137:89–106.

Monteith, J. L. 1972. Solar-radiation and productivity in tropical ecosystems. *Journal of Applied Ecology* 9:747–766.

Monteith, J. L. 1977. Climate and the efficiency of crop production in Britain. *Philosophical Transactions of the Royal Society of London B* 281:277–297.

Monteith, J. L. 1994. Validity of the correlation between intercepted radiation and biomass. *Agricultural and Forest Meteorology* 68:213–220.

Myneni, R. B., S. Hoffman, Y. Knyazikhin, et al. 2002. Global products of vegetation leaf area and fraction absorbed PAR from year one of MODIS data. *Remote Sensing of Environment* 83:214–231.

Myneni, R. B., R. R. Nemani, and S. W. Running. 1997. Estimation of global leaf área index and absorbed PAR using radiative transfer models. *IEEE Transactions on Geoscience and Remote Sensing* 35:1380–1393.

Myneni, R. B., and D. L. Williams. 1994. On the relationship between fAPAR and NDVI. *Remote Sensing of Environment* 49:200–11.

Nakaji, T., H. Oguma, and Y. Fujinuma. 2006. Seasonal changes in the relationship between photochemical reflectance index and photosynthetic light use efficiency of Japanese larch needles. *International Journal of Remote Sensing* 27:493–509.

Nakaji, T., T. Takeda, Y. Fujinuma, and H. Oguma. 2005. Effect of autumn senescence on the relationship between the PRI and LUE of young Japanese larch trees. *Phyton* 45:535–542.

Nemani, R. R., and S. W. Running. 1989. Estimation of regional surface-resistance to evapotranspiration from NDVI and thermal-IR AVHRR data. *Journal of Applied Meteorology* 28:276–284.

Nichol, C. J., K. F. Huemmrich, T. A. Black, et al. 2002. Sensing of photosynthetic-light-use efficiency of boreal forest. *Agricultural and Forest Meteorology* 101:131–142.

Nichol, C. J., J. Lloyd, O. Shibistova, et al. 2002. Remote sensing of photosynthetic-light-use efficiency of a Siberian boreal forest. *Tellus B* 54:677–687.

Niinemets, U., A. Cescatti, and R. Christian. 2004. Constraints on light interception efficiency due to shoot architecture in broad-leaved Nothofagus species. *Tree Physiology* 24:617–630.

Niinemets, U., and L. Sack. 2006. Structural determinants of leaf light-harvesting capacity and photosynthetic potentials. *Progress in Botany* 67:385–419.

Norby, R. J., J. Ledford, C. D. Reilly, et al. 2004. Fine-root production dominates response of a deciduous forest to atmospheric CO_2 enrichment. *Proceedings of the National Academy of Sciences of the United States of America* 101:9689–9693.

Nouvellon, Y., D. Lo Seen, S. Rambal, et al. 2000. Time course of radiation use efficiency in a shortgrass ecosystem: Consequences for remotely sensed estimation of primary production. *Remote Sensing of Environment* 71:43–55.

Paruelo, J. M., H. E. Epstein, W. K. Lauenroth, and I. C. Burke. 1997. ANPP estimates from NDVI for the central grassland region of the US. *Ecology* 78:953–958.

Paruelo, J. M., M. Oesterheld, C. M. Di Bella, et al. 2000. Estimation of primary production of subhumid rangelands from remote sensing data. *Applied Vegetation Science* 3:189–195.

Patel, N. R. 2006. Investigating relations between satellite derived land surface parameters and meteorological variables *Geocarto International* 21:47–53.

Pereira, J. S., J. A. Mateus, L. M. Aires, et al. 2007. Net ecosystem carbon exchange in three contrasting Mediterranean ecosystems: The effect of drought. *Biogeosciences* 4:791–802.

Piñéiro, G., M. Oesterheld, and J. M. Paruelo. 2006. Seasonal variation in aboveground production and radiation use efficiency of temperate rangelands estimated through remote sensing. *Ecosystems* 9:357–373.

Pitman, J. I. 2000. Absorption of photosynthetically active radiation, radiation use efficiency and spectral reflectance of bracken [Pteridium aquilinum (L.) Kuhnl] canopies. *Annals of Botany* 85:101–111.

Potter, C., S. Klooster, A. Huete, and V. Genovese. 2007. Terrestrial carbon sinks for the United States predicted from MODIS satellite data and ecosystem modeling. *Earth Interactions* 11:1–21.

Potter, C. S., J. T. Randerson, C. B. Field, et al. 1993. Terrestrial ecosystem production—a process model-based on global satellite and surface data. *Global Biogeochemical Cycles* 7:811–841.

Prince, S. D., S. J., Goetz, and S. N. Goward. 1995. Monitoring primary production from Earth observing satellites. *Water, Air and Soil Pollution* 82:509–522.

Ruimy, A., B. Saugier, and G. Dedieu. 1994. Methodology for the estimation of terrestrial net primary production from remotely sensed data. *Journal of Geophysical Research* 99:5263–5283.

Running, S. W., D. D. Baldocchi, D. P. Turner, S. T. Gower, P. S. Bakwin, and K. A. Hibbard. 1999. A global terrestrial monitoring network integrating tower fluxes, flask sampling, ecosystem modeling and EOS satellite data. *Remote Sensing of Environment* 70:108–127.

Running, S. W., and R. R. Nemani. 1988. Relating seasonal patterns of the AVHRR vegetation index to simulated photosynthesis and transpiration of forests in different climates. *Remote Sensing of Environment* 24:347–367.

Running, S. W., R. R. Nemani, F. A. Heinsch, M. S. Zhao, M. Reeves, and H. Hashimoto. 2004. A continuous satellite-derived measure of global terrestrial primary production. *BioScience* 54:547–560.

Running, S. W., L. Queen, and M. Thornton. 2000. The Earth observing system and forest management. *Journal of Forestry* 98:29–31.

Russell, G., P. G. Jarvis, and J. L. Monteith. 1989. Absorption of radiation by canopies and stand growth. In *Plant canopies: Their growth, form and function*, eds. G. Russell, B. Marshall, and P. G. Jarvis, 21–39. Cambridge, UK: Cambridge University Press.

Sakai, T. 2005. Microsite variation in light availability and photosynthesis in a cool-temperate deciduous broadleaf forest in central Japan. *Ecological Research* 20:537–545.

Salazar, M. R., B. Chaves, J. W. Jones, and A. Cooman. 2006. A simple potential production model of Cape gooseberry (*Physalis peruviana* L.). *Acta Hortic* 718:105–112.

Schwalm, C. R., T. A. Black, B. D. Amiro, et al. 2006. Photosynthetic light use efficiency of three biomes across an east–west continental-scale transect in Canada. *Agricultural and Forest Meteorology* 140:269–286.

Seaquist, J., L. Olsson, and J. Ardö. 2003. A remote sensing-based primary production model for grassland biomes. *Ecological Modelling* 169:131–155.

Seaquist, J. W., L. Olsson, J. Ardo, and L. Eklundh. 2006. Broad-scale increase in NPP quantified for the African Sahel, 1982–1999. *International Journal of Remote Sensing* 27:5115–5122.

Sims, D., and J. Gamon. 2002. Relationships between leaf pigment content and spectral reflectance across a wide range of species, leaf structures and developmental stages. *Remote Sensing of Environment* 81:337–354.

Sims, D., H. Luo, S. Hastings, W. Oechel, A. Rahman, and J. Gamon. 2006. Parallel adjustments in vegetation greenness and ecosystem CO_2 exchange in response to drought in a southern California chaparral ecosystem. *Remote Sensing of Environment* 103:289–303.

Sims, D. A., A. F. Rahman, V. D. Cordova, et al. 2005. Midday values of gross CO_2 flux and light use efficiency during satellite overpasses can be used to directly estimate eight-day mean flux. *Agricultural and Forest Meteorology* 131:1–12.

Still, C. J., J. Randerson, and I. Fung. 2004. Large-scale plant light-use efficiency inferred from the seasonal cycle of atmospheric CO_2. *Global Change Biology* 10:1240–1252.

Storkey, J. 2006. A functional group approach to the management of UK arable weeds to support biological diversity. *Weed Research* 46:513–522.

Tracol, Y., E. Mougin, P. Hiernaux, and L. Jarlan. 2006. Testing a Sahelian grassland functioning model against herbage mass measurements. *Ecological Modelling* 193:437–446.

Tsubo, M., S. Walker, and H. O. Ogindo. 2005. A simulation model of cereal–legume intercropping systems for semi-arid regions I. Model development. *Field Crops Research* 93:10–22.

Tucker, C. J., I. Y. Fung, C. D. Keeling, and R. H. Gammon. 1986. Relationship between atmospheric CO_2 variations and a satellite-derived vegetation index. *Nature* 319:195–199.

Tucker, C. J., W. H. Jones, W. A. Kley, and G. J. Sundstrom. A 3-band hand-held radiometer for field use. *Science* 211:281–283.

Tucker, C. J., C. O. Justice, and S. D. Prince. 1986. Monitoring the grasslands of the Sahel 1984–1985. *International Journal of Remote Sensing* 7:1571–1581.

Tucker, C. J., and P. J. Sellers. 1986. Satellite remote-sensing of primary production. *International Journal of Remote Sensing* 7:1395–1416.

Turner, D. P., W. D. Ritts, W. B. Cohen, et al. 2003. Scaling gross primary production (GPP) over boreal and deciduous forest landscapes in support of MODIS GPP product validation. *Remote Sensing of Environment* 88:256–270.

Turner, D. P., W. D. Ritts, W. B. Cohen, T. K. Maeirsperger, S. T. Gower, and A. Kirschbaum. 2005. Site-level evaluation of satellite-based global terrestrial gross primary production and net primary production monitoring. *Global Change Biology* 11:666–684.

Turner, D. P., W. D. Ritts, W. B. Cohen, et al. 2006. Evaluation of MODIS NPP and GPP products across multiple biomes. *Remote Sensing of Environment* 102:282–292.

Turner, D. P., S. Urbanski, D. Bremer, et al. 2003. Cross-biome comparison of daily light use efficiency for gross primary production. *Global Change Biology* 9:383–395.

Ueyama, M., Y. Harazono, E. Ohtaki, and A. Miyata. 2006. Controlling factors on the inter-annual CO_2 budget at a sub-arctic black spruce forest in interior Alaska. *Tellus B* 58:491–450.

Veroustraete, F., H. Sabbe, and H. Eerens. 2002. Estimation of carbon mass fluxes over Europe using the C-Fix model and Euroflux data. *Remote Sensing of Environment* 83:376–399.

Walcroft, A. S., K. J. Brown, W. S. F. Schuster, et al. 2005. Radiative transfer and carbon assimilation in relation to canopy architecture, foliage area distribution and clumping in a mature temperate rainforest canopy in New Zealand. *Agricultural and Forest Meteorology* 135:326–339.

Xiao, X. M. 2006. Light absorption by leaf chlorophyll and maximum light use efficiency. *IEEE Transactions on Geoscience and Remote Sensing* 44:1933–1935.

Xiao, X. M., Q. Y. Zhang, B. Braswell, et al. 2004. Modeling gross primary production of temperate deciduous broadleaf forest using satellite images and climate data. *Remote Sensing of Environment* 91:256–270.

Zhang, Q. Y., X. M. Xiao, B. Braswell, E. Linder, F. Baret, and B. Moore. 2005. Estimating light absorption by chlorophyll, leaf and canopy in a deciduous broadleaf forest using MODIS data and a radiative transfer model. *Remote Sensing of Environment* 99:357–371.

Zhao, M., F. A. Heinsch, R. R. Nemani, and S. W. Running. 2005. Improvements of the MODIS terrestrial gross and net primary production global data set. *Remote Sensing of Environment* 95:164–175.

参 考 文 献

Andrade, F. H., S. A. Uhart, and A. Cirilo. 1993. Temperature affects radiation use efficiency in maize. *Field Crops Research* 32:17–25.

Archibold, O. W. 1995. *Ecology of world vegetation*. London: Chapman & Hall.

Asrar, G., M. Fuchs, E. T. Kanemasu, and J. L. Hatfield. 1984. Estimating absorbed photosynthetic radiation and leaf area index from spectral reflectance in wheat. *Agronomy Journal* 76:300–306.

Choudhury, B. J. 1987. Relationships between vegetation indices, radiation absorption, and net photosynthesis evaluated by a sensitivity analysis. *Remote Sensing of Environment* 22:209–233.

Cook, B. D., P. V. Bolstad, J. G. Martin, et al. 2008. Using light-use and production efficiency models to predict photosynthesis and net carbon exchange during forest canopy disturbance. *Ecosystems* 11:26–44.

Drolet, G. G., E. M. Middleton, K. F. Huemmrich, F. G. Hall, and H. A. Margolis. 2008. Regional mapping of gross light-use efficiency using MODIS spectral indices. *Remote Sensing of Environment* 112:3064–3078.

Fensholt, R., I. Sandholt, M. S. Rasmussen, S. Stisen, and A. Diouf. 2006. Evaluation of satellite based primary production modelling in the semi-arid Sahel. *Remote Sensing of Environment* 105:173–188.

Field, C. B., R. B. Jackson, and H. A. Mooney. 1995. Stomatal responses to increased CO_2—Implications from the plant to the global scale. *Plant Cell and Environment* 18:1214–1225.

Filella, I., J. Peñuelas, L. Llorens, and M. Estiarte. 2004. Reflectance assessment of seasonal and annual changes in biomass and CO_2 uptake of a Mediterranean shrubland submitted to experimental warming and drought. *Remote Sensing of Environment* 90:308–318.

Fisher, B., R. K. Turner, and P. Morling. 2009. Defining and classifying ecosystem services for decision making. *Ecological Economics* 3:643–653.

Gamon, J. A., C. B. Field, M. L. Goulden, et al.1995. Relationships between NDVI, canopy structure, and photosynthesis in 3 Californian vegetation types. *Ecological Applications* 5:28–41.

Garbulsky, M. F., J. Peñuelas, D. Papale, et al. 2010. Patterns and controls of the variability of radiation use efficiency and primary productivity across terrestrial ecosystems. *Global Ecology and Biogeography* 19:253–267.

Gower, S. T., C. J. Kucharik, and J. M. Norman. 1999. Direct and indirect estimation of leaf area index, *f(APAR)*, and net primary production of terrestrial ecosystems. *Remote Sensing of Environment* 70:29–51.

Grace, J., C. Nichol, M. Disney, L. T. Quaife, and P. Bowyer. 2007. Can we measure terrestrial photosynthesis from space directly, using spectral reflectance and fluorescence? *Global Change Biology* 13:1484–1497.

Gu, L., D. Baldocchi, S. B. Verma, et al. 2002. Advantages of diffuse radiation for terrestrial ecosystem productivity. *Journal of Geophysical Research: Atmospheres* 107:2–23.

Hilker, T., N. C. Coops, M. A. Wulder, T. A. Black, and R. G. Guy. 2008. The use of remote sensing in light use efficiency based models of gross primary production: A review of current status and future requirements. *Science of the Total Environment* 404:411–423.

Ito, A., and T. Oikawa. 2007. Absorption of photosynthetically active radiation, dry-matter production, and light-use efficiency of terrestrial vegetation: A global model simulation. *Elsevier Oceanography Series* 73:335–359; 503–505.

Jenkins, J. P., A. D. Richardson, B. H. Braswell, S. V. Ollinger, D. Y. Hollinger, and M. L. Smith. 2007. Refining light-use efficiency calculations for a deciduous forest canopy using simultaneous tower-based carbon flux and radiometric measurements. *Agricultural and Forest Meteorology* 143:64–79.

Kiniry, J. R., J. A. Landivar, M. Witt, T. J. Gerik, J. Cavero, and L. J. Wade. 1998. Radiation-use efficiency response to vapor pressure deficit for maize and sorghum. *Field Crops Research* 56:265–270.

Los, S. O., G. J. Collatz, P. J. Sellers, et al. 2000. A global 9-yr biophysical land surface dataset from NOAA AVHRR data. *Journal of Hydrometeorology* 1:183–199.

MA (Millennium Ecosystem Assessment). 2005. *Ecosystems and human well-being: The assessment series (four volumes and summary)*. Washington, DC: Island Press.

Maselli, F., M. Chiesi, M. Moriondo, L. Fibbi, M. Bindi, and S. W. Running. 2009. Modelling the forest carbon budget of a Mediterranean region through the integration of ground and satellite data. *Ecological Modelling* 220:330–342.

McNaughton, S. J., M. Oesterheld, D. A. Frank, and K. J. Williams. 1989. Ecosystem-level patterns of primary productivity and herbivory in terrestrial habitats. *Nature* 341:142–144.

Monteith, J. L. 1972. Solar radiation and productivity in tropical ecosystems. *Journal of Applied Ecology* 9:747–766.

Nouvellon, Y., D. L. Seen, S. Rambal, et al. 2000. Time course of radiation use efficiency in a shortgrass ecosystem: Consequences for remotely sensed estimation of primary production. *Remote Sensing of Environment* 71:43–55.

Paruelo, J. M., M. F. Garbulsky, J. P. Guerschman, and E. G. Jobbágy. 2004. Two decades of Normalized Difference Vegetation Index changes in South America: Identifying the imprint of global change. *International Journal of Remote Sensing* 25:2793–2806.

Pereira, J. S., J. A. Mateus, L. Aires, et al. 2007. Net ecosystem carbon exchange in three contrasting Mediterranean ecosystems—The effect of drought. *Biogeosciences* 4:791–802.

Piñeiro, G., M. Oesterheld, and J. M. Paruelo. 2006. Seasonal variation in aboveground production and radiation-use efficiency of temperate rangelands estimated through remote sensing. *Ecosystems* 9:357–373.

Potter, C. S. 1993. Terrestrial ecosystem production: A process model based on global satellite and surface data. *Global Biogeochemical Cycles* 7:811–841.

Ruimy, A., P. Jarvis, D. D. Baldocchi, and B. Saugier. 1995. CO_2 fluxes over plant cano-

pies and solar radiation: A review. *Advances in Ecological Research* 26:1–68.
Ruimy, A., L. Kergoat, A. Bondeau, et al. 1999. Comparing global NPP models of terrestrial net primary productivity (NPP): Analysis of differences in light absorption and light-use efficiency. *Global Change Biology* 5:56–64.
Sala, O. E., R. B. Jackson, H. A. Mooney, and R. W. Howarth. 2000. Methods in ecosystem science: Progress, tradeoffs, and limitations. In *Methods in ecosystem science*, eds. O. E. Sala, R. B. Jackson, H. A. Mooney, and R. W. Howarth, 1–3. New York: Springer-Verlag.
Scurlock, J. M. O., W. Cramer, R. J. Olson, W. J. Parton, and S. D. Prince. 1999. Terrestrial NPP: Toward a consistent data set for global model evaluation. *Ecological Applications* 9:913–919.
Sellers, P. J., J. A. Berry, G. J. Collatz, C. B. Field, and E. G. Hall. 1992. Canopy reflectance, photosynthesis, and transpiration. III. A reanalysis using improved leaf models and a new canopy integration scheme. *Remote Sensing of Environment* 42:187–216.
Sellers, P. J., C. J. Tucker, G. J. Collatz, et al. 1994. A global 1-degrees by 1 degrees NDVI data set for climate studies. The generation of global fields of terrestrial biophysical parameters from the NDVI. *International Journal of Remote Sensing* 15:3519–3545.
Still, C. J., J. T. Randerson, and I. Y. Fung. 2004. Large-scale plant light-use efficiency inferred from the seasonal cycle of atmospheric CO_2. *Global Change Biology* 10:1240–1252.
Tong, X. J., J. Li, and L. Wang. 2008. A review on radiation use efficiency of the cropland. *Chinese Journal of Ecology* 27:1021–1028.
Turner, D. P., S. T. Gower, W. B. Cohen, M. Gregory, and T. K. Maiersperger. 2002. Effects of spatial variability in light use efficiency on satellite-based NPP monitoring. *Remote Sensing of Environment* 80:397–405.
Turner, D. P., W. D. Ritts, W. B. Cohen, et al. 2005. Site-level evaluation of satellite-based global terrestrial gross primary production and net primary production monitoring. *Global Change Biology* 11:666–684.
Turner, D. P., S. Urbanski, D. Bremer, et al. 2003. A cross-biome comparison of daily light use efficiency for gross primary production. *Global Change Biology* 9:383–395.
Zhao, Y. M., S. K. Niu, J. B. Wang, H. T. Li, and G. C. Li. 2007. Light use efficiency of vegetation: A review. *Chinese Journal of Ecology* 26:1471–1477.

第七章 亚马孙热带森林生物质燃烧排放估算

Y. E. 岛袋　G. 佩雷拉　F. S. 卡多佐　R. 斯托克勒
S. R. 弗雷塔斯　S. M. C. 库拉

一、简介

生态系统服务是环境生产对人类有用的资源的过程。这类服务是广泛和多样的,并决定着我们的土地、水、食品的质量和人类的健康。生态系统服务利用环境带来了各种各样的好处(国际林业研究中心,2009；Haines-Young and Potschin,2013)。在"生态系统服务级联"中(图7-1),服务既可以体现出社会价值,也可以带来经济价值。大多数生态系统,无论是人工的、半人工的还是完全自然的,都是多功能的,能够带来市场和非市场的利益。图7-1从一个合理的视角,呈现出生态系统服务与人们直接使用和重视的东西之间的关联。这些服务可代表经济活动的投入(供给服务,如提供木材)或为经济活动提供的服务,如吸收和处理废物(调节服务)。社会价值可包括文化意义以及面向大众的道德价值和审美价值。例如,在环境经济核算体系模型中,若我们严格地应用"自然流动"这个术语,人工林地的碳封存和水调节服务将不会被视作环境提供的,而在海恩斯-杨和波钦(Haines-Young and Potschin,2013)提出的更传统的生态系统服务范式下,二者均属环境带来的服务。

关于森林生态系统,费恩赛德(Fearnside,1985)首先提出了管理森林木材和非木材产品的计划。他还将环境服务分为三类:提供生物多样性、提供水资源和避免全球变暖(Fearnside,1997,2000)。

第七章 亚马孙热带森林生物质燃烧排放估算

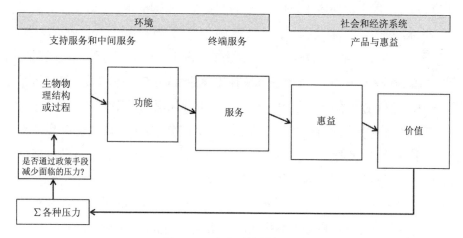

图 7-1 生态系统服务级联

资料来源：Haines-Young, R. H., and M. Potschin. 2013. Common International Classification of Ecosystem Services (CICES): Consultation on Version 4, August-December 2012.

此后，关于生态系统服务的讨论逐渐转向对生态系统服务项目的支付，即政府为流域维护等服务向土地所有者支付津贴。生态系统服务这一概念或有助于描述人类与自然间的联系和人类依赖自然的某些方式。这也是一个挑战，因为人与自然间的联系是复杂的，不同的专家看待它的方式也不同。生态系统服务基于一种新的层次结构，旨在消除类别层次上的重叠和冗余。因此，国际生态系统服务通用分类（Common International Classification of Ecosystem Services, CICES）可被视为严格意义上的分类（Haines-Young and Potschin, 2010, 2013）。

在这一语境下，土地利用和土地覆盖变化被认为是改变环境的主要因素。生物质燃烧被持续用于清除大面积的植被区域，消耗大量的生物质，这一过程中大量的痕量气体和气溶胶被释放到大气中。生物质燃烧对于生态系统服务的影响亦有两面，其利弊需要权衡。一方面，在农业或牲畜的开放空间进行生物质燃烧，可用于控制害虫，或用于回收养分。另一方面，燃烧影响生物多样性模式，改变大气成分和气候调节服务，改变碳循环，影响大气和陆地表面间的能量平衡及生物地球化学和水文循环，并可能损害房屋、基础设施和人类生命。

生物质燃烧改变了地球表面的物理化学和生物特征、大气能量收支和气候系统（Andreae and Merlet, 2001; Ichoku and Kaufman, 2005; Cardoso et al.,

2009)。此外，由于生物质燃烧排放物可被输送到遥远的地区，此等现象从当地尺度外推，并改变了区域和全球尺度的能量平衡（Kaufman et al.，1995；Andreae et al.，2004）。此外，在生物质燃烧过程中排放出部分气体（一氧化碳、二氧化碳、甲烷、非甲烷烃、硝酸等）具有化学活性，并与羟基浓度（OH）相互作用，从而改变氧化效率，并对温室气体之一的对流层臭氧产生影响。

正如此前所指出的，在CICES总结的三个部分中，生物质燃烧以几种方式引发了不利影响。燃烧的后果取决于火灾的规模大小和持久时间长短，部分情况下燃烧甚至会对土壤下的生物多样性、微观和宏观动物及养分循环造成损害，并影响空气质量和大气成分。此外，有一些研究将各区域的呼吸道疾病和其他健康问题的暴发与年度高强度燃烧活动联系起来（Ignotti et al.，2010）。

在亚马孙热带森林中，生物质燃烧经常与农业扩张和生产过程有关，例如森林砍伐、牧场更新和害虫控制（Sampaio et al.，2007；Cardoso et al.，2008；Marengo et al.，2010）。

生物质燃烧期发生在6月~10月的旱季（Crutzen and Andreae，1990）；原因主要是该时期缺乏足量降水，且植被中水分含量较低，因此更易发生火灾（Nobre et al.，1991；Moraes et al.，2004）。此外，燃烧的引发和持续受诸如生物质类型、空气温度、湿度和风速等因素的影响（Freitas et al.，2005；Werf et al.，2006；Fearnside et al.，2009）。

本章的总体目标是展示遥感工具如何在评估生物质燃烧对生态系统服务的直接和间接影响方面提供有价值的信息。特别是，本章着重分析了两种在地球空气质量的数值模型中用于估算和吸收生物质燃烧排放的不同方法。其中，Terra和Aqua卫星平台上的中分辨率成像光谱仪传感器和地球静止环境卫星成像传感器被用于估算亚马孙热带森林生物质燃烧释放到大气中的气溶胶和微量气体的量。

二、生物质燃烧排放估算方法

用于估算气溶胶和痕量气体的传统生物质排放估算方法通常使用的排放因子与燃料负载特性和燃烧生物质干物质量有关（Andreae and Merlet，2001）。此外,虽然不同物种的排放因子已有准确数值,但燃烧效率取决于燃料荷载的水分含量(Chuvieco et al.,2004)和燃烧面积,燃烧面积通常在火灾结束之后很长时间才可得出(Roy et al.,2002;Silva et al.,2005)。最近,已开发出通过环境卫星估算火焰辐射功率来获得燃烧的生物质和火灾排放物的新方法(Silva et al.,2003,2005;Ichoku and Kaufman,2005)。火焰辐射功率(FRP)可定义为在生物质燃烧过程中作为辐射发射的那部分化学能。FRP 的时间积分为火焰辐射能。

理论上,燃烧释放的辐射强度与燃烧的生物质线性相关,并且可能与植被类型无关(Wooster et al.,2005;Freeborn et al.,2008)。此外,卫星对火焰辐射能量(FRE)释放率的测量可与气溶胶光学深度(Aerosol Optical Depth,AOD)和生物质燃烧系数相关联,以测算区域的烟雾排放(Wooster et al.,2003,2005)。此方法允许使用化学传输模型对排放到大气中的气溶胶和痕量气体的浓度进行近实时估算(Chatfield et al.,2002,2003;Freitas et al.,2009)。

三、亚马孙热带雨林生物质分布

亚马孙河流域和其他生态系统的地上生物质总量因采用的测算方法不同而有显著变化。植被中的碳含量和被烧毁地区的碳封存等因素很难计算。萨奇等(Saatchi et al.,2007)估算亚马孙盆地的总生物质为 860 亿吨(Pg),约为每公顷 300 至 400 吨。此外,奥尔森等(Olson et al.,2002)估算亚马孙盆地的地上生物质总量为 $4\sim15 kgC \cdot m^{-3}$。霍顿等(Houghton et al.,1999,2001)和费尔南德斯等(Fernandes et al.,2007)等多项研究分析了南美洲火灾或土地利用和土地覆盖变化所消耗的生物质。然而,生物质估算并未呈现出明显一致性。霍顿等(Houghton et al.,1999)和田等(Tian et al.,1998)估算了土地利用变化对大气的年度碳净通量分别为 $0.2 PgC \cdot y^{-1}$ 和 $0.3 PgC \cdot y^{-1}$。此外,佩雷拉等(Pereira et al.,2011)和凯泽等(Kaiser et al.,2012)使用 FRE 估算了南美洲生物质燃烧火灾总消耗,结果为 $0.28 PgC \cdot y^{-1}$(2002 年)和 $0.33 PgC \cdot y^{-1}$(年平均)。

在南美洲,一些天气系统可改变生物质燃烧排放物的传输,改变空气的化学成分、辐射收支和云的微物理性质(Freitas et al.,2005)。在亚马孙盆地,火焰温度最高可超过 1 600 开尔文,通常在 830~1 440 开尔文之间(Riggan et al.,

2004)。随着生物质燃烧活动释放的高能量强度,排放到大气中的烟羽流可达到地面以上 6 000 米以上。因此,要分析不同尺度下生物质燃烧的影响,就必须(在空间和时间两个层面)估算痕量气体和气溶胶的排放。在这一语境下,本章旨在分析用以估算生物质燃烧排放并将其同化为地球空气质量数值模型的两种方法。

四、材料和方法

(一) 热异常检测

极轨卫星平台 Terra 和 Aqua 上搭载的 MODIS 传感器成像角为 ±55°,高度为海拔 705 千米处。对于不同的卫星,在地面给定点上的传感器过境时间也有所不同:Terra 卫星(其产品命名为 MOD)在其下降轨道上于上午 10:30 和晚上 10:30 穿过赤道,Aqua 卫星(其产品命名为 MYD)在其上升轨道上于下午 01:30 和凌晨 01:30 穿过赤道(Giglio,2005)。

包含火灾和 FRP 信息的产品分别名为 MOD14 和 MYD14。Terra 和 Aqua 卫星上的 MODIS 根据平台的过境时间每天各采集两次数据。MODIS 传感器每日四次的火灾观测有助于促进全球监测火灾过程以及火灾对生态系统、大气和气候的影响。上述产品中使用的科学数据集包括火灾掩码、算法质量、辐射功率,以及利用了考夫曼等(Kaufman et al.,1996)提出的方法描述火点像元属性的多层数据。MODIS 火灾产品在天底点时的空间分辨率为 1 平方千米,在高观测天顶角时可达 3 平方千米,通常提供两到四次对给定地表的观测。

搭载在 GOES 卫星群上的地球静止环境卫星遥感传感器可在电磁光谱的可见光到热红外范围内的五个光谱带中获取信息。热异常检测算法使用以 3.9 微米和 10.7 微米为中心的光谱带(Prins and Menzel,1992),并基于罗伯茨等(Roberts et al.,2005)提出的方法。野火自动生物质燃烧算法(Wildfre Automated Biomass Burning Algorithm,WFABBA)(网址:http://wfabba.ssec.wisc.edu/wfabba.html)是基于 GOES 的火灾探测产品,并利用伍斯特等(Wooster et al.,2005)提出的 FRP 估算方法,如许卫东等(Xu et al.,2010)所述。目前,GOES 计划中有三颗正在运行的卫星,这三颗卫星定期(15~30 分钟)从南美、中美和北美获取信息;METEOSAT 卫星则可用在非洲和欧洲收集信息。对热异常的估算自 1997 年开始,涵盖了 16 年的历史序列。使用低空间分辨率(在天底点时空间分辨率为 4 千米)的地球同步卫星的一个优点是观测频率高,这允许对特定燃烧进行生命周期表征(图 7-2)。然而,一个缺点是,传感器检测到的亚马孙火灾中,5%~10%的火灾使得以 3.9 微米为中心的探测器产生饱和,因此无法对大规模燃烧排放进行估算(Pereira et al.,2009)。当火灾的大小和温度超过传感器捕获电磁辐射的灵敏度时,就会出现饱和像元,使得无法再通过 WFABBA 推出实际 FRP。此外,MODIS 和 GOES 的 FRP 估算存在一些不确定性,如火灾和烟雾的同时观测(误差约为±11%)、云量(±11%)、火焰

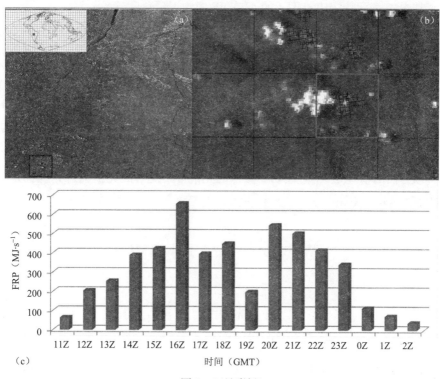

图 7-2(见彩插)

注:(a)为 MODIS/Aqua 传感器影像,1B2G7R 构成,所示为 2007 年 9 月 29 日毁林弧形带中发生的火灾(黑色);(b)为 MYD14 产品探测到且可观测的燃烧,图中正方形为使用卷积掩膜呈现出全部燃烧范围的聚集;(c)描述的是与红色正方形中探测到的燃烧相关的生命周期,并将观测数据用于在整个时间序列上对火灾辐射功率(FRP)进行积分(MJ·s^{-1})。

辐射功的准确性和一致性(±16%)等(Vermote et al.,2009)。

(二) FRP 整合

图 7-3 显示了 FRP 整合方法的流程图。第一步,根据采集时间和信息来源对特定日期的 FRP 数据进行分组。在该步骤中,从 FRP 积分中消除低置信度火点像元(MOD14 和 MYD14 产品的值低于 50%,WFABBA、GOES 产品的标志 4 和 5)。

对于 MODIS 的 FRP 值,使用式 7-1 以使蝴蝶结效应最小化,如费里伯恩等(Freeborn et al.,2011)所述:

$$\text{MODIS}_{\text{FRP}} = \text{FRP}\cos^2(\theta) \qquad 式\ 7-1$$

其中,θ 表示特定火点像元的 MODIS 观测天顶角。

佩雷拉等人(Pereira et al.,2009)分析了 9 年的 WFABBA 数据,发现探测到的火灾中有 6.6% 饱和。因此,饱和百分比可能因季节天气特征和生物质燃烧活动的变化而变化,例如亚马孙河流域的非典型雨季或强烈干旱就有可能对其产生影响。然而,在没有 FRP 估算的情况下去除 GOES 饱和像元并且忽略生物质燃烧的这一类重要事件并不合适,相反应寻找一种替代方案,在中红外(middle-infrared,MIR)辐射率方法的基础上估算 FRP(式 7-2)。

该方法的前提是,中心在 3.9 微米的光谱带中发射的光谱辐射亮度(M_λ)与 FRP 成比例(Wooster et al.,2003、2005):

$$\text{FRP}_{\text{MR}} = \frac{\text{Ag}}{\alpha}\sigma \int_{3.76}^{4.03} M(\lambda,T)_d \lambda - M_b$$

$$M(\lambda,T) = \frac{c_1}{\lambda^5 \left(\exp\left(\frac{c_2}{\lambda T}\right) - 1\right)} \qquad 式\ 7-2$$

其中 Ag 表示 GOES 像元的面积;α 是基于 GOES 的 MIR 光谱带的常数拟合;M_λ 是发射光谱辐射亮度;c_1 和 c_2 是常数(分别为 $3.74 \times 10^8\ \text{W} \cdot \text{m}^{-2}$ 和 $1.44 \times 10^4\ \mu\text{m} \cdot \text{K}$);$\lambda$ 是波长(微米);T 表示温度(开尔文),M_b 是背景发射的辐射亮度(110 兆瓦)。

在修正了 MODIS 燃烧数据的蝴蝶结效应并且在不进行 FRP 估算的情况下给 GOES 饱和像元赋值(第二步)之后,第三步是根据目标 FRP 值的范围值

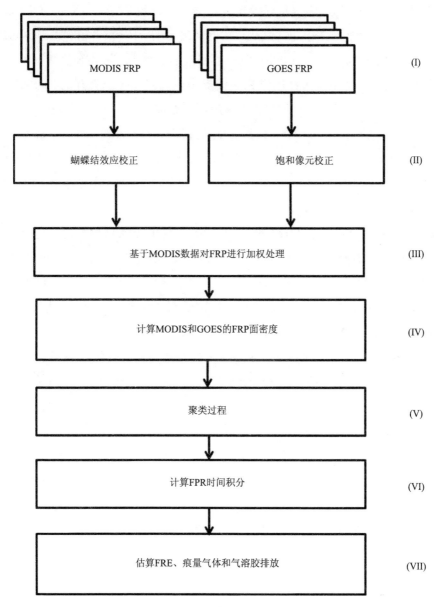

图 7-3

注：该方法的流程图分为七个步骤：(Ⅰ)数据采集；(Ⅱ)蝴蝶结效应和饱和像元校正；(Ⅲ)FRP 调整处理；(Ⅳ)FRP 面密度估算；(Ⅴ)聚类过程；(Ⅵ)FRP 时间积分；(Ⅶ)FRE、痕量气体和气溶胶排放估算。

修正 GOES 的 FRP 值。许卫东等(Xu et al., 2010)比较了 MODIS 和 GOES 在时间差小于 10 分钟、观察天顶角小于 30°的情况下的 FRP 重合值,发现 GOES 的 FRP 值大约有 90%与同时的 MODIS 值相差小于 50%。在本研究中,GOES 卫星估算的 1 000 兆瓦以下的 FRP 值被校正 17%,高于 1 000 兆瓦的 FRP 值被校正 41%(Xu et al., 2010)。该过程还被应用于处理第二代 METEOSAT(Meteosat Second Generation, MSG)上加载的旋转增强可见和红外成像仪(Spinning Enhanced Visible and InfraRed Imager, SEVIRI)收集的数据,但由于空间覆盖的原因,我们决定不在亚马孙生物质燃烧排放估算中使用来自 SEVIRI 的数据。

凯泽等(Kaiser et al., 2012)针对 MODIS 提出的方法被用于对 GOES 和 MODIS 的 FRP 值进行积分。第四步,通过按像元面积加权 FRP 值计算出 MODIS 和 GOES 的 FRP 面密度。在这一步骤中,通过巴西官方的土地利用和土地覆盖图(MMA)得到的水体图被用来校正面密度的最终误差。第五步包括确定聚类过程所需的网格配置。本研究决定在空间分辨率为 0.17°的规则网格上估算 FRE,这一分辨率数值是目前巴西区域大气模型系统模型与耦合化学-气溶胶-示踪输运模型(Coupled Chemistry-Aerosol-Tracer Transport Model-Brazilian Regional Atmospheric Modeling System, CCATT-BRAMS)耦合得出的模拟分辨率。

聚类过程即将来自不同传感器的所有检测到的火灾数据组合在一起。在该步骤中,可根据 CCATT-BRAMS 空间分辨率和网格配置来定义合并 FRP 数据的矩阵的大小。

因此,在由不同卫星 $\xi(\mathrm{lon}, \mathrm{lat})$ 估算的 FRP 面密度值的网格上,运行的 $M \times N$(行×列)大小的卷积掩码 $\eta(\gamma, \kappa)$ 将导致网格($\mathrm{FRP}_{\mathrm{grid}}$)包含在给定时间内所有聚类火灾。

$$\mathrm{FRP}_{\mathrm{grid}(\mathrm{lon}, \mathrm{lat}, t)} = \sum_{\gamma=-\alpha}^{\alpha} \sum_{\kappa=-\beta}^{\beta} \eta(\gamma, \kappa) \xi(\mathrm{lon}+\gamma, \mathrm{lat}+\kappa, t) \quad \text{式 7-3}$$

其中(式 7-3)将聚类网格定义为 $M \times N$ 大小的掩码与影像完全重叠的所有点($\mathrm{lon} \varepsilon [\alpha, M-\alpha]$, $\mathrm{lat} \varepsilon [\beta, N-\beta]$)。在本案例中,本研究开发的算法以 400 平方千米(20 千米×20 千米)的掩码大小贯穿矩阵。据此,Aqua、Terra、GOES-

10、GOES-11、GOES-12、GOES-13 和 GOES-15 卫星检测到的所有火灾的聚类 FRP 温度演变值均被储存。基于每个网格点的 FRP 值及其各自的出现时间，FRE 通过下式(式 7-4)计算(第六步)：

$$\text{FRE}_{\text{grid(lon,lat)}} = \frac{1}{2}\sum_{i=1}^{n}(\text{FRP}_n + \text{FRP}_{n+1})\cdot(T_{n+1} - T_n) \quad \text{式 7-4}$$

其中，$\text{FRE}_{\text{lon,lat}}$ 表示规则网格中特定点的地理位置(经度和纬度)；T 是每次观测的间隔；n 表示此次观测。如此估算，相当于假设在规则网格的一部分中观察到的空间分布可以代表其整体。然而，若两次采集间的间隔大于 4 小时($\Delta T > 14\ 400$ 秒)，则假设有两个或多个独立火灾发生，算法将重启，对 FRP 值重新开始积分($T=0$)。在 FRE 估算之后，将这些数据代入到 CCATT-BRAMS 模型中，计算痕量气体和气溶胶的排放量(第七步)。

(三) CCATT-BRAMS 模型与生物质燃烧估算

CCATT-BRAMS 是为模拟不同尺度下的大气环流而开发的，它基于 BRAMS 数值模型(Freitas *et al.*，2009)。该模型利用相同的时间步长和相同的大气物理和动力学参数，使得痕量气体和气溶胶的输运与大气状态的演化同时进行。一氧化碳和 $\text{PM}_{2.5}$ 粒子的质量守恒方程通过以下(式 7-5)趋势方程计算(Freitas *et al.*，2009)：

$$\frac{\partial \bar{s}}{\partial t} = \underbrace{\left(\frac{\partial \bar{s}}{\partial t}\right)_{\text{adv}}}_{\text{I}} + \underbrace{\left(\frac{\partial \bar{s}}{\partial t}\right)}_{\text{II}} + \underbrace{\left(\frac{\partial \bar{s}}{\partial t}\right)_{\text{deepcony}}}_{\text{III}} + \underbrace{\left(\frac{\partial \bar{s}}{\partial t}\right)_{\text{shallowcony}}}_{\text{IV}} + \underbrace{\left(\frac{\partial \bar{s}}{\partial t}\right)_{\text{chemCO}}}_{\text{V}} + \underbrace{W_{\text{PM}2.5\mu m}}_{\text{VI}} + \underbrace{R}_{\text{VII}} + \underbrace{Q_{\text{pr}}}_{\text{VIII}} \quad \text{式 7-5}$$

其中 s 是网格平均示踪混合比；项(I)是三维运输(平均风的平流)；项(II)表示行星边界层(Planetary Boundary Layer，PBL)中的亚网格扩散；项(III)和项(IV)分别表示深对流和浅对流的亚网格输运；项(V)适用于一氧化碳，其中一氧化碳被视作一种寿命为 30 天的被动示踪剂(Seinfeld and Pandis，1998)，项(VI)是用于 $\text{PM}_{2.5\mu m}$ 粒子的湿法去除；项(VII)用于气体和气溶胶粒子的干沉积；项(VIII)则是与植被燃烧相关的羽流上升机制的源项(Freitas *et al.*，2009；Longo et al.,2010)。

在 CCATT-BRAMS 中,执行燃烧生物质估算的模块被命名为巴西生物质燃烧排放模型(Brazilian Biomass Burning Emission Model,3BEM);更多细节见隆哥等人(Longo et al.,2010)的研究。在该模型中,燃烧的生物量由活植被中存在的碳含量(Olson et al.,2000;Houghton et al.,2001)估算。对于 CCATT-BRAMS 模型中的给定像元,给定气体或气溶胶的总排放源及其在此期间的变化通过 3BEM 模型计算 FRE(式 7-6)并利用燃烧因子和燃烧生物质(式 7-7)计算得出:

$$Q^{[\epsilon]}_{plumerise}(t) = \frac{gf(t)}{p_0 \triangle V} \cdot (EF^{(\epsilon)} \cdot \vartheta \cdot FRE_{grid(lon,lat)}) \qquad 式7-6$$

$$Q^{[\epsilon]}_{plumerise}(t) = \frac{gf(t)}{p_0 \triangle V} \cdot (EF^{[\epsilon]} \cdot BBurned) \qquad 式7-7$$

其中 t 表示时间(秒);gf(t)示以最大排放周期为中心的高斯函数,用于模拟日循环;p_0 与天气状况相关;ΔV 是像元的体积;EF 表示给定物种的排放因子(ϵ);ϑ 是与生物质消耗速率相关的系数($1.37 kg \cdot MJ^{-1}$;Kaiser et al.,2012);*BBurned* 是用 3BEM 模型估算的燃烧生物质的量。

(四) 现场数据和库存对比

阿克里州(Acre State)位于巴西亚马孙河西部(图 7-4 中标记为 1 处),该州有三类保存完好的植物生态区:(1)该州西北部无棕榈树的樟树、灌木和森林地貌区;(2)该州西北和中部两片较小区域、靠近亚马孙州边界的雨林地区,特点为生长在冲积层上、有露生树冠的生物群系(图 7-4 中标记为 2 区域);(3)开阔的雨林是阿克里州主要的植物生态,这类区域遍布全州,有长有棕榈树的冲积层生物群系,长有露生树冠的山麓生物群系,以及树冠均匀露出的低地生物群系和长有棕榈树和竹子的冲积层生物群系。

此外,阿克里州为有人区,尤其是当地人口开垦了供牲畜使用的牧场(畜牧是该州的主要经济活动),在草原及附近地区有少量的次生植被(IBGE,2005)。

阿克里州的燃烧区域地图是根据 2010~2011 年的专题制图仪(Thematic Mapper,TM)/陆地卫星 5 号(Landsat 5)捕获的影像得出的。本研究运用这两年研究区域呈现在所有可用影像中呈现出的燃烧痕迹来进行燃烧面积的估算,

第七章 亚马孙热带森林生物质燃烧排放估算

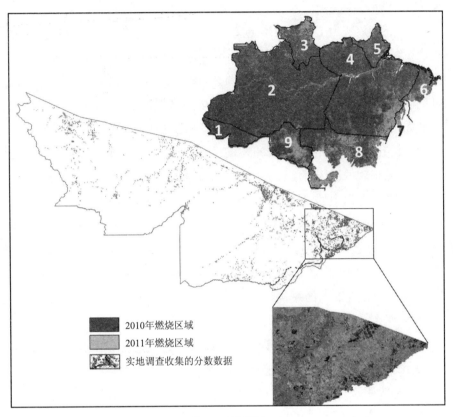

图 7-4(见彩插)

注：巴西部分州及州界；1阿克里州；2亚马孙州；3罗赖马州；4巴拉那州；5阿马帕州；6马拉尼昂州；7托坎廷斯州；8马托格罗索州；9朗多尼亚州；以及2010年(红色)和2011年(黄色)阿克里州燃烧区域地图和现场采集点。

这类影像共占15个条带/行编号，总计56个影像。起初，TM影像被加载到 INPE开发的SPRING4.3.1软件中进行处理(Câmara et al., 1996)。该软件采用多项式模型对所有影像进行几何校正，并用最近邻算法进行插值。

本研究使用影像分割技术对阿克里州的燃烧区域地图进行切割和分类，这一过程即使用相似性阈值(两个区域被认为相似并被定义为单个多边形的最小影像灰度值)和面积阈值(由影像像素数定义的要个性化的最小面积)生成具有均匀光谱特征的多边形。

在本研究中，为了获得燃烧痕迹的最佳分类结果，将相似度和面积阈值分别定义为 12 和 8。

因此，燃烧痕迹由上述多边形定义，并且进行人工影像编辑以使由数字分类过程产生的错分和漏分误差最小化（Almeida-Filho and Shimabukuro，2002）。为了验证燃烧区域图进行了两项实地调查（图 7-4），调查共访问了 33 个地面点，以验证从 TM/Landsat5 影像中得到的燃烧区域地图。此外，由于一些地面点区域无法到达，该地点所采集的样本位于临近主要道路旁。野外采集数据与测绘面积的比较，总体准确率为 93%，遗错分误差为 6%，漏分误差为 1%。此时，遗漏误差出现的原因可解释为植被再生抹去了卫星影像中的燃烧痕迹。

对 2010 年数据中的多边形进行统计分析，结果表明，绘制地图中的燃烧痕迹最小尺寸为 0.01 公顷，最大尺寸为 3 660 公顷，平均值为 15 公顷。若具体来看，大小在 2～6 公顷的多边形占到总地图区域的 54%，而大小在 7～20 公顷的多边形占据 29% 的地图区域。此外，较小的燃烧痕迹区域（0.01～1 公顷）仅占多边形总数的 2%，较大的燃烧痕迹区域（大于 20 公顷）占绘制多边形总数的 13%，这表明在 2010 年，小面积的燃烧发生率最高。

2011 年，燃烧面积的平均值为 14 公顷，最小面积为 0.01 公顷，最大面积为 1 383 公顷，与 2010 年的燃烧痕迹相比，面积较小。此外，1～8 公顷的燃烧面积占多边形总数的 58%，其次是 9～20 公顷的烧伤多边形，占总数的 27%；另外，面积较大的燃烧痕迹（大于 20 公顷）占绘制的多边形总数的 14%，这一分布与 2010 年类似，但研究区域中燃烧的发生率较低。

五、结果与讨论

（一）FRE 分布

图 7-5 展示了根据 MODIS 和 GOES 卫星数据估算的 2000～2011 年亚马孙生物群系 FRE 的空间分布。通常，FRE 值最高的区域位于亚马孙河雨林的边缘，被称为毁林弧形带；它是这个地区的主要森林砍伐和随后的农业扩张发生地。

在诸多地区，巴西亚马孙雨林地带的 FRE 年平均值达到高于 $0.07 MJ \cdot m^{-2}$ 的值。因此，考虑到网格面积（400 平方千米），某些网格点释放的能量应约为

$0.2PJ \cdot y^{-1}$。综合以上数值,我们可估算亚马孙生物群系每年释放的能量大约为400PJ。凯泽等(Kaiser et al.,2012)使用MODIS火灾产品,估算南美地区每月的FRE为100～450PJ;然而,这一数值随着火灾季节和降雨状况而变化。

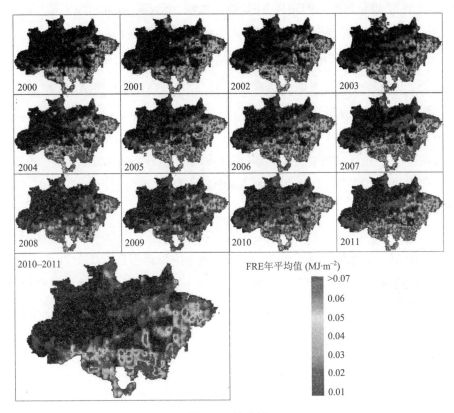

图7-5(见彩插)

注:图为根据每日观测估算的2000～2011年期间亚马孙生物群系火灾辐射能($MJ \cdot m^{-2}$)的空间分布。

在图7-5中,可以验证FRE值是否呈现了经济活动(Ewers et al.,2008)和气候因素(Barlow and Peres,2004;Good et al.,2008)带来的暂时的空间变异性。在亚马孙生物群系中,生物质燃烧主要发生在7～10月的旱季,一些火灾亦会发生在干湿过渡期间。

通过空间变异性分析,2002年、2005年和2007年是火灾频发且生物质消耗较高的年份。在这些年份中,FRE空间上主要分布于巴西的马托格罗索州、帕拉州和朗多尼亚州,这些州亦是巴西农业和畜牧业扩张的前沿地带。利用 FRE 的时间分布数据,可核实燃烧的异常事件,例如(图中用符号✡标记):(1)2005 年发生在阿克里州的异常燃烧,FRE 值大大超过该地区的年平均值;(2)2003 年和 2007 年罗赖马州出现的超高 FRE 值。此外,据观察,在降雨最多的年份,所有地区释放的能量都显著减少(如 2008 年和 2009 年数据显示)。

(二) 亚马孙热带雨林气溶胶和痕量气体排放估算

为了计算释放到大气中的痕量气体和气溶胶,CCATT-BRAMS 模型在其原始形式中使用了隆哥等开发的 3BEM 方法(Longo et al., 2010)。因此,对于由 WFABBA 或 MOD14/MYD14 产品判断为燃烧区域的给定像元,在使用滤波器去除半径为 1km 的重复火灾区域、根据贝尔瓦德(Belward, 1996)的方法确定土地利用和土地覆盖的类型、并更新 MODIS 产品之后,可通过下式(式 7-8)获得排放估算:

$$M^{[\epsilon]} = \alpha_{veg} \cdot \beta_{veg} \cdot EF^{[\epsilon]} \cdot a_{fogo} \qquad 式7-8$$

其中 $M^{[\epsilon]}$ 表示物种排放,$\alpha_{veg}e\beta_{veg}$ 表示地上生物质部分以及表 7-1 中描述的植被燃烧效率,$EF^{[\epsilon]}_{veg}$ 为每个物种的排放因子,a_{fogo} 表示燃烧区。相应地,在 FRP 处理和整合过程中,消耗的生物质以尺寸为 20 千米的网格为单位进行统计。因此,每次检测都会估算土地利用和土地覆盖,并将估算结果应用于每个物种的排放因子。

表 7-1 3BEM 处理燃烧因子与地上生物质的结果

IGBPLULC 图例	燃烧系数	地上生物质($kg \cdot m^{-2}$)
常绿阔叶林	0.5	29.24
落叶阔叶林	0.43	12.14
混交林	0.43	12.14
郁闭灌木丛	0.87	7.40
稀疏灌木丛	0.72	0.86
有林草地	0.45	10.00

续表

IGBPLULC 图例	燃烧系数	地上生物质($kg \cdot m^{-2}$)
稀树草原	0.52	6.20
草地	1.00	0.71
永久湿地	0.40	3.80
农田	0.40	3.80
农田与自然植被镶嵌体	0.40	3.80
裸地或稀疏植物	0.84	1.00

图7-6显示了用3BEM和FRE方法估算的所有亚马孙生物群落的一氧化碳排放估值($10^{-6} kg \cdot m^{-2}$)随时间(2000年1月～2011年12月)的变化。两种方法的对比说明,在一般情况下,两者计算结果总体上是一致的(图7-6(c)),3BEM方法估算的痕量气体和气溶胶值倾向于比FRE方法的计算结果更大一些(t-student, $p>0.05$, $n=144$)。图7-6(a)和(b)分别显示了利用埃弗龙(Efron,1982)最初开发的自助法估算的角度系数和相关性值。使用自助法即按照1.0×10^4的种群数量重建原始曲线,并将该数值用于计算相关参数,从而得出模型估算的置信区间和误差分析,为样本估算提供了精确度的度量。

参见图7-6(c),我们可看到,在火灾少发时期,两个模型得出的数值非常相似,比如在2000年和2001年的10月～12月,FRE的一氧化碳排放估算值比3BEM的估算值低12%,如图7-6(c)顶部的差异图(差值等于3BEM数值减去FRE数值)所示。一般情况下,两种方法的一氧化碳排放估算差异为20%,只有2011年除外,这一年的某些天中FRE估算的一氧化碳排放值大于3BEM估算的一氧化碳排放值,特别是9月和10月的某些天。可能导致这一变化的因素可能是更短的观测间隔和更多观测数据(GOES-13的新数据)及检测算法和FRP估算的变化(Xu et al.,2010)等。

这两种方法的主要区别是,影像东部(托坎廷斯州和马拉尼昂州,图中用符号¤标记)的生物质燃烧排放量增加,表示在亚马孙森林地带和巴西塞拉多地带间的过渡,以及马托格罗索州南部和玻利维亚中部的燃烧排放量增加。在南美洲,生物质燃烧主要与农业活动有关,农业活动具有很强的多变性。因此,燃烧

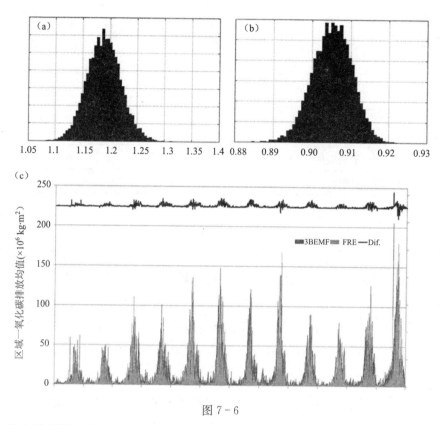

图 7-6

注：(a)角系数分布和(b)自助法中通过 3BEM(x 轴)和 FRP(y 轴)估算的一氧化碳数据的相关性；(c)2000~2011 年期间通过 3BEM 和 FRE 估算的一氧化碳排放平均值。

发生率最高的地区主要位于马托格罗索州、帕拉州和朗多尼亚州的毁林弧形带上。在这些地区，燃烧密度可达到每平方千米 125 个观测点。在过去 16 年中，马托格罗索州是火灾数量最多的地区，具有超过 1.3×10^6 次观测，其次是帕拉州，共有 0.9×10^6 次观测到的燃烧发生(INPE，2012)。

虽然两种方法计算的一氧化碳排放结果在空间分布上有一定的差异，但总体上，模型间呈现出结果的一致性。造成差异的因素包括地面生物量比例、植被燃烧效率和传统方法燃烧面积等。这些变量的表征和估算非常复杂，取决于天气和植被湿度。此外，大面积地区地面生物量的比例不可能是均匀的，并且近实时燃烧区域是遥感数据估算的主要不确定性来源之一(Chuvieco *et al.*，2004；

Yebra et al.,2008)。

从 FRP 得到的痕量气体和气溶胶估算与生物质燃烧直接相关。此外,影响燃烧效率的因素如土壤和植被水分的减少亦会直接影响火灾释放的能量。然而,在这一方法中,误差的产生也与 FRP 的日周期特性和轨道传感器的不同灵敏度有关。此外,如韦尔莫等(Vermote et al.,2009)指出的,还有一些其他因素可能带来不确定性,如生物质燃烧和烟羽的同时观测(近似误差±11%)、FRP 的准确性和一致性(±16%)以及云覆盖率(±11%)等。

(三) 排放模型评估

图 7-7 显示了 2010 年 7 月 15 日～2010 年 11 月 11 日(图 7-7(a))和 2011 年 7 月 15 日至 2011 年 11 月 15 日(图 7-7(b))的一氧化碳平均浓度(ppb),以及同期 MODIS 和 GOES 探测到的燃烧位置(分别为图 7-7(c)和(d))。如图所示,2010 年的火灾发生率高于 2011 年。根据图 7-7(c),燃烧发生率最高的是阿克里州东部地区,与巴西亚马孙河流域森林砍伐评估方案(Program for Deforestation Assessment in the Brazilian Legal Amazonia,PRODES)计算的森林砍伐率最高的地区相吻合。本研究使用 2010 年的数据,绘制了 2 000 平方千米的燃烧区域地图,占阿克里州总面积的 1.3%。同样,数据显示次年(2011 年)火灾发生大幅减少,但空间分布与 2010 年相似,东部地区的火灾发生率最高。燃烧面积减少至 $643km^2$,减少了 68%。导致火灾发生率下降的因素之一是年降水量的差异,根据巴西国家气象局(National Institute of Meteorology,INMET)的数据,2010 年比 2011 年干燥。

此外,还需着重强调的是,减少燃烧面积的另一个根本因素是公共政策的执行,例如加强监测、财产认证和燃烧替代技术。最近,阿克里州创建了一系列名为阿克里州环境森林资产估价政策(Valuation Policy of Acre Environmental Forest Asset,PVAAFA)的项目,这是与政府机构和民间社会组织合作的结果。

根据 2010 年和 2011 年的燃烧区域清单并基于地上活生物质的空间分布,燃烧的总生物质得到估算(Saatchi et al.,2007)。因此,据估算,仅在阿克里州,2010 年和 2011 年分别燃烧了 28.3 百万吨和 8.7 百万吨的生物质。总体上,若将上述结果与 3BEM 和 FRE 估算进行比较,两个模型分别可还原 2010 年燃烧消耗的所有生物质的 82%和 76%,以及 2011 年的 85%和 78%,表明了观测数

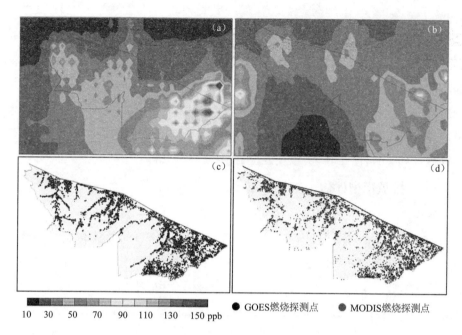

图 7-7（见彩插）

注：基于 CCATT-BRAMS 使用 FRE 方法建立的 2010 年(a)和 2011 年(b)一氧化碳浓度模型；2010 年(c)和 2011 年(d)由 MODIS 和 GOES 探测到的燃烧位置。

据和模型数据的一致性。

六、结论

每一天，诸多人类活动都在损伤、破坏或改造生态系统。生物质燃烧改变了许多对维持生态系统服务和地球生物可持续性至关重要的生态过程。在本研究中，我们在不同的时空尺度上展示了生物质燃烧对生态系统服务的大量直接和间接影响。

在此背景下，遥感产品可为估算火灾探测及其向大气中的排放提供有价值的信息。遥感具有很高的时空重复性，可以用来描述生物质燃烧的生命周期，并将其同化为空气质量的数据模型。

2000~2011 年，亚马孙生物群系的火灾发生率很高，燃烧发生率随降雨异常等有关气候因素以及社会经济过程的变化而变化。在亚马孙生物群系中,生物质燃烧排放的最高值集中在毁林弧形带上,并与森林砍伐和随后的农业扩张有关。在诸多地区,每日 FRE 值可能高于 $0.07\mathrm{MJ} \cdot \mathrm{m}^{-2}$ 或约 $0.2\mathrm{PJ} \cdot \mathrm{y}^{-1}$,大都发生在 7~10 月(旱季)之间,此外在干湿过渡(10~12 月)期间也会发生一些火灾。通过空间变异性分析,2002 年、2005 年和 2007 年是火灾发生率和生物质消耗较高的年份。

对使用燃烧因子和生物质的排放方法和基于 FRE 的方法进行评估后表明,在一般模式下,两种方法呈现出相似性,当用 3BEM 方法估算时,痕量气体和气溶胶的值往往更大一些。从阿克里州获得的库存数据表明,3BEM 和 FRE 方法可分别还原 2010 年消耗的所有生物质的 82% 和 76%,可还原 2011 年观测数据的 85% 和 78%。因此,一些利用 FRP 的新方法可以为生物质燃烧排放估算带来实质性改进,对于实时应用来说尤为如此。然而,在精度发展和新的参数化方面,基于 FRE 的方法仍有很大提高空间,未来应使几种估算中的误差最小化。

致谢

本研究受到了 CAPES、CNPQ(479626-2011-1)和 FAPESP(2010/07083-0,2010/17437-4)提供的资金支持。

参 考 文 献

Almeida-Filho, R., and Y. E. Shimabukuro. 2002. Digital processing of a Landsat-TM time series for mapping and monitoring degraded areas caused by independent gold miners, Roraima State, Brazilian Amazon. *Remote Sensing of Environment* 79:42–50.

Andreae, M., D. Rosenfeld, P. Artaxo, et al. 2004. Smoking rain clouds over the Amazon. *Science* 303:1342–1345.

Andreae, M. O., and P. Merlet. 2001. Emission of trace gases and aerosols from biomass burning. *Global Biogeochemical Cycles* 15:955–966.

Barlow, J. and C. A. Peres. 2004. Ecological responses to El Niño-induced surface fires in central Brazilian Amazonia: management implications for flammable tropical forests. *Philosophical Transactions of the Royal Society* 359:367–380.

Belward, A. 1996. The IGBP-DIS global 1 km land cover data set (DISCover) proposal and implementation plans. IGBP-DIS Working Paper 13.

Câmara, G., R. C. M. Souza, U. M. Freitas et al. 1996. Spring: Integrating Remote Sensing and GIS with Object-Oriented Data Modelling. *Computers and Graphics* 15(6):13–22.

Cardoso, M., C. A. Nobre, D. Lapola et al. 2008. Long-term potential for fires in estimates of the occurrence of savannas in the tropics. *Global Ecology and Biogeography* 17:222–235.

Cardoso, M., C. Nobre, G. Sampaio et al. 2009. Long-term potential for tropical-forest degradation due to deforestation and fires in the Brazilian Amazon. *Biologia (Bratislava)* 64:433–437.

Chatfield, R., Z. Guo, G. Sachse et al. 2002. The subtropical global plume in the Pacific Exploratory Mission-Tropics A (PEM-Tropics A), PEM-Tropics B, and the Global Atmospheric Sampling Program (GASP): How tropical emissions affect the remote Pacific. *Jounal of Geophysical Research* 107:42–78.

Chuvieco, E., D. Cocero, D. Riano et al. 2004. Combining NDVI and surface temperature for the estimation of live fuel moisture content in forest fire danger rating. *Remote Sensing of Environment* 92(3):322–331.

CICES (Common International Classification of Ecosystem Services): Consultation on Version 4, August–December 2012. Available from: http://cices.eu/wp-content/uploads/2012/07/CICES-V43_Revised-Final_Report_29012013.pdf (accessed March 15, 2013).

CIFOR, 2009. Center for International Forestry Research: Realising REDD+: National strategy and policy options, ed. A. Angelsen, 363 p. Bogor, Indonesia.

Crutzen, P. J., and M. O. Andreae. 1990. Biomass burning in the tropics: Impact on atmospheric chemistry and biogeochemical cycles. *Science* 250:1669–1678.

Efron, B. 1982. *The jackknife, the bootstrap, and other resampling plans.* Society of Industrial and Applied Mathematics. CBMS-NSF Monographs, 38, California: Stanford University.

Ewers, R. M., W. F. Laurance, and C. M. Souza Jr. 2008. Temporal fluctuations in Amazonian deforestation rates. *Environmental Conservation* 35:303–310.

Fearnside, P. M. 1985. Uma estrutura para Avaliação de opções de Desenvolvimento Florestal na Amazônia [A Framework for Evaluation Forestry Development options in Amazonia]. Presentation at the 1° Seminário Internacional de Manejo em Floresta Tropical – SEMA/WWF, Serra do Carajás, Pará.

Fearnside, P. M. 1997. Greenhouse gas from deforestation in Brazilian Amazonia: Net committed emissions. *Climate Change* 35:321–360.

Fearnside, P. M. 2000. Global warming and tropical land use change: Greenhouse gas emissions from biomass burning, decomposition and soils in forest conversion, shifting cultivation and secondary vegetation. *Climate Change* 46:115–158.

Fearnside, P. M., C. A. Righi, P. M. L. A. Graça, et al. 2009. Biomass and greenhouse gas emissions from land-use change in Brazil's Amazonian "arc of deforestation": The states of Mato Grosso and Rondônia. *Forest Ecology and Management* 258:1968–1978.

Fernandes, S. D., N. M. Trautmann, D. G. Streets, et al. 2007. Global biofuel use, 1850–2000. *Global Biogeochemical Cycles* 21:GB2019.

Freeborn, P. H., M. J. Wooster, W. M. Hao, et al. 2008. Relationships between energy release, fuel mass loss, and trace gas and aerosol emissions during laboratory biomass fires. *Journal of Geophysical Research Atmospheres* 113:D01301.

Freeborn, P. H., M. J. Wooster, and G. Roberts. 2011. Addressing the spatiotemporal sampling design of MODIS to provide estimates of the fire radiative energy emitted from Africa. *Remote Sensing of Environment* 115:475–498.

Freitas, S. R., K. M. Longo, M. A. F. Silva Dias, et al. 2005. Monitoring the transport of biomass burning emissions in South America. *Environmental Fluid Mechanics* 5:135–167.

Freitas, S. R., K. M. Longo, M. A. F. Silva Dias, et al. 2009. The Coupled Aerosol and Tracer Transport model to the Brazilian developments on the Regional Atmospheric Modeling System (CATT-BRAMS). Part 1: Model description and evaluation. *Atmospheric Chemistry and Physics* 9:2843–2861.

Giglio, L. 2005. MODIS collection 4 active fire product user's guide. Version 2.2. 2005. Available from: http://maps.geog.umd.edu/products/MODIS_Fire_Users_Guide_2.2.pdf (accessed August 3, 2010).

Good, P., J. Lowe, M. Collins, et al. 2008. An objective tropical Atlantic SST gradient index for studies of South Amazon dry season climate variability and change. *Philosophical Transactions of the Royal Society* 363:1761–1766.

Haines-Young, R. H., and M. Potschin. 2010. *Proposal for a common international classification of ecosystem goods and services (CICES) for integrated environmental and economic accounting*. European Environment Agency, New York: Department of Economic and Social Affairs.

Haines-Young, R. H., and M. Potschin. 2013. Common International Classification of Ecosystem Services (CICES): Consultation on Version 4, August–December 2012.

Horowitz, L., S. Walters, D. Mauzerall, et al. 2003. A global simulation of tropospheric ozone and related tracers: Description and evaluation of MOZART, version 2. *Journal of Geophysical Research* 108:47–84.

Houghton, R. A., K. T. Lawrence, J. L. Hackler, et al. 2001. The spatial distribution of forest biomass in the Brazilian Amazon: A comparison of estimates. *Global Change Biology* 7:731–746.

Houghton, R. A., D. L. Skole, C. A. Nobre, et al. 1999. Annual fluxes of carbon from deforestation and regrowth in the Brazilian Amazon. *Nature* 403:301–304.

IBGE (Instituto Brasileiro de Geografia e Estatística). 2005. Vegetação do Estado do Acre. Escala 1:1.000.000. Avaliable from: ftp://geoftp.ibge.gov.br/.../manual_tecnico_vegetacao_brasileira.pdf (accessed 16 March 2013).

Ichoku, C., and Y. J. Kaufman. 2005. A method to derive smoke emission rates from MODIS fire radiative energy measurements. *IEEE Transactions on Geoscience and Remote Sensing* 43:2636–2649.

Ignotti, E., S. S. Hacon, W. L. Junger, et al. 2010. Air pollution and hospital admissions for respiratory diseases in the subequatorial Amazon: A time series approach. *Caderno de Saúde Pública* 26:747–761.

INPE (Instituto Nacional de Pesquisas Espaciais). 2012. Portal do Monitoramento de Queimadas e Incêndios. Available from: http://www.inpe.br/queimadas (accessed November 8, 2012).

Kaiser, J. W., A. Heil, M. O. Andreae, et al. 2012. Biomass burning emissions estimated with a global fire assimilation system based on observed fire radiative power. *Biogeosciences* 9:527–554.

Kaufman, J. B., D. L. Cummings, D. E. Ward, et al. 1995. Fire in the Brazilian Amazon: Biomass, nutrient pools, and losses in slashed primary forests. *Oecologia* 104:397–408.

Kaufman, Y. J., L. Remer, R. Ottmar, et al. 1996. Relationship between remotely sensed fire intensity and rate of emission of smoke: SCAR-C experiment. In *Global biomass burning*, ed. J. Levine, 685–696. Cambridge, MA: MIT Press.

Longo, K. M., S. R. Freitas, M. O. Andreae, et al. 2010. The Coupled Aerosol and Tracer Transport model to the Brazilian developments on the Regional Atmospheric Modeling System (CATT-BRAMS). Part 2: Model sensitivity to the biomass burning inventories. *Atmospheric Chemistry and Physics* 10:5785–5795.

Marengo, J. A., C. A. Nobre, and L. F. Salazar. 2010. Regional climate change scenarios in South America in the late XXI century: Projections and expected impacts. *Nova Acta Leopoldina* 112:251–265.

Moraes, E. C., S. H. Franchito, and V. Brahmananda Rao. 2004. Effects of biomass burning in Amazonia on climate: A numerical experiment with a statistical-dynamical model. *Journal of Geophysical Research* 109:1–12.

Nobre, C. A., P. J. Sellers, and J. Shukla. 1991. Amazonian deforestation and regional climate change. *Journal of Climate* 4:957–988.

Olson, J. S., J. A. Watts, and L. J. Allison. 2000. Major world ecosystem complexes ranked by carbon in live vegetation: A database. Available from: http://cdiac.esd.ornl.gov/ndps/ndp017.html (accessed August 7, 2010).

Olson, J. S., J. A. Watts, and L. J. Allison. 2002. Major world ecosystem complexes ranked by carbon in live vegetation: A database (Revised November 2000). Available from: http://cdiac.esd.ornl.gov/ndps/ndp017.html (accessed November 7, 2012).

Pereira, G., N. J. Ferreira, F. S. Cardozo, et al. 2011. Aerosol and trace gas retrievals from remote sensing fire products. In *Fire detection*, ed. R. P. Bennett, 103–118. New York: Nova Science Publishers.

Pereira, G., S. R. Freitas, E. C. Moraes, et al. 2009. Estimating trace gas and aerosol emissions over South America: Relationship between fire radiative energy released and aerosol optical depth observations. *Atmospheric Environment* 43:6388–6397.

Prins, E. M., and W. P. Menzel. 1992. Geostationary satellite detection of biomass burning in South America. *International Journal of Remote Sensing* 13:2783–2799.

Riggan, P., R. Tissell, R. Lockwood, et al. 2004. Remote measurement of energy and carbon flux from wildfires in Brazil. *Ecological Applications* 14:855–872.

Roberts, G., M. J. Wooster, G. L. W. Perry, et al. 2005. Retrieval of biomass combustion rates and totals from fire radiative power observations: Application to southern Africa using geostationary SEVIRI imagery. *Journal of Geophysical Research* 110:D21111.

Roy, D. P., P. E. Lewis, and C. O. Justice. 2002. Burned area mapping using multi-temporal moderate spatial resolution data—A bi-directional reflectance model-based expectation approach. *Remote Sensing of Environment* 83:263–286.

Saatchi, S., R. Houghton, R. Avala, et al. 2007. Spatial distribution of live aboveground biomass in Amazon Basin. *Global Change Biology* 13:816–837.

Sampaio, G., C. Nobre, M. H. Costa, et al. 2007. Regional climate change over eastern Amazonia caused by pasture and soybean cropland expansion. *Geophysical Research Letters* 34:1–7.

SEEA. 2003. *Handbook of national accounting system of integrated environmental and economic accounting*. United Nations European Commission, International Monetary Fund, Organisation for Economic Co-operation and Development, World Bank, 572 p., New York: United Nations.

Seinfeld, J., and S. Pandis. 1998. *Atmospheric chemistry and physics*. New York: John Wiley & Sons.

Silva, J. M. N., A. C. L. Sa, and J. M. C. Pereira. 2005. Comparison of burned area estimates derived from SPOT-VEGETATION and Landsat ETM+ data in Africa: Influence of spatial pattern and vegetation type. *Remote Sensing of Environment* 96:188–201.

Tian, H., J. M. Melillo, D. W. Kicklighter, et al. 1998. Effect of interannual climate variability on carbon storage in Amazonian ecosystems. *Nature* 396:664–667.

Vermote, E., E. Ellicott, O. Dubovik, et al. 2009. An approach to estimate global biomass burning emissions of organic and black carbon from MODIS fire radiative power. *Journal of Geophysical Research* 114:205–227.

Werf, G. R., J. T. Randerson, L. Giglio, et al. 2006. Interannual variability in global biomass burning emissions from 1997 to 2004. *Atmospheric Chemistry and Physics* 6:3423–3441.

Wooster, M. J., G. Roberts, G. Perry, et al. 2005. Retrieval of biomass combustion rates and totals from fire radiative power observations: Calibration relationships between biomass consumption and fire radiative energy release. *Journal of Geophysical Research* 110:D24311.

Wooster, M. J., B. Zhukov, and D. Oertel. 2003. Fire radiative energy for quantitative study of biomass burning: Derivation from the BIRD experimental satellite and comparison to MODIS fire products. *Remote Sensing of Environment* 86:83–107.

Xu, W., M. Wooster, G. Roberts, et al. 2010. New GOES imager algorithms for cloud and active fire detection and fire radiative power assessment across North, South and Central America. *Remote Sensing of Environment* 114:1876–1895.

Yebra, M., E. Chuvieco, and D. Riano. 2008. Estimation of live fuel moisture content from MODIS images for fire risk assessment. *Agricultural and Forest Meteorology* 148:523–536.

第三部分

生物多样性相关生态系统服务

第八章 基于对地观测的物种多样性评价和监测

N. 费尔南德斯

一、简介

物种丧失和种群数量减少是生态系统过程及对人类的生态服务保护中最重要的威胁之一(Chapin et al., 2000)。而在不久的将来,这些威胁还会增加,即使是最保守的评估也表明,自白垩纪事件以来,当前的灭绝率也是史无前例的(Barnosky et al., 2011),这也支持了人类活动正在触发地球上的第六次物种大灭绝的观点。例如,最近对动物物种状况的评估表明,在世界自然保护联盟(International Union for Conservation of Nature, IUCN)目录的物种中,19%的脊椎动物和26%的无脊椎动物受到威胁,其中,每年有大量物种升级成为了高风险类别(Hoffmann et al., 2010; Collen et al., 2012)。总的来说,由于种群的丧失,物种多样性正在进一步降低。

"地球生命力指数"是基于1 400多个脊椎动物物种数据得出的种群数量趋势指标,其自1970年以来已下降了约30%,但在如热带等地区观察到的降幅甚至是全球下降率的两倍(Collen et al., 2009)。与此同时,其他类群数量也在普遍下降,如传粉昆虫在维持野生植物群落和作物生产方面发挥着至关重要的作用,其数量的下降严重影响了生态过程(Potts et al., 2010; Cameron et al., 2011)。尽管绝大多数物种、种群和生态系统仍未得到评估,但科学界普遍认为,这些生态过程在不同的生物类群间普遍存在,影响着全球生态系统(Dobson, 2005)。阻止物种多样性的下降具有至关重要的作用,对于其作为生态系统过程的调节器，以及为生态系统服务，乃至其本身均有益处(Mace et al., 2012)。此外，物种数量和种群丰富度决定了供给服务中的食物供给和遗传资源供给，以及调节服务中的疾病控制和授粉，与此同时，物种多样性对于支持服务中初级生产和养分循环的重要性也得到了越来越多的认可(Chapin et al., 2000)。

尽管在过去几十年中研究取得了巨大进展，但无论在区域还是全球尺度，我们量化和预测生物多样性变化的能力都非常有限(Balmford and Bond, 2005)。特别是，我们还需要投入大量的研究来了解物种及其多样性和数量如何应对环境变化，包括气候变化、土地利用变化、采伐和生物入侵等。基于卫星影像的对地观测则在此方面大有可为。尽管遥感是 21 世纪初才开始广泛应用于物种环境研究的，但其仍是过去 30 年来环境研究方法上最大的进步之一。基于克尔和奥斯特洛夫斯基(Kerr and Ostrovsky, 2003)和特纳等人(Turner et al., 2003)的早期文献综述，卫星技术在生物多样性科学和保护方面的应用可概括为三个主要领域：(1)生物、物种和群落组成的直接探测；(2)物种栖息地的种类、类型和范围的分类；(3)生态系统功能的测定，例如作为环境异质性和生物多样性指标的初级生产力。在过去十年中，尽管通过遥感研究生物多样性仍存在一些重要挑战，但这些领域都取得了巨大进展(Pettorelli et al., 2011；Pfeifer et al., 2012)。数据采集成本大幅下降，一些卫星产品甚至是免费的，但高成本仍阻碍着高光谱影像等产品的广泛使用。计算机的处理能力和专用软件都有了很大的发展，但应用现代分析技术需要高度专业化的人才。

最后，生态学家和生物保护学家已逐渐将卫星数据应用于研究中，但他们仍未接受充分的遥感实践和理论培训，以便充分利用快速提升的技术和产品等潜力进行相应研究(Cabello et al., 2012)。

本章概述了光学遥感卫星在陆地生物多样性探测、分析和监测中的应用。对地观测应用的广泛开发(并非详尽无遗)，目的是使其更贴近生物多样性管理和保护的行动者和实践者。出于进行生物多样性评估的目的，重点在于物种水平和物种组合的研究，这对生态系统服务的提供具有特别重要的影响(Pereira and Cooper, 2006)。本章讲述了对地观测系统如何协助进行物种远程探测、分布模拟以及理解生物物理环境变化如何影响种群和物种组合。

二、从太空寻找物种

从太空中直接探测物种及其多样性仍是对地观测系统最大的挑战之一。基于遥感的调查方案可以提供一种全面、系统和可重复的方法来绘制物种分布和量化多样性。其可远距离提供传统实地调查难以涉及的多时相信息，甚至是历史信息。但有两大重要障碍，一是相对于生物的个体、群体和群落的大小，大多数传感器的空间分辨率较低（图8-1）；二是受到光学特征相似物种间的光谱分离技术限制。在物种入侵评估领域中，物种遥感已有了很大提升（Huang and Asner, 2009）。及时监测并绘制侵入物种图对于防止其传播至关重要。为此，入侵后地表反射特性的变化是能够特别有效证明入侵生物存在的指标。

一些入侵物种可能表现出非常显著的信号，这些信号可在比生物个体更大的空间分辨率中探测到。例如，美国宇航局的中分辨率成像光谱仪（MODIS）的标称分辨率为250米，可以监测到亚利桑那州沙漠中由外来草本植物雷曼氏画眉草（拉丁名：*Eragrostis lehmanniana*）引起的植被绿度异常升高（Huang and Geiger, 2008）。

由于入侵物种在寒冷季节为应对罕见的降雨产生了新的组织，而本地的草本植物则处于衰老或休眠状态，这使得绘制受影响区域的地图成为可能。实地调查的野外生物量与植被绿度增加量间的相关性更是有助于预测物种丰度。同样的，北美大盆地上旱雀麦（拉丁名：*Bromus tectorum*）的物种入侵动态的回顾性研究，已有研究利用Landsat影像，通过分析经过增强的植被对降水的响应得出结论：旱雀麦比其他禾本科和灌木更早变绿。1973～2001年的Landsat影像均观察到了这一特征，从而用于绘制物种扩张图，并基于过去入侵状况和景观特征间的关系构建未来传播风险的预测模型（Bradley and Mustard, 2006）。总之，植物独特的物候特征非常有助于利用相对简单的技术和广泛可用的卫星产品跟踪其时空入侵状况。

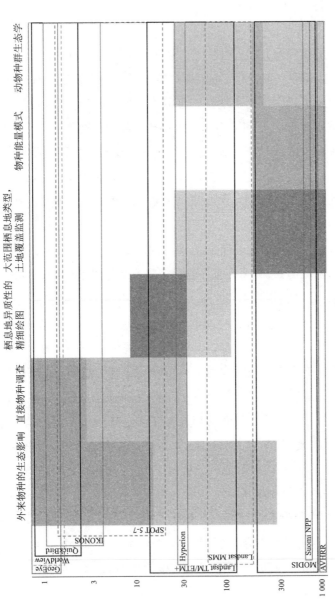

图 8-1（见彩插）

注：生物多样性研究的空间分辨率要求和与选定的卫星传感器比较概述。图中空框包含 GeoEye、WorldView、QuickBird、IKONOS、SPOT 和 Landsat ETM+ 每个卫星系列在像底点（NADIR）的分辨率范围以全色波段的最高分辨率。在列中，标色区域为各项应用所需的生态系统候分辨率，而深色区域为更合适的分辨率。在某些情况下，如动物群体研究，低分辨率数据的适用性差是由于数据采集频率低而无法开展详细的生态系统候分析，而不是由于空间分辨率不足。同时，也并非越高的分辨率开展研究就越有利，例如在大尺度物种-能量关系的研究案例中，物种分布数据通常以比大多数传感器提供的分辨率更低的分辨率记录，这是由于使用低分辨率产品可节省大量处理环节。

还有其他方法利用物种独特的光谱特性来进行监测。柽柳(拉丁名：*Tamarix ramosissima*)已蔓延到了美国西南部和墨西哥的广大地区,这不仅威胁了当地的植物群落,减少了动物种群的丰富度和多样性,还改变了河岸生态系统的水文状况。当地生态保护的首要目标就是控制和根除柽柳,并非常鼓励对受侵害地区进行卫星遥感监测的尝试(Morisette et al.，2006)。基于 Landsat TM5 影像 30 米分辨率的光谱分类,该物种在无叶期的低冠层反射率已成为对其进行监测的一种手段。柽柳的茎比其他河岸物种都暗,使其在近红外吸收波长上能与当地植被区分开来。一种基于 Landsat TM5 波段 4 和波段 5 的简单阈值法在识别受侵害的大面积地块上有着非常高的准确性($\geqslant 90\%$)(Groeneveld and Watson，2008)。而通过更复杂的光谱分解技术,完整景观中部分覆盖的柽柳范围可以被更精确地解译出来(Silvan-Cardenas and Wang，2010)。通过对多景 Landsat ETM+进行亚像素物种覆盖量化,可实现低密度林分的高精度探测(约 70%)。这些林分的探测对于早期的生物入侵管理非常重要,而这也是开展控制和根除措施最有效的阶段。

空间分辨率接近或优于 1 米的对地观测产品,有时被称为"超空间"影像,在直接监测物种方面具有更大的潜力(图 8-1)。显然,将数据与研究物种的个体大小匹配,将改善其在空间上的识别度,并减少由于像素中包含不同物种导致的像素特征混杂的问题。例如,研究表明,在测算美国科罗拉多河流域中的柽柳分布时,2.5 米分辨率的 QuickBird 数据相较于粗颗粒度的 Landsat TM 和 Hyperion 数据表现更优,其结果只有航拍数据与其相当(Carter et al.，2009)。

类似的,超空间影像使得在湿地进行的物种级别的入侵植物监测成为可能,而较粗分辨率的数据往往起不到效果(Laba et al.，2008)。然而,空间分辨率的提升也存在一些问题,可能影响调查精度。例如,更高的分辨率可能会由于监测对象内部反射率的异质性而增加特征变异(Nagendra and Rocchini，2008)。例如,植物冠层中的阴影差异,或是如树皮与树叶,新陈代谢旺盛与衰退的组织等的不同光学性质,都可能导致特征变异性的提升。因此,尽管在物种监测中将数据粒度缩小至小于或接近研究物种的个体大小很有发展前景,但高分辨率数据也存在着分类不确定性等的缺陷。

高光谱遥感(Underwood et al.，2003)允许从诸多较窄的、连续的光谱波段获取信息,这些波段覆盖了相对较宽的电磁光谱部分,尤其是可见近红外和中红外部分。该特性提供了巨大的潜力,可以根据生物化学和结构成分的差异(包括色素成分、水和干物质含量以及氮浓度)来区分生物体。由此,树木冠层的结构和化学特征可用于区分细微差异并构建高精度光谱特征(Ustin et al.,2004)。高光谱遥感对于绘制以低密度或分散出现在异源群落中的物种有着巨大的潜力(He et al.,2011);然而,将植被的生物化学和结构特征转化为分辨物种仍极具挑战性。一项基于航拍影像的夏威夷亚热带森林研究发现,一组外来树种表现出不同于本地树种的光谱特征,而且还可区分树种是否具有固氮功能(Asner et al.,2008b)。由于树冠的衰老、缝隙、阴影以及地形因素导致了物种分类中较大的不确定性,基于这种方式来绘制树种分布图也并不容易。通过使用激光雷达获得森林冠层的高光谱和三维数据的复杂组合有可能克服这些困难。因此,可根据森林光谱和结构特性及植被改变森林三维结构的独特方式来绘制夏威夷森林的树种图(Asner et al.,2008a)。这一结果对于监测入侵植物对生态系统结构、生物多样性和功能的长期影响具有启示。

高光谱除应用于植物入侵领域之外,还成为了绘制物种多样性分布图的普遍手段。例如,在澳大利亚昆士兰的一个小区域,使用这些技术对三种红树林物种进行了调查,得到的分布图总体准确率超过了 75% (Kamal and Phinn, 2011)。然而,生成地图的精度在很大程度上取决于分类算法,为了充分发挥高光谱影像的潜力,仍需要更好的方法来改进分类(Heumann,2011)。

目前,大多数高光谱研究均基于航空机载传感器(He et al.,2011)。该数据的获取成本昂贵,逻辑复杂,并且空间覆盖率低。地球观测-1 卫星(Earth Observing -1,EO-1)Hyperion 目前是唯一的航天高光谱数据源,为 30 米中等分辨率。尽管如此,一些物种调查仍能从中获益良多。在秘鲁亚马孙森林中进行的一项研究发现,使用 Hyperion 影像可对大冠树木进行分类分离(Papes et al.,2010)。判别分析考虑了旱季和雨季的季节光谱差异,准确地分离出 3 个树种和 2 个附加属。然而,由于空间分辨率和混合光谱,在较高树木多样性的区域定位物种仍具有挑战性(Carter et al.,2009)。此外,由于大气吸收信号导致的低信噪比,Hyperion 数据的分析工作也是较为困难的(Pengra et al.,2007)。

对动物物种的遥感探测显然比对植物的探测更具挑战性,因为动物是活动的,而且其位置又无法被预测。尽管如此,遥感已能够探测特定种群。例如,海冰上遗留的粪便污渍的特殊反射特性有助于使用 Landsat ETM+影像调查全部南极海岸线上的帝企鹅聚居地(Fretwell and Trathan,2009)。通过对影像的视觉检视,和简单地从红色可见波段中减去蓝色波段,就有可能找到以前未曾发现的帝企鹅聚居地。通过 QuickBird 影像用监督分类等方法更深入地调查,能够从企鹅群中监测个体企鹅,并通过提取雪、阴影和海鸟粪等信号信息来估算其的大小(Fretwell et al.,2012)。

三、土地覆盖和物种生态位的对地观测

由于土地利用变化对物种栖息地具有破坏性,其被认为是下个世纪陆地生态系统生物多样性丧失的主要原因(Sala,2001)。在此等背景下,对地观测系统应通过提供准确和空间上一致的土地覆盖数据,以预测物种在不断变化的环境中的命运,从而在评估和减轻物种损失方面发挥关键作用。

土地覆盖制图实际上是遥感最普遍的应用,尤其是在生物多样性保护领域(图8-2)。1970年代和1980年代发射的第一批 Landsat 和 NOAA 卫星推动了土地覆盖制图的发展(Defries and Townshend,1999)。最近,基于长期数据获取,新的中高分辨率卫星及影像分类方法的进展推动了土地覆盖制图成为了地球观测系统的核心研究领域(Giri,2012)。分类方法通常是基于卫星影像的光谱和纹理特性的差异将陆地表面分类为零碎的、离散的土地覆盖类型,既可使用单独时间节点数据,也可使用多时间序列数据来研究时序变化。由此得到的地图已逐渐成为了诸多生态应用的底版数据,其中包括陆地维管植物(Nagendra,2001)和陆地动物(Leyequien et al.,2007)的生物多样性研究。这为从濒危物种保护规划到生物多样性状况评估等广泛领域的分析工作提供了帮助。

此外,在生物多样性监测方面强调了区域和全球土地覆盖监测的必要性,其旨在指导和评估《生物多样性公约》(Pereira and Cooper,2006)中的跨国生物多样性协议。

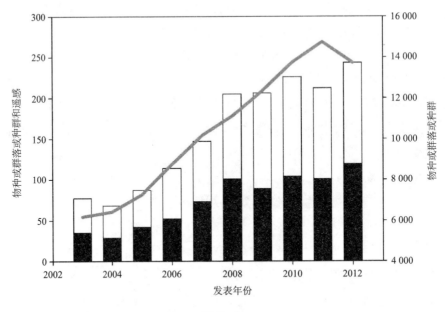

图 8-2

注:在斯高帕斯数据库(Scopus)中搜索的各年份发表的出版物数量。实线为标题、关键词和摘要中经常使用了物种或群落或种群等术语的出版物数量。整体柱状图显示为标题、关键词和摘要中经常使用了物种或群落或种群和遥感等术语的出版物数量。其中,黑色柱状部分为术语涉及土地覆盖和栖息地结构的出版物数量,白色柱状部分为其他出版物数量。需要注意的是左右纵向坐标轴的数量比例是不相同的。在近 10 年期间,可以看到关于生物多样性的相关出版物大幅增加。而这些出版物中,使用遥感的相关研究数量也在显著增加,尽管其占总出版物数量的比例增幅较小(从 2003 年的 1.25%增至 2012 年的 1.77%)。

(一) 物种分布模型

如土地覆盖图等从遥感中提取的产品是用于分析物种与生境关系的有用信息源。在地方、区域和全球尺度,及时、准确生产土地覆盖数据集均已取得了重大进步(Giri,2012),在不久的将来,更先进的卫星和技术的出现必将加强土地覆盖制图。土地覆盖数据已被证明是研究物种生境选择和生境利用、物种分布的环境约束、预测物种在较远地区出现及监测土地利用和土地利用变化影响等方面的基础资料。然而,目前的土地覆盖数据并不总是符合物种环境研究的要求,这取决于诸如地图性质(如空间和时间分辨率、准确性和图例)、物种的生态特征及物种栖息地评估的具体目的等因素。

作为保护生物学的一个分支学科,保护生物地理学对认识调查物种的空间分布、变化及其驱动因素在生物多样性管理决策中的重要性做出了重要贡献(Whittaker et al., 2005)。它还提供了一个从全球到区域尺度研究物种分布的理论和分析框架(Richardson and Whittaker, 2010)。物种分布预测模型(Species Distribution Models, SDMs)得到了广泛应用,SDMs 是一种推断物种及物种多样性如何受到全球环境变化影响的方法。SDMs 旨在揭示特定物种或种群的出现与环境特征间的联系,目标是预测栖息地对物种的适宜性,并描述可表征的生态位(在 Hutchinsonian 概念中;Franklin, 2010)。这些模型基于对物种存在与否的数据,或仅存在数据与描述气候、植被、人类活动及与物种生态需求相关的其他生物和非生物因素的变量关系进行分析(Elith and Leathwick, 2009)。此外,物种分布预测模型还可用于预测环境变化对物种的潜在影响,并支撑例如设计保护区网络、制定物种保护计划和防治入侵物种等保护策略。例如欧洲的 CORINE (Coordination of Information on the Environment, CORINE,环境信息整合)土地覆盖清单等通用产品,被用在了物种分布的生物地理模型中,用于补充完全基于气候数据的传统生物气候方法(Luoto et al., 2007)。CORINE 是一个由欧盟所属的欧洲环境署整合的通用数据集,它将欧洲土地覆盖至少分为 44 类。在芬兰,一项对 88 种鸟类分布的研究表明,将 CORINE 数据应用到生物气候模型中,将鸟类空间分布预测的准确性显著提高至近 90%(Luoto et al., 2007)。类似的方法也已被成功用于模拟新苏格兰的无尾目类动物和海龟的分布(Tingley and Herman, 2009),证实了土地覆盖对于确定气候适宜但栖息地不适宜地区的重要性。然而,对更大尺度的研究却质疑土地覆盖对预测物种分布的重要性。一项欧洲范围内对所有陆生脊椎动物和大约 20% 的植物进行的分析表明,在 50 千米的分辨率下,气候变化是土地覆盖和物种分布的主要驱动力,而加入土地覆盖变量通常不会改善模型性能(Thuiller et al., 2004)。有学者认为,空间分布的决定因素是分层级的,在大尺度上,气候决定了物种分布,在小尺度上,则是土地覆盖决定了物种分布(Pearson et al., 2004)。然而,很少有研究去评估气候与土地覆盖的相对重要性。显然,需要对更多种类的物种进行更多的评估,以更好地了解环境变化对物种多样性和分布的影响(Heikkinen et al., 2006)。

区域土地覆盖数据对于评估世界上大多数区域的绝大多数物种来说仍是稀缺的,大量的研究需要根据研究物种提供相应的土地覆盖和植被信息。这类数据的生产是有益的,因为它使物种的生态需求和环境预测因子之间联系更紧密,例如,可以更好地划定相关植被类型(Fernández et al., 2003)。戈特沙尔克等人(Gottschalk et al., 2005)综述了112篇关于鸟类分布的文献,发现大多数研究使用了基于研究物种的先验知识预设分类方案,对卫星影像中的植被进行分类。Landsat影像是第一批可供广大学者使用的卫星数据集,其在近80%的研究中得到了应用。相比之下,可能是由于空间分辨率较低和波段数量较少,甚高分辨率辐射计的数据仅用在了20%的研究中(图8-1),土地覆盖分类从仅有2类到70类以上,分类结果的精度也有很大差异。通过捕捉更多细节,更高的空间和专题分辨率,可以揭示更多的物种与栖息地间的关系,而分类较少的小比例尺地图不太可能检测到如小图斑、线性结构和异质景观等重要的栖息地成分(Gottschalk et al., 2005)。

选择具有生态重要性和生态相近性的预测因子有助于模型的统计稳健性,提高模型外推的准确性,并降低将无关变量纳入建模过程的可能性(Elith and Leathwick, 2009)。然而,将细节改进到一定程度后,再改进可能并不总能再有所提升:一项对比使用CORINE数据和基于Landsat的特定植被分类数据对西班牙54种鸟类分布的分析研究表明,除了与河岸地带等特殊栖息地密切相关的少数物种,补充分类信息对预测准确性没有显著贡献(Seoane et al., 2004)。同样的研究发现,提高CORINE数据的空间分辨率并不能提高鸟类SDMs的精度。因此,考虑到生产土地覆盖数据集所需的高成本和专业性,通用对地观测产品可能是诸多应用中高性价比的替代方案。这个解决方案取决于研究物种的特定生态特征、研究区的空间尺度以及如准确性、专题图例和空间分辨率等地图特征。令人惊讶的是,只有少数研究对环境数据源在SDMs背景下的总体适用性进行了评估,与之形成鲜明对比的是,学界在对于物种数据准确性和分辨率以及统计建模方法选择等因素评估的重视(Franklin, 2010)。

(二) 全球物种评估

在过去几年中,已制定了若干对地观测实施计划,旨在提供关于土地利用和土地覆盖的全球的、空间连续的信息(表8-1)。虽然是基于不同的遥感数据来源和不同的方法,其均聚焦于描述全球的植被类型,其基本目的均是为了满足大量用户的需求。成果均可在互联网上免费获取。

表 8-1 全球土地覆盖产品

产品	传感器	卫星数据周期	分类类型	标称空间分辨率(km)	类型数	整体面积加权精度[a](%)
IGBP-Discover	NOAA-AVHRR	1992~1993年	非监督分类	1.1	17	66.9
UMD Global Land Cover	NOAA-AVHRR	1992~1993年	监督分类	1.1	14	65
GLC2000	SPOT-VEGETATION	1999~2000年	非监督分类	1.5	22[b]	68.6
MCD12Q1	EOS-MODIS	年度数据,2001年至今	监督分类	0.5	17、14、11、9[c]	75
GlobCover	ENVISAT-MERIS	2005~2006年和2009年	非监督分类	0.3	22[b]	67.1

注:a 各数据集精度评估使用了不同的方法,数据集之间的精度数据不能直接比较。IGBP-DISCover:使用 Landsat 和 SPOT 影像进行专家影像解译来验证(Scepan et al.,1999)。GLC2000:基于使用 Landsat 影像、航空摄影影像和专题地图的参考数据分层随机抽样的定量评估(Mayaux et al.,2006)。MCD12Q1:对第五个产品版本通过监督分类的交叉验证分析进行评估(Friedl et al.,2010)。GlobCover 2005:比克伦等人(Bicheron et al.,2006)。

b 指全球产品的分类数。

c 分别为类型 1、2、3 和 4 的分类方案的分类数。

(1) 国际地圈生物圈计划数据集(IGBP-DISCover)首次以 1 千米分辨率绘制陆地表面地图。它是基于 17 种类别的 IGBP 土地覆盖分类方案。1992 年 4 月~1993 年 3 月期间的 AVHRR 的全球月度归一化差异植被指数也被用于制作地图(Loveland et al.,2000)。一个相同类型的数据集则由马里兰大学(UMD)基于相同的数据集,并使用 14 个类别的不同分类方案生产出来(Hansen et al.,2000)。

(2) 全球土地覆盖 2000(Global Land Cover 2000,GLC2000)是由欧洲委员会联合研究中心协调国际研究合作伙伴制作的一张分辨率为 1 千米的 2000 年全球地图。该地图基于 SPOT 4-VEGETATION 影像将 19 种不同的区域产品组合成一个数据集。

其包括两套产品：一个是按照联合国土地覆盖分类体系（LCCS）将区域产品统一分为 22 类全球土地覆盖数据集；以及根据每个区域土地覆盖进行分类优化的区域地图（Bartholome and Belward，2005）。GLC2000 数据库是被用作《千年生态系统评估》的核心数据集，以确定生态系统间的边界。

（3）MODIS 土地覆盖产品系列 5（MODIS Land Cover Product Collection 5, MCD12Q1）包含一套以 500 米像素分辨率生成的自 2001 年以来每个公历年的数据集。分类为使用分布于全球近 2 000 个训练样本的监督分类算法（Friedl et al.，2010）。每年的数据均以五种不同的土地覆盖分类方案进行分类，其中包括 IGBP（土地覆盖类型 1,17 个类别）和 UMD（类型 2,14 个类别）的分类方案。这些方案使得可以与从 AVHRR 提取的产品进行比较。

（4）GlobCover 是来自不同国际机构的一项联合倡议，旨在制作包括 22 种土地覆盖类型的全球地图，300 米的分辨率也使其成为最详细的全球产品（Bicheron et al.，2006）。其是利用 MERIS 卫星数据生产的。其专题图例与 GLC2000 兼容。此外，还以相同的分辨率提供了 11 个区域数据集，其二级图例最多为 51 个类别。共编制了两张全球覆盖地图，一张为 2005~2006 年期间，一张为 2009 年。

这些产品对全球物种多样性和分布的评估有很大帮助。其中一个例子是最近的哺乳动物研究，首次分析了几乎所有已知的陆地物种的栖息地适宜性（Rondinini et al.，2011）。根据 IUCN 红名单中所述的生境偏好，GlobCover 数据集被用于构建一个根据土地覆盖数据、海拔和水文信息来区分当地对每个物种的高、中、低适宜性的模型。成果地图描绘了物种的潜在分布，并与 IUCN 数据集进行了比较，该数据集的物种地理范围是诸多全球生物多样性评估的参考。潜在分布图中预测的已知哺乳动物的出现地，比 IUCN 中的地理范围更准确，这表明 IUCN 地图经常高估物种所占据的范围。这一结果体现了在生物多样性保护评估中考虑地理范围环境异质性的重要性。例如，除了 IUCN 的地理范围外，如再考虑生境偏好和土地覆盖类型的分布，评估的非洲物种保护所需的面积将增加两倍（Rondinini et al.，2005）。

全球土地覆盖监测对于预测物种在变化环境中的命运也是至关重要的。根据《千年生态系统评估》的气候和土地利用预测，对鸟类受环境变化的影响进行全球评估得出，到 2100 年，所有 8 750 个陆生物种中 10%～20% 的物种数量将减少超过 50%（Jetz et al.，2007）。绝大多数的物种是在热带国家里发现的，而在这些国家，土地覆盖变化是最严重的威胁。类似地，对两栖动物多样性的预测显示，土地覆盖变化是热带中美洲、南美洲和非洲物种的主要威胁，并将加剧对其他地区气候变化的影响（Hof et al.，2011）。

然而，目前一般用途的土地覆盖产品仍有不足，需要在生物多样性研究的背景下特别考虑如下问题：

（1）土地覆盖数据的不确定性可能会严重影响物种分布分析结果。虽然对不同全球产品评估的总体准确性都相对较高（表 8-1），但在相互比较数据集时，个别类别和土地覆盖的总体分布存在显著差异。例如，两个最新和最高分辨率的全球地图，即 GlobCover 和 MODIS MCD12Q1，农田的分布差异超过了 35%，而森林的差异只有 10% 左右（Fritz et al.，2011）。土地覆盖图的使用通常不考虑其准确性；然而，从底图数据传递的误差明显影响了物种分布预测模型的可靠性。因此，在任何特定应用中使用这些产品都需要对错误分类的敏感性进行谨慎检查。

（2）专题分辨率低可能会妨碍某些相关生境的发现，并限制可能的适宜和不适宜的土地覆盖类别间的区分。例如，可从土地覆盖信息中描述欧亚棕熊的概率分布，因为森林的分布和破碎程度决定了物种栖息地的适宜性（Fernández et al.，2012）。然而，一般土地覆盖产品并不区分当地森林和不适合该物种的伐木种植园。

（3）空间分辨率可能会影响并限制生境异质性的识别，这强调了需要将建模工作的目标和空间尺度与现有信息数据相匹配。例如，物种可能受到栖息地变化的影响，其空间尺度比土地覆盖地图所记录的更为精细。

（4）由于不同的分类体系、分辨率和类别定义，不同的土地覆盖图很难进行比较，从而影响了我们分析物种栖息地时序变化的能力。

现阶段已经开发了一些方法，来比较不同分类的图例，并找出专题不确定性高的区域（Fritz and See，2008），但如前文所述，为了避免将土地覆盖变化与分类差异混淆起来,必须进行谨慎检查。

四、测定生态系统在动物生态和保护中的作用

卫星传感器提供的光谱信息经常被用于模拟与地表能量和物质平衡相关的生态系统过程和特性,包括净初级生产力的动态变化、养分的动态变化及将热量分成潜通量和显通量。这个过程最终描述了生态系统中生命活动的能量特性,同时确定了异养生物可用的资源量。最近,越来越多的研究通过遥感的方式描述生态系统功能,旨在了解维持物种种群和分类多样性的生态过程(Pettorelli et al.,2011;Cabello et al.,2012)。

(一) 生态系统功能与物种丰富度的关系

物种能量假说预测物种丰富度与当地的可用能量直接相关(Wright,1983)。它由以下的观点支撑:(1)生物个体的密度是可用能量运转的结果;(2)物种丰富度由该地区可支撑的个体总数决定。可用能量在本文中被定义为产生物种可用资源的比率。这个比率应该考虑用每个物种的各自的能量需求和限制来估算。例如,蒸散发量被用作植物可用能量的相对量度,因为它提供了有关可用于光合作用的入射太阳能总量和由降水量、蒸发量、蒸腾量和土壤蓄水量确定的水分限制的信息(Wright,1983)(见第十八章)。对于动物来说,物种丰富度既与净初级生产力相关,也直接或间接地与如温度、降水、潜在蒸散量和实际蒸散量等气候协变量相关(Currie et al.,2004)。物种能量假说的支持证据仍在研讨之中,而最近一些研究通过遥感估算可用能量的方式加入了这一讨论。

赫尔伯特和哈斯克尔(Hurlbert and Haskell,2003)通过分析北美初级生产力与鸟类物种丰富度间的季节关系,并假设 NDVI 指数与鸟类可获得的食物资源呈正相关,发现:(1)栖息物种数量与最小月度 NDVI 指数(评价可获得的最少食物资源量指标)呈正相关;(2)迁移物种的数量与 6 月 NDVI 指数和最小月度 NDVI 指数间的差值(测度繁殖季节生产力的季节波动指标)呈正相关;(3)迁移物种的比例与历年 6 月 NDVI 指数波动比例比(繁殖季节生产力与可利用的最少生产力之间的相对关系指标)呈正相关。所有这些结果都支持了鸟类会根据资源的波动而自主选择栖息地的假设,而海洋环境产生力决定了特定时段

内可以共同生存于当地的物种总数量。然而,菲利普斯等人(Phillips et al.,2010)发现,北美的鸟类物种数量与能量之间的关系不是线性的,而是符合单峰分布的:二者关系的斜率在生产力较低的地区为正,而在能量较高的环境中变平并变为负。这种模式的一个可能性解释为,在低产环境中增加初级生产力会使生境变得复杂,并增加食物产量(如种子和无脊椎动物)。在一定程度上来说,生产力并不代表多样性的关键限制因素,而高环境生产力可能涉及生境结构复杂性的降低,导致某些具有更高竞争优势的物种总体丰富度降低。

理解跨生产力梯度的物种与能量关系,对于生物多样性保护是高度相关的。生态系统在关系的表征和强度上存在差异,因此,需要制定不同的管理策略(Phillips et al.,2010),但这些方面的问题很少得到解决。值得注意的是,遥感估算的初级生产力可通过提供纯气候变量无法提供的生态系统能量平衡状态、变异性和变化的指标,为监测和预测气候变化对分类丰富度的影响做出重大贡献(Ivits et al.,2013)。而关于生态系统能量与如稀有物种的数量和物种均匀度等,除了物种丰富度以外的其他物种保护指标间的关系,还需要进一步的研究。例如,在北美洲,稀有鸟类物种与 NDVI 指数的关系比与常见物种的弱,这与增加能源利用率可以降低这些物种灭绝风险的预测相矛盾(Evans et al.,2006)。

利用如 NDVI 指数等植被指数作为可用能量的指标是基于植被指数与植被吸收的光合有效辐射量的直接关系。后者是度量单位面积二氧化碳的最大潜在吸收量的指标(Goetz and Prince,1998),它已被用作代表不同生态系统中 NPP 指数的指标(如 Paruelo et al.,1998)。

然而,实际生产率也受到温度、水资源可利用量和其他资源的可利用量等生产效率项的限制(Sellers et al.,1995)。这些因素的作用效果均被归纳在了蒙蒂斯第一生产量的光利用效率模型里(Monteith,1972)(另见第六章):

$$GPP = \varepsilon \times F_{PAR} \times PAR \qquad 式8-1$$

其中:GPP 为初级生产力总量,PAR 为入射的光合有效辐射($\mu mol \cdot m^{-2} s^{-1}$),FPAR 为吸收的光合有效辐射量,$\varepsilon$ 为描述吸收的光合有效辐射量转化为生物量的比例的效率项($gC \cdot MJ^{-1}$)。

根据这个方程,仅使用 NDVI 指数来评估能源可用性的研究假设,不同生态系统和不同时序间光合有效辐射量转化为生物量(ε)的差异可忽略不计。在诸多未完全达到潜在植物光合能力(如由于环境和生理限制)的生态系统中,这一假设使得估算能量可用性时出现了显著偏差(见第三章)。然而,在物种和能量关系的背景下,违反这一假设的重要性应进行充分研讨。

为了克服这些问题,MODIS 现在提供全球 NPP 和 GPP(MOD17)指数。MOD17 基于光使用效率算法,该算法根据气象数据和植被类型来估算效率项,试图提供更直接的生产力估算方式。首个将 NDVI 指数与 MODIS 的 GPP 和 NPP 指数进行比较的物种多样性研究发现,在北美稀疏和密集植被区,后者与的鸟类丰富度具有更好的相关性,而当所有植被类型一起分析时,后者的优势就并不明显了(Phillips et al.,2008)。在植被稀疏的地区、土壤背景反射率有显著噪声的地区和 NDVI 较饱和的植被密集地区,NDVI 也表现出了较差的性能(Huete et al.,2002)。这些发现要求在物种-能源模型中更普遍地使用直接生产力评估结果。然而,MOD17 也并非万灵丹,在不加鉴别地使用本产品和类似产品之前,应仔细考虑一些不确定性。MODIS 中用于估算 GPP 和 NPP 指数的算法是基于从较粗分辨率数据计算得到的气象和植被参数,这也将引入额外的观测误差。此外,其是基于考虑气象和植被类型对效率项的影响假设。一项比较北美 9 个生态系统中 MOD17 中的 GPP 指数与涡流协方差通量数据的研究发现(Sims et al.,2006),其与增强植被指数相比没有显著提升,EVI 是一种通过增加蓝色波段以获得对植被结构特征的敏感性来减少背景信号和饱和度问题的指数(Huete et al.,2011)。这些结论表明,EVI 可提供至少与更复杂的完全依赖于遥感数据的 MOD17 一样好的生产率评估结果。

最近,又有几种新的专门根据遥感数据模拟生产力的方法被开发出来了。可使用遥感数据而非气象再分析数据来估算抑制 ε 的环境压力(Hilker et al.,2008)。基于来自 MODIS 的 EVI 数据和陆地表面温度(Land Surface Temperature,LST)数据的 GPP 模型,与北美 11 个涡流协方差塔点的碳通量数据吻合得很好(Sims et al.,2008)。其他尝试集中在基于光化学植被指数(PRI)用卫星数据直接估算辐射利用效率上,该指数体现指示植被光合效率的类胡萝卜素色素的化学状态(Gamon et al.,1997)(见第三章)。对已发表文献的分析发现,根据涡流协方差数据计算,MODIS 的 PRI 数据占生态系统层面的辐射利用效率变化的 53%~67%(Garbulsky et al.，2011)。因此，提出了另一版本的蒙蒂斯方程，其中效率项可由 PRI 数据代替：

$$CO_2 吸收 = f(PRI \times F_{PAR}) \times PAR \qquad 式 8-2$$

虽然这些方法仍需检验，但其在改进利用遥感数据进行生产力估算方面具有巨大潜力，而且随时可用于生物多样性研究。

(二) 野生动物种群的生态保护

生态系统功能的描述指标，如初级生产力速率，提供了与种群动态直接相关的时空变化过程的关键信息(图 8-3)，尽管种群生态学家最近才将遥感应用于测定这些关键的种群驱动因素。最近的综述文献突出了 NDVI 指数(Pettorelli et al.，2011)和更普遍的卫星生态系统功能估算(Cabello et al.，2012)，在种群和物种层面收集保护规划有用的生态信息方面的优势。这些综述文献都强调了生态系统功能描述指标对于更好地理解和预测物种栖息地和种群统计的重要性。这些方面是保护生物学的核心，保护生物学旨在保护和恢复处于危险中的种群所需的生态条件。例如，植被指数被用于模拟濒危物种的栖息地特征，如西

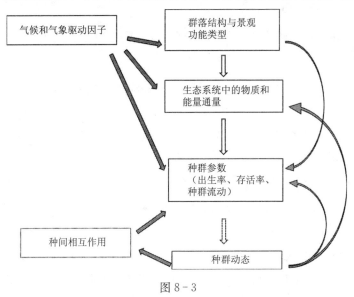

图 8-3

注：物种种群动态的环境控制因素示意图。生态系统功能的遥感指数可用于表征影响种群统计参数的物质和能量通量。反过来，这些通量是由自上而下的非生物控制因素(如气候、景观地貌)和生物多样性反馈确定的。

班牙北部的欧洲棕熊(Wiegand et al.，2008)。熊繁殖频率更高的高质量栖息地与高边际 NDVI 季节模式相对应,这种模式与物种追寻波动的可用能量的生态适应性是一致的。

相比之下,质量下降的地区,即那些繁殖较少、没有繁殖但物种频繁出现或零星存在的地区,专业化和边际化的 NDVI 指数特征逐步依次下降。对于濒临灭绝的大熊猫(Viña et al.，2008),NDVI 物候变量也有助于描述物种的栖息地,因为其与食物资源的可获取性及人类干扰引起的栖息地变化有关。这些发现对于将保护工作集中在最有价值的栖息地和开发更敏锐的栖息地遥感监测计划具有重要意义。

根据资源限制假说(Dunning and Brown，1982)预测生态系统生产力与种群丰度、种群统计之间存在正相关,根据此假说,生态系统功能的遥感指数也有助于描述种群对环境变化的动态响应。在使用 NDVI 时间序列数据作为"生态质量"度量的建模研究中,高置信度地预测了肯尼亚北部草原象的受孕率(Rasmussen et al.，2006)。该指数也可用于预测大象的繁殖物候和种群征召(Wittemyer et al.，2007)。

生产力和生殖指数间的关系往往有一段时间的滞后,因为能量可获得性对生殖的影响可能在受孕之前起作用(Rasmussen et al.，2006)。对这种滞后的分析使预测应对先前生态系统状态的种群数量变化并及时做出保护决策成为可能。

濒危的埃及秃鹫(Grande et al.，2009)和白鹳(Schaub et al.，2005)的存活状况均与植被绿度相关;在越冬地区,所有年龄段的存活状况与季节 NDVI 指数呈现正相关。这一模式似乎与低产量年份的食物短缺有关,可能与鹳可获取的小型野生猎物和昆虫较少,以及秃鹫可获取的野生和家庭的食草动物较少有关。有趣的是,越冬时的物种存活状况与 NDVI 指数的相关性可以解释在鹳繁殖地观察到的种群数量同步性。

穆勒等人(Mueller et al.，2011)发现,有蹄类动物中的不同物种的种群数量动态变化情况与植被生产力的大尺度时空变化保持同步,而初级生产力动态变化的不可预测性与同一物种不同个体之间活动缺乏协调性有关。他们的结果表明,在不可预测的环境中,物种可能倾向于采取游牧行为,而迁移行为能更好地适应高度动态但可预测的生产力模式。了解动物应对环境生产力的不同活动,对于采取以下保护策略具有重要的意义:(1)维护迁移动物的关键走廊、越冬和产犊区;(2)为活动范围更难预测的游牧动物提供大型连片的栖息地地块(Mueller et al., 2011)。

在本章概述的大多数研究中,需要高频率的数据采集来跟踪初级生产力的变化,因为种群数量对环境波动的反应可能发生在非常短的时间范围内。在这方面,非常适合使用 AVHRR 和 MODIS 传感器分别以 1 千米和 250 米分辨率获取的每日数据,虽然为了提升获取相关植被信号的概率,这些影像数据会以 8~16 天进行合成。这种分辨率已被证明适用于有关种群动态的大尺度研究,尽管它可能限制遥感指数在更小尺度的资源可用性研究中的应用(如生活在小栖息地地块或高异质性环境中的物种)。

除了植被指数外,生态系统功能遥感为描述环境变化提供了更多可能性,这在保护研究中基本上被忽视了(Cabello et al., 2012)。

除了上述 NPP 和 GPP 指数估算外,基于地表能量平衡的描述还可提供非常有用的信息,例如功能异质性(Fernández et al., 2010)、由于降解而导致的生态系统功能损失(Garcia et al., 2008)和火灾对生态系统的干扰(Mildrexler et al., 2009),这些信息可能有助于物种栖息地评估和监测。

五、结论

物种多样性本身就是一种核心生态系统服务,它通过调节生态系统过程来维持诸多其他服务,包括初级生产、养分和水循环、气候调节等。因此,物种多样性的评估和保护应成为生态系统及其服务管理的一个组成部分。本文概述了遥感通过直接探测、模拟分布和集聚及分析其在时空上对环境变化的响应,为物种研究提供了诸多机会。然而,对地观测系统的潜力肯定比迄今为止已开发的要大得多。

目前可用的数据和技术仍十分有限,无法解决物种多样性评估和保护中的一些关键问题。而传感器的空间分辨率和覆盖范围、光谱覆盖范围、数据获取成本及标准化流程缺失,都阻碍了物种和种群调查更广泛地实施。大多数全球和区域土地覆盖数据集不能（也并非设计为）提供评估物种与环境之间关系所需的关于特定土地覆盖和植被类别的更精确数据。物种对生态系统物质和能量通量的响应研究依赖于生态系统过程的遥感指标，而这些指标仍需要进一步改进。

为了解决上述的一些不足，目前正在安排新的卫星任务。HyspIRI 任务（http://hyspiri.jpl.nasa.gov）旨在从太空获取新的 60 米分辨率的高光谱数据，以支持植被特征和功能的有关研究。Landsat 8-LDCM 于 2013 年 2 月发射，通过两个机载传感器收集数据，提供 9 个波段的光谱信息数据，其中 1 个波段为 15 米空间分辨率的全色通道信息，8 个波段为 30 米空间分辨率的短波波段信息；并还提供了 2 个 100 米空间分辨率的短波波段信息（http://Landsat.gsfc.nasa.gov/about/ldcm.html）。这一任务提供的数据可与先前的 Landsat 卫星数据的相衔接，因此可以继续支持众多生物多样性学科中的广泛应用，特别是绘制土地覆盖和其他物种栖息地特征的地图。这些数据集对于监测生境变化和预测物种在变化环境中的命运极具价值。然而，利用长时间序列对地观测数据对生境变化的研究仍然非常少。

最后，在过去十年中，将遥感数据和技术与物种、种群和群落结合的研究数量显著增长，但遥感数据和技术仍未系统地用于生物多样性评估、监测和保护规划。因此，增加遥感系统知识在这些领域的传播，可更好地推动对地观测技术在生物多样性研究中的应用。

致谢

作者感谢 M. 德利贝斯、N. 霍宁、M. 加西亚和对早期版本的手稿提出建设性意见的两位匿名审稿人。安达卢西亚自治区政府通过卓越研究项目 RNM-6685 提供了财政支持。

参 考 文 献

Asner, G. P., R. F. Hughes, P. M. Vitousek, et al. 2008a. Invasive plants transform the three-dimensional structure of rain forests. *Proceedings of the National Academy of Sciences of the United States of America* 105:4519–4523.

Asner, G. P., M. O. Jones, R. E. Martin, D. E. Knapp, and R. F. Hughes. 2008b. Remote sensing of native and invasive species in Hawaiian forests. *Remote Sensing of Environment* 112:1912–1926.

Balmford, A., and W. Bond. 2005. Trends in the state of nature and their implications for human well-being. *Ecology Letters* 8:1218–1234.

Barnosky, A. D., N. Matzke, S. Tomiya, et al. 2011. Has the Earth's sixth mass extinction already arrived? *Nature* 471:51–57.

Bartholome, E., and A. S. Belward. 2005. GLC2000: A new approach to global land cover mapping from Earth observation data. *International Journal of Remote Sensing* 26:1959–1977.

Bicheron, P., M. Leroy, C. Brockmann, et al. 2006. GLOBCOVER: A 300 m global land cover product for 2005 using ENVISAT MERIS time series. *Proceedings of the Recent Advances in Quantitative Remote Sensing Symposium*. Valencia, Spain: Universidad de Valencia, 538–542.

Bradley, B. A., and J. F. Mustard. 2006. Characterizing the landscape dynamics of an invasive plant and risk of invasion using remote sensing. *Ecological Applications* 16:1132–1147.

Cabello, J., N. Fernández, D. Alcaraz-Segura, et al. 2012. The ecosystem functioning dimension in conservation: Insights from remote sensing. *Biodiversity and Conservation* 21:3287–3305.

Cameron, S. A., J. D. Lozier, J. P. Strange, et al. 2011. Patterns of widespread decline in North American bumble bees. *Proceedings of the National Academy of Sciences of the United States of America* 108:662–667.

Carter, G. A., K. L. Lucas, G. A. Blossom, et al. 2009. Remote sensing and mapping of tamarisk along the Colorado River, USA: A comparative use of summer-acquired hyperion, Thematic Mapper and QuickBird Data. *Remote Sensing* 1:318–329.

Chapin, F. S., E. S. Zavaleta, V. T. Eviner, et al. 2000. Consequences of changing biodiversity. *Nature* 405:234–242.

Collen, B., M. Böhm, R. Kemp, and J. E. M. Baillie. 2012. *Spineless: Status and trends of the world's invertebrates*. London: Zoological Society of London.

Collen, B., J. Loh, S. Whitmee, L. Mcrae, R. Amin, and J. E. M. Baillie. 2009. Monitoring change in vertebrate abundance: The living planet index. *Conservation Biology* 23:317–327.

Currie, D. J., G. G. Mittelbach, H. V. Cornell, et al. 2004. Predictions and tests of climate-based hypotheses of broad-scale variation in taxonomic richness. *Ecology Letters* 7:1121–1134.

Defries, R. S., and J. R. G. Townshend. 1999. Global land cover characterization from satellite data: From research to operational implementation? *Global Ecology and Biogeography* 8:367–379.

Dobson, A. 2005. Monitoring global rates of biodiversity change: Challenges that arise in meeting the Convention on Biological Diversity (CBD) 2010 goals. *Philosophical Transactions of the Royal Society—Biological Sciences* 360:229–241.

Dunning, J. B., and J. H. Brown. 1982. Summer rainfall and winter sparrow densities—A test of the food limitation hypothesis. *The Auk* 99:123–129.

Elith, J., and J. R. Leathwick. 2009. Species distribution models: Ecological explanation and prediction across space and time. *Annual Review of Ecology Evolution and Systematics* 40:677–697.

Evans, K. L., N. A. James, and K. J. Gaston. 2006. Abundance, species richness and energy availability in the North American avifauna. *Global Ecology and Biogeography* 15:372–385.

Fernández, N., M. Delibes, F. Palomares, and D. J. Mladenoff. 2003. Identifying breeding habitat for the Iberian lynx: Inferences from a fine-scale spatial analysis. *Ecological Applications* 13:1310–1324.

Fernández, N., J. M. Paruelo, and M. Delibes. 2010. Ecosystem functioning of protected and altered Mediterranean environments: A remote sensing classification in Doñana, Spain. *Remote Sensing of Environment* 114(1):211–220.

Fernández, N., N. Selva, C. Yuste, H. Okarma, and Z. Jakubiec. 2012. Brown bears at the edge: Modeling habitat constrains at the periphery of the Carpathian population. *Biological Conservation* 153:134–142.

Franklin, J. 2010. *Mapping species distributions. Spatial inference and prediction.* Cambridge, UK: Cambridge University Press.

Fretwell, P. T., M. A. LaRue, P. Morin, et al. 2012. An emperor penguin population estimate: The first global, synoptic survey of a species from space. *Plos One* 7:e33751.

Fretwell, P. T., and P. N. Trathan. 2009. Penguins from space: Faecal stains reveal the location of emperor penguin colonies. *Global Ecology and Biogeography* 18:543–552.

Friedl, M. A., D. Sulla-Menashe, B. Tan, et al. 2010. MODIS Collection 5 global land cover: Algorithm refinements and characterization of new datasets. *Remote Sensing of Environment* 114:168–182.

Fritz, S., and L. See. 2008. Identifying and quantifying uncertainty and spatial disagreement in the comparison of Global Land Cover for different applications. *Global Change Biology* 14:1057–1075.

Fritz, S., L. See, I. McCallum, et al. 2011. Highlighting continued uncertainty in global land cover maps for the user community. *Environmental Research Letters* 6:044005.

Gamon, J. A., L. Serrano, and J. S. Surfus. 1997. The photochemical reflectance index: An optical indicator of photosynthetic radiation use efficiency across species, functional types, and nutrient levels. *Oecologia* 112:492–501.

Garbulsky, M. F., J. Penuelas, J. Gamon, Y. Inoue, and I. Filella. 2011. The photochemical reflectance index (PRI) and the remote sensing of leaf, canopy and ecosystem radiation use efficiencies. A review and meta-analysis. *Remote Sensing of Environment* 115:281–297.

Garcia, M., C. Oyonarte, L. Villagarcia, S. Contreras, F. Domingo, and J. Puigdefabregas. 2008. Monitoring land degradation risk using ASTER data: The non-evaporative fraction as an indicator of ecosystem function. *Remote Sensing of Environment* 112:3720–3736.

Giri, C. (ed.). 2012. *Remote sensing of land use and land cover: Principles and applications*. Boca Raton, FL: CRC Press.

Goetz, S. J., and S. D. Prince. 1998. Modeling terrestrial carbon exchange and storage: Evidence and implications of functional convergence in light use efficiency. *Advances in Ecological Research* 28:57–92.

Gottschalk, T. K., F. Huettmann, and M. Ehlers. 2005. Thirty years of analysing and modelling avian habitat relationships using satellite imagery data: A review. *International Journal of Remote Sensing* 26:2631–2656.

Grande, J. M., D. Serrano, G. Tavecchia, et al. 2009. Survival in a long-lived territorial migrant: Effects of life-history traits and ecological conditions in wintering and breeding areas. *Oikos* 118:580–590.

Groeneveld, D. P., and R. P. Watson. 2008. Near-infrared discrimination of leafless saltcedar in wintertime Landsat TM. *International Journal of Remote Sensing* 29:3577–3588.

Hansen, M. C., R. S. Defries, J. R. G. Townshend, and R. Sohlberg. 2000. Global land cover classification at 1km spatial resolution using a classification tree approach. *International Journal of Remote Sensing* 21:1331–1364.

He, K. S., D. Rocchini, M. Neteler, and H. Nagendra. 2011. Benefits of hyperspectral remote sensing for tracking plant invasions. *Diversity and Distributions* 17:381–392.

Heikkinen, R. K., M. Luoto, M. B. Araujo, R. Virkkala, W. Thuiller, and M. T. Sykes. 2006. Methods and uncertainties in bioclimatic envelope modelling under climate change. *Progress in Physical Geography* 30:751–777.

Heumann, B. W. 2011. Satellite remote sensing of mangrove forests: Recent advances and future opportunities. *Progress in Physical Geography* 35:87–108.

Hilker, T., N. C. Coops, M. A. Wulder, T. A. Black, and R. D. Guy. 2008. The use of remote sensing in light use efficiency based models of gross primary production: A review of current status and future requirements. *Science of the Total Environment* 404:411–423.

Hof, C., M. B. Araujo, W. Jetz, and C. Rahbek. 2011. Additive threats from pathogens, climate and land-use change for global amphibian diversity. *Nature* 480:516–U137.

Hoffmann, M., C. Hilton-Taylor, A. Angulo, et al. 2010. The impact of conservation on the status of the world's vertebrates. *Science* 330:1503–1509.

Huang, C. Y., and G. P. Asner. 2009. Applications of remote sensing to alien invasive plant studies. *Sensors* 9:4869–4889.

Huang, C. Y., and E. L. Geiger. 2008. Climate anomalies provide opportunities for large-scale mapping of non-native plant abundance in desert grasslands. *Diversity and Distributions* 14:875–884.

Huete, A., K. Didan, T. Miura, E. P. Rodriguez, X. Gao, and L. G. Ferreira. 2002. Overview of the radiometric and biophysical performance of the MODIS vegetation indices. *Remote Sensing of Environment* 83:195–213.

Huete, A., K. Didan, W. van Leeuwen, T. Miura, and E. Glenn. 2011. MODIS vegetation indices. In *Land remote sensing and global environmental change*, eds. B. Ramachandran, C. O. Justice, and M. J. Abrams, 579–602. New York: Springer-Verlag.

Hurlbert, A. H., and J. P. Haskell. 2003. The effect of energy and seasonality on avian species richness and community composition. *American Naturalist* 161:83–97.

Ivits, E., M. Cherlet, W. Mehl, and S. Sommer. 2013. Ecosystem functional units characterized by satellite observed phenology and productivity gradients: A case study for Europe. *Ecological Indicators* 27:17–28.

Jetz, W., D. S. Wilcove, and A. P. Dobson. 2007. Projected impacts of climate and land-use change on the global diversity of birds. *Plos Biology* 5:1211–1219.

Kamal, M., and S. Phinn. 2011. Hyperspectral data for mangrove species mapping: A comparison of pixel-based and object-based approach. *Remote Sensing* 3:2222–2242.

Kerr, J. T., and M. Ostrovsky. 2003. From space to species: Ecological applications for remote sensing. *Trends in Ecology and Evolution* 18:299–305.

Laba, M., R. Downs, S. Smith, et al. 2008. Mapping invasive wetland plants in the Hudson River National Estuarine Research Reserve using quickbird satellite imagery. *Remote Sensing of Environment* 112:286–300.

Leyequien, E., J. Verrelst, M. Slot, G. Schaepman-Strub, I. M. A. Heitkonig, and A. Skidmore. 2007. Capturing the fugitive: Applying remote sensing to terrestrial animal distribution and diversity. *International Journal of Applied Earth Observation and Geoinformation* 9:1–20.

Loveland, T. R., B. C. Reed, J. F. Brown, et al. 2000. Development of a global land cover characteristics database and IGBP DISCover from 1 km AVHRR data. *International Journal of Remote Sensing* 21:1303–1330.

Luoto, M., R. Virkkala, and R. K. Heikkinen. 2007. The role of land cover in bioclimatic models depends on spatial resolution. *Global Ecology and Biogeography* 16:34–42.

Mace, G. M., K. Norris, and A. H. Fitter. 2012. Biodiversity and ecosystem services: A multilayered relationship. *Trends in Ecology and Evolution* 27:19–26.

Mayaux, P., H. Eva, J. Gallego, et al. 2006. Validation of the global land cover 2000 map. *IEEE Transactions on Geoscience and Remote Sensing* 44:1728–1739.

Mildrexler, D. J., M. S. Zhao, and S. W. Running. 2009. Testing a MODIS Global Disturbance Index across North America. *Remote Sensing of Environment* 113:2103–2117.

Monteith, J. L. 1972. Solar-radiation and productivity in tropical ecosystems. *Journal of Applied Ecology* 9:747–766.

Morisette, J. T., C. S. Jarnevich, A. Ullah, et al. 2006. A tamarisk habitat suitability map for the continental United States. *Frontiers in Ecology and the Environment* 4:11–17.

Mueller, T., K. A. Olson, G. Dressler, et al. 2011. How landscape dynamics link individual-to population-level movement patterns: A multispecies comparison of ungulate relocation data. *Global Ecology and Biogeography* 20:683–694.

Nagendra, H. 2001. Using remote sensing to assess biodiversity. *International Journal of Remote Sensing* 22:2377–2400.

Nagendra, H., and D. Rocchini. 2008. High resolution satellite imagery for tropical biodiversity studies: The devil is in the detail. *Biodiversity and Conservation* 17:3431–3442.

Papes, M., R. Tupayachi, P. Martinez, A. T. Peterson, and G. V. N. Powell. 2010. Using hyperspectral satellite imagery for regional inventories: A test with tropical emergent trees in the Amazon Basin. *Journal of Vegetation Science* 21:342–354.

Paruelo, J. M., E. G. Jobbágy, O. E. Sala, W. K. Lauenroth, and I. Burke. 1998. Functional and structural convergence of temperate grassland and shrubland ecosystems. *Ecological Applications* 8:194–206.

Pearson, R. G., T. P. Dawson, and C. Liu. 2004. Modelling species distributions in Britain: A hierarchical integration of climate and land-cover data. *Ecography* 27:285–298.

Pengra, B. W., C. A. Johnston, and T. R. Loveland. 2007. Mapping an invasive plant, *Phragmites australis*, in coastal wetlands using the EO-1 Hyperion hyperspectral sensor. *Remote Sensing of Environment* 108:74–81.

Pereira, H. M., and H. D. Cooper. 2006. Towards the global monitoring of biodiversity change. *Trends in Ecology and Evolution* 21:123–129.

Pettorelli, N., S. Ryan, T. Mueller, et al. 2011. The Normalized Difference Vegetation Index (NDVI): Unforeseen successes in animal ecology. *Climate Research* 46:15–27.

Pfeifer, M., M. Disney, T. Quaife, and R. Marchant. 2012. Terrestrial ecosystems from space: A review of Earth observation products for macroecology applications. *Global Ecology and Biogeography* 21:603–624.

Phillips, L. B., A. J. Hansen, and C. H. Flather. 2008. Evaluating the species energy relationship with the newest measures of ecosystem energy: NDVI versus MODIS primary production. *Remote Sensing of Environment* 112:3538, 4381–4392.

Phillips, L. B., A. J. Hansen, C. H. Flather, and J. Robison-Cox. 2010. Applying species-energy theory to conservation: A case study for North American birds. *Ecological Applications* 20:2007–2023.

Potts, S. G., J. C. Biesmeijer, C. Kremen, P. Neumann, O. Schweiger, and W. E. Kunin. 2010. Global pollinator declines: Trends, impacts and drivers. *Trends in Ecology and Evolution* 25:345–353.

Rasmussen, H. B., G. Wittemyer, and I. Douglas Hamilton. 2006. Predicting time specific changes in demographic processes using remote-sensing data. *Journal of Applied Ecology* 43:366–376.

Richardson, D. M. and R. J. Whittaker. 2010. Conservation biogeography—Foundations, concepts and challenges. *Diversity and Distributions* 16:313–320.

Rondinini, C., M. Di Marco, F. Chiozza, et al. 2011. Global habitat suitability models of terrestrial mammals. *Philosophical Transactions of the Royal Society—Biological Sciences* 366:2633–2641.

Rondinini, C., S. Stuart, and L. Boitani. 2005. Habitat suitability models and the shortfall in conservation planning for African vertebrates. *Conservation Biology* 19:1488–1497.

Sala, O. E. 2001. Price put on biodiversity. *Nature* 412:34–36.

Scepan, J., G. Menz, and M. C. Hansen. 1999. The DISCover validation image interpretation process. *Photogrammetric Engineering and Remote Sensing* 65:1075–1081.

Schaub, M., W. Kania, and U. Koppen. 2005. Variation of primary production during winter induces synchrony in survival rates in migratory white storks *Ciconia ciconia*. *Journal of Animal Ecology* 74:656–666.

Sellers, P. J., B. W. Meeson, F. G. Hall, et al. 1995. Remote sensing of the land surface for studies of global change models, algorithms, experiments. *Remote Sensing of Environment* 51:3–26.

Seoane, J., J. Bustamante, and R. Diaz-Delgado. 2004. Are existing vegetation maps adequate to predict bird distributions? *Ecological Modelling* 175:137–149.

Silvan-Cardenas, J. L., and L. Wang. 2010. Retrieval of subpixel *Tamarix* canopy cover from Landsat data along the Forgotten River using linear and nonlinear spectral mixture models. *Remote Sensing of Environment* 114:1777–1790.

Sims, D. A., A. F. Rahman, V. D. Cordova, et al. 2006. On the use of MODIS EVI to assess gross primary productivity of North American ecosystems. *Journal of Geophysical Research-Biogeosciences* 111:G04015.

Sims, D. A., A. F. Rahman, V. D. Cordova, et al. 2008. A new model of gross primary productivity for North American ecosystems based solely on the enhanced vegetation index and land surface temperature from MODIS. *Remote Sensing of Environment* 112:1633–1646.

Thuiller, W., M. B. Araujo, and S. Lavorel. 2004. Do we need land-cover data to model species distributions in Europe? *Journal of Biogeography* 31:353–361.

Tingley, R., and T. B. Herman. 2009. Land-cover data improve bioclimatic models for anurans and turtles at a regional scale. *Journal of Biogeography* 36:1656–1672.

Turner, W., S. Spector, N. Gardiner, M. Fladeland, E. Sterling, and M. Steininger. 2003. Remote sensing for biodiversity science and conservation. *Trends in Ecology and Evolution* 18:306–314.

Underwood, E., S. Ustin, and D. DiPietro. 2003. Mapping nonnative plants using hyperspectral imagery. *Remote Sensing of Environment* 86:150–161.

Ustin, S. L., D. A. Roberts, J. A. Gamon, G. P. Asner, and R. O. Green. 2004. Using imaging spectroscopy to study ecosystem processes and properties. *Bioscience* 54:523–534.

Viña, A., S. Bearer, H. M. Zhang, Z. Y. Ouyang, and J. G. Liu. 2008. Evaluating MODIS data for mapping wildlife habitat distribution. *Remote Sensing of Environment* 112:2160–2169.

Whittaker, R. J., M. B. Araujo, J. Paul, R. J. Ladle, J. E. M. Watson, and K. J. Willis. 2005. Conservation biogeography: Assessment and prospect. *Diversity and Distributions* 11:3–23.

Wiegand, T., J. Naves, M. F. Garbulsky, and N. Fernandez. 2008. Animal habitat quality and ecosystem functioning: Exploring seasonal patterns using NDVI. *Ecological Monographs* 78:87–103.

Wittemyer, G., H. B. Rasmussen, and I. Douglas-Hamilton. 2007. Breeding phenology in relation to NDVI variability in free-ranging African elephant. *Ecography* 30:42–50.

Wright, D. H. 1983. Species-energy theory—An extension of species-area theory. *Oikos* 41:496–506.

第九章 利用遥感对国家公园网络的生态系统服务功能多样性和碳保护战略进行评估

J. 卡贝洛　P. 劳伦　A. 雷耶斯　D. 阿尔卡拉斯-塞古拉

一、简介

保护实践面临着促进人类福祉的发展和保护生物多样性的挑战。截至目前，这两个目标仍然存在差异，但现在，生物保护学家正在推动向从更广阔的视角看待保护生物多样性的模式转变(Mace et al., 2012)。仅以生物多样性保护为基础的保护战略目前被认为是不切合实际的，其目标也无法实现，例如，欧洲和全球 2010 年生物多样性目标的失败就表明了这一点。这项在 2010 年底前防止生物多样性下降的战略已被 2020 年和 2050 年新的、更广泛的目标（《生物多样性公约》缔约方会议第十届会议，2010 年）所取代，其中还包括遏制生态系统服务退化，并对其进行充分评估，作为对人类福祉的重要贡献。这种新的保护目标视角提供了在当前全球变化的情况下促进可持续发展的可能性，同时又不忽视人类与生态系统的相互作用。

生态系统服务概念是指人类从生态系统中获得的利益(MA, 2005)，它代表了将生物多样性保护与人类福祉相结合的步骤方法的关键。尽管在过去的十年中，生态系统服务的科学取得了相当大的进步，但将其纳入保护实践仍然面临着重大挑战。首先，生物多样性和生态系统服务概念的复杂性导致了在生态系统评估中对生物多样性的混淆。正如梅斯等人(Mace et al., 2012)所报告的那样，迄今为止开发的方法在将生物多样性等同于生态系统服务（生态系统服务视角）

和将生物多样性本身视为生态系统服务(保护视角)之间交替进行。尽管生态系统服务视角(生物多样性等同于生态系统服务),意味着其中一个领域的管理会自动增强另一个领域的发展(例如,生态系统与生物多样性的经济学,或 TEEB; http:www.tee-bweb.org),保护视角(生物多样性作为一种生态系统服务)反映了生物多样性的内在价值。其次,必须对提供生态系统服务的地区进行测绘,以量化其提供的保护效益,并探索其与最大限度地提高生物多样性区域的一致性。保护视角强调生态系统管理和政策决策的必要性,重点将生物多样性和生态系统服务的优先级相结合(Chan et al.,2006;Turner et al.,2007;Naidoo et al.,2008;Egoh et al.,2009)。然而,一些研究表明,专为保护生物多样性而设计的保护优先级可能无法保持最佳的生态系统服务水平,反之亦然(Naidoo et al.,2008)。因此,生物多样性和生态系统服务之间的权衡和协同效应是管理和政策决策的关键。虽然权衡通常意味着对生物多样性保护的强烈限制,但协同效应可以为自然保护和人类福祉提供重要的"双赢"机会(Cowling et al.,2008;Egoh et al.,2009;Strassburg et al.,2009;de Groot et al.,2010;Reyers et al.,2012)。

生态系统服务领域的科学发展导致将净初级生产力视为支持生态系统服务交付的基本生态过程(Richmond et al.,2007;Egoh et al.,2008;Fisher et al.,2009)。特别是,人们对全球气候变化的关注越来越集中在净初级生产力的作用上。光合作用固定的二氧化碳与自养呼吸损失的二氧化碳之间的差异是碳封存服务中的最重要组成部分之一。因此,在生物多样性和气候变化危机的背景下,分析生物多样性和碳保护成效之间的一致性成为管理者最重要的优先级之一(Harvey et al.,2009;Thomas et al.,2013)。目前需要在全球范围内实施共同协议,以制定相关的生物多样性和碳保护战略(Harvey et al.,2009)。

卫星影像是在指定生态系统尺度上量化生物多样性和碳封存的区域模式的重要工具(Cabello et al.,2012)。遥感工具在光谱植被指数如归一化差异植被指数和增强植被指数的获得上得到成功应用。这些指数是植被截获的光合有效辐射比例的线性估计(Wang et al.,2004),是碳收益的主要控制因素(Monteith,1981)(见第六章)。这些指数不仅可以估计生态系统的碳收益(Huete et al.,2002),而且还可以得出与关键生态过程密切相关的功能属性,例如初级生产(Ruimy et al.,1994)、季节性(Potter and Brooks,1998)和物候学(Reed et al.,1994)。这些功能属性已成功地用于识别生态系统功能类型(Ecosystem Functional Types，EFTs)(Paruelo et al.，2001；Alcaraz-Segura et al.，2006，2010)。生态系统功能类型(EFTs)是一组具有功能多样性特征的生态系统，这些功能多样性特征与生物群和大气之间物质与能量交换动力学相关(例如初级生产、从植物到动物的营养转移、营养循环、水动力学和热传递)。生态系统功能类型为环境因素提供了特定协调反应，并被用作景观尺度的生物实体，以量化区域多样性(Alcaraz-Segura et al.，2013)。

现有的保护区网络可以通过将生物多样性以外的生态系统服务(如碳收益)纳入其设计和管理中，为公众获得更多的附加值(Eigenbrod et al.，2009；Sutherland et al.，2009)。在实践中，虽然经常评估保护区网络在最大化生物多样性方面的有效性(Rodrigues et al.，2004；Langhammer et al.，2007)，但很少有案例真正将其他生态系统服务纳入此类评估。本项研究分析了当前国家公园网络如何在国家层面上保护生态系统功能的多样性和碳收益。我们选择国家公园作为研究案例，因为它们是代表一个国家主要自然系统的最佳范例。我们首先绘制了每个国家的生态系统功能类型空间异质性图。然后，我们根据其生态系统功能类型的代表性(以及相反的保护差距)和稀有性以及碳收益，估计了公园保护多样性的最大化贡献。

最后，我们分析了整个网络中生态系统功能多样性和碳收益之间的空间一致性，以找出葡萄牙、西班牙和摩洛哥多样性和碳保护战略之间的权衡或协同效应(表9-1)。这些国家有一些相同的生物多样性和环境特征，但保护政策不同。此外，尽管葡萄牙和西班牙有着悠久的自然保护历史，并积极响应欧洲的保护政策，但在摩洛哥，关于生物多样性和生态系统服务的先验知识仍然匮乏。

表9-1 葡萄牙、西班牙和摩洛哥的国家公园网络

国家	代码	国家公园	年份	环境特征	IUCN[a]
葡萄牙	1	佩内达—格雷斯	1971	地中海和欧洲西伯利亚过渡区(主要是橡树林、荒原)	II
西班牙	2	欧罗巴山	1918/1995	大西洋森林(温带落叶森林、欧洲西伯利亚草原、山地和亚高山灌木林)	II
	3	奥尔德萨	1918/1982	高山针叶林和欧洲西伯利亚草原	II
	4	艾格斯托尔特斯	1955/1996	高山针叶林和欧洲西伯利亚草原	II

续表

国家	代码	国家公园	年份	环境特征	IUCN[a]
西班牙	5	蒙弗拉圭	2007	地中海橡树林和灌木丛	V
	6	加贝内罗斯	1995	有地中海森林和灌木丛的广阔低地	II
	7	多尼亚纳	1969/2004	盐沼、海岸沙丘、马基群落和地中海常绿矮灌丛	II
	8	内华达山脉	1999	地中海橡树林、灌木丛和草原与地中海中高山脉相连	V
摩洛哥	9	塔拉斯曼塔	2004	摩洛哥冷杉林(冷杉)、阿特拉斯雪松林和黑松林	NR
	10	胡塞马	2004	马基群落和地中海常绿矮灌丛,高海岸悬崖	NR
	11	塔兹卡	1950	橡树林和阿特拉斯雪松林	NR
	12	伊芙兰	2004	阿特拉斯雪松林和混交林	V
	13	海尼夫拉	2008	阿特拉斯雪松林	NR
	14	大阿特拉斯山	2004	阿特拉斯雪松林、橡树林和杜松林	NR
	15	图卜卡勒	1942	从地中海以下到北非最高峰的强烈海拔梯度,橡树林和雪松林,灌木丛和草原	V
	16	苏斯一马塞	1991	海岸沙丘和湿地	NA
	17	伊里奇	1994	沙漠植被(阿沙西娅瑞蒂安娜干草原和稀树草原)、临时湿地和沙丘	NR
	18	赫尼菲斯	2006	湿地、盐滩、海岸沙丘和沙漠	NA

资料来源:世界保护区数据库。http://www.wdpa.org。

注:NA=不适用;NR=未报告。

a 根据世界保护区数据库成立的国际自然保护联盟。

二、方法

(一) 从卫星衍生的增强植被指数中识别生态系统功能类型及量化碳收益

生态系统功能类型的识别和碳收益的量化基于从中分辨率成像光谱仪(MODIS)影像获得的增强植被指数。

我们使用了 2001~2010 年的 MOD13C1 产品,该产品生成全球影像最长需要 16 天,其空间分辨率为 0.05°×0.05°(赤道约 25 平方千米)。光谱植被指数的时间动态已被广泛用于表征生态系统的初级生产力、季节性和物候学(Pettorelli et al.,2005)。按照阿尔卡拉斯-塞古拉等人(Alcaraz-Segura et al.,2013)解释的方法,我们使用以下三个属性来识别生态系统功能类型。首先,利用年均增强植被指数作为年度初级生产力的线性估计量,这是生态系统功能最综合的描述之一(Virginia and Wall, 2001)。年均增强植被指数也被用作碳收益量化指标。其次,增强植被指数季节变化系数(EVI-sCV)被用来描述季节性或生长季节与非生长季节碳收益之间的差异。第三,最大增强植被指数日期(Date of Maximum EVI,DMAX)被用作生长季节的物候指标(Paruelo et al.,2001; Pettorelli et al.,2005; Alcaraz-Segura et al.,2006)。众所周知,这三个增强植被指数指标可捕捉增强植被指数时间序列中的大部分变化(Alcaraz-Segura et al.,2006,2011)。

使用阿尔卡拉斯-塞古拉等人(Alcaraz-Segura et al.,2013)提出的方法,将三个增强植被指数指标的范围划分为四个间期,可能导致 $4×4×4=64$ 个生态系统功能类型。对于最大增强植被指数日期,这四个间期与温带生态系统一年中的四个季节相一致。对于年均增强植被指数和增强植被指数季节变化系数,我们计算了每年直方图的第一、第二和第三个四分位数。然后将间歇期之间的限制设置为每项指标的每四分位数的 10 年中位数。按照帕鲁埃洛等人(Paruelo et al.,2001)基于两个字母和一个数字提出的术语,为每个生态系统功能类型分配三个字符的代码。年均增强植被指数代码的第一个字母(大写)从"A"到"D",表示生产率从低到高(不断提高)。第二个字母(小写)显示增强植被指数季节变异系数,从"a"到"d"表示季节性从高到低(递减)。这些数字指的是增强植被指数最大值的季节（1～4：春、夏、秋、冬）。生态系统功能类型的定义和编码，仅基于生态系统功能的描述，并可以用于生态解释。一旦在每个指标的四个间期中设定了限制，我们将应用这些限制每年生成十张生态系统功能类型地图（2001～2010 年）。

由于多年来生态系统功能类型图的变化较小，最后我们利用年际中位数对 2001～2010 年生态系统功能异质性进行了总结。

（二）国家公园网络功能多样性评估

在每个国家，我们的研究仅限于自然区域（荒地）的生态系统功能类型。这一限制是基于人类影响指数（HIi）（1 千米×1 千米空间分辨率的数据集），该指数集通过整合人类存在的八个指标来评估人类对生态系统的直接影响（Sanderson et al.，2002）。人类影响指数的范围是 0～64。0 代表无人为影响，64 代表最强人为影响。因此，我们的分析只使用平均人为影响指数低于 20 的 MODIS 像素。自然生态系统的阈值为人类影响指数低于 20，因为这是在国家公园中发现的最大人类影响指数。此外，我们使用正射影像（orthophotos）和谷歌地球（Google Earth）对像素选择的自然性进行了双重检查。在每个国家公园中，我们仅使用在国家公园范围内面积超过 75% 的 MODIS 像素。

功能分析基于丰富性、代表性和稀有性。这些标准在保护生物学中被广泛应用，以量化国家保护区最大限度地提高国家生物多样性的能力。每个网络在代表各国生态系统功能类型多样性方面的有效性（代表性）被评估为国家公园网络中生态系统功能类型数量与该国所有自然区域中生态系统功能类型总数的简单比率（Cabello et al.，2008）。因此，我们还确定了国家层面生态系统功能类型保护方面的差距。每个国家公园中生态系统功能类型组成的稀有性（奇异性或独特性）被计算为公园中生态系统功能类型相对稀有性的总和。首先，根据以下等式估算每种生态系统功能类型的绝对稀有性：

$$\text{Rarity EFT}_i = (\text{AEFT}_{max} - \text{AEFT}_i)/\text{AEFT}_{max} \qquad 式 9-1$$

其中，AEFT_{max} 是最丰富的 EFT（生态系统功能类型）所占据的区域，AEFT_i 是被评估的 EFT（生态系统功能类型）的区域。该指数在 0 到 1 的范围内对稀有性进行从低到高分级。然后求出每个国家所有生态系统功能类型的绝对稀有性之和,并计算每个生态系统功能类型的相对稀有性占整个国家的百分比(Rarity EFTi [%] = Rarity EFTi ×100/Σ Rarity EFTs)。此外,为了将公园视为稀有或普通,每个国家的所有自然像素都被绘制在由年均增强植被指数和增强植被指数季节变异系数定义的功能空间中。

(三) 网络中功能多样性与碳收益的一致性

为了探讨网络中功能多样性与碳收益之间的空间一致性,我们绘制了各国国家公园的生态系统功能类型丰富度(生态系统功能多样性估计值)与年均增强植被指数(碳收益估计值)之间的成对关联图。通过该项分析,我们能够确定哪些公园最能代表生物多样性和保护战略的"双赢"机会,以及哪些公园只是最大限度地达到了这两种战略中的一种。

三、结果

(一) 葡萄牙、西班牙和摩洛哥自然区域生态系统功能类型的空间格局

图 9-1 中的生态系统功能类型图综合描述了 2001~2010 年葡萄牙、西班牙和摩洛哥自然区域(荒地)生态系统功能空间格局的综合特征。

在葡萄牙,年均增强植被指数、增强植被指数季节变异系数和增强植被指数最大值日期的 64 种组合中,有 58 种组合被发现(图 9-2 和表 9-2),且大多数的生态系统功能类型同样丰富。最丰富的生态系统功能类型呈现为春季和夏季最大值。高产生态系统位于该国西北象限地中海和欧洲西伯利亚生物地理区域间的过渡地带(从洛萨山和埃什特雷拉山脉向北),具有较低的季节性,春季和夏季达到极大值。生产力较低的生态系统功能类型位于地中海地区的东北部(杜罗河流域)和东南部(瓜迪亚纳河流域)。这些地区的季节性随秋季极大值的增加而降低至中等,随冬季极大值的增加而升高。几乎所有可能的生产力与季节性组合都存在于冬季和春季最大值的生态系统中。

在 64 种可能的生态系统功能类型组合中,西班牙有 63 种(图 9-2 和表 9-2),尽管春季和夏季最大值的生态系统功能类型更为丰富。在欧洲西伯利亚(北部和西北部)和地中海(中部、南部和东部)的生物地理区域之间形成明显的对比

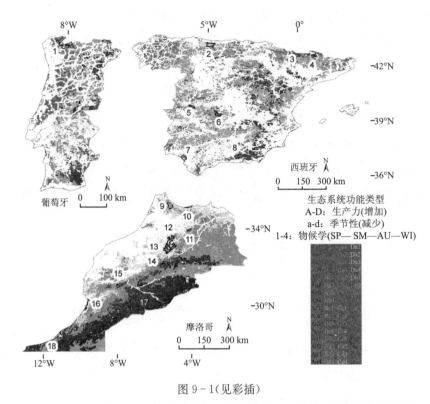

图 9-1（见彩插）

注：葡萄牙、西班牙和摩洛哥生态系统功能类型的空间格局。生态系统功能类型按国家分类。因此，用于确定生产力、季节性和最大类别的数值范围因国家而异。

(Alcaraz-Segura et al., 2006)。总体而言，欧洲西伯利亚地区和地中海山区的生产力较高。在与主要河流相关的流域，生产力较低，例如，东部的埃布罗、北部的杜埃罗及拉曼查内陆平原。东南部的半干旱地区生产力最低。季节性在欧洲西伯利亚山脉（比利牛斯山脉和欧罗巴山）的高峰地区较大，在地中海地区、河流流域、湿地（多尼亚纳、埃布罗河三角洲和瓦伦西亚阿尔布费拉）和伊比利亚东南部的半干旱地区最大。全国大部分地区在春季和夏季出现增强植被指数最大值。欧洲西伯利亚生态系统的特征是夏季出现明显的最大值。在地中海地区，高山、湿地和水域（河岸地区）也出现了夏季最大值。西班牙其他地区有两种主要的物候模式。地中海山脉在秋季至初冬达到最大值，而半干旱地区、河流流域和大陆平原则有春季最大值。

第九章 利用遥感对国家公园网络的生态系统服务功能多样性和碳保护战略……

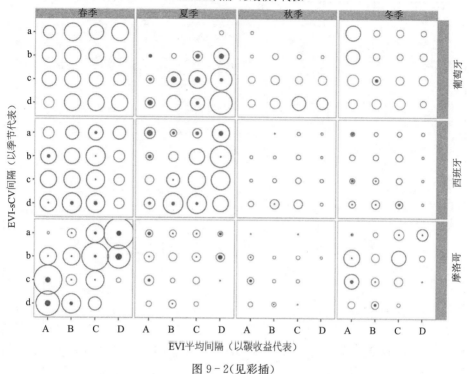

图 9-2（见彩插）

注：基于生态系统功能类型分类，对葡萄牙、西班牙和摩洛哥国家公园网络进行差距分析。生态系统功能类型按国家分类。因此，用于确定碳收益（从"A"增加至"D"）、季节性（从"a"减少至"d"）和物候期的数值范围因国家而异。外圆表示每个生态系统功能类型对应的自然区域所占国家的百分比（即人类影响指数小于20）。内实心圆表示每个生态系统功能类型所占用的国家公园网络的百分比。

表 9-2 基于生态系统功能多样性和碳收益的国家公园网络保护战略评价

国家	国家公园	保护战略				双赢公园[a]
		代表性（生态系统功能类型数量）	稀有程度		碳收益（增强植被指数平均值）	
			平均值	累积(%)		
葡萄牙	网络	10/58[b] (17.2%)	0.789	18.4	1/1[c] (100%)	—
	佩内达—格雷斯	10	0.789±0.201	18.4	0.264±0.034	

续表

国家	国家公园	保护战略				双赢公园[a]
		代表性(生态系统功能类型数量)	稀有程度		碳收益(增强植被指数平均值)	
			平均值	累积(%)		
西班牙	网络	24/63 (38.1%)	0.596±0.254	44.8	6/7(86%)	4/7
	艾格斯托尔特斯	2	0.833±0.094	3.6	0.162±0.028	
	加贝内罗斯	6	0.361±0.302	4.7	0.270±0.015	
	多尼亚纳	8	0.599±0.266	9.0	0.166±0.042	
	蒙弗拉圭	2	0.178±0.042	0.8	0.245±0.013	
	奥尔德萨	3	0.841±0.075	5.5	0.170±0.091	
	欧罗巴山	6	0.562±0.351	7.3	0.288±0.066	
	内华达山脉	8	0.798±0.297	13.9	0.166 8±0.062	
摩洛哥	网络	31/57 (54.4%)	0.613±0.202	66.3	5/10(50%)	1*/10
	胡塞马	4	0.559±0.434	4.6	0.159	
	大阿特拉斯山	6	0.687±0.314	8.4	0.098	
	伊夫兰	3	0.456±0.467	2.9	0.219	
	伊里基	3	0.422±0.346	2.6	0.060	
	赫尼菲斯	15	0.784±0.256	24.0	0.073	
	海尼夫拉	3	0.783±0.303	4.8	0.224	
	苏斯—马塞	4	0.822±0.193	6.7	0.064	
	塔拉斯曼塔	4	0.587±0.463	4.8	0.277	
	塔兹卡	2	0.217±0.307	0.9	0.194	
	图卜卡勒	4	0.810±0.171	6.6	0.074	

注：a显示高功能多样性(生态系统功能类型)和碳收益的公园数量；
b国家公园网络中包含的生态系统功能类型数量超过该自然区域的生态系统功能类型总数；
c显示生态系统功能类型平均值超过国家中位数的公园数量。

在64种可能的生态系统功能类型组合中,摩洛哥有57种(图9-2和表9-2)。摩洛哥的生产力低于葡萄牙或西班牙。在生产力较高的北部和西北部地区(受地中海气候和大西洋影响)与生产力较低的南部和东南部地区(受撒哈拉沙漠气候影响)之间形成明显的对比。最高的年均增强植被指数出现在以下三个地区,分别是靠近大西洋海岸(马莫拉森林)附近的一些地区、阿特拉斯山脉北部地区以及出现夏季最大值的塔拉瑟姆塔讷国家公园(这三个地区都位于地中海最西部的林地生态区)。最低年均增强植被指数出现在瓦尔扎扎特省、塔塔省和坦坦省(北撒哈拉草原生态区)的阿特拉斯山脉以南。北部和西北部地区具有高季节性和春季最大值的特征,南部和东南部地区具有低季节性和冬季与春季最大值的特征。增强植被指数最大值的生态系统功能类型在冬季与春季最丰富。

(二) 国家公园网络的代表性和稀有性

国家公园网络的代表性和稀有性在这三个国家之间有所不同(表9-2)。在葡萄牙公园网络中只有一个国家公园,很少有机会将葡萄牙的保护战略与其他国家进行比较。然而,该网络在国家生态系统功能多样性中的代表性(17.2%)和累积稀有性(18.4%)最低(表9-2)。

因此,尽管包括地中海和欧洲西伯利亚地区橡树林和荒原的佩内达-格雷斯公园足够大,且足以体现该国生态系统功能类型异质性的重要部分,并显示出相对较高的平均稀有性(0.789±0.201),但该国在生态系统保护方面仍存在许多差距(图9-2)。该公园几乎占该国稀有程度的20%;然而,该公园的大多数像素显示了葡萄牙常见的年均增强植被指数和增强植被指数季节变异系数的功能组合(图9-3)。该公园只包括六个生态系统功能类型夏季最大值和一个生态系统功能类型冬季最大值,而与季节性或春季或秋季最大值相对应的生态系统功能类型必须在其他保护区中出现,以满足更多的生物多样性优先级(图9-2)。

西班牙网络涵盖了63个生态系统功能类型中的24个(表9-2和图9-2),更倾向于包含稀有生态系统(占累积稀有性的44.8%),而非代表国家功能异质性(占代表性的38.1%)。内华达山脉公园和多尼亚纳公园是对网络代表性贡献最大的两个公园,前者沿海拔梯度保护着从森林到草原的多种地中海生态系统,后者拥有盐沼、海岸和非森林生态系统,每个生态系统包括8个生态系统功能类型。欧罗巴山和加贝内罗斯、大西洋和地中海森林分别有6个生态系统功能类型,功能多样性略低。奥尔德萨拥有3个生态系统功能类型,特别是艾格斯托尔特斯和蒙弗拉圭,通常只有两个生态系统功能类型的森林地区,就生态系统功能类型丰富度而言,是最差的公园。该网络的平均稀有性为0.596,为了达到保护目标,公园之间的差异被认为是重要的(±0.254)。该网络占国家自然生态系统功能稀有性的44.8%。稀有性与低生产力(年均增强植被指数)生态系统以及中等季节性(增强植被指数季节变异系数)有关(图9-3)。虽然平均稀有性最高的公园为艾格斯托尔特斯(0.833 ± 0.094)和奥尔德萨(0.841 ± 0.075),但累积稀有性最高的公园为内华达山脉(13.9%)和多尼亚纳(9%)。在该网络中,生态系统功能类型保护的差距是指具有秋季增强植被指数最大值的生态系统(图9-2)。

摩洛哥网络覆盖了大部分生态系统功能类型(57个生态系统功能类型中有31个)(表9-2和图9-2)。在该网络中,与功能多样性相关的代表性和积累稀有性占比最高,即具有54.4%的代表性和66.3%的积累稀有性。赫尼菲斯公园拥有湿地、盐滩、海岸沙丘以及沙漠,同时拥有15个生态系统功能类型,是生态系统功能类型最丰富的公园。其他的包括从大阿特拉斯山的6个生态系统功能类型到塔兹卡的2个生态系统功能类型。虽然该网络的稀有性(0.613 ± 0.202)与西班牙相似,但其累积稀有性却是三个国家中最高的(66.3%)。累积稀有性最高的是赫尼菲斯公园(24%),最低的是塔兹卡地区。总体而言,稀有性与生产力极低和中、高季节性的生态系统有关(图9-3)。

然而,在赫尼菲斯,高稀有性与物候有关,因为其像素与生产力和季节性的共同组合有关(图9-3),但有四个物候类别显示差异。在该网络中,差距主要出现在夏季和秋季增强植被指数最大的生态系统中(图9-2)。

(三) 国家公园网络的碳收益

佩内达—格雷斯是葡萄牙唯一的国家公园,其生产力非常高,平均碳收益高于该国的中位数(表9-2和图9-4)。在西班牙,86%的公园的碳收益高于该国的中位数(表9-2和图9-4)。生态系统最具生产力的国家公园(年均增强植被指数高于该国的中位数)是欧罗巴山、加贝内罗斯以及蒙弗拉圭(拥有广阔的森林)和多尼亚纳(拥有与沙丘和湿地相连的灌木丛和植被)。奥尔德萨和内华达山脉处于该国碳收益的中位数,而艾格斯托尔特斯(拥有森林、灌木丛和山地湿

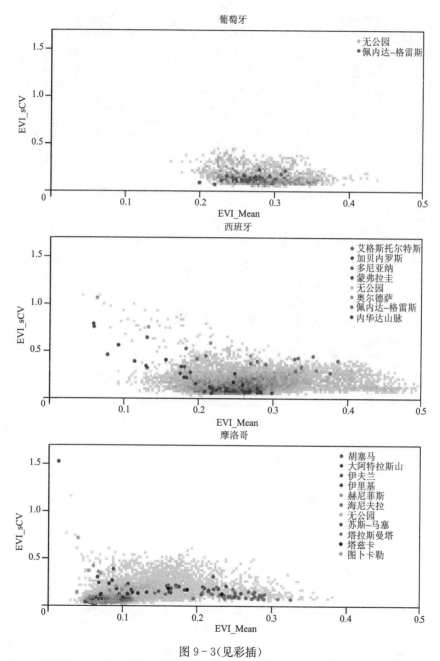

图 9-3（见彩插）

注：通过年均增强植被指数作为碳收益和增强植被指数空间变异系数的估计值进行功能表征，增强植被指数空间变异系数是葡萄牙、西班牙和摩洛哥及其国家公园自然植被季节性的描述。

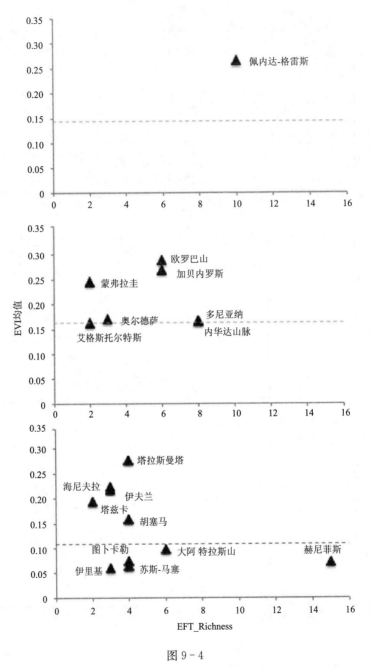

图 9-4

注：国家公园网络功能多样性与碳收益之间的空间一致性分析。虚线表示每个国家增强植被指数平均值的中位数。

地)的增强植被指数平均值最低。另一方面,摩洛哥网络中50%的公园碳收益高于国家中位数(表9-2和图9-4)。生产力最高的公园(增强植被指数平均值高于国家中位数)是塔拉斯曼塔、海尼夫拉、伊夫兰和塔兹卡,它们拥有典型的地中海针叶林(冷杉、雪松和松树林),以及主要拥有地中海灌木丛的胡塞马。生产力最低的国家公园是伊里基、苏斯-马塞、赫尼菲斯、图卜卡勒和大阿特拉斯山,它们都位于撒哈拉沙漠的生物地理区域,主要拥有灌木丛、湿地和沙漠植被。总体而言,摩洛哥国家公园网络对生产性生态系统的重视程度低于葡萄牙和西班牙。

(四) 国家公园网络多样性与碳收益的空间一致性

这三个国家的生态系统功能类型丰富度和碳收益之间的一致性存在差异。在葡萄牙,佩内达-格雷斯有10个生态系统功能类型,与其他两个国家的结果相比,这是一个很高的数字,而且还显示出了高碳收益。在西班牙网络中,内华达山脉和多尼亚纳是生态系统功能类型丰富度最高的两个公园(拥有8个生态系统功能类型),也属于中低碳收益地区,它们的碳收益高于全国的中位数(表9-2和图9-4)。同样拥有相对较高的生态系统功能类型丰富度(拥有6个生态系统功能类型)的欧罗巴山和加贝内罗斯的碳收益最高。蒙弗拉圭的情况并非如此,虽然其碳收益相对较高,但却只有2个生态系统功能类型。艾格斯托尔特斯和奥尔德萨这两个公园的生态系统功能类型(分别为2个和3个)丰富度较低同时碳收益也相对较低。最后,在摩洛哥大阿特拉斯山,特别是赫尼菲斯公园,它们的生态系统功能类型丰富度相对较高(分别为6个和15个),但它们的碳收益却低于该国的中位数。其余的公园,均少于4个生态系统功能类型,其碳收益变化很大。

四、讨论

全球变化和生态过程与人类福祉之间的联系强调了将生态系统服务纳入保护目标的必要性。为此,寻求保护工作和生态系统服务之间的空间一致性已成为一项新的挑战。我们现在需要大量的关于生物多样性和生态系统服务如何与现有保护区网络相关联的知识。任何此类评估都必须对保护区的生物多样性和生态系统服务水平进行评估。然而,由于生物多样性种类繁多和可用的环境信息数量大以及与生态系统服务水平制图相关的错误,导致这项工作可能比预期的更为缓慢,也会更为复杂。如前所述,可以通过卫星影像中获得的光谱信息转换为生态系统功能变量来克服这些困难。首先(通过识别生态系统功能类型),景观多样性测量值可以被估计出来并作为多样性指标,因为它们对环境控制的反应与物种多样性相同(Alcaraz-Segura et al.,2013)(见第八章)。其次,卫星获取的信息最大限度地减少了与大区域生态系统服务空间变异性量化相关的误差(Cabello et al.,2012)。在这一基础上,我们发现了三个不同国家的每个保护区的碳收益水平,以此作为一个必须纳入新保护模式的重要例子。我们直接根据各部分陆地区域的光谱特征计算其空间变异性,避免对不同的土地覆盖一概而论(Eigenbrod et al.,2010)。

根据葡萄牙、西班牙和摩洛哥的区域生态系统功能类型模式,在这些国家的国家公园网络中,我们确定了与生物多样性和生态系统服务优先级相关的三个保护目标(代表性、稀有性和碳收益)的差距、偏差和空间一致性。所研究的国家公园在不同程度上符合这些目标,而且在任何情况下,就生态系统功能多样性和碳收益服务而言,都不是双赢公园的明智选择。在葡萄牙,佩内达—格雷斯是该国唯一的公园,对基于代表性(17.2%)和稀有性(18.4%)保护战略的响应不明显。如果该公园符合碳保护战略,就可以被视为双赢公园(表9-2和图9-4)。然而,该网络还应包括其他保护区,以填补在功能多样性方面发现的众多空白(图9-2)。从碳收益的角度来看,高生产力生态系统在这个国家有很好的代表性(图9-1)。然而,完全基于碳收益的保护战略可能不包括重要的生物多样性特征(Naidoo et al.,2008),例如分布在该国一半以上地区的中低生产力生态系统(图9-1)。

西班牙国家公园网络在不同层面上对这三项保护战略作出了回应,它是最符合碳保护战略的网络,可能是因为它主要包括森林生态系统;其80%的公园的碳收益高于本国的中位数。与生态系统功能类型的代表性相比,该网络更倾向于稀有性(表9-2和图9-3)。内华达山脉公园有八个生态系统功能类型(表9-2),以及在生产力(年均增强植被指数)、季节性(增强植被指数季节变异系数)(图9-3)和物候学方面具有罕见的生态系统功能区域,在丰富性、代表性和稀有性方面为该国生态系统功能多样性做出了重大贡献。就生态系统功能多样性和碳收益而言,西班牙的系统是最明显的双赢候选公园网络(图9-4)。欧罗巴山和加贝内罗斯公园的碳收益最高,多尼亚纳和内华达山脉的生态系统功能多样性最高。尽管如此,这里似乎出现了一种权衡,因为功能多样性或生物多样性(隐性)的增加,对应于多尼亚纳和内华达山脉不同类型植被的镶嵌,与欧罗巴山与加贝内罗斯森林相比,可能会导致碳收益减少。

在摩洛哥,国家公园网络具有最高的生态系统功能多样性(53%的代表性和49.5%的累积稀有性)(表9-2)和碳收益(50%的公园)标准。特别是赫尼菲斯公园,包含湿地、盐滩、海岸沙丘和沙漠,以及与之相关的高度多样化的动物群(http:ma.chm-cbd.net)。该公园的生态系统功能类型丰富度最高,碳收益最低(图9-4),是温带和干旱地区生物多样性和碳收益之间缺乏空间一致性的一个例子。在热带地区以外,高生态系统功能多样性(以及隐性生物多样性)与独特的环境特征或生态交错带环境有关(Alcaraz-Segura et al., 2013),最大限度地降低了森林地区对生物多样性的影响,而森林地区通常具有高碳收益。此外,塔拉斯曼塔包括最具代表性的摩洛哥针叶林(冷杉、雪松和松树),拥有四个生态系统功能类型,其生态系统功能类型丰富度为全国中位数,可以被视为最接近双赢方案的公园。

这项对国家公园网络的评估提升了当前保护工作的价值。它还提供了当前气候变化减缓战略(如碳封存)如何影响保护区的信息(Thomas et al., 2012)。然而,为了设计双赢网络,需要对多个生态系统服务进行更广泛的分析(Bennett et al., 2009)。正如我们所展示的,在非热带国家,高生物多样性可能集中在非森林生态系统中,或者根本没有任何双赢的地区。因此,在仅限碳保护的战略下,生物多样性可能面临风险(CBD-UNEP, 2011)。此外,如国家公园等保护区可以为管理全球变化和人类福祉提供其他生态系统服务(除生物多样性和碳封存)的庇护。例如,在摩洛哥,为保护水资源而建立了一些山地公园,而另一些公园的建立则为科学研究提供了场所,并在实践中发展了生态旅游(http://ma.chm-cbd.net)。考虑到生态系统服务评估的多维性(见第二十章),从生物多样性、碳或其他服务的角度来看,保护政策应该推动有助于人类福祉的保护区的发展。

五、结论

正如我们所展示的,在考虑生态系统服务保护目标的新保护模式下,遥感工具可用于分析当前国家公园网络。将光谱信息转化为植被指数使我们能够量化国家生物多样性指标和关键生态系统服务，如碳收益。基于利用光谱植被指数得出的生态系统功能特征，我们确定了葡萄牙、西班牙和摩洛哥国家公园网络中的代表性、稀有性和碳收益之间的偏差、协同作用和权衡。在所有情况下，这些网络都会显示出保护方面的差距，这对葡萄牙尤为重要。此外，生态系统功能多样性和碳保护战略之间缺乏明确的协同作用，这需要考虑整个网络范围内的双赢方案，而不仅仅是针对个别公园。这一战略在热带地区以外可能特别重要，因为在那些地区，高生物多样性和其他生态系统服务可能与非森林地区有关。

致谢

来自匿名审稿人的意见提升了最终文稿的质量。资助来自于ERDF（FEDER）、西班牙-外国边境跨境合作计划（POXTEFEX-Transhabitat）、安达卢西亚地区政府（Junta de Andalucía GLOCHARID and SEGALERT Projects，P09-RNM-5048）和科学与创新部（Project CGL2010-22314）。我们感谢黛博拉·富尔道尔对于英文的修订。

参考文献

Alcaraz-Segura, D., E. H. Berbery, S. J. Lee, and J. M. Paruelo. 2011. Use of ecosystem functional types to represent the interannual variability of vegetation biophysical properties in regional models. *CLIVAR Exchanges* 17:23–27.

Alcaraz-Segura, D., E. Liras, S. Tabik, J. Paruelo, and J. Cabello. 2010. Evaluating the consistency of the 1982–1999 NDVI trends in the Iberian Peninsula across four time-series derived from the AVHRR Sensor: LTDR, GIMMS, FASIR, and PAL-II. *Sensors* 10:1291–1314.

Alcaraz-Segura, D., J. M. Paruelo, and J. Cabello. 2006. Identification of current ecosystem functional types in the Iberian Peninsula. *Global Ecology and Biogeography* 15:200–212.

Alcaraz-Segura, D., J. M. Paruelo, H. E. Epstein, and J. Cabello. 2013. Environmental and human controls of ecosystem functional diversity in temperate South America. *Remote Sensing* 5:127–154.

Bennett, E. M., G. D. Peterson, and L. J. Gordon. 2009. Understanding relationships among multiple ecosystem services. *Ecology Letters* 12:1394–1404.

Cabello, J., D. Alcaraz-Segura, A. Altesor, M. Delibes, and E. Liras. 2008. Funcionamiento ecosistémico y evaluación de prioridades geográficas en conservación [Ecosystem functioning and assessment of conservation geographical priorities]. *Ecosistemas* 17:53–63.

Cabello, J., N. Fernández, D. Alcaraz, et al. 2012. The ecosystem functioning dimension in conservation biology: Insights from remote sensing. *Biodiversity and Conservation* 21:3287–3305.

CBD–UNEP (Convention on Biological Diversity–United Nations Environment Programme). 2011. *REDD-plus and biodiversity.* CBD Technical Series No. 59. Montreal, Canada: CBD–UNEP.

Chan, K. M. A., M. R. Shaw, D. R. Cameron, E. C. Underwood, and G. C. Daily. 2006. Conservation planning for ecosystem services. *PLoS Biology* 4:14.

Convention on Biological Diversity, Conference of the Parties, 2010. The strategic Plan for Biodiversity 2011–2020 and the Aichi Biodiversity Targets.

Cowling, R. M., B. Egoh, A. T. Knight, et al. 2008. An operational model for mainstreaming ecosystem services for implementation. *Proceedings of the National Academy of Sciences of the United States of America* 105:9483–9488.

de Groot, R. S., R. Alkemade, L. Braat, L. Hein, and L. Willemen. 2010. Challenges in integrating the concept of ecosystem services and values in landscape planning, management and decision making. *Ecological Complexity* 7:260–272.

Egoh, B., B. Reyers, M. Rouget, M. Bode, and D. M. Richardson. 2009. Spatial congruence between biodiversity and ecosystem services in South Africa. *Biological Conservation* 142:553–562.

Egoh, B., B. Reyers, M. Rouget, D. M. Richardson, D. C. Le Maitre, and A. S. van Jaarsveld. 2008. Mapping ecosystem services for planning and management. *Agriculture, Ecosystems and Environment* 127:135–140.

Eigenbrod, F., B. J. Anderson, P. R. Armsworth, et al. 2009. Ecosystem service benefits of contrasting conservation strategies in a human-dominated region. *Proceedings of the Royal Society—Biological Sciences* 276:8.

Eigenbrod, F., P. R. Armsworth, B. J. Anderson, et al. 2010. The impact of proxy-based methods on mapping the distribution of ecosystem services. *Journal of Applied Ecology* 47:8.

Fisher, B., R. K. Turner, and P. Morling. 2009. Defining and classifying ecosystem services for decision making. *Ecological Economics* 68:643–653.

Harvey, C. A., B. Dickson, and C. Kormos. 2009. Opportunities for achieving biodiversity conservation through REDD. *Conservation Letters* 3:53–61.

Huete, A., K. Didan, T. Miura, E. P. Rodriguez, X. Gao, and L. G. Ferreira. 2002. Overview of the radiometric and biophysical performance of the MODIS vegetation indices. *Remote Sensing of Environment* 83:195–213.

Langhammer, P. F., M. I. Bakarr, L. A. Bennun, et al. 2007. *Identification and gap analysis of key biodiversity areas: Targets for comprehensive protected area systems.* Gland, Switzerland: IUCN.

MA (Millennium Ecosystem Assessment). 2005. Ecosystems and human well-being: Desertification Synthesis. Washington, DC: Island Press.

Mace, G. M., K. Norris, and A. H. Fitter. 2012. Biodiversity and ecosystem services: A multilayered relationship. *Trends in Ecology and Evolution* 27:19–26.

Monteith, J. L. 1981. Climatic variation and the growth of crops. *Quarterly Journal of the Royal Meteorological Society* 107:749–774.

Naidoo, R., A. Balmford, R. Costanza, et al. 2008. Global mapping of ecosystem services and conservation priorities. *Proceedings of the National Academy of Sciences of the United States of America* 105:5.

Paruelo, J. M., E. G. Jobbagy, and O. E. Sala. 2001. Current distribution of ecosystem functional types in temperate South America. *Ecosystems* 4:683–698.

Pettorelli, N., J. O. Vik, A. Mysterud, J. M. Gaillard, C. J. Tucker, and N. C. Stenseth. 2005. Using the satellite-derived NDVI to assess ecological responses to environmental change. *Trends in Ecology and Evolution* 20:503–510.

Potter, C. S., and V. Brooks. 1998. Global analysis of empirical relations between annual climate and seasonality of NDVI. *International Journal of Remote Sensing* 19:2921–2948.

Reed, B. C., J. F. Brown, D. Vanderzee, T. R. Loveland, J. W. Merchant, and D. O. Ohlen. 1994. Measuring phenological variability from satellite imagery. *Journal of Vegetation Science* 5:703–714.

Reyers, B., S. Polasky, H. Tallis, H. A. Mooney, and A. Larigauderie. 2012. Finding common ground for biodiversity and ecosystem services. *BioScience* 62:503–507.

Richmond, A., R. K. Kaufmann, and R. B. Myneni. 2007. Valuing ecosystem services: A shadow price for net primary production. *Ecological Economics* 64:454–462.

Rodrigues, A. S. L., H. R. Akçakaya, S. J. Andelman, et al. 2004. Global gap analysis: Priority regions for expanding the global protected-area network. *BioScience* 54:1092–1100.

Ruimy, A., B. Saugier, and G. Dedieu. 1994. Methodology for the estimation of terrestrial net primary production from remotely sensed data. *Journal of Geophysical Research* 99:5263–5283.

Sanderson, E. W., M. Jaiteh, M. A. Levy, K. H. Redford, A. V. Wannebo, and G. Woolmer. 2002. The human footprint and the last of the wild. *BioScience* 52:891–904.

Strassburg, B. B. N., A. Kelly, A. Balmford, et al. 2009. Global congruence of carbon storage and biodiversity in terrestrial ecosystems. *Conservation Letters* 3:98–105.

Sutherland, W., W. Adams, R. Aronson, et al. 2009. One hundred questions of importance to the conservation of global biological diversity. *Conservation Biology* 23:557–567.

Thomas, C. D., B. J. Anderson, A. Moilanen, et al. 2013. Reconciling biodiversity and carbon conservation. *Ecology Letters* 16:39–47. doi:10.1111/ele.12054.

Turner, W. R., K. Brandon, T. M. Brooks, R. Costanza, G. A. B. Da Fonseca, and R. Portela. 2007. Global conservation of biodiversity and ecosystem services. *BioScience* 57:868–873.

Virginia, R. A., and D. H. Wall. 2001. Principles of ecosystem function. In *Encyclopedia of biodiversity*, ed. S. A. Levin, 345–352. San Diego, CA: Academic Press.

Wang, Q., J. Tenhunen, N. Q. Dinh, M. Reichstein, T. Vesala, and P. Keronen. 2004. Similarities in ground- and satellite-based NDVI time series and their relationship to physiological activity of a Scots pine forest in Finland. *Remote Sensing of Environment* 93:225–237.

第十章 利用对地观测数据在流域尺度上分析河岸植被对河流生态完整性的影响

T. 托莫斯　K. 范洛伊　P. 科苏特　B. 维伦纽夫　Y. 苏雄

一、简介

在过去的十年里,河流生态系统已被确定是对人类福祉最基本的生态系统服务(Postel and Carpenter,1997;Aylward et al.,2005)。河流生态系统提供了文化服务(如娱乐、旅游、生存价值)、调节服务(如水质维护、洪水缓冲、侵蚀控制)和支持服务(如在营养物循环、捕食者—被捕食者关系、生态系统恢复能力和生物多样性维护中的作用)。这些服务直接和间接地对人类福祉做出了巨大贡献(Aylward et al.,2005)。尽管我们对生物多样性、生态功能和生态系统服务供给之间的联系尚且知之甚少(Mertz et al.,2007),但是事实证明,维持河流生态系统完整性可以支持对河流生态系统服务的保护。

河岸带作为陆地生态系统和水生生态系统的交界面,是河流边缘含有独特植被的地带。植被有树木、木本灌木、草本植物/非禾本草本植物、牧草和莎草等。河岸带对生态系统功能的生态重要性已得到广泛认可(如 Naiman and Décamps,1997;Naiman et al.,2005;Shearer and Xiang,2007)。其重要性一方面体现在,河岸植被对于河流生态系统的营养状态和食物链(有机物残体)、温度(为水生群落提供荫蔽)以及栖息地(稳固河岸、提供木质碎屑)都起到了调节作用,而这些是河流生态系统的重要组成。另一方面,它对农业和城市污染扩散(如营养物质、沉积物、农药)起到了缓冲作用。

因此，维护和恢复正常的河岸缓冲条件可能有助于保护河流生态系统的完整性，进而益于河流生态系统服务的正常维持。

一个流域内的河流功能过程在时间和空间上以多个尺度相互作用，并与人类活动密切相关（Allan，2004；Wang，2006；Johnson and Host，2010）。人类活动产生的压力综合作用且不规则地分布在流域内。因此，需要更准确地对河岸植被条件和人类活动压力的生物响应（反映河流生态系统功能的完整性）（Hynes，1970；Karr，1993）在多个空间尺度上进行阐释并提供可靠的科学信息，以便优先考虑和设计流域的有效河岸缓冲区。然而，关于河岸植被影响的报告在文献中存在分歧，一些作者解释并指出了河岸地区土地覆盖（Riparian Area Land Cover，RALC）空间信息的局限性（如 Frimpong et al.，2005；Roy et al.，2007；Wasson et al.，2010）。

长期以来，由于小尺度区域的局限（Aguiar and Ferreira，2005）以及利用航测解析法耗时且成本高昂（Coulter et al.，2000），在广阔领域上精确绘制 RALC 仍是无法想象的。20 世纪 70 年代，卫星影像提供了更广阔的视野（如 Landsat 卫星 30 米分辨率影像和 SPOT 卫星 20 米影像），但中等空间分辨率（MRS）影像，除了大型洪泛平原（如亚马孙洪泛平原 2 700 平方千米）等一些非常明确的场景以外并不能特别详细地捕获到河岸生态系统（Mertes et al.，1995）。考虑到河岸生态系统的空间范围和土地覆盖类型的多样性，中等空间分辨率影像似乎不足以详细描述 RALC（Müller，1997；Hollenhorst et al.，2006；Tormos et al.，2011）。然而，大多数分析 RALC 与生物反应之间关系的研究都是基于中等空间分辨率影像衍生出的土地覆盖图。

自 20 世纪 90 年代中期以来，传感器和平台技术的进步提供了地面单元尺寸在 0.6～4 米的宽视角影像，其质量与航摄相片相当，同时具有多光谱信息、稳定的几何结构和一致的观测条件的额外优势（Goetz，2006）。然而，在 20 世纪 90 年代末，与大多数高空间分辨率和甚高空间分辨率（Very High Spatial Resolution，VHSR）制图研究一样，RALC 制图研究使用传统基于像素的技术，在准确、快速处理此类影像中包含的丰富且详细的辐射信息方面遇到了困难（如 Müller，1997；Neale，1997）。

在过去的十年里，随着计算机科学和地理信息系统（GIS）技术的进步，特别是面向对象的影像分析,通过提供有效的计算机辅助分类技术,彻底改变了高空间分辨率和 VHSR 影像的处理方式,其结果已经接近人工解译的质量,同时速度更快、成本更低、可重复性更高(如 Durieux et al., 2007; Tiede et al., 2010; Dupuy et al., 2012)。这些改进能够对大范围河岸地区的精细尺度和可靠空间信息进行探索与生产,如戈茨等人(Goetz et al., 2003)和约翰森等人(Johansen et al., 2008)。然而,这种大规模生态系统的研究没有考虑到影像可用性方面的限制,也没有对河岸地区主要的人类压力(农业和城市地区)和植被条件进行描述。

为了解决所提出的问题,本章介绍了以下领域的新方法:(1)利用高空间分辨率和 VHSR 对地观测数据绘制大范围 RALC 图;(2)在多个空间尺度上分析河岸植被条件和人类压力的生物响应。

二、研究区域

本研究集中于诺曼底河网(法国西北部)石灰岩大陆架水文生态区(HER9 地区;图 10-1(a))的部分区域。

研究区包含了塞纳河下游支流的全部流域:厄尔河(5 935 平方千米)、埃普特河(1 490 平方千米)、伊顿河(1 300 平方千米)和昂代勒河(740 平方千米),以及里勒河(2 310 平方千米)、图克河(1 605 平方千米)、迪沃河(1 573 平方千米)和贝蒂讷河(307 平方千米)等数条直接流入大海的河流。由 6 000 千米以上河网排水的 297 个次流域组成,面积共计 25 000 平方千米。根据斯特拉赫排序法(1957),其中 48% 为一级河网、20% 为二级河网、18% 为三级河网、6% 为四级河网、2% 为五级河网、1% 为六级河网、5% 为七级河网(即塞纳河)。来自国家监测网络的 155 个地点可以获得本研究区域河流的生态完整性。

HER9 地区是一个海拔 200 米以下的平原,基底由沉积岩组成,主要为板状碳酸盐岩。这些岩石在地表渗透性方面表现出不同的特征,导致排水网络的密度差异,或硬度和抗侵蚀性的差异,从而在研究区域局部产生更明显的地貌(丘陵和海岸)。农业活动较活跃的地区主要集中在田间作物(如油籽、混合农业、牧草)和牲畜上,这些地区河岸带结构复杂,落叶林和河流附近草地交替。这种农

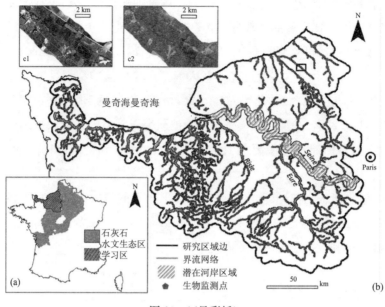

图 10-1（见彩插）

注：研究区域（诺曼底沿海地区）介绍。(a)石灰岩大陆架法国水文生态区；(b)河流网络、潜在河岸区域和生物监测点区域；(c)对地观测数据图示：(c1)航拍影像数据（0.5 米/像素）和(c2)SPOT5 XS 影像数据（10 米/像素）。

业景观夹杂着稀疏城市地区。

三、数据集和方法

（一）数据集

1. 遥感和辅助数据

遥感和辅助数据根据其对法国土地管理人员总体可用性和成本效益来选择的。通过对不同数据的经济性比较分析，收集了两种类型的高分辨率和甚高分辨率、多光谱遥感数据：正射航空相片和 SPOT5 卫星影像。这些影像的参数和预处理总结见表 10-1。正射相片（0.5 米像素）可以提供检测河流沿岸狭窄、零碎的地表覆盖类型所需的纹理信息（Müller,1997）（图 10-1(c1)）。正射航空相片由法国国家地理研究所（IGN）制作，每 5 年更新一次，用于法国各部门。为避免大量的云覆盖影响，通常在春季或夏季采集数据信息。正射影像之间进行辐射量均衡化处理在相同条件下进行，以减少不同航空影像间的辐射测量不均匀性(Paparoditis et al.，2006)。为获得对区分植被类别至关重要的近红外波段信息(Johansen and Phinn，2006)，还采集了 SPOT5 XS 影像(10 米像素)(图 10-1(c2))。

表 10-1 对地观测数据特征及预处理

影像数据	幅宽	空间分辨率	光谱分辨率	采集日期	图块数量	制作人	预处理
正射影片	5 千米	0.5 米	B,G,R	2003 年夏季，2006 年	2455	IGN ©	剪切成河岸区域并镶嵌
SPOT5 XS	60 千米	10 米	G,R,NIR,SWIR	2003 年夏季，2007 年	11	SPOTimage ©存档影像	TOA 辐射校正，剪切成河岸区域并镶嵌

注：B=蓝色；G=绿色；R=红色；NIR=近红外；SWIR=短波红外。

除影像外，还收集了分米级和米级精度的空间主题数据(辅助数据)(表 10-2)。我们收集了 IGN 制作的道路网络地理数据库(BDRoute©，分米级精度)，因为道路的光谱特性不均匀，特别难以从遥感影像中自动提取。我们使用了 IGN 制作的法国水文数据库(BDCarthage©，米级精度)中的地表水文实体(河流、湖泊和水库)。利用数值包裹寄存器(NPR，米级精度)提高了草地与作物的识别能力。NPR 提供了欧洲共同农业政策福利补贴区块(相邻地块)的空间信息。

表 10-2 常见土地覆盖类型的对应关系

一般土地覆被类型(6 类)	从 OBIA 分类流程获得的类别(25 个类别)	数据来源
C1. 水体	水道	A&B
	湖泊和水库	C
	海洋	C
C2. 农业区	耕地	A&B
	一年生作物(小麦、油菜、玉米等)	A&B&E
	牧草	E
	葡萄园	A&B
	果树	E

续表

一般土地覆被类型(6类)	从 OBIA 分类流程获得的类别(25 个类别)	数据来源
C3. 城市地区	连续城市结构	C
	人工土	A&B
	工业或商业单位	C
	道路和铁路网及相关土地	C
	采矿场	C
	运动休闲设施	C
	当地道路	D
	部门道路	D
	国道	D
	高速公路	D
C4. 树木植被	树木植被	A&B
C5. 草本植物、灌木植被	永久草地	E
	临时草地	E
	半生草本和灌木植被	A&B
C6. 半自然裸土	裸露的土壤和稀疏的植被	A&B
	裸土	A&B
	沙子	A&B

注：文献中用于研究对河流生态完整性影响的类型和土地覆盖类别，此等类别是从面向对象的影像分析(Object-Based Image Analysis, OBIA)分类流程中提取的。分类采用的数据：A 正射影像；B SPOT5 XS；C CORINE 土地覆盖数据库；D BDRoute；E BDCarthage；和 F 数字包裹登记册。

区块边界每年由农民在正射航空相片上进行影像解译和数字化。最后，利用 1990 年和 2000 年获得的 Landsat 和 SPOT 卫星影像目视解译，获得了 CO-RINE(环境信息协调计划)土地覆盖数据库(CLC, 分米级精度)。对影像的解译是基于覆盖在 1/100 000 卫星影像拷贝的透明投影片上进行的(Bossard et al., 2000)。CLC 中要素的最小面积为 25 公顷，是同质区域或可识别结构的多土地覆盖类型混合区域。该分类基于分级的标准命名法，包括三个级别：1 级的 5 大土地覆盖类别(1：人造地表；2：农业区；3：森林和半自然区；4：湿地；5：水体)、15

种土地覆盖类别(2级)和44种土地覆盖类型(3级)。虽然CLC对于描述河流沿线狭窄和碎片化物体的特征不够精确,但它适用于提取河流流域走廊内的大型物体,特别是大型人造地表(表10-2)。

2. 生物数据

从1992~2002年期间研究区的国家监测网络中提取了大型底栖无脊椎动物数据集(155个站点,1 038个样本;见图10-1(b))。在此期间,每个站点的样本数量是变化的。2002年后,随着《水框架指令》(Water Framework Directive,WFD)的实施,监测过程才趋于稳定。这些站点平均采样6.7次(约每两年一次),极限情况分别是采样2次(4个站)和10次(44个站)。我们还注意到,有些站点每年只取样一次(主要在春季)。

在法国,各区域环境委员会使用全球标准化(l'Indice Biologique Global Normalisé,IBGN)指数(AFNOR,1992)检查生物监测点。IBGN是两个指标的组合:分类群总数(14个科级)和动物学指标群(代表39个指示分类群的存在/不存在),分为9个敏感性等级。该指数对污染(包括有毒物质)和一般退化(包括栖息地改变)很敏感。模型中使用了每个站点在1992~2002年期间的IBGN平均值。

(二) 河岸地区土地覆盖分类

1. 分类流程概述

我们设计了一个由六个类别组成的分类模型(表10-2)。"C1:水体"和"C6:半自然裸土"是划定河床的必要条件;"C2:农业区"和"C3:城市区域"被认为是河流生态状况的主要变化原因(Allan,2004);以及"C4:树木植被"和"C5:半自然草本和灌木植被",构成了维持生物多样性和调节非点源污染的河流走廊景观的主要自然要素(Naiman et al.,2005)。

为此,我们使用了托尔莫斯等人(Tormos et al.,2012)设计的面向对象的影像分析流程。本流程的主要目标是在不同地理环境(即不同地形、气候和地质条件)下,对大型河岸地区绘制高分辨率RALC地图。该过程主要依靠管理人员或相关人员提供的多源空间数据。第二个目标是验证RALC图的准确性。图10-2显示了该过程的不同阶段的流程图。

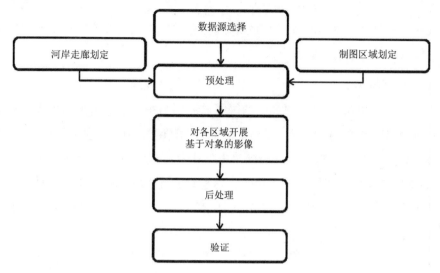

图 10-2 河岸地区土地覆盖分类流程图
资料来源:Tormos, T., P. Kosuth, et al., 2012.

该过程分为七个阶段。第一阶段选择输入数据。第二阶段划定了潜在的河岸走廊区域(使用水文网络周围的缓冲区,其宽度取决于 Strahler 河流分级法的顺序;我们研究区域的结果如图 10-1(b)所示)。第三阶段对同质区域(地理和影像层面)进行绘图。以下三个阶段分别用于 RALC 分类,包括预处理(阶段四)、对每个绘制区域进行 OBIA 分析(阶段五)和后处理(阶段六)阶段。最后一个阶段对生成的 RALC 图进行验证(Tormos et al.,2012)。

2. 为河岸地区土地覆盖设计的面向对象的影像分析

面向对象的影像分析是一种适用于高分辨率情况,即当像素明显小于所考虑的对象时的技术(Blaschke,2010)。它解决了基于像素的分类精度低及分类结果"椒盐"效应的问题(如 Latty and Hoffer,1981;Kressler et al.,2003;Durieux et al.,2007)。在第一阶段中,识别分割过程并建立同质区域(图斑或影像对象),以划定不同尺度下的对象。在第二阶段中,使用光谱及空间信息(例如纹理、形状和环境特征)将分类过程应用于上述对象,以增加对光谱特征相似的不同土地覆盖类型的区分。大多数比较OBIA和像素分类方法的研究显示,OBIA具有更高的准确性(如Carleer and Wolff,2006;Cleve et al.,2008;Myint et al.,2011)。

托莫斯等人(2012)设计的OBIA方法的独创性在于:(1)在分类中使用专题空间信息;(2)分层的影像对象网络;(3)分类树的构建;(4)分类规则定义。

(1) 专题空间数据包含了评估RALC的可靠信息。因此,这是OBIA流程利用的第一项信息。根据该信息创建了第一级分类("主题"级)。研究区域也根据专题数据实体边界进行分割(它构成了分层影像对象网络的第一层,图10-3)。然后,使用布尔规则将得到的影像对象归类或不归类于专题类。

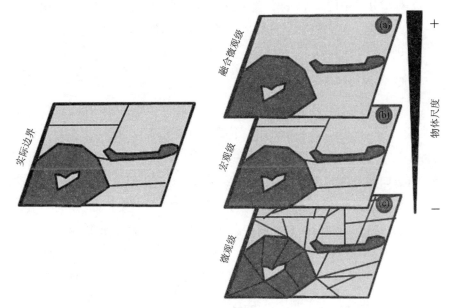

图10-3 为在河岸区域内提取不同大小的对象而设计的分割过程的示意图

注:(a)微观级分割能够检测河岸景观中狭窄的和破碎的对象,若分割过程在较粗糙的尺度实现,则此等对象将被忽略。(b)和(c)宏观级和融合的微观级对于划分中等大小的对象是必要的,同时保持已在微观级划定和分类的独立对象。为此,需要依据微观级的副本创建融合的微观级,并且在微观级对相同类的相邻对象进行合并。然后,基于微观级和融合的微观级的边界范围来创建宏观级。

(2) 为了提取河岸景观中不同大小的地物(从狭窄且零散的物体,如沿岸沙堤、孤立树木植被或微小不透水的表面,到大型地物,如农业地块和连续城市区域等),设计了由三个层次组成的分割过程(图10-3)。在我们的案例中,第一层分割("微观"层)是基于正射航空相片(具有最高空间分辨率的影像)提供的"专题"级边界范围。"宏观"级别使用了来自 SPOT5XS 卫星的影像信息。因此,我们的 OBIA 是基于由四个分割层级组成的分层影像对象网络。

(3) 分类树的构建可以归类为自上而下(即低到高层次)的影像解译。根据现有的数据来源,树从最容易提取的类别开始,到兴趣类别(例如"水面"与"地表",然后是"地表""高植被的土壤""低植被土壤"……)。当不能在当前分割层级上可靠地分离和明确地定义更多的类别时,分类树的构建就停止了。

(4) 对于分类中的每个决策,基于由专家知识、试错运行或单纯目视判别选择的一个或多个相关的光谱、空间和背景特征,使用模糊或清晰的隶属函数来制定决策规则。

3. 分类自动化和准确性评估

首先根据影像采集日期将研究区域划分为相同类型的制图区域(流程的第三阶段)。然后,根据 OBIA 方法在一个试点制图区域上设计一个主规则集。使用的特征是:(1)辐射值(逐波段计算斑块/对象内像素的平均值);(2)纹理值,特别是所有方向的灰度熵差向量(Haralick et al., 1973),从正射影像的蓝色波段到不同类型人工区域到无植被土壤;(3)软件提供的上下文特征(如到一个类别的距离,超大地物的存在);(4)亮度指数、归一化植被指数等标准指标。

例如,我们使用 NDVI,它来自 SPOT5 XS 红色和近红外波段。众所周知,它可正确区分城市地区的植被和非植被区域(Carleer and Wolff, 2006)。最后,在调整分类规则时,对每个制图区域运用主规则集,并在考虑区域特性的基础上,将一些新分类添加到分类树中。研究使用了 eCognition 软件开发者版本和服务器版本进行主规则集制定和分类规则调整。

为了评估准确性,根据文献中用于研究河流生态完整性影响的普通土地覆盖类型对类别进行分类后,我们计算出了混淆矩阵(表 10-2)。鉴于用于分类的特征是在 OBIA 中的对象尺度上计算的,因此选择了地物或多边形作为选取控制数据的采样单元(Tiede et al., 2006; Grenier et al., 2008)。然而,混淆矩阵是使用所选控制对象的面积(以 0.50 米像素表示)来计算的,因为土地覆盖图是由多级 OBIA 方案的实施产生的,并且包含从小到大的不同大小的对象。为了建立一个专用于从遥感数据导出的分类规则的混淆矩阵,我们没有对辅助数据中的对象进行采样。最终,我们从研究区域的影像信息(表10-2)中随机收集了50个样本。考虑到研究区域(和制图区域)的大小,对照样本现场采集数据将是非常耗时的。根据朱等人(Zhu et al.,2000)的建议,以空间分辨率最高的影像进行目视解译并作为控制对象。为了保持影像解译的客观性,解译过程中没有查看分类地图。

(三) 土地覆盖与河流生态完整性间的大尺度关系模型

1. 模型概述

从对地观测数据得到的土地覆盖图可作为计算空间(或景观)指标的基础,为量化人类压力在河流生态系统不同功能尺度(从流域到局部尺度)和不同河岸植被条件的影响提供了直接方法。其可与描述河流生态完整性的诸多河流生物指标相关联,以更好地理解人类压力影响(如 Allan *et al.*,1997;Sponseller *et al.*,2001;Wasson *et al.*,2010)。

我们设计了一个统计模型来研究河流生态完整性之间的关系,利用大型底栖无脊椎动物生物指标(见生物数据一节)和在此类研究中广泛使用的对三种空间尺度上进行估算的土地覆盖空间指标(Land Cover Spatial Indicator,LCSI)进行评估(Allan,2004)。这三个空间尺度分别是:上游流域、上游河岸走廊(对应于生物监测点上游河网河岸走廊的空间)及局部河岸走廊(对应于生物监测站点上游河网3千米沿线的空间)(图10-4)。为了更好地定位影响源,使用了根据(1)生物监测点上游纵向距离(仅针对局部河岸走廊尺度)和(2)河流横向距离(针对上游河岸和局部河岸尺度)变化的缓冲区。

2. 河流生态完整性评价

通过将观测值与国家河流类型的参考值相除,IBGN 指数值被转变为生态质量比(Ecological Quality Ration,EQR)值——这与水框架指令中的参考数值存在偏差。参考数值是从参考地点数据得出的。选择参考地点的标准符合 WFD 相互校准过程(参考版2003年版)中商定的标准,且 EQR 值与国家一级用于确定 WFD 生态状况的 EQR 值相当。

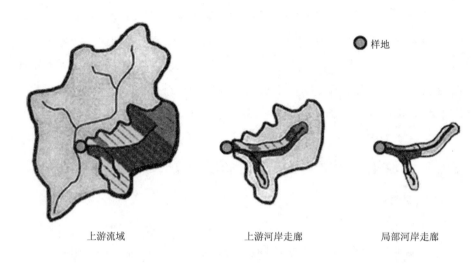

上游流域　　　　　上游河岸走廊　　　　局部河岸走廊

图 10-4

注：描述了在给定生物监测点将土地覆盖与河流完整性的生物测量指标联系起来所使用的三个空间尺度：上游流域、上游河岸走廊（对应于整个上游段的河岸走廊）和局部河岸走廊（对应于上游段最长可达3千米的河岸走廊）。（修改自 Morley, S. A., and J. R. Karr, 2002.）

3. 空间指标

空间指标是通过对划定区域的景观结构属性进行聚合来确定的（Gergel et al., 2002）。每种土地覆盖类型的比例被用来估算某种土地覆盖类别的压力强度（Tormos et al., 2011）。

因此，本研究建立了诸多 LCSI，这些数值会随土地覆盖类别和空间面积而变化。

研究计算出了每个生物监测点在流域尺度上的土地覆盖指标，流域是根据对法国国家地理研究所 BDALTI©（250米/像素）数字地形模型的分析划定的。使用 CLC 数据库分析了该尺度下的土地覆盖百分比。使用第三级 CLC 对每个流域的土地覆盖进行了描述。

在上游河岸走廊根据宽度（两侧与河流的横向距离）不同建立了15个缓冲区：0～50米之间10个缓冲区，间距5米；50～300米之间5个缓冲区，间距50米（图 10-5）。

在局部河岸走廊根据长度（上游曲线距离生物监测点）和宽度（两侧河流横

向距离)不同建立了150个缓冲区:确定了10个长度类别,从0～500米间距为100米,从500～3 000米间距为500米(图10-6(a))。然后,对于每个长度类别,确定15个宽度间隔,0～50米间的间距为5米,50～300米间的间距为50米(图10-6(b))。

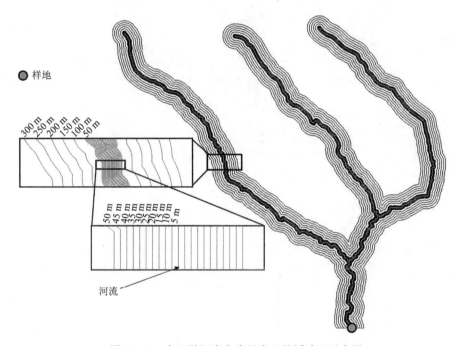

图10-5 在上游河岸走廊尺度上的缓冲区示意图

缓冲区使用ESRI©GIS软件中基于Python的GIS工具(空间分析、网络分析和动态分割工具)开发的算法进行确定。根据VHSR对地观测得出的地图,连同辅助数据,用于评估此等尺度下的土地覆盖。我们计算了每个缓冲区内所有可提取土地覆盖类别的百分比。

4. 统计方法

首先,在每个尺度上选择一个缓冲区,在该缓冲区中,给定的土地覆盖类别对EQR-IBGN指数的影响最为显著。通过比较不同候选缓冲区空间指标和生物学反应间的二元变量相关性,选择出了确定河岸尺度内对于特定土地覆盖分类最具影响力的缓冲区(Johnson and Covich, 1997; Sponseller *et al.*,

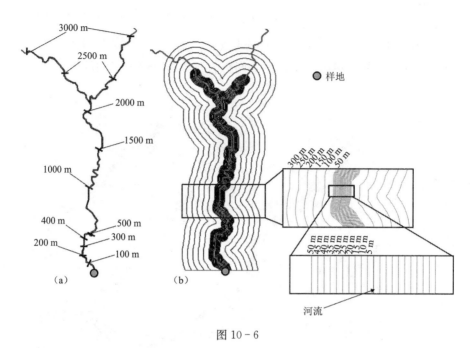

图 10-6

注：在局部河岸走廊尺度上的缓冲区示意图：(a)不同长度和(b)确定长度(2 500 米)不同宽度缓冲区案例。

2001)。

第二步，在第一步中选择的空间指标和流域尺度上的空间指标中，确定出对生物响应影响最大的 LCSI 组合。与华森等人(Wasson et al.,2010)一样，我们使用了偏最小二乘(PLS)回归模拟了土地覆盖对无脊椎底栖动物指数(EQR-IBGN)的影响。PLS 是多元线性回归(MLR)(Wold,1966,1982)的扩展。PLS 回归允许通过指定明确数量的和共线性的预测变量来研究复杂问题(Wold et al.,2001)。

我们在 XLStat 2006 软件(AddinSoft,巴黎,法国)中使用了 NIPALS 算法来开展了 PLS 回归分析。采用沃尔德(1982)、马滕斯(2000)所推荐的 jack-knifing 技术(Efron and Gong,1983)估算了系数的标准误差和置信区间。模型中仅保留了根据 jack-knifing 法与生物指标有显著关系的 LCSI。

四、结果

(一) 河岸地区土地覆盖地图

通过正射航空相片和 SPOT5 XS 影像的面向对象的影像分析结合辅助数据获得了详细的 RALC 图(总共 25 个类别;见表 10-2)。图 10-7 中可看到生成的地图图例,六个类别的分类用于评估地图的准确性。

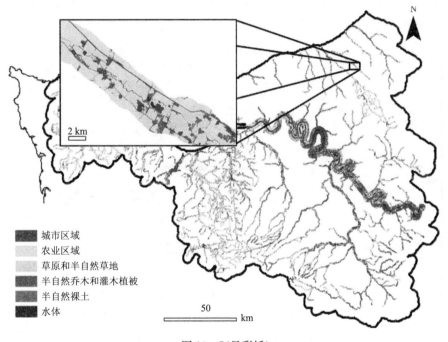

图 10-7(见彩插)

注:研究区河岸区域土地覆盖图按分类学分为六类以评价地图质量。

混淆矩阵(文章未给出)显示,地物分类获得了较高的精度:在诺曼底沿海地区(5 600 平方千米)河岸区域上 85% 的像素得以被正确地分类。对于"C1:水体"、"C2:农业区"和"C4:森林区域"获得了高度准确的结果(用户和生产者的准确度均大于 90%),而对于"C3:城市区域"和"C6:半自然裸土"类别获得了最差的结果(用户和生产者的准确度均小于 70%)。

为处理 1 000 平方千米的河岸走廊影像解译,制图区域之间规则阈值的调整时间和处理时间估计为 16 天。

(二) 土地覆盖效应

EQR-IBGN 均值 38% 的变异性可以用三个尺度(上游流域、上游河岸和局部河岸)中的 14 个 LCSI 组合来解释。图 10-8 说明了重要 LCSI 对平均 EQR-IBGN 的贡献。我们可以看到,在三种尺度上,与平均 EQR-IBGN 呈负相关的土地覆盖类别与呈正相关的土地覆盖类别之间有明显的区别。人工表面("非连续城市"、"工业区"和"道路"等各类型)在流域尺度和河岸尺度上有明显而强烈的负面效应。

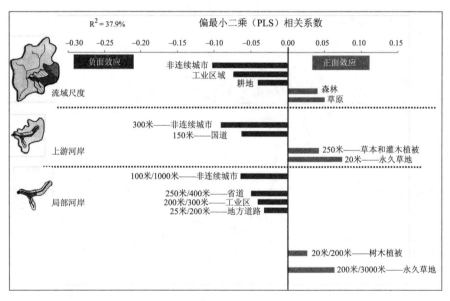

图 10-8

注:平均 EQR-IBGN 和统计上显著的(根据 Jack-knifing 法)LCSI 变量间的标准化偏最小二乘(PLS)回归系数,该变量来源于流域尺度的 Corine 土地覆盖(2000 年版本)和河岸尺度(上游和局部河岸尺度)的河岸地区土地覆盖图。

特别是"非连续城市"类别是影响大型无脊椎底栖动物群落的主要因素。流域尺度上的农业区(耕地)对大型无脊椎动物群落也有负面影响,但不如人工表面显著。草地和林区在三个尺度上的正面效应是明显的。"永久草地"在两个河岸尺度上都是有益的影响,在上游河岸尺度上的缓冲区要窄得多(河两岸20米的缓冲条带)。在监测点附近(上游200米)河流两侧20米的条带上,森林区域在流域和局部河岸尺度上具有有益效果。

五、讨论

(一) 绘制广域河岸地区土地覆盖地图

首先,分类规则对于所获得的准确性是有效的,但这些结果必须谨慎对待。一方面,分类精度可能会受到训练样本量的影响(Foody et al.,2006);这些结果必须使用不同数量的控制样本数据集进行验证(如 Durrieu et al.,2007)。另一方面,六种分类融合的效率可能被高估,根据遥感数据得到的最终分类来计算混淆矩阵是很有意义的。此外,使用精确方法(Sauro and Lewis,2005)或KHAT统计量(Congalton,1991)来量化分类的置信区间对于我们的分析是有用的。

提出的 OBIA 过程是完全可操作的(即可重复、易于转移、在广泛领域快速适用的)。它可以提供广域尺度的河岸土地覆盖和土地利用的可靠精细信息。为描述人类压力对河流生态完整性的影响提供了工具。尽管研究区域的景观具有多样性,但 OBIA 在不改变分类特征的情况下,可在所有制图区域上使用相同的分割参数和相同的分类层次结构,操作的主要任务是调整用于定义类别的不同特征的阈值。这主要是 OBIA 使用自上而下的方法来构建分类树(将空间要素划分为更细的单元)和使用"基于知识的规则"分类技术。自上而下的方法促进了简单规则的使用,这些规则很容易地移植到其他制图区域,并便于新的操作人员使用该方法。此外,给定制图区域的特定类别可被集成到分类树中而不影响分类树的整体构造,或可在用户的完全控制下被细化(Lucas et al.,2007; Tiede et al.,2010)。

此外,虽然基于知识的规则分类技术可以很容易地分发至集群主机上,从而大大减少了计算时间,但监督分类技术不能被分发,因为它需要作为整体处理映射区域以收集具有空间代表性的训练样本。由于模糊逻辑概念的出现——一种处理不同数据源之间冲突的强大技术(Benz et al.,2004)——新的数据源可以很容易地与初始数据源进行集成和结合。例如,在不久的将来,基于激光遥感的垂直信息将在广阔领域上可用且用户能够负担得起,集成该数据能够(1)利用建筑物高度信息来解决城市区域和半自然裸土间的分类误差;(2)更好地描述河岸植被类别(如 Antonarakis et al.,2008;Geerling et al.,2009;Johansen et al.,2010)。

OBIA 显然是一种在相关尺度上处理河岸地区空间信息以推断与河流生态完整性关系的合适方法。然而,尽管这种解译方法准确性较高,但获得特定时间或时间段的 RALC 图,并将这些信息与生物评估相关联仍是一个挑战。显然,这取决于该区域影像数据的可用性[①]。

此外,必须在 OBIA 流程中进行一些改进,以便对每个主题空间数据的几何和语义信息进行可靠和灵活的匹配。

(二) 大规模河岸植被对河流生态完整性的影响

我们对土地覆盖(不包括河岸植被)和生物响应之间的关系的研究结果与现有文献没有分歧。城市化(不连续的城市、工业区和道路)和农业区(耕地)对大型无脊椎动物群落的影响是公认的(如 Roth et al.,1996;Wang et al.,1997;Paul and Meyer,2001;Sponseller et al.,2001;Wang and Kanehl,2003;Allan,2004;Wasson et al.,2010)。流域尺度和河岸走廊尺度上由于人类占领所产生的多重直接和间接压力改变了生态系统的结构和功能。河岸走廊中城市化和农业活动的存在由于限制了再定殖过程,特别是成年大型无脊椎动物的再定殖过程,导致了水生系统的碎片化(Petersen et al.,1999,2004)。同样,诸多作者也强调了半自然区(森林和草原)对流域尺度上大型无脊椎动物群落的有益影响(如 Roth et al.,1996;Wang et al.,1997;Wasson et al.,2010)。这些地类减少了弥漫性污染,其可能因为在流域中所占的位置(Gergel,2005)而成为污染的"下水池"。

关于河岸植被对生物完整性影响的权重,文献中的结果是有差异的。显然,研究是在不同的自然和仿生化背景下进行的,但大多数研究使用的是中等空间分辨率卫星影像,不适合可靠地估计河岸缓冲区内不同覆盖类型的表面积。在

---

① 以本研究为例,我们一开始很难在研究区域利用合理成本获得具有近红外信息的 VHSR 卫星影像镶嵌图。自 2011 年以来,GEOSUD 计划 (http://www.geosud.teledetection.fr)面向科学界/土地管理的公共和私人参与者、研究人员和决策者免费提供了法国全域的 Rapideye 卫星影像(5 米像素;蓝色、绿色、红色、红边和近红外光谱波段)。

没有河岸空间信息约束的情况下（使用了基于 VHSR 生成的地图），本研究清楚地指出了上游河岸尺度中森林和永久草地的积极影响（影像强度相当于耕地，尤其是草地），从而佐证了在局部尺度上的研究。沿河的农用草地对于防止农业和城市地区的污染扩散方面可以起到过滤作用（如 Lyons *et al.*, 2000; Lin *et al.*, 2004）。森林植被有助于减少扩散污染（通过植物吸收），但也在调节温度和营养状态方面发挥了关键作用；此外，它直接为大型无脊椎底栖动物提供了食物和栖息地异质性（如 Hachmoller *et al.*, 1991; Maridet *et al.*, 1997）。

此外，通过对局部和上游河流走廊尺度上多个缓冲区的分析，我们能够确定森林和草地最具影响力的条带宽度。研究结果为管理者优先制定河岸地区生态修复策略提供了重要的信息。我们的研究结果表明，应努力开展管理河流植被的政策和生态修复战略决定，将河岸缓冲区扩大到河流边缘 20 米条带宽度的最低阈值。

因此，河岸区域内精细尺度空间信息似乎是正确理解河流生态完整性与人类压力、河岸条件之间关系的一个条件。由此可设想模型的进一步改进及其在管理策略中的应用：(1) 可以引入新的空间指标，包括可能影响河流生态完整性（Weller *et al.*, 2011）的河岸走廊植被斑块空间配置（宽度、均匀性、碎片化程度）；(2) 分析其他生物响应和营养物质响应；(3) 考虑河流和集水群的大小。例如，波特等人（Potter *et al.* 2005）表明，大型无脊椎底栖动物对源头上存在河岸植被的反应更大。此等改进对于在流域尺度上正确量化河岸地区提供的河流生态系统服务是必要的。

六、结论

在本章中，我们引入新近的遥感处理方法来获得广域范围内河岸带的精细尺度空间信息，以分析其与河流生态完整性的关系。该方法基于 OBIA 方法，并结合了 VHSR 对地观测数据和米级辅助数据。

它在法国诺曼底沿海地区（5 600 平方千米的河岸地区）的实施证明了它的效率、可转移性和适用性。通过整合获得的空间信息，可以反映大型无脊椎动物群落的响应，量化不同缓冲区的河岸土地覆盖对河流生态完整性的影响。结果表明，在河流两侧 20 米缓冲区内，河岸林地和草地植被对河流的局部和整个上游网络都产生了有利影响。

从这些结果中,我们得出这样的结论,正如王(2006)、约翰逊和霍斯特(2010)在他们的综述中所倡导的,我们正确理解河流生态完整性与人类压力和河岸植被间的大规模关系的能力,似乎高度依赖于广泛区域空间信息的可用性和处理结果。需要进行进一步的研究,以提供精细尺度的空间信息,解决当前河流生态研究中有关生物响应非线性、模型不确定性、空间自相关、遗留效应和自然变异性的问题(Allan,2004;Wang,2006;Johnson and Host,2010)。但可以肯定的是,利用这些精细尺度的信息,可以探索与河岸植被的机理影响相关的新的可靠的空间指标(例如连续性、宽度和均匀性;Tormos *et al.*,2011;Weller *et al.*,2011),以便更好地支持管理者制定有效的河岸生态恢复策略,维持河流生态完整性和生态系统服务。此外,基于这些新的探索,可以用示意教学地图对流域尺度上的河流生态系统服务状况进行表示(Pert *et al.*,2010)。

致谢

本研究是通过法国科研部授予第一作者的博士学位资助实现的。这项工作部分是在 ONEMA 和 Irstea 间的研究合同框架下实现的。遥感数据的获取得益于法国国家航天机构法国空间研究中心 ISIS 计划的支持(SPOT-5 影像,SPOT 影像分发),IGN 的机载正射相片则由法国农业部提供。

参 考 文 献

AFNOR (Association Française de NORmalisation). 1992. Essai des eaux. Détermination de l'Indice Biologique Global Normalisé (IBGN). *Association Française de Normalisation - norme homologuée T 90-350*:1–8.

Aguiar, F. C., and M. T. Ferreira. 2005. Human-disturbed landscapes: Effects on composition and integrity of riparian woody vegetation in the Tagus River basin, Portugal. *Environmental Conservation* 32:30–41.

Allan, J. D. 2004. Landscapes and riverscapes: The influence of land use on stream ecosystems. *Annual Review of Ecology, Evolution, and Systematics* 35:257–284.

Allan, J. D., D. Erickson, and J. Fay. 1997. The influence of catchment land use on stream integrity across multiple spatial scales. *Freshwater Biology* 37:149–161.

Antonarakis, A. S., K. S. Richards, and J. Brasington. 2008. Object-based land cover classification using airborne LiDAR. *Remote Sensing of Environment* 112:2988–2998.

Aylward, B., J. Bandyopadhyay, J. C. Belausteguigotia, et al. 2005. Freshwater ecosystem services. In *Ecosystems and human well-being: Policy responses*, eds. R. L. K. Chopra, R. Leemans, P. Kumar, and H Simons, 215–255. Washington, DC: Island Press.

Benz, U. C., P. Hofmann, G. Willhauck, I. Lingenfelder, and M. Heynen. 2004. Multi-resolution, object-oriented fuzzy analysis of remote sensing data for GIS-ready information. *ISPRS Journal of Photogrammetry and Remote Sensing* 58:239–258.

Blaschke, T. 2010. Object based image analysis for remote sensing. *ISPRS Journal of Photogrammetry and Remote Sensing* 65:2–16.

Bossard, M., J. Feranec, and J. Otahel. 2000. *CORINE land cover technical guide—Addendum 2000*. Technical report No 40. Copenhagen: European Environment Agency.

Carleer, A. P., and E. Wolff. 2006. Urban land cover multi-level region-based classification of VHR data by selecting relevant features. *International Journal of Remote Sensing* 27:1035–1051.

Cleve, C., M. Kelly, F. R. Kearns, and M. Moritz. 2008. Classification of the wildland-urban interface: A comparison of pixel- and object-based classifications using high-resolution aerial photography. *Computers, Environment and Urban Systems* 32:317–326.

Congalton, R. G. 1991. A review of assessing the accuracy of classifications of remotely sensed data. *Remote Sensing of Environment* 37:35–46.

Coulter, L., D. Stow, A. Hope, et al. 2000. Comparison of high spatial resolution imagery for efficient generation of GIS vegetation layers. *Photogrammetric Engineering and Remote Sensing* 66:1329–1335.

Dupuy, S., E. Barbe, and M. Balestrat. 2012. An object-based image analysis method for monitoring land conversion by artificial sprawl use of RapidEye and IRS data. *Remote Sensing* 4:404–423.

Durieux, L., E. Lagabrielle, and A. Nelson. 2007. A method for monitoring building construction in urban sprawl areas using object-based analysis of Spot 5 images and existing GIS data. *ISPRS Journal of Photogrammetry and Remote Sensing* 63:399–408.

Durrieu, S., T. Tormos, P. Kosuth, and C. Golden. 2007. Influence of training sampling protocol and of feature space optimization methods on supervised classification results. *Paper presented at the International Geoscience and Remote Sensing Symposium (IGARSS)*. 23–28 July, Barcelona, Spain.

Efron, B., and G. Gong. 1983. A leisurely look at the bootstrap, the jackknife, and cross validation. *American Statistician* 37:36–48.

Foody, G. M., A. Mathur, C. Sánchez-Hernández, and D. S. Boyd. 2006. Training set size requirements for the classification of a specific class. *Remote Sensing of Environment* 104:1–14.

Frimpong, E. A., T. M. Sutton, K. J. Lim, et al. 2005. Determination of optimal riparian forest buffer dimensions for stream biota-landscape association models using multimetric and multivariate responses. *Canadian Journal of Fisheries and Aquatic Sciences* 62:1–6.

Geerling, G. W., M. J. Vreeken-Buijs, P. Jesse, A. M. J. Ragas, and A. J. M. Smits.

2009. Mapping river floodplain ecotopes by segmentation of spectral (CASI) and structural (LiDAR) remote sensing data. *River Research and Applications* 25:795–813.

Gergel, S. E. 2005. Spatial and non-spatial factors: When do they affect landscape indicators of watershed loading? *Landscape Ecology* 20:177–189.

Gergel, S. E., M. G. Turner, J. R. Miller, J. M. Melack, and E. H. Stanley. 2002. Landscape indicators of human impacts to riverine systems. *Aquatic Sciences* 34:118–128.

Goetz, S. J. 2006. Remote sensing of riparian buffers: Past progress and future prospects. *Journal of the American Water Resources Association* 42:133–143.

Goetz, S. J., R. Wright, A. J. Smith, E. Zinecker, and E. Schaub. 2003. IKONOS imagery for resource management: Tree cover, impervious surfaces and riparian buffer analyses in the mid-Atlantic region. *Remote Sensing of Environment* 88:195–208.

Grenier, M., S. Labrecque, M. Benoit, and M. Allard. 2008. Accuracy assessment method for wetland object-based classification. *Paper presented at the Geographic Object Based Image Analysis (GEOBIA), ISPRS Commission IV*. August 5–8, 2008, Calgary, AB, Canada.

Hachmoller, B., R. A. Matthews, and D. F. Brakke. 1991. Effects of riparian community structure, sediment size, and water quality on the macroinvertebrate communities in a small, suburban stream. *Northwest Science* 65:125–132.

Haralick, R. M., K. Shanmugam, and I. Dinstein. 1973. Textural features for image classification. *IEEE Transactions on Systems, Man, and Cybernetics Society* 3:610–621.

Hollenhorst, T., G. Host, and L. Johnson. 2006. Scaling issues in mapping riparian zones with remote sensing data: Quantifying errors and sources of uncertainty. In *Scaling and uncertainty analysis in ecology*, eds. J. Wu, B. Jones, H. Li, and O. Loucks, 275–295. Dordrecht, The Netherlands: Springer.

Hynes, H. B. N. 1970. *The ecology of running waters*. Liverpool, UK: Liverpool University Press.

Johansen, K., and S. Phinn. 2006. Linking riparian vegetation spatial structure in Australian tropical savannas to ecosystem health indicators: Semi-variogram analysis of high spatial resolution satellite imagery. *Canadian Journal of Remote Sensing* 32:228–243.

Johansen, K., S. Phinn, J. Lowry, and M. Douglas. 2008. Quantifying indicators of riparian condition in Australian tropical savannas: Integrating high spatial resolution imagery and field survey data. *International Journal of Remote Sensing* 29:7003–7028.

Johansen, K., S. Phinn, and C. Witte. 2010. Mapping of riparian zone attributes using discrete return LiDAR, QuickBird and SPOT-5 imagery: Assessing accuracy and costs. *Remote Sensing of Environment* 114:2679–2691.

Johnson, L. B., and G. E. Host. 2010. Recent developments in landscape approaches for the study of aquatic ecosystems. *Journal of the North American Benthological Society* 29:41–66.

Johnson, S. L., and A. P. Covich. 1997. Scales of observation of riparian forests and distributions of suspended detritus in a prairie river. *Freshwater Biology* 37:163–175.

Karr, J. R. 1993. Defining and assessing ecological integrity: Beyond water quality. *Environmental Toxicology and Chemistry* 12:1521–1531.

Kressler, F. P., Y. S. Kim, and K. T. Steinnocher. 2003. Object-oriented land cover classification of panchromatic KOMPSAT-1 and SPOT-5 data. *Geoscience and Remote Sensing* 19:263–269.

Latty, R. S., and R. M. Hoffer. 1981. Computer-based classification accuracy due to the spatial resolution using per-point versus per-field classification techniques. *Proceedings of the 7th symposium on machine processing of remotely sensed data with special emphasis on range, forest, and wetlands assessment, IEEE Computer Society*, June 23–26, 1981, Indiana, USA: 384–393.

Lin, C. H., R. N. Lerch, H. E. Garrett, and M. F. George. 2004. Incorporating forage grasses in riparian buffers for bioremediation of atrazine, isoxaflutole and nitrate in Missouri. *Agroforestry Systems* 63:91–99.

Lucas, R., A. Rowlands, A. Brown, S. Keyworth, and P. Bunting. 2007. Rule-based classification of multi-temporal satellite imagery for habitat and agricultural land cover mapping. *ISPRS Journal of Photogrammetry and Remote Sensing* 62:165–185.

Lyons, J., S. W. Trimble, and L. K. Paine. 2000. Grass versus trees: Managing riparian areas to benefit streams of central North America. *Journal of the American Water Resources Association* 36:919–930.

Maridet, L., M. Phillippe, J. G. Wasson, and J. Mathieu. 1997. Seasonal dynamics and storage of particulate organic matter within bed sediment of three streams with contrasted riparian vegetation. In *Groundwater/surface water ecotones: Biological and hydrological interactions and management options*, eds. J. Gibert, J. Mathieu, and F. Fournier, 68–74. Cambridge, UK: Cambridge University Press.

Martens, H., and M. Martens. 2000. Modified jack-knife estimation of parameter uncertainty in bilinear modelling by partial least squares regression (PLSR). *Food Quality and Preference* 11:5–16.

Mertes, L. A. K., D. L. Daniel, J. M. Melack, B. Nelson, L. A. Martinelli, and B. R. Forsberg. 1995. Spatial patterns of hydrology, geomorphology, and vegetation on the floodplain of the Amazon River in Brazil from a remote sensing perspective. *Geomorphology* 13:215–232.

Mertz, O., H. M. Ravnborg, G. L. Lövei, I. Nielsen, and C. C. Konijnendijk. 2007. Ecosystem services and biodiversity in developing countries. *Biodiversity and Conservation* 16:2729–2737.

Morley, S. A., and J. R. Karr. 2002. Assessing and restoring the health of urban streams in the Puget Sound Basin. *Conservation Biology* 16:1498–1509.

Müller, E. 1997. Mapping riparian vegetation along rivers: Old concepts and new methods. *Aquatic Botany* 58:437.

Myint, S. W., P. Gober, A. Brazel, S. Grossman-Clarke, and Q. Weng. 2011. Per-pixel vs. object-based classification of urban land cover extraction using high spatial resolution imagery. *Remote Sensing of Environment* 115:1145–1161.

Naiman, R. J., and H. Décamps. 1997. The ecology of interfaces: Riparian zones. *Annual Review of Ecology and Systematics* 28:621–658.

Naiman, R. J., H. Décamps, and M. E. McClain. 2005. *Riparia: Ecology, conservation, and management of streamside communities*. Boston, MA: Elsevier Academic.

Neale, C. M. U. 1997. Classification and mapping of riparian systems using airborne multispectral videography. *Restoration Ecology* 5(4 Suppl.):103–112.

Paparoditis, N., J. P. Souchon, G. Martinoty, and M. Pierrot-Deseilligny. 2006. High-end aerial digital cameras and their impact on the automation and quality of the production workflow. *ISPRS Journal of Photogrammetry and Remote Sensing* 60:400–412.

Paul, M. J., and J. L. Meyer. 2001. Streams in the urban landscape. *Annual Review of Ecology and Systematics* 32:333–365.

Pert, P. L., J. R. A. Butler, J. E. Brodie, et al. 2010. A catchment-based approach to mapping hydrological ecosystem services using riparian habitat: A case study from the wet tropics, Australia. *Ecological Complexity* 7:378–388.

Petersen, I., Z. Masters, A. G. Hildrew, and S. J. Ormerod. 2004. Dispersal of adult aquatic insects in catchments of differing land use. *Journal of Applied Ecology* 41:934–950.

Petersen, I., J. H. Winterbottom, S. Orton, et al. 1999. Emergence and lateral dispersal of adult Plecoptera and Trichoptera from Broadstone Stream, UK. *Freshwater Biology* 42:401–416.

Postel, S., and S. Carpenter. 1997. Freshwater ecosystems services. In *Nature's services—Socieltal dependance on natural ecosystems*, ed. G. C. Daily, 195–214. Washington, DC: Island Press.

Potter, K. M., F. W. Cubbage, and R. H. Schaberg. 2005. Multiple-scale landscape predictors of benthic macroinvertebrate community structure in North Carolina. *Landscape and Urban Planning* 71:77–90.

REFCOND. 2003. Rivers and lakes—Typology, reference conditions and classification systems. Common implementation strategy for the Water Framework Directive. In *Guidance document no. 10*, ed. Office for Official Publications of the European Communities. Luxembourg: European Communities.

Roth, N. E., J. D. Allan, and D. L. Erickson. 1996. Landscape influences on stream biotics integrity assessed at multiple spatial scales. *Landscape Ecology* 11:16.

Roy, A. H., B. J. Freeman, and M. C. Freeman. 2007. Riparian influences on stream fish assemblage structure in urbanizing streams. *Landscape Ecology* 22:385–402.

Sauro, J., and J. R. Lewis. 2005. Estimating completion rates from small samples using binomial confidence intervals: Comparisons and recommendations. *Proceedings of the Human Factors and Ergonomics Society Annual Meeting. September 2005, Orlando, USA*, vol. 49 no. 24: 2100–2103.

Shearer, K. S., and W. N. Xiang. 2007. The characteristics of riparian buffer studies. *Journal of Environmental Informatics* 9:41–55.

Sponseller, R. A., E. F. Benfield, and H. M. Valett. 2001. Relationships between land use, spatial scale and stream macroinvertebrate communities. *Freshwater Biology* 46:1409–1424.

Strahler, A. N. 1957. Quantitative analysis of watershed geomorphology. *Transactions of the American Geophysical Union* 38:913–920.

Tiede, D., S. Lang, and C. Hoffmann. 2006. Supervised and forest type-specific multiscale segmentation for a one-level-representation of single trees. *Paper presented at the 1st International Conference on Object Based Image Analysis (OBIA)*. July 4–5, 2006, Salzburg, Austria.

Tiede, D., S. Lang, D. Hölbling, and P. Füreder. 2010. Transferability of OBIA rulesets for IDP camp analysis in Darfur. *Paper presented at Geographic Object-Based Image Analysis (GEOBIA)*, Ghent, Belgium, 29 June–2 July.

Tormos, T., P. Kosuth, S. Durrieu, S. Dupuy, B. Villeneuve, and J. G. Wasson. 2012. Object-based image analysis for operational fine-scale regional mapping of land cover within river corridors from multispectral imagery and thematic data. *International Journal of Remote Sensing* 33:4603–4633.

Tormos, T., P. Kosuth, S. Durrieu, B. Villeneuve, and J. G. Wasson. 2011. Improving the quantification of land cover pressure on stream ecological status at the riparian scale using high spatial resolution imagery. *Physics and Chemistry of the Earth* 36:549–559.

Wang, L. 2006. Introduction to landscape influences on stream habitats and biological assemblages. In *Landscape influences on stream habitats and biological assemblages*, eds. R. M. Hughes, L. Wang, and P. W. Seelbach, 1–23. Bethesda, MD: American Fisheries Society.

Wang, L., and P. Kanehl. 2003. Influences of watershed urbanization and instream habitat on macroinvertebrates in cold water streams. *Journal of the American Water Resources Association* 35:1181–1196.

Wang, L., J. Lyons, P. Kanehl, and R. Gatti. 1997. Influences of watershed land use on habitat quality and biotic integrity in Wisconsin streams. *Fisheries* 22:6–12.

Wasson, J. G., B. Villeneuve, A. Iital, et al. 2010. Large scale relationships between basin and riparian land cover and ecological status of European rivers: Examples with invertebrate indices from France, Estonia, Slovakia and United Kingdom. *Freshwater Biology* 55:1465–1482.

Weller, D. E., M. E. Baker, and T. E. Jordan. 2011. Effects of riparian buffers on nitrate concentrations in watershed discharges: New models and management implications. *Ecological Applications* 21:1679–1695.

Wold, H. 1966. Estimation of principal components and related models by iterative least squares. In *Multivariate analysis*, ed. P. R. Krishnaiaah, 391–420. New York: Academic Press.

Wold, H. 1982. Soft modeling: The basic design and some extensions. In *Systems under indirect observation*, eds. K. G. Jö Reskog and H. Wold, 589–591. New York: North–Holland.

Wold, S., M. Sjostrom, and L. Eriksson. 2001. PLS-regression: A basic tool of chemometrics. *Chemometrics and Intelligent Laboratory Systems* 58:109–130.

Zhu, Z., L. Yang, S. V. Stehman, and R. L. Czaplewski. 2000. Accuracy assessment for the U.S. Geological Survey Regional Land-Cover Mapping Program: New York and New Jersey region. *Photogrammetric Engineering and Remote Sensing* 66:1425–1435.

第四部分

水循环相关生态系统服务

第十一章 遥感水文生态系统服务评价

C. 卡瓦略-桑托斯　B. 马科斯　J. 埃斯皮尼亚·马克斯
D. 阿尔卡拉斯-塞古拉　L. 海因　J. 洪拉多

一、社会和水文服务

水在生态系统运转、支撑生化循环、支持生物有机体及其生长以及在地球上创造水生栖息地等方面发挥着重要作用(Chapin et al.,2002)。此外,人类和社会依靠生态系统提供水文服务和由此产生的效益(MA,2003)。根据所产生的效益,可确定两种主要类型的水文服务(图 11-1):(1)供水:包括家庭用水、灌溉用水和工业用水、水力发电、淡水产品、运输,以及娱乐和精神效益;(2)减轻水害:包括降低洪水的数量和严重程度,减少土壤侵蚀和沉积物沉积,以及减轻滑坡(Brauman et al.,2007)。这两种服务可按照三个维度评估:(1)数量(即水的总量)、(2)时段(即水的季节分布)和(3)质量(与污染物的清除、分解及沉积物的捕集有关)(Brauman et al.,2007;Elmqvist et al.,2009)。

生态系统中涉及水的生物物理学结构和过程决定了水文服务的供应(图 11-1)。生态系统的运转率决定了为人们提供潜在服务的能力(Haines-Young and Potschin,2010)。提供服务的内在能力本质上以生态系统功能的形式存在,独立于人类的选择,只有当人们使用或感受到此等功能的好处时,服务才会实现(Fisher et al.,2009)。通过一项服务,可产生多重收益并转化为福利所得——以经济、生态和社会评估为主题(Ansink et al.,2008;de Groot et al.,2010)。举例说,渗透的水量将补充地下水储备,增加蓄水能力(性质和功能;图 11-1)。一旦

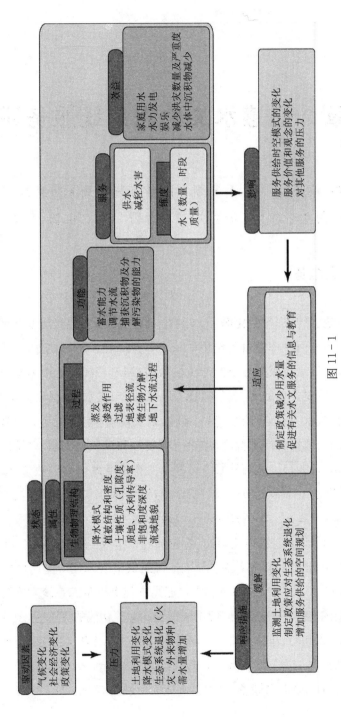

图 11-1

注：社会生态系统范围内提供水文服务的框架(基于 Rounsevell, M. D. A., et al., 2010; Haines-Young, R. and M. Potschin, 2010.)。

人们使用此类水,供水服务就转化为经济效益,比如水可供家庭消费使用。在过去的几十年里,人们对水问题的担忧与日俱增,尤其重视干旱和半干旱地区的水资源短缺问题(van Beek et al.,2011)。水的可用性由每个区域生物物理条件所决定,在气候变化的背景下,水文服务带来的降损影响往往会恶化(Brauman et al.,2007)。

生态变化的外部驱动因素如气候、社会经济和政治变化可能会影响水文服务的提供(图11-1)。此等驱动因素可能会影响那些直接影响生态系统状态的内部压力,如土地利用变化或需水量增加(Rounsevell et al.,2010)。此等压力将冲击水文服务的提供,并最终影响相应的利益。然而,水的综合评价和规划不仅必须考虑水的供应,还必须考虑水的需求(de Roo et al.,2012),因此需要政府和社会做出回应。为应对过度用水,减少水消耗的政策是一种重要的响应方式。为缓解压力,通过如土地利用变化或降水模式等方法监测内部压力,将有助于设计维持水文服务完整性的响应措施(图11-1)。

面对日益增加的水资源压力和生态系统中水流调节的紧张情况,可持续的水资源管理至关重要。联合国水机制里约+20峰会(UN-Water for Rio+20)的报告说明了这一点,该报告指出,"绿色经济"的成功取决于水资源的可持续管理、供水和适当的卫生服务(UN,2011)。苏等人(Su et al.,2012)建议促进科学知识分享,并在各国之间提供水循环方面的能力建设和技术转让,包括使用遥感,特别是使用卫星遥感技术来改善水资源管理。因此,遥感的使用对于支持水管理至关重要,特别是通过允许评估和监测水平衡要素以及在空间和时间上提供水文服务的生态系统状况(Wagner et al.,2009)。

本章旨在概述如何利用遥感技术支持可持续水资源管理。我们着重研究了如何利用遥感,特别是卫星遥感技术,分析和监测与水循环和提供水文服务相关的生态系统成分,以及遥感如何支持水文建模。我们的分析基于对遥感和水文文献的回顾及我们在水文服务空间建模方面的经验。

二、水文服务和水循环

千年生态系统评估框架强调了人类福祉与自然系统水循环间的密切关系

(MA，2003)。说明这一联系的最佳方法是利用水循环概念模型，该模型报告了整个地球水库(即海洋、冰盖、冰川、含水层、河流、湖泊、土壤和大气)的持续水流(Fetter，2001)。

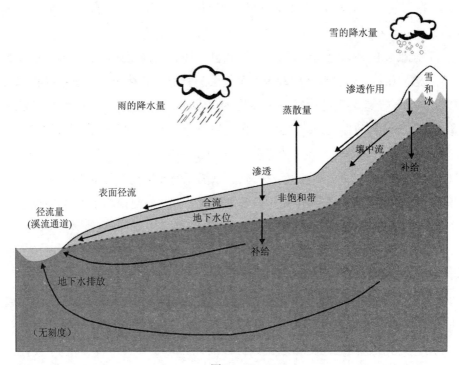

图 11-2

注：地表附近的水循环。(概念模型基于 Fitts，C. R.，2002.)

水循环中涉及的主要物理过程是蒸发、蒸散、凝结、降水、下渗、地表径流、壤中流、径流量和地下径流(Fitts，2002)。

上一节所述的水文服务(供水和减轻水损害)与发生在或接近陆地表面的水循环部分密切相关(图 11-2)。全球水循环的这一阶段可简明地描述如下(Fitts，2002；van Brahana，2003)：(1)降水产生的水到达地面并渗入或产生地表径流(取决于地形条件、植被覆盖、土壤质地和结构，及自然或人为不可渗透层的存在，及其他因素)；(2)一旦进入非饱和带，水可能会在水平方向(相互流动)下循环，经过一段较短的路径后出现在斜坡或河岸上；(3)水可能会沿着地下水位的方向向下循环,并对含水层进行补给;(4)在这里,地下水通过饱和介质循环,最终排入河流(基流)。

供水服务首先取决于降水量和蒸散量。之后,入渗和地表径流的平衡决定了后续径流量的时间。若大部分来自降水的水由于缓坡、高渗透性土壤和足够的植被覆盖而渗入,含水层的回灌率可能会增加,从而导致全年更大的河流基流。在此等情况下,供水服务将受益于有利于渗透的非饱和带特征,产生更多的地下水资源(如可通过井利用)及更及时地分布的地表水资源(如可通过水坝利用)(Lal,2000;Fetter,2001)。除其他因素外,水质在很大程度上取决于非饱和带发生的自然衰减(Fetter,1999)。事实上,由于物理化学和生物过程的作用,污染物的浓度随着水渗过不饱和带而降低。

非饱和带的性质也紧紧控制着有关水毁减灾的水文服务。同时,非饱和带的结构和水文地质学特征是控制渗透和地下水补给的关键问题(Espinha Marques et al.,2011)。植被、土壤类型和地形的综合作用在很大程度上决定了洪水和土壤侵蚀的发生。例如,稀疏的植被与质地细密的土壤(黏土或淤泥)相结合,可能会促进地表径流产生并减少水体的渗透,从而增加此类自然灾害的风险。此外,在地形陡峭的高度异质非饱和带(关于水力传导)可能会由于暴雨引发山体滑坡(Fernandes et al.,2004)。

三、生态系统遥感对提供水文服务的作用

在过去的几十年里,有大量关于遥感和水文科学的评论和特刊发表(Kite and Pietroniro,1996;Rango and Ahlam,1998;Pietroniro and Prowse,2002;Schmugge et al.,2002;van Dijk and Renzullo,2011;Fernández-Prieto et al.,2012;Su et al.,2012)。最近,一些尝试结合遥感和生态系统服务(所有类别)的文章也已发表(Feng et al.,2010;Ayanu et al.,2012)。此节将探讨用于水文服务研究的遥感,特别是用于观测和监测不同地球水库(大气层、低温圈、地表水、土壤/地面和植被)中的水(液态、固态和气态)的卫星遥感技术和传感器。

表11-1描述了一些用于评估与供水和(或)水灾害减轻服务相关的单个水要素所需的卫星产品。虽然本文是一篇很全面的综述,但其并没有描述所有可

用的卫星产品。

(一) 供水

1. 大气层

在大气中,水蒸气、云和降水(降雨和雪)是影响供水服务提供的重要水元素。水蒸气和云吸收并发射红外辐射(IR),云还反射可见光辐射(Visible Radiation,VIS),这有助于地球的能量平衡(Su et al.,2012)。微波辐射计用于测量发射的辐射,而光学仪器用于测量可见光辐射、红外辐射和近红外辐射。观测云和间接降水(VIS 和 IR)背后的理由是:与地球表面相比,云反射率越高温度越低(Tapiador et al.,2012)。

搭载在提供大气层产品的泰拉(Terra)卫星和水(Aqua)卫星上的中分辨率成像光谱仪和米堤欧卫星 8 号(METEOSAT-8)上的改进型自旋可见光和红外成像仪(SEVIRI)都被用于使用改进的算法在高空间分辨率下获取云特性(Kidd et al.,2009)。利用激光探测和测距(LIDAR)以及使用电磁波探测和测距(Radio Detection and Ranging,RADAR)来获取云的物理属性,最先进的测量来自美国宇航局(NASA-CSA)于 2006 年发射的 Cloudsat 和 Calypso 卫星(Su et al.,2012)。对于水汽观测(全水柱、水汽剖面和对流层上层湿度),舒尔茨等人(Schulz et al.,2009)和苏等人(Su et al.,2012)对一些反演算法和传感器进行了全面描述。例如,气候监测卫星应用设施(CM-SAF)使用 METEOSAT 卫星和美国国家海洋和大气管理局(NOAA)运行卫星上的若干仪器来提供云和水蒸气参数(Schulz et al.,2009)。

微波技术可更好地测量降水速率,这是由于雨粒子的强相互作用和对云覆盖的相对不敏感(Michaelides et al.,2009;Tang et al.,2009)。基德等人(Kidd et al.,2009)、米海利季斯等人(Michaelides et al.,2009)和塔皮亚多等人(Tapiador et al.,2012)详尽地描述了降水卫星产品。多传感器技术,如热带降水测量任务(Tropical Rainfall Measuring Mission,TRMM)拥有用于被动微波检索的 TRMM 微波成像仪(TMI)和用于主动微波的第一台星载测雨雷达(PR),从 1997 年到现在,每隔三小时提供一次降水估算,并经常用于全球气候模型(Tapiador et al.,2012)。

表 11-1 测量水循环元素的传感器和卫星的示例

水循环	服务	类型	产品/传感器	人造卫星	空间分辨率	时间分辨率	成本	期限	机构/来源
水蒸气	供水	光学	来自全球臭氧层监测实验(GOME/SCIAMACHY)的水蒸气柱总量	ERS	80~40千米；60~30千米	3 天	有	1995 年至今	EUMETSAT;http://Gome. aeronomie. be
			中分辨率成像光谱仪的水蒸气柱总量	ENVISAT	300 米	3 天	有	2005 年至今	欧空局；http://www.envirport. org/meris
			气象卫星可见光红外成像仪(MVIRI)	METEO-SAT	5 千米	30 分钟	有	1977 年至今	EUMETSAT; http://www.eumetsat. int
		被动微波	来自大气红外探测器(AIRS)的水汽总剖面和云产品	Aqua	3~14千米	每天	免费	2002 年至今	NASA;http://airs. jpl.nasa. gov
			来自专用传感微波成像仪(Special Microwave/Imager, SSM/I)的总可降水量产品	NOAA 极地轨道环境卫星星座(POES)	15~20千米	4 小时	是的	1987 年至今	NESDIS;http://www.ncdc. noaa. gov
云	供水量和供水时间；减轻水害(防雨防汛)	光学	来自中分辨率成像光谱仪的大气产品(云温度、高度、发射率等)	Terra	1 千米	一天一两次	免费	2000 年至今	nasa;http://modis-atmos.gsfc. nasa. gov
			红外大气探测干涉仪(IASI)的水汽剖面、云顶温度和压力产品	METOP-A	1~25千米	3 天	是的	2007 年至今	EUMESAT;http://www. eumetsat. int

续表

水循环	服务	类型	产品/传感器	人造卫星	空间分辨率	时间分辨率	成本	期限	机构/来源
云	供水量和供水时间;减轻水害(防雨防汛)	光学	改进型甚高分辨率辐射计的云产品	NOAA 的 POES 星座	25 千米	每周一次	免费	1982 年至今	NESDIS;http://noaasis.noaa.gov
				Cloudsat	2.5 千米	太阳同步轨道	收费	2006 年至今	NASA-CSA;http://cloudsat.atmos.colostate.edu
		微波雷达	云物理属性	云-气溶胶激光雷达和红外探路者卫星观测 (CALIPSO)	0.3～5 千米	太阳同步轨道	收费	2006 年至今	NASA-CSA;http://www-calipso.larc.nasa.gov
降水	供水量和供水时间;减轻水害(防雨防汛)	光学	雨水估算器-GOES 地球同步卫星	NOAA 的 POES 星座	4 千米	3 小时	免费	2002 年至今	NESDIS;http://www.star.nesdis.NOAA.gov
			MODIS 大气产品-36 通道 VIS/IR 传感器	Terra 和 Aqua	1 千米	每天	免费	2000 年至今 2002 年至今	nasa;http://modis-atmos.gsfc.nasa.gov

续表

水循环	服务	类型	产品/传感器	人造卫星	空间分辨率	时间分辨率	成本	期限	机构/来源
降水	供水量和供水时间;减轻水害(防雨防汛)	光学	自旋增强可见光红外成像仪(SEVIRI)	METEOSAT-8	1千米	15分钟	收费	2002年至今	EUMESAT;http://www.eumetsat.int
		被动微波	改进型微波扫描辐射计(AMSR-e)	Aqua	25千米	次日	免费	2002~2011年	NASA;http://www.ghcc.msfc.nasa.gov/AMSR
			TRMM微波成像仪(TMI)	热带地区的热带降雨测量任务卫星	250米~5千米	3小时,每天,每周,每月	免费	1997年至今	NASA+JAXA;http://trmm.gsfc.nasa.gov
		主动微波	空间降水雷达	TRMM	250米~5千米	3小时,每天,每周,每月	免费	1997年至今	NASA+JAXA;HTTP://pmm.nasa.gov/node/162
		多卫星/传感器及算法	人工神经网络-云分类系统数据库	TRMM和计量站	4千米	3小时,每天,每周,每月	免费	2000年至今	chrs;http://chrs.web.uci.edu/persiann
			气候变化技术(气候预测中心)	地球卫星红外数据+被动微波AMSR-e和TMI	8千米	30分钟至每天	免费	2002年至今	NWS;http://www.cpc.ncep.NOAA.gov

续表

水循环	服务	类型	产品/传感器	人造卫星	空间分辨率	时间分辨率	成本	期限	机构/来源
冰雪	供水量（冰川）；水损害缓解（海平面上升，洪水）	光学	AVHRR的雪产品	NOAA的POES星座	25千米	每周一次	免费	1966年至今	NESDIS; http://www.nsoft.class.NOAA.gov
			MODIS雪积–归一化差分雪指数（Normalized Difference Snow Index, NDSI）	Terra和Aqua	500米~1千米	每日、每周和每月合成	免费	2000年至今	nsIDC+DAAc; http://modis-snow-ice.gsfc.nasa.gov
		被动微波	来自AMSR-e的雪水当量（SWE）	Aqua	25千米	每天	免费	2002~2011年	Vua+NASA+NSIDC; http://nsidc.org
			来自CryoSat任务的北极冰图	CryoSat-2	5~10千米（250米）	30天	收费	2010年至今	欧空局; http://www.esa.int
		主动微波	RADARSAT冰监测	RADARSAT-1/3	500千米	每天	收费	1995~2013年；2007年至今	CSA+MDA; http://www.asc-csa.gc.ca/eng/satellite/radarsat

续表

水循环	服务	类型	产品/传感器	人造卫星	空间分辨率	时间分辨率	成本	期限	机构/来源
冰雪	供水量(冰川);水损害缓解(海平面上升,洪水)	多卫星/传感器及算法	来自太空的全球陆地冰测量冰川数据库	先进星载热发射和反射辐射计(Advanced Spaceborne Thermal Emission and Reflection Radiometer, ASTER)结合卫星影像和其他冰川监测地理信息系统数据的数据库	150米	16天	收费	2005年至今	NSIDC+USGS; http://www.glims.org
			GlobSnow	SMMR、SSM/I和AMSR-e 结合地面数据	25千米	每天,每周,每月	收费	1979年至今	欧空局; http://www.globsow.info
			交互式多传感器冰雪测绘系统	NOAA/POES-VHRR;MODIS;ME-TEOSAT	4~24千米	每天,每周	免费	1997年至今	NIC-NOAA/NESDIS; http://www.natic.NOAA.gov/ims/ims.html
蒸散	供水和干旱监测	多卫星/传感器及算法	MOD 16 ET 产品来自MODIS土地覆盖、LAI/FPAR和全球地表气象学(GMAO)	Terra	250米~1千米	16天	免费	2001年至今	NASA; http://modis.gsfc.nasa.gov

续表

水循环	服务	类型	产品/传感器	人造卫星	空间分辨率	时间分辨率	成本	期限	机构/来源
蒸散	供水和干旱监测	多卫星/传感器及算法	地面能量平衡算法(Surface Energy Balance Algorithm for Land,SEBAL)使用不同种类的信息(例如,表面温度、半球反射率和NDVI)	例如,METEOSAT和NOAA-AVHRR	根据研究结果	根据研究结果	免费	1998年至今	Bastiaanssen et al., 1998年
地表水	供水(数量、时间和质量);减轻水害	雷达高度计	Jason-2 海洋地形	海洋表面地形任务(OSTM)	2千米	10天	收费	2008年至今	NASA、CNES、EUMETSAT、NOAA; http://www.nasa.gov
		光学	使用可见光波段传感器(如SPOT)的地表水波段	SPOT	5~25米	26天	收费	1986年至今	CNES; http://www.cnes.fr
		光学	MERIS 水质波段	ENVISAT	300米	2~3天	收费	2005年至今	ESA; https://earth.esa.int
	供水数量和时间;减轻水害(滑坡和洪水)	主动微波	高级散射计(Advanced Scatterometer,ASCAT)	METOP-A	25千米	每天	收费	2006年至今	EUMETSAT+ESA; http://manat.star.nesdis.noaa.gov
			先进合成孔径雷达	ENVISAT	1~5千米	每周一次	收费	2005年至今	ESA; https://earth.esa.int
土壤湿度			AMSR-e	Aqua	25千米	次日	免费	2002~2011年	Vua+NASA+NSIDC; http://nsidc.org/data/amsre

续表

水循环	服务	类型	产品/传感器	人造卫星	空间分辨率	时间分辨率	成本	期限	机构/来源
土壤湿度	供水数量和时间;减轻水害(滑坡和洪水)	多卫星/传感器及算法	土地参数反演模型(Land Parameter Retrieval Model,LPRM)	AMSR-e、Nimbus SMMR、trmm tmi,SSM/I	50千米	次日	免费	1978年至今	vua;http://www.falw.vu/~jeur/LPRM/pubs.htm
		被动微波	孔径合成微波成像辐射计的土壤水分	土壤水分和海洋盐度卫星(SMOS)	50千米	3天	免费	2009年至今	欧空局;http://www.esa.int
		光学	可见光波段(植被识别点)和NDVI	Landsat/IKONOS 和MODIS	250米	10~30天复合材料	免费(Ikonos 是)	1972年至今;1999年至今	NASA
			NIR波段(地下水排放温度)	Landsat-5/7	15~60米	16天	免费	1999~出席	NASA;http://landsat.usgs.gov
地下水	供水量	重力测量学	重力反演与气候实验(GRACE)地球微重力模型	GRACE	400~500千米	30天	收费	2002~出席	NASA + DLR;http://www.csr.utexas.edu/grace
			重力场和恒稳态海洋环流探测器(Gravity Field and Steady-State Ocean Circulation Explorer,GOCE)地球重力场和大地水准面模型	GOCE	100千米	10~30天	收费	2009~出席	欧空局;http://www.esa.int

注:这里描述的大多数卫星/产品可很好地用于除指定卫星/产品之外的其他水循环用途。

利用人工神经网络从遥感信息中估算降水（Precipitation Estimation from Remotely Sensed Information using Artifcial Neural Networks, PERSIANN-CCS）是一种将低空极地轨道卫星和地球静止红外影像结合起来的降水算法。此外，将记录与测雨观测值进行比较，并用被动微波雨量值进行调整，以获得高度准确的雨量估算值（Soroshian et al.，2000；Sahoo et al.，2011）。国际全球降水测量任务（GPM）卫星将于2014年发射，其携带先进的仪器，将利用GPM卫星星座，成为TRMM的替代者，为太空降水测量设定新的标准（Su et al.，2012；Tapiador et al.，2012）。

2. 冰冻圈

对冰雪的观测是很重要的，因为流域冷水储存的时间决定了不同的径流和河流流量，影响供水时段和减轻水害（海平面上升和洪水）。与降雨相比，不同的物理性质需要不同的传感技术（Schmugge et al.，2002）。考虑到雪和其他类型土地覆盖间的反射率对比，VIS和NIR波段被用于绘制雪图。即便如此，由于雪和云之间存在相似性，使得区分二者或在获取雪深信息中存在一定的局限性（Wagner et al.，2009）。在中分辨率光学影像产品中，冰雪测图的最佳来源是具有高时间、高空间分辨率的MODIS雪产品和NOAA极地轨道平台星座（POES）上的改进型甚高分辨率辐射计产品。

从低层地面发射的微波辐射（Microwave Radiation，MR）被雪层中的雪粒向诸多不同方向散射。反过来，微波辐射可受到积雪的一些性质的影响，例如颗粒的尺寸，可能受到被动和主动微波观测的影响（Schmugge et al.，2002）。在辐射传输过程中，雪水当量（SWE）是一个对被动微波信号敏感的参数。同样，全球降雪监测（GlobSnow）产品将地面数据与被动观测相结合，使用扫描多通道微波辐射计（SSMR）、专用传感器微波成像仪（SSM/I）和先进微波扫描辐射计（AMSR-e；Luojus et al.，2010）来监测降雪范围和雪水当量。主动微波仪器提供具有更高空间分辨率的冰雪地图，例如最近发射的CryoSat任务，空间分辨率为5千米（Drinkwater et al.，2003）。

然而，被动和主动微波方法有一些限制，影响反演，即与光学观测相比，时间和空间分辨率更粗糙，并且对积雪微物理性质的敏感性很强（Tang et al.，2009）。多传感器产品的组合，如交互式多传感器冰雪测图系统（IMS），从

NOAA-AVHRR、Terra/Aqua MODIS 和 METEOSAT 卫星收集信息,是克服此等限制并生成大规模雪数据的替代方案(Helfrich et al.,2007)。

3. 地表水

地表水包含内陆水和海水,同时包括它们的储存和排放。地表水的观测有助于评价供水服务和水质,其中卫星观测是非常重要的。水在近红外和中红外吸收能量,而土壤和植被在此等波长反射能量(Pietroniro and Prowse,2002)。与此类似,光学影像已用于绘制洪泛区、湖泊和水库的地图,但是伴随存在掩盖底层水体的云相关的限制,对水体量化有限制(Alsdorf and Lettenmaier,2003)。为克服这一点,雷达测高法被广泛用于监测水面水位,并与高程一起计算水量(Alsdorf and Lettenmaier,2003;Su et al.,2012)。20 世纪 90 年代初,雷达测高仪,如 ERS-1 和 TOPEX-Poseidon,开始提供大的湖泊和水库的信息;现在则由正在运行的第二代 ERS-2、ENVISAT、GFO 和 Jason-2(Tang et al.,2009)继续提供这些信息。未来,地表水和海洋地形(SWOT)任务将克服当前测高仪器的空间和温度分辨率限制,并包括轨道之间缺失的水体(Tang et al.,2009;Su et al.,2012)。事实上,单靠卫星观测无法测量水流流量和流速(Tang et al.,2009)。估算河流流量的一种方法是使用遥感水力信息,除雷达测高外,可选择是否与水文模型结合(见第十一章,第五节)(Bjerklie et al.,2003,2005)。最后,空间重力测量任务——重力反演和气候实验(GRACE)卫星,提供了水量的测量,其与降水和蒸散数据相结合是近实时估算大陆淡水排放量的有用工具(Syed et al.,2010)。

监测水质的遥感技术发展始于 20 世纪 70 年代初。大多数研究评估了光谱特性和水质参数间的经验关系(Ritchie et al.,2003)。影响水质的因素可分为:(1)改变反射太阳能谱和(或)从地表水发射热辐射的因素,此等因素可使用遥感技术进行测量,如悬浮沉积物/浊度(如 Potes et al.,2011)、藻类(即叶绿素、胡萝卜素)(Carvalho et al.,2010;Song et al.,2012)、溶解有机物(DOM;例如 Del Castillo and Miller,2008;Jørgensen et al.,2011)、油类(如 Jha et al.,2008)、水生维管植物(如 Santos et al.,2009;Ward et al.,2012)和热释放(如 Alcantara et al.,2010);(2)通过测量其他水质参数(对能谱敏感)间接推断出来的物质,如大多数化学物质(即营养素、杀虫剂、金属)(Hadibarata et al.,2012)和病原体

(Tran et al.,2010)。多光谱影像传感器,如 Envisat 卫星上的中分辨率成像光谱仪,为悬浮沉积物、叶绿素和其他水质参数提供了潜在应用的光谱波段(Ayanu et al.,2012)。与不易获得的现场测量相比,使用卫星观测工具识别和监测地表水质问题的优势在于数据具有空间和时间覆盖(Ritchie et al.,2003)。

4. 土壤和地面

地表以下的水是位于非饱和生根区的土壤水分,以及饱和区的地下水(图 11-2)。在 VIS 和 IR 波段工作的光学传感器可通过区分水平衡方程元素间接推断土壤水分(Su et al.,2012)(见第十四章)。然而,由于干燥土壤的介电常数特性(其在存在水的情况下改变),通过主动和被动探测在微波波段下更精确地测量土壤水分(Schmugge et al.,2002)。主动微波传感器(局部到区域尺度的 SAR 和用于全球监测的散射计)发射电磁脉冲,并捕获从地球散射回的测量电磁能量(Su et al.,2012)。一个例子是来自高级散射仪的全球土壤湿度产品,该产品安装在 EUMESAT 的 METOP-A 卫星上,从 2006 年运行至今,空间分辨率为 25 千米(Bartalis et al.,2007)。反过来,被动微波传感器测量从地球表面发射的辐射。获得土壤水分信息的算法有几种。一个是土地参数检索模型(Land Parameter Retrieval Model,LPRM),其结合了从 1978 年到现在的历史数据集的全球土壤湿度模型(Owe et al.,2008)。然而,工作在 C 波段和更长波长的被动传感器在植被覆盖丰富的地区受到限制(Tang et al.,2009)。为了克服这一限制,2009 年发射的土壤湿度和海洋盐度(SMOS)卫星任务每三天以 50 千米的空间分辨率进行观测(Albergel et al.,2011)。土壤水分的近实时观测的优点是:(1)更好地了解水循环;(2)揭示其对气候变化的影响;(3)改进对洪水和干旱等自然灾害的预测(Su et al.,2012)。

地下水是水循环的最后一个受益于卫星技术的组成部分,但由于其显著的季节性和年际变化性,使得对其监测非常重要。卫星测量的植被分布、地形、温度、土壤湿度和重力已被用于收集有关地下水存在的信息(Becker,2006)(见第十三章)。重力测量基于水在地球不同部位中的重新分配原理,其重分配过程随重力场而变化(Su et al.,2012)。同样,地下水由 GRACE 测量,GRACE 于 2002 年发射,是一个双星任务,旨在绘制地球重力场的静态和时变分量(Ramillien et al.,2008)。Grace 通过分离其他部位(土壤水分、海洋、蒸散)的影响,提供了对地下水储存变化的高精度测量的方法(Rodell et al.,2006;Llovel et al.,2010)。最后,于2009年发射的重力场和稳态海洋环流探测器(Gravity Field and Steady-State Ocean Circulation Explorer,GOCE),正以无与伦比的精度绘制地球引力图,是对GRACE测量的补充。

5. 植被

从局部到全球尺度下,植被含水量(Vegetation Water Content,VWC)的估算是了解环境中水流的核心,是干旱和火灾监测的重要变量。考虑到现场测量能够可靠地与VWC的光谱反射率关联,遥感技术为VWC评估(近红外和短波红外的反射率)提供了一种即时且无损的方法(Wu et al.,2009)。然而,此等类型的方法需要进一步改进,以解释观察到的叶结构、叶干物质、冠层结构和叶面积指数的影响(Zarco-Tejada et al.,2003)。

利用各种植被水分指数(Yilmaz et al.,2008)估算了冠层水分含量,此等植被水分指数由归一化差异水分指数(Normalized Difference Water Index,NDWI)、归一化差异红外指数(Normalized Difference Infrared Index,NDII)、最大差异水分指数(Maximum Difference Water Index,MDWI)和水分带指数(Water Band Index,WBI)组成,并已被证明适用于VWC的估算(Chen et al.,2005)。最近,研究人员通过基于辐射传输模型的遥感技术,如PROSPECT和SAILH,探索了VWC估算的方法模型(Jacquemoud et al.,2009;Suárez et al.,2009)。

(二) 水毁减灾

供水部分第一节所述的所有卫星和传感器也可用于监测与水相关的灾害。尤其是降水,是与水相关的危险触发因素。因此,为进行准确的风险评估,在全球范围内估算精确的降水量是很重要的(Tapiador et al.,2012)。与地表观测相比,卫星平台对气象现象的观测提供了更准确的视图,在气候评估和极端事件预测方面具有优势(Kidd et al.,2009)。可使用光学影像(如Landsat系列)在不同时期探测洪水和相关损害。微波传感器(SRTM、MODIS、AMSR-e)可用来重建地表动力学历史,有助于预测洪水、海岸淹没和滑坡等先前事件(特别是更具破坏性的事件)造成的危害(Tralli et al.,2005;Syitski et al.,2012)。监测冰盖和融化过程对于跟踪海平面上升和可能的海岸淹没和侵蚀过程很重要。CryoSat（2010年至今）和ICESat（2003～2010年）等任务提供了确定冰盖质量平衡所需的多年海拔高程数据。

可使用光学影像（如QuickBird）识别滑坡，并可以使用收集地貌和土壤水分信息的RADAR和InSAR技术进行预测（Tralli et al.,2005; van Westen et al.,2008）。较高时间和空间分辨率卫星可用于预警滑坡易发性增强变化，并有助于了解导致边坡崩塌的过程（Wasowski et al.,2010）。

干旱是由于蒸散量和温度增加、降水不足和土壤水分减少而导致缺水造成的。一些指数使用水循环的不同元素来推断关于干旱严重程度的信息（植被、土壤湿度和蒸散量；Su et al.,2003）。这在干旱和半干旱生态系统中尤其突出，干旱是控制生态系统生产力年际和年内变化的主要因素，从而控制生态系统提供的效益（Hein et al.,2011）。已有不同的卫星数据被收集，以支持风险评估。例如，达特茅斯洪水观测站（Dartmouth Flood Observatory）提供了全球水资源数据库和洪水及干旱的关键指数，可用于全球和区域风险评估（如全球洪水风险；Jongman et al.,2012）。

四、水文服务的驱动因素和压力的遥感

遥感对于监测可能影响提供水文服务的驱动因素和压力非常重要，包括淡水资源的安全及在气候变化背景下减轻水危害。气候变化是生态系统功能和服务的全球驱动力，与降水模式的变化及其他因素有关（图11-1）。因此，气候监测对水资源管理至关重要，自20世纪70年代末以来，已从METEOSAT和NOAA系列卫星上收集了高精度的长期降水数据集（Kidd et al.,2009）。土地利用变化，从城市扩张到农田废弃或森林经营的集约化，是水文服务的另一个重要压力。目前，利用QuickBird和IKONOS等卫星影像，可以在非常高的空间和时间精度上监测土地利用变化（Rogan and Chen,2004）。

五、遥感数据与水文模型的集成

水文模型的使用使管理者能够了解河流集水区对大气强迫条件的响应,这对于更准确的水资源管理和水灾害预测和缓解非常重要(Xie and Zhang,2010)。

空间数据对更复杂的基于物理的分布式水文模型的需求不断增加,再加上更复杂的遥感产品的出现,增加了水文学家对遥感应用的兴趣(Kite and Pietroni r,1996;Pietroniro and Prowse,2002)。特别是,遥感数据带来了改进水文模型校准和验证的新机会(Montanari et al.,2009)。对水文模拟可能有用的遥感产品清单包括提供有关降水、土地利用、土壤湿度、排放量和蒸散量的数据(Pietroniro and Prowse,2002)。

使用遥感产品的优点可扩展到其具有全区域覆盖的近乎实时的数据可用性,这使得即使在空间和时间上缺乏地面观测的区域也可进行水文建模(Grimes,2008)。以接近实时的特征为重点,水文模型可使用被称为数据同化的一系列技术来丰富时间连续性和动态信息(Walker and Houser,2005)。此等技术合并模型和观测值,解决来自不同强制条件和参数化的不确定性问题,提高了模型性能(Xie and Zhang,2010)。尽管遥感在不同的空间尺度上提供了连续不断的最新的测量方式,但它仍依赖于地面观测来开发、校准和验证算法(Tang et al.,2009)。遥感数据可用于估算影响流域或区域中水过程的水文重要生物地貌变量(地形、土地覆盖/使用、土壤数据)及水文状态变量(例如降水)(Pietroniro and Prowse,2002)。

(一)水文生物地貌变量

水文生物地貌变量是物理和分布式水文模型的空间输入数据。地形数据用于水文模型的划定和离散化步骤(Arnold and Fohrer,2005)。此等类型的数据以高精度分辨率收集,主要来自 RADAR,但也来自短波传感器。最流行的可以免费获得的资源是 NASA 的航天飞机雷达地形任务(Shuttle Radar Topography Mission,SRTM)和先进星载热发射和反射辐射仪获取的全球数字高程模型(Global Digital Elevation Model,GDEM)。此外,土地覆盖/使用和土壤数据对于分布式水文模型尤其重要,在分布式水文模型中,水文响应单元受到土地覆盖和土壤特征的空间变异性的影响(Pietroniro and Prowse,2002)。有免费的土地覆盖数据集可以使用,其基于卫星影像的解译。全球土地覆盖2000(1∶5 000 000)可在美国地质调查局网站上查阅。对于欧洲国家,CORINE 土地覆盖(1∶100 000)使用卫星影像解译(Landsat、SPOT-4、SOPT-5、IRS-P6、Liss iii)来生成土地覆盖产品,该产品在1990年、2000年和2006年可用,并对所有欧洲国家进行通用分类(EEA 1997)。最后,关于土壤信息,联合国粮食及农业组织(Food and Agricultural Organization,FAO)提供的低分辨率"世界数字土壤地图"也可免费获得。

(二) 水文状态变量

从卫星观测中得到的水文状态变量已被引入应用,以补充甚至取代水文模型的现场输入(Tang et al.,2009)。一些主要来自卫星传感器的气候数据、蒸散量、土壤水分、蓄水和排放的例子,用于校准和验证水文模型(van Dijk and Renzullo,2011)。蒸散是全球能源预算和水文循环间的主要联系变量(Smith and Choudhury,1990)。卫星遥感提供日常观测,如植被、能量和地表温度,用于估算蒸散量(Courault et al.,2005)(见第十八章)。该过程采用了两种方法:(1)基于能量平衡的物理模型;(2)将蒸散与整个生长期植被指数测量值联系起来的经验模型(Zhang et al.,2009)。基于第一种方法的示例是广泛使用的模型算法——陆地表面能量平衡算法(Surface Energy Balance Algorithm for Land,SEBAL;Bastiaanssen et al.,1998)。SEBAL 被引入水文模型的校准过程中,与传统的基于地面的数据算法用于蒸散计算相比,其结果更好(Immerzeel and Droogers,2008)。

卫星衍生的水文状态变量在测量不佳的流域中是必不可少的,在此等流域中,水文地面数据的可用性对校准过程是一个挑战(Milzow et al.,2011)。其中的一些变量已成功地用于若干水文建模工作(Kite and Pietroniro,1996;Fernández-Prieto et al.,2012)。降水是水文模拟中最重要的输入,一些卫星任务及其衍生产品已对降水进行了估算,如 TRMM。在一些地区,例如亚马孙盆地,TRMM 卫星降雨数据的性能与雨量计观测获得的数据相当(Collischonn et al.,2008)。然而,中国最近的一项研究表明,与使用雨量计观测数据相比,使用

TRMM 数据有利于每月径流模拟,但不适用于每日模拟。因此,需要进一步发展基于卫星的每日降雨量估算算法(Li and Zhang,2012)。陆地上的卫星雷达测高仪允许检索小型和狭窄水体的数据,这些数据可使用评级曲线方法转换成流量,以校准和验证水文模型(Leon et al.,2006;Calmant et al.,2008)。结合地面排放数据,它可在更精细的时间和空间尺度上改进排放估算(Michailovsky and McEnnis,2012)。

(三) 遥感在水土评估工具中的应用

在耦合模型输出中,地面测量和遥感地球观测的结合是更精确地解决水平衡方程的有趣方法(Tang et al.,2009)(见第十二章)。结合这种方法广泛使用的水文模型是土壤和水评估工具(Soil and Water Assessment Tool,SWAT),该工具由美国农业部研究服务部(Arnold and Fohrer,2005)在 20 世纪 90 年代早期开发。SWAT 被开发用于预测土地管理对水资源的影响,按月或日的时间步长执行模拟排放程序。它是一个分布式的、基于物理的模型。

引入遥感数据来校准和验证的原始 SWAT 的改进已展开(Gassman et al.,2007)。还引入了提高模型可靠性的数据同化技术,例如使用土壤水分的技术(Han et al.,2012)。托宾和本尼特(Tobin and Bennett,2012)比较了美国六个流域不同降水产品(一种来自测雨观测,另一种来自卫星数据)的效率,产生了按次月时间步长的径流,在 TRMM 和地面降水数据之间观察到相当的性能。那拉辛汗等人(Narasimhan et al.,2005)应用 NOAA-AVHRR 传感器导出的归一化植被指数(NDVI)来验证 SWAT 模拟的土壤水分,作为传统径流校准和验证的补充。结果表明,模拟土壤水分可作为半干旱条件下作物胁迫的良好指标。在对南非一条河流的研究中,最初的 SWAT 代码引入了一些变化,允许结合雷达测高仪、降水产品、SAR 表面土壤湿度和 GRACE 总储水量变化数据。尽管降水产品对降水量的估算有所不同,但在降水差异较大的时期,利用地表土壤水分和总储水量可确定可能的误差(Milzow et al.,2011)。

六、结论和观点

本章详尽地回顾了地球观测在水文服务的评估、管理和监测方面的诸多优点。首先,与地面测量相比,它大大提高了分布式信息的空间质量。同样,它可更好地覆盖地面条件较差的偏远地区。其次,它可提供近乎实时的信息,这在预测自然灾害和启动应急计划(例如 TRMM)方面尤其有用,在降水警报方面已经高度发展。最后,遥感的使用使得评估水文服务的三个方面(即数量、时间和质量)成为可能。考虑到水循环时间覆盖信息的一致性和频繁性,可改进水循环要素的年际变化和季节变化的研究(Schulz et al.,2009)。自 20 世纪 70 年代初以来,已在全球尺度上获得陆地卫星和其他影像(用于更详细地分析植被和土地利用模式)。

然而,在参数化方面及观察结果的费用方面仍存在一些限制。卫星收集的水文信息需要强有力的参数化和验证,以提高水文研究的准确性和一致性,最好使用地基测量(Tang et al.,2009;Hein et al.,2011)。虽然地球观测的成本可能很高,这取决于其的空间和时间分辨率,但使用卫星观测似乎比水文参数观测的传统方法具有更高的成本效果(Dreher et al.,2000;Pietroniro and Prowse,2002)。

未来,越来越多的更复杂的卫星任务将为观测、分析和监测水循环的不同组成部分创造新的机会(Fernández-Prieto et al.,2012)。此外,遥感应用,特别是卫星产品,在水文模拟方面的整合取得了显著进展,提高了产品的空间和时间分辨率。剩下的挑战是开发/改进卫星算法,使其适应水文模型的常规及进一步的数据同化过程。此外,此等卫星产品应根据每个地区的环境特征进行校准和验证。

总之,对供水和水毁减灾服务的评估从遥感技术和数据,特别是卫星产品中受益很大。在空间上明确和准实时的基础上,卫星产品有助于提高对提供水文服务的过程和功能的理解。此外,还可评估和监测影响水循环和提供水文服务的驱动因素和压力,从而有助于对水文资源和服务进行更有力和有效的管理。

致谢

本研究得到了葡萄牙科学和技术基金会(Fundação para a Ciência e a Tecnologia)为考迪亚·卡瓦略-桑托斯的博士学位授予(SFRH/BD/66260/2009)

提供的资助。乔奥·洪拉多通过项目"ECOSENSING"(PTDC/AGR-AAM/104819/2008)获得了该基金会的支持。布鲁诺·马尔科斯(Bruno Marcos)通过项目"MoBiA"(PTDC/AAC-AMB/114522/2009)获得了该基金会的支持。

参 考 文 献

Albergel, C., E. Zakharova, J. C. Calvet, et al. 2011. A first assessment of the SMOS data in southwestern France using in situ and airborne soil moisture estimates: The CAROLS airborne campaign. *Remote Sensing of Environment* 115:2718–2728.

Alcantara, E. H., J. L. Stech, J. A. Lorenzzetti, et al. 2010. Remote sensing of water surface temperature and heat flux over a tropical hydroelectric reservoir. *Remote Sensing of Environment* 114:2651–2665.

Alsdorf, D. E., and D. P. Lettenmaier. 2003. Geophysics. Tracking fresh water from space. *Science* 301:1491–1494.

Ansink, E., L. Hein, and K. Per Hasund. 2008. To value functions or services? An analysis of ecosystem valuation approaches. *Environmental Values* 17:489–503.

Arnold, J. G., and N. Fohrer. 2005. SWAT2000: Current capabilities and research opportunities in applied watershed modelling. *Hydrological Processes* 19:563–572.

Ayanu, Y. Z., C. Conrad, T. Nauss, M. Wegmann, and T. Koellner. 2012. Quantifying and mapping ecosystem services supplies and demands: A review of remote sensing applications. *Environmental Science & Technology* 46:8529–8541.

Bartalis, Z., W. Wagner, V. Naeimi, et al. 2007. Initial soil moisture retrievals from the METOP-A Advanced Scatterometer (ASCAT). *Geophysical Research Letters* 34:L20401.

Bastiaanssen, W. G. M., H. Pelgrum, J. Wang, et al. 1998. A remote sensing surface energy balance algorithm for land (SEBAL): Part 2: Validation. *Journal of Hydrology* 212:213–229.

Becker, M. W. 2006. Potential for satellite remote sensing of ground water. *Ground Water* 44:306–318.

Bjerklie, D. M., S. L. Dingman, C. J. Vorosmarty, C. H. Bolster, and R. G. Congalton. 2003. Evaluating the potential for measuring river discharge from space. *Journal of Hydrology* 278:17–38.

Bjerklie, D. M., D. Moller, L. C. Smith, and S. L. Dingman. 2005. Estimating discharge in rivers using remotely sensed hydraulic information. *Journal of Hydrology* 309:191–209.

Brauman, K. A., G. C. Daily, T. K. Duarte, and H. A. Mooney. 2007. The nature and value of ecosystem services: An overview highlighting hydrologic services. *Annual Review of Environment and Resourses* 32:67–98.

Calmant, S., F. Seyler, and J. F. Crétaux. 2008. Monitoring continental surface waters by satellite altimetry. *Surveys in Geophysics* 29:247–269.

Carvalho, G. A., P. J. Minnett, L. E. Fleming, V. F. Banzon, and W. Baringer. 2010. Satellite remote sensing of harmful algal blooms: A new multi-algorithm method for detecting the Florida Red Tide (*Karenia brevis*). *Harmful Algae* 9:440–448.

Chapin III, F. S., P. A. Matson, and H. A. Mooney. 2002. *Principles of terrestrial ecosystem ecology*. 1st ed. New York: Springer.

Chen, D., J. Huang, and T. J. Jackson. 2005. Vegetation water content estimation for corn and soybeans using spectral indices derived from MODIS near- and shortwave infrared bands. *Remote Sensing of Environment* 98:225–236.

Collischonn, B., W. Collischonn, and C. E. M. Tucci. 2008. Daily hydrological modeling in the Amazon basin using TRMM rainfall estimates. *Journal of Hydrology* 360:207–216.

Courault, D., B. Seguin, and A. Olioso. 2005. Review on estimation of evapotranspiration from remote sensing data: From empirical to numerical modeling approaches. *Irrigation and Drainage Systems* 19:223–249.

de Groot, R. S., R. Alkemade, L. Braat, L. Hein, and L. Willemen. 2010. Challenges in integrating the concept of ecosystem services and values in landscape planning, management and decision making. *Ecological Complexity* 7:260–272.

Del Castillo, C. E., and R. L. Miller. 2008. On the use of ocean color remote sensing to measure the transport of dissolved organic carbon by the Mississippi River Plume. *Remote Sensing of Environment* 112:836–844.

de Roo, A., F. Bouraoui, P. Burek, et al. 2012. *Current water resources in Europe and Africa*. Italy: Joint Research Center.

Dreher, J., F. Gampe, M. Kirchebner, et al. 2000. Hydrological services: The need for the integration of space based information. *Proceedings of the International Symposium GEOMARK 2000*, Paris. 10–12 April 2000, 119–127.

Drinkwater, M. R., R. Francis, G. Ratier, and D. J. Wingham. 2003. The European space agency's earth explorer mission CryoSat: Measuring variability in the cryosphere. *Annals of Glaciology* 39:313–320.

EEA (European Environment Agency). 1997. *CORINE Land Cover—Technical guide, European Commission, Brussels* pp. 1–130. Retrieved from http://www.eea.europa.eu/publications/COR0-landcover.

Elmqvist, T., E. Maltby, T. Barker, et al. 2009. Biodiversity, ecosystems and ecosystem services (chapter 2). In *(TEEB) The Economics of Ecosystems and Biodiversity: Ecological and Economic Foundations*, ed. P. Kumar, 41–111. London: Earthscan.

Espinha Marques, J., J. Samper, B. Pisani, et al. 2011. Evaluation of water resources in a high-mountain basin in Serra da Estrela, Central Portugal, using a semi-distributed hydrological model. *Environmental Earth Sciences* 62:1219–1234.

Feng, X., B. Fu, X. Yang, and Y. Lü. 2010. Remote sensing of ecosystem services: An opportunity for spatially explicit assessment. *Chinese Geographical Science* 20:522–535.

Fernandes, N. F., R. F. Guimarães, R. A. T. Gomes, B. C. Vieira, D. R. Montgomery, and H. Greenberg. 2004. Topographic controls of landslides in Rio de Janeiro: Field evidence and modeling. *Catena* 55:163–181.

Fernández-Prieto, D., P. van Oevelen, Z. Su, and W. Wagner. 2012. Advances in Earth observation for water cycle science. *Hydrology and Earth System Sciences* 16:543–549.

Fetter, C. W. 1999. *Contaminant hydrogeology*. 2nd ed. Upper Saddle River, NJ: Prentice Hall.

Fetter, C. W. 2001. *Applied hydrology*. 4th ed. Upper Saddle River, NJ: Prentice Hall.

Fisher, B., R. K. Turner, and P. Morling. 2009. Defining and classifying ecosystem services for decision making. *Ecological Economics* 68:643–653.

Fitts, C. R. 2002. *Groundwater science*. London: Academic Press.

Gassman, P. W., M. R. Reyes, C. H. Green, and J. G. Arnold. 2007. The soil and water assessment tool: Historical development, applications, and future research directions. *Transactions of the ASABE* 50:1211–1250.

Grimes, D. I. F. 2008. An ensemble approach to uncertainty estimation for satellite-based rainfall estimates. *Hydrological Modelling and the Water Cycle* 63:145–162.

Hadibarata, T., F. Abdullah, A. R. M. Yusoff, R. Ismail, S. Azman, and N. Adnan. 2012. Correlation study between land use, water quality, and heavy metals (Cd, Pb, and Zn) content in water and green lipped mussels *Perna viridis* (Linnaeus.) at the Johor Strait. *Water, Air, & Soil Pollution* 223:3125–3136.

Haines-Young, R., and M. Potschin. 2010. The links between biodiversity, ecosystem services and human well-being. In *Ecosystem ecology: A new synthesis*, eds. D. Raffaelli and C. Frid, 110–139. BES Ecological Reviews Series. Cambridge, UK: Cambridge University Press.

Han, E., V. Merwade, and G. C. Heathman. 2012. Implementation of surface soil moisture data assimilation with watershed scale distributed hydrological model. *Journal of Hydrology* 416–417:98–117.

Hein, L., N. de Ridder, P. Hiernaux, R. Leemans, A. de Wit, and M. Schaepman. 2011. Desertification in the Sahel: Towards better accounting for ecosystem dynamics in the interpretation of remote sensing images. *Journal of Arid Environments* 75:1164–1172.

Helfrich, S. R., D. McNamara, B. H. Ramsay, T. Baldwin, and T. Kasheta. 2007. Enhancements to, and forthcoming developments in the Interactive Multisensor Snow and Ice Mapping System (IMS). *Hydrological Processes* 21:1576–1586.

Immerzeel, W. W., and P. Droogers. 2008. Calibration of a distributed hydrological model based on satellite evapotranspiration. *Journal of Hydrology* 349:411–424.

Jacquemoud, S., W. Verhoef, F. Baret, et al. 2009. PROSPECT+SAIL models: A review of use for vegetation characterization. *Remote Sensing of Environment* 113:S56–S66.

Jha, M. N., J. Levy, and Y. Gao. 2008. Advances in remote sensing for oil spill disaster management: State-of-the-art sensors technology for oil spill surveillance. *Sensors* 8:236–255.

Jongman, B., P. J. Ward, and J. C. J. H. Aerts. 2012. Global exposure to river and coastal flooding: Long term trends and changes. *Global Environmental Change* 22:823–835.

Jørgensen, L., C. A. Stedmon, T. Kragh, S. Markager, M. Middelboe, and M. Søndergaard. 2011. Global trends in the fluorescence characteristics and distribution of marine dissolved organic matter. *Marine Chemistry* 126:139–148.

Kidd, C., V. Levizzani, J. Turk, and R. Ferraro. 2009. Satellite precipitation measurements for water resource monitoring. *Journal of the American Water Resources Association* 45:567–579.

Kite, G. W., and A. Pietroniro. 1996. Remote sensing applications in hydrological modelling. *Hydrological Sciences Journal* 41:563–591.

Lal, R. 2000. Rationale for watershed as a basis for sustainable management of soil and water resources. In *Integrated watershed management in the global ecosystem*, ed. R. Lal, 2–16. Boca Raton, FL: CRC Press.

Leon, J. G., S. Calmant, F. Seyler, et al. 2006. Rating curves and estimation of average water depth at the Upper Negro River based on satellite altimeter data and modeled discharges. *Journal of Hydrology* 328:481–496.

Li, X. H., and Q. Zhang. 2012. Suitability of the TRMM satellite rainfalls in driving a distributed hydrological model for water balance computations in Xinjiang catchment, Poyang lake basin. *Journal of Hydrology* 426–427:28–38.

Llovel, W., M. Becker, A. Cazenave, J. F. Cretaux, and G. Ramillien. 2010. Global land water storage change from GRACE over 2002–2009. Inference on sea level. *Comptes Rendus—Geoscience* 342:179–188.

Luojus, K., J. Pulliainen, M. Takala, and J. Lemmetyinen. 2010. *Global snow monitoring for climate research*. European Space Agency Study Contract Report ESRIN Contract 21703/08/I-EC.

MA (Millenium Ecosystem Assessment). 2003. *Ecosystems and human well-being: A framework for assessment*. Washington, DC: Island Press.

Michaelides, S., V. Levizzani, E. Anagnostou, P. Bauer, T. Kasparis, and J. E. Lane. 2009. Precipitation: Measurement, remote sensing, climatology and modeling. *Atmospheric Research* 94:512–533.

Michailovsky, C. I., and S. McEnnis. 2012. River monitoring from satellite radar altimetry in the Zambezi River basin. *Hydrology and Earth System Sciences* 16:2181–2192.

Milzow, C., P. E. Krogh, and P. Bauer-Gottwein. 2011. Combining satellite radar altimetry, SAR surface soil moisture and GRACE total storage changes for hydrological model calibration in a large poorly gauged catchment. *Hydrology and Earth System Sciences* 15:1729–1743.

Montanari, M., R. Hostache, P. Matgen, G. Schumann, L. Pfister, and L. Hoffmann. 2009. Calibration and sequential updating of a coupled hydrologic–hydraulic model using remote sensing-derived water stages. *Hydrology and Earth System Sciences* 13:367–380.

Narasimhan, B., R. Srinivasan, J. G. Arnold, and M. Di Luzio. 2005. Estimation of long-term soil moisture using a distributed parameter hydrologic model and verification using remotely sensed data. *Transactions of the ASAE* 48:1101–1113.

Owe, M., R. de Jeu, and T. Holmes. 2008. Multisensor historical climatology of satellite-derived global land surface moisture. *Journal of Geophysical Research* 113:17.

Pietroniro, A., and T. D. Prowse. 2002. Applications of remote sensing in hydrology. *Hydrological Processes* 16:1537–1541.

Potes, M., M. J. Costa, J. C. B. da Silva, A. M. Silva, and M. Morais. 2011. Remote sensing of water quality parameters over Alqueva Reservoir in the south of Portugal. *International Journal of Remote Sensing* 32:3373–3388.

Ramillien, G., J. S. Famiglietti, and J. Wahr. 2008. Detection of continental hydrology and glaciology signals from GRACE: A review. *Surveys in Geophysics* 29:361–374.

Rango, A., and I. S. Ahlam. 1998. Operational applications of remote sensing in hydrology: Success, prospects and problems. *Hydrological Sciences Journal* 43:947–968.

Ritchie, J., P. Zimbra, and J. Everitt. 2003. Remote sensing techniques to assess water quality. *Photogrammetric Engineering and Remote Sensing* 69:695–704.

Rodell, M., J. Chen, H. Kato, J. S. Famiglietti, J. Nigro, and C. R. Wilson. 2006. Estimating groundwater storage changes in the Mississippi River basin (USA) using GRACE. *Hydrogeology Journal* 15:159–166.

Rogan, J., and D. M. Chen. 2004. Remote sensing technology for mapping and monitoring land-cover and land-use change. *Progress in Planning* 61:301–325.

Rounsevell, M. D. A., T. P. Dawson, and P. A. Harriuson. 2010. A conceptual framework to assess the effects of environmental change on ecosystem services. *Biodiversity and Conservation* 19:2823–2842.

Sahoo, A. K., M. Pan, T. J. Troy, R. K. Vinukollu, J. Sheffield, and E. F. Wood. 2011. Reconciling the global terrestrial water budget using satellite remote sensing. *Remote Sensing of Environment* 115:1850–1865.

Santos, M. J., S. Khanna, E. L. Hestir, et al. 2009. Use of hyperspectral remote sensing to evaluate efficacy of aquatic plant management. *Invasive Plant Science and Management* 2:216–229.

Schmugge, T. J., W. P. Kustas, J. C. Ritchie, T. J. Jackson, and A. Rango. 2002. Remote sensing in hydrology. *Advances in Water Resources* 25:1367–1385.

Schulz, J., P. Albert, H. D. Behr, et al. 2009. Operational climate monitoring from space: The EUMETSAT satellite application facility on climate monitoring (CM-SAF). *Atmospheric Chemistry and Physics* 9:1687–1709.

Smith, R. C. G., and B. J. Choudhury. 1990. Relationship of multispectral satellite data to land surface evaporation from the Australian continent. *International Journal of Remote Sensing* 11:2069–2088.

Song, K., D. Lu, L. Li, S. Li, Z. Wang, and J. Du. 2012. Remote sensing of chlorophyll-*a* concentration for drinking water source using genetic algorithms (GA)–partial least square (PLS) modeling. *Ecological Informatics* 10:25–36.

Sorooshian, S., K. L. Hsu, G. Xiaogang, H. V. Gupta, B. Imam, and D. Braithwaite. 2000. Evaluation of PERSIANN system satellite-based estimates of tropical rainfall. *Bulletin of the American Meteorological Society* 81:2035–2046.

Su, Z., R. A. Roebeling, J. Schulz, et al. 2012. Observation of hydrological processes using remote sensing. In *Treatise on water science*, ed. P. Wilderer, vol. 2:351–399. Oxford: Academic Press.

Su, Z., A. Yacob, J. Wen, et al. 2003. Assessing relative soil moisture with remote sensing data: Theory, experimental validation, and application to drought monitoring over the North China Plain. *Physics and Chemistry of the Earth, Parts A/B/C* 28:89–101.

Suárez, L., P. J. Zarco-Tejada, J. A. J. Berni, V. González-Dugo, and E. Fereres. 2009. Modelling PRI for water stress detection using radiative transfer models. *Remote Sensing of Environment* 113:730–744.

Syed, T. H., J. S. Famiglietti, D. P. Chambers, J. K. Willis, and K. Hilburn. 2010. Satellite-based global-ocean mass balance estimates of interannual variability and emerging trends in continental freshwater discharge. *Proceedings of the National Academy of Sciences of the United States of America* 107:17916–17921.

Syvitski, J. P. M., I. Overeem, G. R. Brakenridge, and M. Hannon. 2012. Floods, floodplains, delta plains—A satellite imaging approach. *Sedimentary Geology* 267–268:1–14.

Tang, Q., H. Gao, H. Lu, and D. P. Lettenmaier. 2009. Remote sensing: Hydrology. *Progress in Physical Geography* 33:490–509.

Tapiador, F. J., F. J. Turk, W. Petersen. 2012. Global precipitation measurement: Methods, datasets and applications. *Atmospheric Research* 104–105:70–97.

Tobin, K. J., and M. E. Bennett. 2012. Temporal analysis of soil and water assessment tool (SWAT) performance based on remotely sensed precipitation products. *Hydrological Processes* 27:505–514.

Tralli, D. M., R. G. Blom, V. Zlotnicki, A. Donnellan, and D. L. Evans. 2005. Satellite remote sensing of earthquake, volcano, flood, landslide and coastal inundation hazards. *ISPRS Journal of Photogrammetry and Remote Sensing* 59:185–198.

Tran, A., F. Goutard, L. Chamaille, N. Baghdadi, and D. Lo Seen. 2010. Remote sensing and avian influenza: A review of image processing methods for extracting key variables affecting avian influenza virus survival in water from Earth observation satellites. *International Journal of Applied Earth Observation and Geoinformation* 12:1–8.

UN (United Nations). 2011. *Water in a green economy*. UN-Water. Retrieved from www.unwater.org.

van Beek, L. P. H., Y. Wada, and M. F. P. Bierkens. 2011. Global monthly water stress: 1. Water balance and water availability. *Water Resources Research* 47:W07517.

Van Brahana, J. 2003. Hydrological cycle. In *Encyclopedia of water science*, eds. B. A. Stewart and T. A. Howell, 412–414. New York: Marcel Dekker.

van Dijk, A. I. J. M., and L. J. Renzullo. 2011. Water resource monitoring systems and the role of satellite observations. *Hydrology and Earth System Sciences* 15:39–55.

van Westen, C. J., E. Castellanos, and S. L. Kuriakose. 2008. Spatial data for landslide susceptibility, hazard, and vulnerability assessment: An overview. *Engineering Geology* 102:112–131.

Wagner, W., N. E. C. Verhoest, R. Ludwig, and M. Tedesco. 2009. Remote sensing in hydrological sciences. *Hydrology and Earth System Sciences* 13:813–817.

Walker, J., and P. Houser. 2005. Hydrologic data assimilation. In *Advances in water science methodologies*, ed. U. Aswathanarayana, 230. Boca Raton, FL: CRC Press.

Ward, D. P., S. K. Hamilton, T. D. Jardine, et al. 2012. Assessing the seasonal dynamics of inundation, turbidity, and aquatic vegetation in the Australian wet-dry tropics using optical remote sensing. *Ecohydrology* 6:312–323.

Wasowski, J., C. Lamanna, G. Gigante, and D. Casarano. 2010. High resolution satellite imagery analysis for inferring surface–subsurface water relationships in unstable slopes. *Remote Sensing of Environment* 124:1–14.

Wu, C., Z. Niu, Q. Tang, and W. Huang. 2009. Predicting vegetation water content in wheat using normalized difference water indices derived from ground measurements. *Journal of Plant Research* 122:317–326.

Xie, X., and D. Zhang. 2010. Data assimilation for distributed hydrological catchment modeling via ensemble Kalman filter. *Advances in Water Resources* 33:678–690.

Yilmaz, M. T., E. R. Hunt, and T. J. Jackson. 2008. Remote sensing of vegetation water content from equivalent water thickness using satellite imagery. *Remote Sensing of Environment* 112:2514–2522.

Zarco-Tejada, P. J., C. A. Rueda, and S. L. Ustin. 2003. Water content estimation in vegetation with MODIS reflectance data and model inversion methods. *Remote Sensing of Environment* 85:109–124.

Zhang, J., Y. Hu, X. Xiao, et al. 2009. Satellite-based estimation of evapotranspiration of an old-growth temperate mixed forest. *Agricultural and Forest Meteorology* 149:976–984.

第十二章　用于生态系统服务评估的水文模型的遥感数据同化

J. 埃雷罗　A. 米拉尔斯　C. 阿吉拉尔　F. J. 博内特　M. J. 波罗

一、简介

古代文明已观察、研究和描述了所谓的水循环(Biswas,1970)。但直到19世纪,测量能力达到了获取大量相关数据的时候,水文学作为一门独立科学的地位才得到巩固。这引出了20世纪各种合理化的假说、结论和模型,最终带来了复杂水文模型的发展和应用。在过去的几十年里,对地球表面的观测实现了从地面到太空的飞跃,将人们对尺度效应的关注由利用对点位的测量表征连续三维系统(升尺度)转移到对遥感数据及其产品的降尺度。在舒尔茨和恩格曼(Schultz and Engman,2000)、施穆格等(Schmugge et al.,2002)或苏等(Su et al.,2011)的著作中,可找到遥感技术在水文观测和建模中应用的相关例子。

系统性的水文模拟,无论其尺度如何,都是对水质及生态系统动态性分析和模拟的首要和具有重要影响的基础。然而,水文模拟的准确性是计算全链条相互作用和演进过程中生态变量不确定性的最重要来源之一。瓦格纳等(Wagner et al.,2009)指出,需要基于这类新的观测手段、对宏观尺度过程的捕捉以及对相关不确定性的量化来改进建模策略,这也是水文学家和科学家普遍面临的主要挑战。在这种情况下,数据同化方法在促进遥感在水文和其他科学中的应用方面发挥着关键作用。

二、水文模型和生态系统服务量化

随着社会认识到生态系统服务的重要性,人们对数学模型作为提供决策过程必要信息的工具作用越来越感兴趣。在过去 30 年中,这些模型的使用有所增加,这主要是由于信息和数据采集技术(卫星影像、航空摄影、远程数据传输、数字高程模型等)的普及和计算机能力的提升。

为适应不同的过程和系统,大量模型已经被开发出来。因此,我们可区分生物地球化学模型,如 CANDY(Franko et al.,1995)和 ICBM(Andren et al.,2004);陆地植被模型,如 TRIFFID(Cox,2001)和 YieldSafe(Van der Werf et al.,2007);碳循环模型,如 Hybrid(Friend et al.,1997)、CenW(Kirschbaum,1999)和 ASPECTS(Rasse et al.,2001);水文模型,如 TOPMODEL(Beven and Kirkby,1979)或 SWAT(Arnold et al.,1998;Neitsch et al.,2005)。

这些模型中的每一个都给出了与系统或子系统相关联的结果,考虑到了过程的时空尺度、可用的初始信息及决策所需的最终结果。即使在当下,广泛评估中涉及的所有系统和过程的复杂性限制了全局建模的可能。最近,一些工作已开始以耦合或嵌入的方式在单个模型中集成不同的模型,从而允许量化与系统相对应的服务,此等系统虽然相互关联,但传统上是单独建模的。

水文循环是将自然概念化为一组相互连接的系统(大气—地球—海洋)的一个明显例子,相互串联的物理过程(降水、融雪、径流、渗透、地下流动、含水层贡献等)在此系统中发生。这些物理过程直接制约着形态、化学、生物、经济和社会行为,并对社会服务产生明显影响。因此,使用建模方法评估生态系统服务而开展的诸多工作,以某种方式被构造为基于水文建模的子程序,这并非巧合事件(Band et al.,1991;Tague and Band,2001;Yates et al.,2005)。水文循环涉及的过程允许对其组成部分(如可用水量、土壤流失量、降雪量或地下水储存)进行离散量化,从而实现对增益和损失的测量。对有形资源(水、雪、沉积物、养分等)的努力量化,通常与社会实践获得的结果相去甚远。这是由于分析中包含的非有形资源(如文化或美学价值),增加了评估过程的复杂性。一些作者(Porras et al.,2008;Carpenter et al.,2009)提到,水文模型的结果与服务量化一样,这主要是由于难以估算水循环和用户需求的效益。在未来的生态系统服务的水文建模研究中,这方面的研究有待进一步完善。

一般而言,在使用水文模型量化生态系统服务时,可确定两种不同的方法(Bellamy et al.,2011;Vigerstol and Aukema,2011):(1)传统水文模型,需要对结果进行进一步处理或后处理,才能评估最终的量化结果;(2)通过不同方法或模型的组合开发的综合生态系统模型,其最终产出是服务的量化结果及其空间分布。

几十年来,水文模型一直用于估算与水循环相关的过程,例如对洪水(见第十七章)、可用水资源、土壤损失的评估等。严格地说,早在科学界和社会确立生态系统服务的概念之前,水文模型就一直在为决策者提供过程评估的相关信息。这些模型已从集中/聚集式和基于事件的模型,如 HEC-1(USACE 1982)、TR-20(USCS 1982),发展到具有物理基础、离散或连续模拟结果的复杂模型,如 WMS(Dellman et al.,2002)、MIKE-SHE(Abbott et al.,1986)或 WiMMed(Polo et al.,2009;Herrero et al.,2010)。从生态系统服务量化的角度来看,尽管集中式模型和基于事件的模型需要较少的信息和计算能力,但其适用范围有限。相反,如前所述,基于物理的和离散式模型不仅可以估算水平衡,还可以估算水文变量的空间分布,并且可以量化水文过程及其相互作用。

水土保持评价工具模型具有半经验半离散的基础,在诸多研究中得到广泛的应用,特别是在生态系统服务功能量化方面。其应用包括水文资源可用性评估(Notter et al.,2012)、流域污染物分布(Prochnow et al.,2008;Schilling and Wolter,2009)、不同气候场景评估(Stone et al.,2001)、流域管理最佳实践(Gassman et al.,2007)、山区雪中的蓄水和融雪供给(Herrero et al.,2005)及土壤流失评估(Shen et al.,2009)等。然而,它的主要局限在于许多过程中的经验方法和空间离散化,因为它是基于均匀的水文响应空间单位(Hydrologic Response Units,HRU)聚合后得出的结果,有时不包括许多重要的过程。

值得肯定的是,尽管不同工作提到了由于高计算、大数据需求及需要校准大量参数而存在局限性,但是基于物理和分布式的生态系统服务量化模型仍具有重要作用。需要指出的是,遥感的贡献及其在相对较短的时间内为模型提供分布信息的能力(如土地覆盖类型、土壤湿度、LAI 叶面积指数),这使得模型的初始配置和获得的结果(积雪演化、洪涝面积、土壤流失等)得到最终验证。此外,还可以建立遥感数据和不同水文参数间的关系,以便从这些观测结果中获得新的信息(Kite and Pietroniro,1996;Chen et al.,2005;Liu and Li,2008;Feng et al.,2010;Su et al.,2010;Aguilar et al.,2012)。专栏12-1和专栏12-2提供了这方面的例子。

专栏12-1 陆地卫星影像中的地表反射率和植被覆盖率的年度和季节变化:同化水文模型作为输入变量

地球表面的能量和水平衡是蒸发和蒸腾共同作用的过程。地表和土壤上层的有效水分是此平衡中能量不足或过剩的源头;在水分缺乏条件下,植被通过呼吸作用直接将水分从根区输送到大气,并通过气孔控制蒸腾速率来改变土壤蒸发状况。在实践中,很难区分植被区域的蒸腾速率和蒸发速率,而蒸散量是用于描述水汽蒸发的术语。此外,由于植被的存在,大气边界层和表层土壤层中局部和区域的能量和水的平衡也密切相关:根区决定了土壤平衡中作为控制量的有效土壤深度;植被覆盖及其种类、密度、结构组成了一个粗略的三维层,在大气和地表之间的动量、能量和水的流动传递中起着基本作用;植被类型及活力和密度也影响入射太阳能反射回大气的比例,即反照率;植被的气生结构可保留一定比例的降雨,这些降雨被蒸发回大气而非渗入土壤,这是水分收支中的截留项。在空间基础上,植被的表面密度由覆盖率估算,它是经过水平投影后植被覆盖地形表面的占比,范围从0(裸土)到1(完全被植被覆盖的土壤)。因此,覆盖率是将植被涉及的能量和水平衡从单位植被面积换算到单位表面积的指数(图12-1)。

对于中到大尺度的水文分析,覆盖率变化的量化具有重要意义。可使用不同的产品来计算归一化差异植被指数与覆盖率之间的关系(Curran,1981;Sellers,1989;Bannari et al.,1995),例如中分辨率成像光谱辐射计(MODIS)数据,但在不均匀地区由于显著的尺度效应,其空间分辨率对正确信息的量化存在一定限制。Landsat(陆地卫星)影像,由于其具有稳定的空间分辨率,已被广泛使用(Ramsey et al.,2004)于流域和区域尺度的NDVI值估算。

图 12-1(见彩插)计算

注:瓜达尔费奥河流域植被分类(左)和覆盖率分布示例(右),2005年3月。

为了量化西班牙南部的瓜达尔费奥河流域这一复杂沿海山区的截流损失,迪亚斯-古铁雷斯(Díaz-Gutiérrez,2007)将基于加什(Gash,1979)和鲁特等(Rutter et al.,1971,1975)方法的截流模型与通过分析2002~2005年期间每年4~6幅Landsat5TM影像以及7幅Landsat7ETM+影像的NDVI季节特征得到的植被覆盖率的图集相结合;并证明采用一种具有稳定状态和演化周期的简单插值算法就能模拟生成连续的以"天"为步长的时间序列产品(Polo et al.,2011)。

将这个时间序列产品同化到水文模型"地中海环境流域综合管理"(Watershed integrated Management for Mediterranean environments,WiMMed)中,就可以估算流域沿线的降雨截留损失(图12-2),并为管理者提供了一种有效工具,以评估不同的作物选择方案、野火效应和流域尺度上截留比例变化引起的干旱后果。

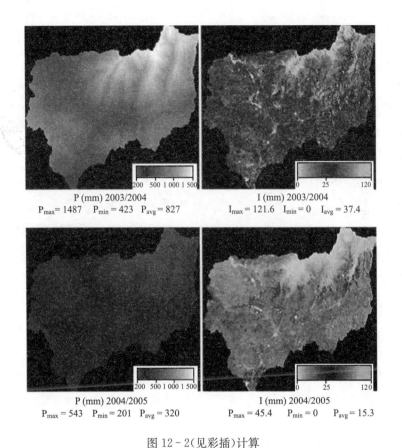

图 12-2（见彩插）计算

注：瓜达尔费奥河流域连续两个水文年降水（P，左）和截流（I，右）

专栏 12-2　与沉积物供应相关的三角洲沿海地区变化（瓜达尔费奥案例研究）

由于推移质侵蚀过程对大坝淤积、河床颗粒稳定性和河口动力学具有高度影响作用，因此显著影响了河流动力和河流管理的频次。沉积物侵蚀对社会和经济具有不同影响，有时甚至相反，这表明需要更复杂的方法来了解此过程及其产生的相关影响。

> 瓜达尔费奥河流域(西班牙南部)展示了储存在干流和干涸支流(河渠)到地中海入海口的大量冲积沉积物。2004 年建造的 Rules 大坝,容量为 110hm³,给上游和下游,尤其是三角洲地区,带来了重大环境变化的风险。
>
> 对位于主河道的两个控制点的监测,以及对遥感信息的分析作为补充,不仅可估算由于淤积而导致的水库库容损失,还可估算对下游三角洲的动态影响。通过综合利用 1956~2008 年期间的正射影像和卫星图像(Redde Información Ambiental de Andalucía,2010)以及 2003~2011 年期间的野外水深测量,估算了 2002 年以来的淤积率,估算泥沙淤积量为 200 万立方米。
>
> 水库中大量的物质沉积导致三角洲自 2005 年以来一直在退化。河流的输入实际上已经停止,在海岸的强烈风暴中造成海滩物质的不平衡流失,这些物质正在消退。在正常情况下,洋流和破碎波所产生的海岸侵蚀作用保持同一个方向,并将沉积物从三角洲输送到近岸区域,主要是沙滩和港口(Ávila,2007)。直接影响可以从该旅游区额外支出的人工育滩费用。在 2004~2009 年期间人工育滩 11.21 万立方米,相关费用为 728.65 万欧元(Ruiz de Almirón,2011)。可以在地中海流域的许多特定的三角洲和入海口中发现这些动态影响。

第二种方法与水文模型和生态系统服务量化集成形成的模型相关,无论是否与水有关。这些模型正处于最初发展阶段,尽管还没有完全对比,但其具有广泛应用前景。生态系统服务和权衡综合评估(Integrated Valuation of Ecosystem Services and Tradeoffs,InVEST)模型(Tallis et al.,2011)对当前或未来情景下不同生态系统服务的数量和价值进行评估。结果可以在与水直接相关或不直接相关的服务(水库水电生产、水库淤积防治、水质净化、生物多样性可持续性、碳储存、木材生产评估)的分布地图上呈现。大量应用案例给出了生物多样性损失(Goldman et al.,2012;Reyers et al.,2012)以及当前和未来土壤管理方案(Daily et al.,2009;Nelson et al.,2009;Johnson et al.,2012)。在这种情况下,可发现水文过程的简化及一些重要因素(如地下水资源)缺失会导致一些局限(第十三章)。遵循这一理念,其他模型基于贝叶斯网络方法建立输入数据和预期不同服务间的关系。生态系统人工智能服务(Artifcial Intelligence for Ecosystem Services,ARIES)模型(Villa et al.,2009)就是这种一个例子,该模型已应用于不同气候情景的评估。两类模型的更多细节见专栏20-1。

三、水文建模和遥感

地球上水和能量收支的结合,决定了环境条件的基本局部规则,大部分的物理、化学、生物和生态规则对其高度依赖。

此外,水是沉积物、营养物、污染物等在不同尺度生态系统流动的主要途径之一。生物地球化学循环与生态系统动力学中水循环相似,因为它界定了系统和子系统、内部的联系和交换、外部的不可抗力和相关响应以及相互作用。想要量化生物地球化学通量,必须首先估算水和能量的通量。生物地球化学循环在物理上和数学上都依赖于水文循环,因此,需要接收和传递水文循环的质量水平及其结果相关的不确定性。这一事实证明了水文模型在生态系统服务量化中的重要性和影响。

用于模拟水文循环的计算机模型不仅能够对可预见未来情景中的行为进行预测,还能够加深对当前时间特定区域中的行为的理解。为了利用当今遥感在空间分辨率上的能力,必须使用分布式模型。此外,为了充分利用现有的水文知识来描述不同的发生过程,并以足够的精度估算不同的水道和相互作用,这类模型应该是具有物理基础的。具有这两个特征的模型必须能够估算水文循环中显著数量的组成部分和水通量,否则将是无法可靠量化的。

有诸多广泛使用的分布式和基于物理的水文模型,例如TOPMODEL、DHSVM(Wigmosta et al.,1994)或SHE,这只是几个例子。尽管有相似之处,但这些模型在解决某些特定水文过程的方式有所不同。例如,部分模型重视地下径流或地表径流,将其作为应对较为潮湿或干燥环境的战略实施的一部分(这类方法通常与模型最初开发应用的地区有关)。连续模拟的时间步长从按天计算到按小时计算不等。WiMMed是一个由西班牙格拉纳达大学和科尔多瓦大学开发的分布式且基于物理过程的水文模型(Polo et al.,2009;Herrero et al.,2010)。该模型最初是为地中海山区环境设计的,特别是用于处理存在低纬度的融雪的极端变化的山区盆地：陡峭的地形、气象梯度、植被覆盖的异质性、具有优先地形地下流动的山区含水层、季节性的暴雨和干燥期、大量沉积物产生和常年过度干旱。在这种高度异构和快速变化的环境中,使用物理方法求解与此过程相关的问题尤其重要。这对于积雪水文学尤其重要(见第十五章),只有通过能量和质量平衡方法才能正确模拟。与其他分布式模型一样,遥感在 WiMMed 水文模拟中发挥着重要作用。

在建模过程中使用遥感技术可以识别具有某些物理意义的参数,同时也可以对整个区域或水文单元上的水进行描述。这些参数可作为输入数据并入模型中,可被认为是恒定或随时间变化的。分布式模型只有以卫星影像信息相同的分辨率模拟水循环,才能充分利用这些数据。在此情况下,数据同化和单元模型结果与状态变量的远程测量间的比较将是直接的。以 Landsat 卫星多光谱数据为例,利用多光谱数据获得信息,并将其纳入一个分布模型,为利用所有可用信息,该模型所应达到的最小空间分辨率为 30 米×30 米。过去,所需的高性能计算能力被认为是使用这一模型的主要限制,但现在该限制已不复存在。

分布式水文模型可同化不同来源和不同方法获得的数据(Houser *et al.*,1998,2012;Reichle *et al.*,2002;de Lannoy *et al.*,2011;Malik *et al.*,2012)。例如,WiMMed 将由多光谱数据计算 NDVI 得到的两个直接结果作为输入：植被覆盖率 f_v 和地表反照率 α(Díaz-Gutiérrez,2007)。第一个参数表示每个像素被植被覆盖比例,而 α 是地形的短波反射率,这两个参数都是多光谱源空间分辨率的有效值。当使用陆地卫星专题制图仪(Landsat Thematic Mapper)数据时,若没有云层干扰,可将 15 天作为时间间隔。接下来可对得到的 f_v 或 α 图进行插值,以获得每日分布的序列,并将序列作为模型的直接输入,以估算土壤或积雪中的降雨截留或能量收支。在其他变量中,也可计算出参考蒸散量(第十八章)和实际蒸散量(Aguilar *et al.*,2010)。这是将从遥感数据获得的物理参数直接同化为模型输入数据的例子。

相反,水文模型也可使用遥感信息获得的状态变量估算值来校准和验证其性能。积雪分布(第十五章)和土壤湿度是相关参考案例(第十四章)。同化过程可采用基于卡尔曼滤波的方法,如数据同化研究试验台(Data Assimilation Research Testbed,DART)模型(Anderson *et al.*,2009),并可结合针对包括源数据质量和不确定性的校正技术。一个更简单的方法是直接插入,在给定状态没有达到令人满意的调整程度时,用"测量"值替换变量的模拟值,如 WiMMed 使用航空或遥感获得积雪数据(Pimentel et al.,2012)(专栏 12-3)。

可利用可见光图像或多光谱图像来检测积雪的存在,通过使用诸如归一化差异积雪指数(Hall et al.,1995)进行近红外反射分析。同样,积雪的水量可通过激光雷达、地面伽马射线衰减数据(de Walle and Rango,2008)来检测,土壤湿度可以通过微波或多光谱分析以及伽马辐射测量(Carroll,1981;Wang and Qu,2009)来估算。

专栏 12-3　通过遥感获得的积雪图对融雪模型进行校准和验证

埃雷罗等(Herrero et al.,2011)在瓜达尔费奥河流域(西班牙内华达山脉附近)应用水文模型时,使用 WiMMed 模型以小时为尺度绘制了积雪覆盖扩张及积雪水量图。在 30 米的空间分辨率下,将模型结果与陆地卫星在足够能见度下特定时间获得的积雪覆盖图进行直接比较。比较为每个像素定义了四种可能的组合:(1)具有模拟和测量的积雪的像素;(2)没有模拟和测量的积雪的像素((1)和(2)是正确的情况);(3)有测量但没有模拟的积雪的像素;(4)没有测量但在模拟中覆盖了积雪的像素。通过从上述四种可能的组合推导出不同的指标来实现校准-验证过程的准确性。

图 12-3 展示了模拟积雪和测量积雪图逐像素的比较。在对 LandsatTM7 图像处理之后获得测量结果,同时用 WiMMed 模型的积雪模块运行模拟。正确的像素是绿色,错误则是透明的,而蓝色和红色像素代表模型对积雪的低估和高估。识别的偏差主要是由于模拟过程中每个像素的温度和降水分配不正确,这通常是山地盆地水文模拟不确定性的主要来源,因为山地盆地的梯度变化非常明显,气象网络的覆盖范围通常不足(这甚至比物理模型本身的固有局限性更重要)。

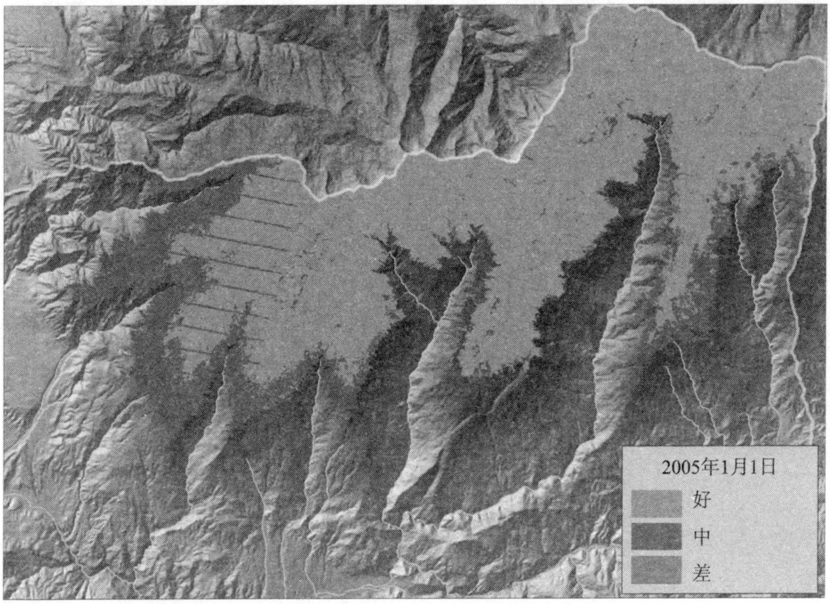

图 12-3（见彩插）

注：2005年1月1日用 WiMMed 模拟和 Landsat TM7 测量的积雪的逐像素对比。

四、水质监测和遥感

地表水体可提供极为与众不同的供应、调节和文化生态系统服务，如供水、渔业生产、运输、娱乐等。地表水体对于诸多关键物种的生存也是至关重要的，因为这些物种利用其生存、喂养和繁殖。由于人类活动压力和自然不可抗力间不断变化的相互作用，水体正不断变化。因此，地表水系统同时具有可恢复性和脆弱性（Ji,2008）。

当河流流过流域时，其集中了水、沉积物、营养物和被排放的污染物。其他地表水系统，如水库和河口，实现对水和相关污染物、沉积物、营养物、毒性物质等来自上游的物质的过滤。土地利用类型的骤然变化可能增加污染物的排放或导致水体的过度开发，往往会给下游带来常见的环境问题，如富营养化、栖息地丧失、藻华发育、鱼类和野生动物数量下降、海水入侵、水利基础设施处淤积等。因此,对水资源的管理和规划不仅要求正确评估水量,而且要求正确评估水质。

对水质的第一个关切集中在健康和卫生问题上。控制措施主要是通过管道或明渠将工业排放废水接入污水处理厂进行处理(Engman and Gurney,1991)。但近年来,非点源污染已成为大众关注和科学研究的主题。非点源污染被认为是暴雨径流的一部分,因此对这类污染的识别、测量和控制可能非常复杂。流域尺度的综合模型将数量和质量标准结合,构成了水资源管理的重要工具。这类模型旨在精确描述影响整个流域水和泥沙通量的水文和侵蚀过程,以及影响水质的物理、化学和生物过程。在非监测区域,尽管总是需要进行实地测量以实现一定程度的校准,但基于物理的模型可以估算流域响应程度。

在流域尺度上再现非点源污染过程细节的水文模型有 ANSWERS、SWMM、AGNPS、HSPF、GLEAMS、SWRRBWQ、CREAMS、SWAT 等。然而由于涉及大量参数,这类模型运行和校准的过程大多非常复杂。此外,由于缺乏可用的测量值或历史数据记录不够,无法在所需时间尺度上确定其动态过程和演变过程,因此难以实现对非点源排放的量化。因此,遥感在水质评价中是极具价值的数据来源,特别是考虑到研究所需的非点源污染的空间分布性质和广阔的空间尺度(Engman and Gurney,1991)。在区域尺度上,只有通过遥感才能实现经济上可行的高频监测。此外,如今卫星对全球范围的覆盖实现了对偏远和难以接近地区的水质的估算,也为建模人员提供了没有地面测量时期的历史水质研究数据(Hellweger et al.,2004)。

在非点源污染模型的应用中,遥感在三个层面构成了极具价值的数据源:作为输入数据、状态变量以及用于模型估算值的校准和验证的测量数据。作为输入数据和状态变量而言,遥感数据通常用于获取对径流潜在水质有很大影响的地表覆盖、土地利用、地形和土壤类型相关的数据。作为模型输出或水质指标,遥感允许推导出水面浊度、悬浮沉积物、叶绿素浓度、有色溶解有机物和水体温度的估算值。这些水质特征可用作更具体的污染问题(例如富营养化水平)的指标,并与非点源模型输出有关(Engman and Gurney,1991)。

浊度是一种光学效应,它与沉积物和其他有机质的总浓度有关。在通过能够量化光合作用过程的传感器获得水生生态系统的状态,叶绿素 a 是监测水生种群(主要是浮游植物)的关键指标。有色溶解有机物(Colored Dissolved Organic Matter,CDOM)是动植物分解过程的产物。所有这些水质变量都可通过直接取水采样的常规技术进行现场测量(Salama et al.,2012)。然而,受水体接收载荷和系统的流体力学的影响,这些变量在空间和时间上有很大的变化。因此,点样本监测通常是不充分的,因为它较为耗时并且仅代表有限的空间和时间域。在此情况下,遥感源提供了非常有价值的空间和时间数据,因为几乎能够实现大面积区域的即时监测(Hadjimitsis et al.,2006;Budhiman et al.,2012)。

光谱的可见光和近红外光的反射率可作为评价水面或近水面水质的指标。水质指标还可通过热红外进行估算,即直接测量发射的能量(Engman and Gurney,1991)。通常情况下,尽管间接测量浊度的 Secchi 深度与蓝色和绿色波段(450～600 纳米)的反射率间的相关性远低于其与红色波段(600～690 纳米有色溶解有机物,Colored Dissolved Organic Matter,CDOM)的反射率间的相关性,但沉积物在所有可见光波段均呈现高反射率。然而,光谱响应受到水系统性质的强烈影响。在受淡水排放影响的水体中,浊度与红色波段反射率之间存在高度相关性。在淡水排放低、沉积物载量不明显的沿海水域,可以通过叶绿素 a 估算的浮游植物浓度对反射率的影响更大(Hellweger et al.,2004;Lane et al.,2007)。

水质变量可根据基于测量和观测间的回归分析的经验方法远程量化。另一种选择是使用半分析方法,该方法应用描述观测光谱和水成分浓度之间关系的水光学模型(Salama et al.,2012)。就传感器和平台而言,需要通过所需数据的分辨率来确定选择,而所需数据则取决于待估算变量和研究区域的空间范围。这就是从卫星数据推导出的相对较小规模的河流、湖泊和河口水域的水质变量应用到公海和一些沿海地区常常受限的原因(Salama and Su,2010,2011;Shen et al.,2010;Budhiman et al.,2012)。在河流监测中,重点在于使用激光雷达、高度计和机载高光谱数据,因为即使是对大型河流,能提供所需光谱和空间分辨率的卫星传感器数量也是有限的(Salama et al.,2012)。例如,ENVISAT 的 300 米空间分辨率,即使是对于最大的河流,其分辨率也过于粗糙,而对于 Landsat 数据,有限数量的可见光波段和粗糙的光谱分辨率是主要限制因素

(Dekker and Peters,1993)。

然而,Landsat一直是湖泊水质监测卫星图像的主要来源,因为具有30米的良好空间分辨率。在文献中,有许多关于内陆湖泊的研究,这些研究已经提出了根据陆地卫星数据估算悬浮固体、浊度、叶绿素a、盐度和温度的表达式(Lathrop,1992;Baban,1993;Mayo et al.,1995;Hadjimitsis et al.,2006;Wang et al.,2006)。然而,对于河口、三角洲和潟湖等涉及较大空间范围的海岸系统,可以使用来自中等空间分辨率的卫星传感器的数据,如MODIS(Hellweger et al.,2004;Chen et al.,2007),以及来自中等分辨率成像光谱仪MERIS(Matthews et al.,2010)的数据。当然,也有许多研究将更高空间分辨率的卫星数据应用于水质参数的估算,例如Landsat(Lavery et al.,1993;Hellweger et al.,2004;Kabbara et al.,2008;Wang and Xu,2008;Bustamante et al.,2009)或EO-1 ALI(Chen et al.,2009)。与应用水-光学模型的研究不同(Salama et al.,2012),这些研究的主要局限性在于其大多数都是针对特定地点的。然而,它们允许对水体的状况进行大规模评估,而在许多情况下,这种评估无法通过其他方式获得(Engman and Gurney,1991)。未来,预计将有更多的研究来充分了解这些水体光学特性的可变性(Budhiman et al.,2012)。

五、结论

遥感技术无疑是流域、区域乃至全球尺度水文建模的强大数据来源,是生态系统服务评估、制图和量化的必要基础。事实上,数据可用性和质量的不断提升,为建模者提供了将同化技术包含在建模环境本身中的方案。直接估算在水文过程中相关的地形特性的空间分布,以及在卫星频率下计算不同状态变量,这两种方法都为科学家和技术人员提供了指定位置上的离散空间信息,这些信息可以被认为在给定的传感器的空间分辨率尺度上是连续的。因此,尺度问题构成了一个需要考虑的重要问题。卫星资源当前和未来的发展,及已在进行的利用多个卫星各自较高的空间或时间分辨率的优势来组合来自多个卫星数据的工作,构成了技术人员和科学家的可靠视野。过去几十年里,获得详细分布信息的可能性极大地增加和扩大了人类观察地球和模拟过程的能力,而在未来仍将如此。

在水质模拟方面,遥感提供了有关地形和土壤性质空间分布的有用信息。这些数据可用作水文模型的输入,在流域尺度上再现非点源污染过程。然而,遥感数据在水质建模中的主要应用领域是估算水体中的水质参数。这些信息可用于水质模型的校准和验证。而尺度问题和水体光学特性的复杂性将仍然构成了未来研究的主线。

参 考 文 献

Abbott, M. B., J. C. Bathurst, J. A. Cunge, P. E. O'Connel, and J. Rasmussen. 1986. An introduction to the European Hydrological System—Systeme Hydrologique "SHE", 1: History and philosophy of a physically based distributed modelling system. *Journal of Hydrology* 87:45–59.

Aguilar, C., J. Herrero, and M. J. Polo. 2010. Topographic effects on solar radiation distribution in mountainous watersheds and their influence on reference evapotranspiration estimates at watershed scale. *Hydrology and Earth System Sciences* 14:2479–2494.

Aguilar, C., J. C. Zinnert, M. J. Polo, and D. R. Young. 2012. NDVI as an indicator for changes in water availability to woody vegetation. *Ecological Indicators* 23:290–300.

Anderson, J., T. Hoar, K. Raeder, et al. 2009. The Data Assimilation Research Testbed: A community facility. *Bulletin of the American Meteorological Society* 90:1283–1296.

Andren, O., T. Kätterer, and T. Karlsson. 2004. ICBM regional model for estimations of dynamics of agricultural soil carbon pools. *Nutrient Cycling in Agroecosystems* 70:231–239.

Arnold, J. G., R. Srinivasan, R. S. Muttiah, and J. R. Williams. 1998. Large area hydrologic modeling and assessment Part I: Model development. *Journal of American Water Resources Association* 34:73–89.

Ávila, A. 2007. *Procesos de múltiple escala en la evolución de la línea de costa* [Multiple scale processes in the coastline evolution]. PhD diss., Environmental Flow Dynamics Research Group, University of Granada, Spain.

Baban, S. M. 1993. Detecting water-quality parameters in the Norfolk broads, UK, using Landsat imagery. *International Journal of Remote Sensing* 14:1247–1267.

Band, L., D. Peterson, S. Running, et al. 1991. Forest ecosystem processes at the watershed scale: Basis for distributed simulation. *Ecological Modeling* 56:171–196.

Bannari, A., D. Morin, F. Bonn, and A. R. Huete. 1995. A review of vegetation indices. *Remote Sensing Reviews* 12:335–357.

Bellamy, P., M. Camino, J. Harris, R. Corstanje, I. Holman, and T. Mayr. 2011. Monitoring and modelling ecosystem services: A scoping study for the ecosystem services pilots. Natural England Commissioned Report NECR073, Cranfield.

Beven, K., and M. Kirkby. 1979. A physically-based variable contributing area model of basin hydrology. *Hydrologic Science Bulletin* 24:43–69.

Biswas, A. K. 1970. *History of hydrology*. Amsterdam: North Holland Publishing Co.

Budhiman, S., M. S. Salama, Z. Vekerdy, and W. Verhoef. 2012. Deriving optical properties of Mahakam Delta coastal waters, Indonesia, using in situ measurements and ocean color model inversion. *ISPRS Journal of Photogrammetry and Remote Sensing* 68:157–169.

Bustamante, J., F. Palacios, R. Díaz-Delgado, and D. Aragonés. 2009. Predictive models of turbidity and water depth in the Doñana marshes using Landsat TM and ETM+ images. *Journal of Environmental Management* 90:2219–2225.

Carpenter, S., H. Mooney, J. Agard, et al. 2009. Science for managing ecosystem services: Beyond the Millennium Ecosystem Assessment. *Proceedings of the National Academy of Sciences of the United States of America* 106:1305.

Carroll, T. R. 1981. Airborne soil moisture measurement using natural terrestrial gamma radiation. *Soil Science* 132:358–366.

Chen, J., X. Chen, W. Ju, and X. Geng. 2005. Distributed hydrological model for mapping evapotranspiration using remote sensing inputs. *Journal of Hydrology* 305:15–39.

Chen, S., L. Fang, L. Zhang, and W. Huang. 2009. Remote sensing of turbidity in seawater intrusion reaches of Pearl River Estuary—A case study in Modaomen waterway, China. *Estuarine, Coastal and Shelf Science* 82:119–127.

Chen, Z., C. Hu, and F. Muller-Karger. 2007. Monitoring turbidity in Tampa Bay using MODIS/Aqua 250-m imagery. *Remote Sensing of Environment* 109:207–220.

Cox, P. M. 2001. *Description of the "TRIFFID" dynamic global vegetation model*. Technical Note 24. Berks: Hadley Centre.

Curran, P. 1981. Multispectral remote sensing for estimating vegetation biomass and productivity. In *Plants and the daylight spectrum*, ed. H. Smith, 65–99. London: Academic Press.

Daily, S., J. Polasky, P. M. Goldstein, et al. 2009. Ecosystem services in decision making: Time to deliver. *Frontiers in Ecology and the Environment* 7:21–28.

Dekker, A., and S. Peters. 1993. The use of the Thematic Mapper for the analysis of eutrophic lakes—A case study in the Netherlands. *International Journal of Remote Sensing* 14:799–821.

De Lannoy, G. J. M., R. H. Reichle, K. Arsenault, et al. 2011. Multi-scale assimilation of AMSR-E snow water equivalent and MODIS snow cover fraction observations in northern Colorado. *Water Resources Research* 48:W01522.

Dellman, P. N., C. E. Ruiz, C. T. Manwaring, and E. J. Nelson. 2002. *Watershed modeling system hydrological simulation program; watershed model user documentation and tutorial*. Vicksburg, MS: Engineer Research and Development Center, Environmental Lab.

DeWalle, D. R., and A. Rango. 2008. *Principles of snow hydrology*. Cambridge, UK: Cambridge University Press.

Díaz-Gutiérrez, A. 2007. *Series temporales de vegetación para un modelo hidrológico distribuido* [Temporal series of vegetation for a distributed hydrological model]. Monografías 2007. Spain: Grupo de Hidrología e Hidráulica Agrícola, University of Córdoba.

Engman, E. T., and R. J. Gurney. 1991. *Remote sensing in hydrology*. London: Chapman & Hall.

Feng, B., B. Fu, X. Yang, and Y. Lü. 2010. Remote sensing of ecosystem services: An opportunity for spatially explicit assessment. *Chinese Geographical Science* 20:522–535.

Franko, U., B. Oelschlagel, and S. Schenk. 1995. Simulation of temperature, water and nitrogen dynamics using the model CANDY. *Ecological Modelling* 81:213–222.

Friend, A. K., R. G. Stevens, M. G. R. Knox, and A. Cannell. 1997. Process-based terrestrial biosphere model of ecosystem dynamics (Hybrid v3.0). *Ecological Modelling* 95:249–287.

Gash, J. H. C. 1979. An analytical model of rainfall interception by forests. *Quarterly Journal of the Royal Meteorological Society* 105:43–55.

Gassman, P. W., M. R. Reyes, C. H. Green, and J. G. Arnold. 2007. The soil and water assessment tool: Historical development, applications, and future research directions. Working Paper 07-WP 443. Center for Agricultural and Rural Development, Iowa State University, Ames, Iowa.

Goldman, R. L., S. Benitez, T. Boucher, et al. 2012. Water funds and payments for ecosystem services: Practice learns from theory and theory can learn from practice. *Oryx* 46:55–63.

Hadjimitsis, D., M. Hadjimitsis, C. Clayton, and B. Clarke. 2006. Determination of turbidity in Kourris Dam in Cyprus utilizing Landsat TM remotely sensed data. *Water Resources Management* 20:449–465.

Hall, D. K., G. A. Riggs, and V. V. Salomonson. 1995. Development of methods for mapping global snow cover using moderate resolution imaging spectroradiometer data. *Remote Sensing of Environment* 54:127–140.

Hellweger, F. L., P. Schlosser, U. Lall, and J. K. Weissel. 2004. Use of satellite imagery for water quality studies in New York Harbor. *Estuarine, Coastal and Shelf Science* 61:437–448.

Herrero, J., C. Aguilar, A. Millares, et al. 2010. *WiMMed. User Manual v1.1.* Granada, Spain: Grupo de Dinámica Fluvial e Hidrología (University of Córdoba) and Grupo de Dinámica de Flujos Ambientales (University of Granada).

Herrero, J., M. J. Polo, and M. A. Losada. 2005. Modelo SWAT aplicado a la cuenca del río Guadalfeo. Balance hidrológico dentro de un modelo de gestión [Assessment of SWAT model in the Guadalfeo River basin: hydrological balance in a management model]. *VI Simposio del Agua en Andalucía (SIAGA)*. IGME. 237–248.

Herrero, J., M. J. Polo, and M. A. Losada. 2011. Snow evolution in Sierra Nevada (Spain) from an energy balance model validated with Landsat TM data. Proc. SPIE 8174, *Remote Sensing for Agriculture, Ecosystems, and Hydrology XIII*, 817403.

Houser, P. R., G. J. M. De Lannoy, and J. P. Walker. 2012. *Hydrologic data assimilation, approaches to managing disaster—Assessing hazards, emergencies and disaster impacts*, ed. J. Tiefenbacher. Rijeka, Croatia: InTech.

Houser, P. R., W. J. Shuttleworth, J. S. Famiglietti, H. V. Gupta, K. H. Syed, and C. Goodrich. 1998. Integration of soil moisture remote sensing and hydrologic modeling using data assimilation. *Water Resources Research* 34:3405–3420.

Ji, Z. G. 2008. *Hydrodynamics and water quality: Modeling rivers, lakes and estuaries.* Hoboken, NJ: John Wiley & Sons.

Johnson, K. A., S. Polasky, E. Nelson, and D. Pennington. 2012. Uncertainty in ecosystem services valuation and implications for assessing land use tradeoffs: An agricultural case study in the Minnesota River Basin. *Ecological Economics* 79:71–79.

Kabbara, N., J. Benkhelil, M. Awad, and V. Barale. 2008. Monitoring water quality in the coastal area of Tripoli (Lebanon) using high-resolution satellite data. *ISPRS Journal of Photogrammetry and Remote Sensing* 63:488–495.

Kirschbaum, M. U. F. 1999. CenW, a forest growth model with linked carbon, energy, nutrient and water cycles. *Ecological Modelling* 181:17–59.

Kite, G., and A. Pietroniro. 1996. Remote sensing applications in hydrological modeling. *Hydrological Science* 41:563–591.

Lane, R. R., J. W. Day, B. D. Marx, E. Reyes, E. Hyfield, and J. N. Day. 2007. The effects of riverine discharge on temperature, salinity, suspended sediment and chlorophyll-a in a Mississippi Delta estuary measured using a flow-through system. *Estuarine, Coastal and Shelf Science* 74:145–154.

Lathrop, R. G. 1992. Landsat Thematic Mapper monitoring of turbid inland water quality. *Photogrammetric Engineering and Remote Sensing* 58:465–470.

Lavery, P., C. Pattiaratchi, and P. Hick. 1993. Water quality monitoring in estuarine waters using the Landsat Thematic Mapper. *Remote Sensing of Environment* 46:268–280.

Liu, X., and J. Li. 2008. Application of SCS model in estimation of runoff from small watershed in loess plateau of China. *Chinese Geographical Science* 18:235–241.

Malik, M. J., R. van der Velde, Z. Vekerdy, and Z. Su. 2012. Assimilation of satellite observed snow albedo in a land surface model. *Journal of Hydrometeorology* 13:1119–1130.

Matthews, M. W., S. Bernard, and K. Winter. 2010. Remote sensing of cyanobacteria-dominant algal blooms and water quality parameters in Zeekoevlei, a small hypertrophic lake, using MERIS. *Remote Sensing of Environment* 114:2070–2087.

Mayo, M., A. Gitelson, Y. Z. Yacobi, and Z. Ben-Avraham. 1995. Chlorophyll distribution in Lake Kinneret determined from Landsat Thematic Mapper data. *International Journal of Remote Sensing* 16:175–182.

Neitsch, S. L., J. G. Arnold, J. R. Kiniry, and J. R. Williams. 2005. *Soil and water assessment tool theoretical documentation*, version 2005. Temple, TX: Grassland, Soil and Water Research Service.

Nelson, E., G. Mendoza, J. Regetz, et al. 2009. Modeling multiple ecosystem services, biodiversity conservation, commodity production, and tradeoffs at landscape scales. *Frontiers in Ecology and the Environment* 7:4–11.

Notter, B., H. Hurni, U. Wiesmann, and K. C. Abbaspour. 2012. Modelling water provision as an ecosystem service in a large East African river basin. *Hydrology and Earth System Sciences* 16:69–86.

Pimentel, R., J. Herrero, and M. J. Polo. 2012. Terrestrial photography as an alternative to satellite images to study snow cover evolution at hillslope scale. *Proceedings of SPIE - The International Society for Optical Engineering*, vol 8531, art 85310Y.

Polo, M. J., A. Díaz-Gutiérrez, and M. P. González-Dugo. 2011. Interception modeling with vegetation time series derived from Landsat TM data. *Proceedings of SPIE - The International Society for Optical Engineering*, vol 8174, art. 817403.

Polo, M. J., J. Herrero, C. Aguilar, et al. 2009. WiMMed, a distributed physically-based watershed model (I): Description and validation. In *Environmental hydraulics: Theoretical, experimental and computational solutions, IWEH09*, eds. P. A. López-Jiménez, V. S. Fuertes-Miquel, P. L. Iglesias-Rey, et al., 225–228. London: CRC Press.

Porras, I., M. Grieg-Gran, and N. Neves. 2008. *All that glitters: A review of payment for watershed services in developing countries*. London: International Institute for Environment and Development.

Prochnow, S. J., J. D. White, T. Scott, and C. D. Filstrup. 2008. Multi-scenario simulation analysis in prioritizing management options for an impacted watershed system. *Ecohydrology & Hydrobiology* 8:3–15.

Ramsey, R. D., J. R. Wright, and C. McGinty. 2004. Evaluating the use of Landsat 30m Enhanced Thematic Mapper to monitor vegetation cover in shrub-steppe environments. *Geocarto International* 19:39–47.

Rasse, D. P., L. François, M. Aubinet, et al. 2001. Modelling short-term CO_2 fluxes and long-term tree growth in temperate forests with ASPECTS. *Ecological Modelling* 141:35–52.

Red de Información Ambiental de Andalucía. 2010. Ortofotografía Digital Histórica de Andalucía 1956–2008: Medio siglo de cambios en Andalucía [Historical digital ortophotography of Andalusia 1956–2008: half a century of change in Andalusia]. Seville, Spain: Junta de Andalucía.

Reichle, R. H., B. Dennis McLaughlin, and D. Entekhabi. 2002. Hydrologic data assimilation with the ensemble Kalman filter. *Monthly Weather Review* 130:103–114.

Reyers, B., S. Polasky, H. Tallis, H. Mooney, and A. Larigauderie. 2012. Finding a common ground for biodiversity and ecosystem services. *BioScience* 62:503–507.

Ruiz de Almirón, C. 2011. *Modelo conceptual para el control de salinidad en la Charca de Suárez mediante un sistema de compuertas* [Conceptual model for the control of salinity in "Charca Suárez" using a floodgate system]. Master thesis, Environmental Flows Dynamics Research Group, University of Granada, Spain.

Rutter, A. J., K. A. Kershaw, P. C. Robins, and A. J. Morton. 1971. A predictive model of rainfall interception in forests. I. Derivation of the model from observations in a plantation of Corsican pine. *Agricultural Meteorology* 9:367–384.

Rutter, A. J., A. J. Morton, and P. C. Robins. 1975. A predictive model of rainfall interception in forests. II. Generalization of the model and comparison with observations in some coniferous and hardwood stands. *Journal of Applied Ecology* 12:367–380.

Salama, M. S., M. Radwan, and R. van der Velde. 2012. A hydro-optical model for deriving water quality variables from satellite images (HydroSat): A case study of the Nile River demonstrating the future Sentinel-2 capabilities. *Physics and Chemistry of the Earth*, Parts A/B/C:224–232.

Salama, M. S., and Z. Su. 2010. Bayesian model for matching the radiometric measurements of aerospace and field ocean color sensors. *Sensors* 10:7561–7575.

Salama, M. S., and Z. Su. 2011. Resolving the subscale spatial variability of apparent and inherent optical properties in ocean color match-up sites. *IEEE Transactions on Geoscience and Remote Sensing* 49:2612–2622.

Schilling, K. E., and C. F. Wolter. 2009. Modeling nitrate-nitrogen load reduction strategies for the Des Moines River, Iowa, using SWAT. *Journal of Environmental Management* 44:671–682.

Schmugge, T. J., W. P. Kustas, J. C. Ritchie, T. J. Jackson, and A. Rango. 2002. Remote sensing in hydrology. *Advances in Water Resources* 25:1367–1385.

Schultz, G. A., and E. T. Engman, eds. 2000. *Remote sensing in hydrology and water management*. Berlin: Springer-Verlag.

Sellers, P. J. 1989. *Theory and applications of optical remote sensing*. New York: Wiley-Interscience.

Shen, F., M. S. Salama, Y. X. Zhou, J. F. Li, Z. Su, and D. B. Kuang. 2010. Remote-sensing reflectance characteristics of highly turbid estuarine waters—A comparative experiment of the Yangtze River and the Yellow River. *International Journal of Remote Sensing* 31:2639–2654.

Shen, Z. Y., Y. W. Gong, Y. H. Li, Q. Hong, L. Xu, and R. M. Liu. 2009. A comparison of WEPP and SWAT for modeling soil erosion of the Zhangjiachong watershed in the Three Gorges reservoir area. *Agricultural Water Management* 96:1435–1442.

Stone, M. C., R. H. Hotchkiss, C. M. Hubbard, T. A. Fontaine, L. O. Mearns, and J. G. Arnold. 2001. Impacts of climate change on Missouri River basin water yield. *Journal of the American Water Resources Association* 37:1119–1129.

Su, Z., R. A. Roebeling, I. Holleman, et al. 2011. Observation of hydrological processes using remote sensing. In *Treatise on water science*, ed. P. Wilderer, vol. 2:351–399. Oxford: Academic Press.

Su, Z., J. Wen, and W. Wagner. 2010. Advances in land surface hydrological processes: Field observations, modeling and data assimilation: Preface. *Hydrology and Earth System Sciences* 14:365–367.

Tague, C. L., and L. E. Band. 2001. Evaluating explicit and implicit routing for watershed hydro-ecological models of forest hydrology at the small catchment scale. *Hydrological Processes* 15:1415–1439.

Tallis, H. T., T. Ricketts, A. D. Guerry, et al. 2011. *InVEST 2.3.0 user's guide*. Stanford, CA: The Natural Capital Project.

USACE (U.S. Army Corps of Engineers). 1982. Hydrologic analysis of ungaged watersheds with HEC-1. Davis, CA: Hydrologic Engineering Center.

USSCS (U.S. Soil Conservation Service). 1982. *Project formulation, hydrology*. Technical release No. 20. Washington, DC: USSCS.

Van der Werf, K., P. J. Keesman, A. R. Burgess, et al. 2007. Yield-SAFE: A parameter-sparse process-based dynamic model for predicting resource capture, growth and production in agroforestry systems. *Ecological Engineering* 29:419–433.

Vigerstol, K. L., and J. E. Aukema. 2011. A comparison of tools for modeling freshwater ecosystem services. *Journal of Environmental Management* 92:2403–2240.

Villa, F., M. Ceroni, K. Bagstad, G. Johnson, and S. Krivovet. 2009. ARIES (ARtificial Intelligence for Ecosystem Services): A new tool for ecosystem services assessment, planning, and valuation. *11th International BIOECON Conference on Economic Instruments to Enhance the Conservation and Sustainable Use of Biodiversity*. Venice, Italy.

Wagner, W., N. E. C. Verhoest, R. Ludwig, and M. Tedesco. 2009. Remote sensing in hydrological sciences. *Hydrology and Earth System Sciences* 13:813–817.

Wang, F., L. Han, H. T. Kung, and R. B. Van Arsdale. 2006. Applications of Landsat-5 TM imagery in assessing and mapping water quality in Reelfoot Lake, Tennessee. *International Journal of Remote Sensing* 27:5269–5283.

Wang, F., and Y. J. Xu. 2008. Development and application of a remote sensing-based salinity prediction model for a large estuarine lake in the US Gulf of Mexico coast. *Journal of Hydrology* 360:184–194.

Wang, L., and J. J. Qu. 2009. Satellite remote sensing applications for surface soil moisture monitoring: A review. *Frontiers of Earth Science in China* 3:237–247.

Wigmosta, M. S., L. Vail, and D. P. Lettenmaier. 1994. A distributed hydrology-vegetation model for complex terrain. *Water Resources Research* 30:1665–1679.

Yates, D., J. Sieber, D. R. Purkey, and A. Huber-Lee. 2005. A demand, priority and preference-driven water planning model: Part 1, model characteristics. *Water International* 30:487–500.

第十三章 基于卫星绿度异常和时间动力学的地下水生态系统依赖探测

S. 孔特雷拉斯 D. 阿尔卡拉斯-塞古拉 B. 斯坎伦 E.G. 乔巴吉

一、简介

地下水依赖型生态系统(Groundwater-dependent Ecosystems, GDEs)在人类发展中发挥着关键作用,在降雨率较低的地区尤其重要,它提供了广泛的生态系统服务,如为野生动物栖息地和生物多样性热点提供物理支持,控制洪水和侵蚀,调节营养循环或为认知发展提供避难所性质景观(de Groot et al., 2002; Chen et al., 2004; Eamus et al., 2005; Bergkamp and Katharine, 2006; Ridolfi et al., 2007)。在过去的十年中,关于 GDEs 的生态学和功能的研究越来越受到科学界和陆地景观管理者的关注。尽管这类生态系统具有很高的内在价值,但由于与地下水资源的水文联系被破坏,诸多生态系统受到了严重影响。地下水开采和耗竭速度过快通常加速了这种破坏,例如西班牙的 Las Tablas de Daimiel 和多尼亚纳国家保护区(Llamas, 1988; Muñoz-Reinoso and García-Novo, 2005);澳大利亚西南部的斯旺沿海平原(Groom et al., 2000);美国莫哈维沙漠和大盆地沙漠(Patten et al., 2008);美国的圣佩德罗河(Stromberg et al., 1996)。破坏也有可能因为疏浚或人工改道改变了河道或湿地的形态(Ellery and McCarthy, 1998)或由于气候因素导致水平衡变化而引起(Murray-Hudson et al., 2006)。因此,要评估 GDEs 提供的生态服务就必须更好地理解 GDEs 的功能和用水量(Murray et al., 2006; Brauman et al., 2007),并在未来土地利用和气候变化情景下制定协调人类活动、生态系统保护及其潜在水文权衡的适应性管理框架(MacKay,2006;Barron et al.,2012)。

GDEs 是需要地下水流入以维持其当前结构和功能,并提供生态系统服务的生态系统(Hatton and Evans,1997;Murray et al.,2003;Eamus et al.,2006)。GDEs 可能表现出一种强制性的依赖关系,需要恒定的地下水存在,或是一种兼性的依赖关系,使其功能适应地下水可用性的波动(Murray et al.,2003;Bertrand et al.,2012)。根据含水层-生态系统的交互关系,GDEs 包括(Eamus et al.,2006;Eamus,2009):(1)洞穴和地下水生生态系统,包括岩溶含水层和裂隙岩体系统;(2)依赖于地下水永久或间歇地表出露的生态系统,包括底流河流、泉水、湿地或泥炭地,以及河口、海洋海岸线生态系统;(3)依赖于地下水地下存在形式的生态系统,亦称"陆地 GDEs"或潜水植物生态系统(Richardson et al.,2011)。其他研究也从功能角度提出了对 GDEs 进行分类的土壤学、形态学、水文和生物地球化学标准(Bertrand et al.,2012)。

为了保持 GDEs 的生态完整性和服务提供,GDEs 需要基于以下知识来制定水资源分配计划和适应性管理策略:(1)类型和空间分布;(2)定量需水量;(3)对地下水状况的自然和人类扰动的抵抗力和恢复力。通常采用许多方法和技术来完成这三个方面的衡量评估,包括遥感、水平衡分析、水文地质建模、示踪和同位素研究、生理生态学测量、根系表征和水生动物群取样(见 Richardson 等人2011 年的综合)。

使用归一化差异植被指数或增强植被指数等卫星指数跟踪植被的光合作用和绿度活动,为描述河岸、湿地和陆地 GDEs 的功能提供了相对便宜和有效的方法(Bradley and Mustard,2008;Barron et al.,2012)。这些光谱指标与半干旱地区的地上净初级生产力和蒸散量密切相关(Running and Nemani,1988;Paruelo et al.,1997;Jobbágy et al.,2002;Nagler et al.,2005;Guerschman et al.,2009)(见第十八章)。当不受地下水和侧向流入资源的影响时,半干旱地区的 ANPP 和蒸散量年速率主要受降水控制,其次受辐射强迫及其与降水量输入的季节耦合影响(Specht,1972;Specht and Specht,1989;Ellis and Hatton,2008;Palmer et al.,2010)。由于地下水为陆地生态系统提供的水资源在时间上比降雨更可靠,因此与非地下水生态系统相比,GDEs 中的年度蒸散量和 ANPP 应更高且更稳定(Contreras et al.,2011;O'Grady et al.,2011)。野外观察表明,叶面积、

ANPP 和水可用性密切相关,这为使用基于卫星的植被指数来识别和描述 GDEs 提供了支撑,使得量化不同时间尺度下的需水量成为可能(Nagler et al.,2005;Contreras et al.,2011;Devitt et al.,2011;O'Grady et al.,2011)。野外证据还表明,即使存在获得无限地下水资源的途径,GDEs 的生产力也可能受到其他有限资源或过程的强烈限制,例如输入的能量、养分可用性、形态约束或扰动等(Eamus et al.,2000;Do et al.,2008)。

因此,从光谱植被指数的年度总结中反演出的年度初级生产力估算似乎不足以识别和描述陆地 GDEs 的需水量。为了解决这一潜在限制,对季节性绿度时序的补充评估可提供关于生态系统对其环境的功能反应的额外和有价值的信息(Morisette et al.,2009)。在此类研究中,除了初级生产力估算外,季节性和物候特征通常从生态系统年度绿度动态中反演出来,以对生态系统功能类型进行分类和描述,形成生物群和物理环境之间物质和能量交换相似的陆地表面斑块(Paruelo et al.,2001;Alcaraz Segura et al.,2006;Fernández et al.,2010)(见第九章和第十六章)。

本研究旨在评估利用卫星识别确定依赖水流流入型生态系统的方法,并探测湿地和潜水植物生态系统对地下水的依赖类型和程度。该方法包括根据孔特雷拉斯等人(Contreras et al.,2011)计算的年度绿度异常的补充分析,及从年内和年际绿度轨迹中反演出的地表物候指标。这种方法的性能在蒙特卡中部沙漠(阿根廷)的低地势地区进行了测试,该研究区域以前已确定了当地依赖水流流入的生态系统的潜在梯度。最后,将此类型生态系统的代表性样本计算出的生产力、季节性和物候度量与从位于上游灌溉绿洲的利用地表和地下水资源进行维持的站点样本中提取的指标进行了比较。

二、方法

(一) 研究地点

研究区域是覆盖 87 500 平方千米的低地势地区(小于等于海平面上 1 000 米),并在南纬 31 度和南纬 36 度之间延伸至阿根廷的蒙特卡中部沙漠(图 13-1)。该地区西面以安第斯山脉为界,东面以塞拉斯山脉为界。该地区的降水量范围为 150~400 毫米/年,大部分集中在南侧夏季(10 月~次年 3 月),年平均温度范围为 13~19℃。在研究区域最干燥的地区,潜在蒸散量达到 1 400 毫米/年。亚伯拉罕等人(Abraham et al.,2009)详细回顾了蒙特州沙漠的主要生物物理和社会经济特征,维拉格拉等人(Villagra et al.,2009)回顾了土地利用和干扰因素对该沙漠自然生态动态的一些影响。

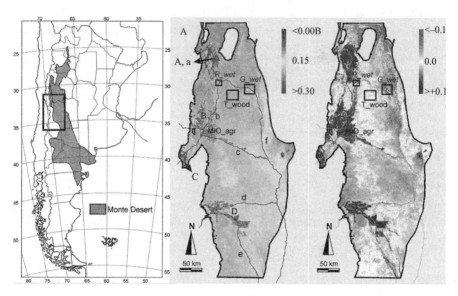

图 13-1(见彩插)

注:2000 年 9 月~2009 年 8 月计算的研究区域 EVI 年平均值(A,左)和 EVI 异常(B,右)。主要河流(小写字母):a 圣胡安河;b 门多萨河;c 图努扬河;d 迪亚曼蒂河;e 阿图埃尔河;f 德萨瓜德罗-萨拉多河。灌溉绿洲大写字母:A 圣胡安绿洲;B 门多萨绿洲;C 上屯垦绿洲;D 圣拉斐尔绿洲。对照区(黑色方块)的样点位于泰尔特卡自然保护区的开阔木豆树属林地和周围(T_wood)、罗萨里奥和瓜纳卡齐湿地系统(分别为 R_wet 和 G_wet)及门多萨绿洲的灌溉作物(MIO_agr)。

该地区有五条主要河流(从北向南依次为:圣胡安、门多萨、图努扬、迪亚曼特和阿图埃尔)穿过,其起源于安第斯山脉。这些河流穿越山脉后,到达中部蒙特卡沙漠的冲积扇和沉积平原,最终汇入德萨瓜德罗-萨拉多水系。安第斯河是该地区山脚下四个大型人工绿洲的主要水源(图 13-1),葡萄园、橄榄和果树是主要作物。这些绿洲约占该地区经济活动的 90%,150 多万人居住在那里。沿着冲积扇的路线,河流注入大量松散含水层,含水层延伸到人工绿洲下游,到达该地区被河流、湖泊和风成起源的沙质平原覆盖的低地势地区。

安第斯山脉山麓的冲积平原和低地平原主要由三种类型的生态系统覆盖：(1)以落叶松属为主的灌木草原；(2)牧豆树的开放植生林（当地称为阿尔加罗瓦）；(3)主要河流沿线的沼泽和湿地。两个最大的湿地系统是罗萨里奥系统——位于门多萨河与圣胡安河汇合之前的最后一段，以及位于圣胡安河末端的瓜纳卡奇系统。

由于灌溉绿洲上游河流的调节和农业的大量引水，门多萨河在门多萨市下游仅具有短暂的水文动态。只有在强降雨之后，河流末端的河岸植被和罗萨里奥湿地才会被供应地表水。圣胡安河则有一个永久性的水文动态，作为瓜纳卡奇湿地的恒定地表水源。然而，在过去的几十年里，与门多萨河流域的情况类似，圣胡安绿洲的水力调节和灌溉农业对湿地系统的流量影响很大。湖泊植被，如加州三棱藨草和香蒲占据了湿地的主导地位，但由于河流的水情变化和河流的补给，柽柳属的外来物种越来越多地入侵这类地区。开阔的牧豆树属林地主要位于土壤有90%砂土以上的冲积平原。林地根据其对地下水资源的依赖而具有不同的结构，并且在泰尔特卡自然保护区和图努扬河的远段（6～15米深度的地下潜水水位）中的增长率和健康状况高于纳库南国家保护区（70～80米深度的地下水位）(Villagra et al., 2005)。在以广泛沙丘系统为主的泰尔特卡地区，沙丘间山谷中的开阔林地发育良好。同位素和水化学剖面研究表明，这些林地对地下水高度依赖(Aranibar et al., 2011; Jobbágy et al., 2011)。开阔的林地和湿地在历史上都为当地居民点和经济提供了木材、泥炭和木炭、食物和水，也为家畜提供了所需的实物支持(Villagra et al., 2009)。

在本研究的框架下，对罗萨里奥(R_wet)和瓜纳卡奇(G_wet)湿地系统及位于泰尔特卡国家保护区和周边(T_wood)的开放牧豆树属林地的植被动态进行了卫星遥感测量。位于门多萨灌溉绿洲(MIO_agr)的灌溉作物的代表性样本也具有同样的植被动态特征（图13-1）。

（二）用于绿度异常估算的气候和卫星数据集

根据孔特雷拉斯等人(Contreras et al., 2011)，我们将"绿度异常"定义为在景观中的任何像素处观察到的年平均绿度与根据当地降水估算的特定地点参考绿度值之间的绝对差异。在本研究中，我们使用EVI作为植被绿度的指标。假定以降水为基础的参考绿度值与年平均降水量（Mean Annual Precipitation，

MAP)线性相关,如下所示:

$$EVI_{ref} = a\,MAP + b \qquad 式13-1$$

其中,EVI_{ref}是基于降水的EVI,a和b是拟合参数,通过对一组参考地点的观测EVI-MAP散点图进行的分位数进行回归分析计算参数。

基于半干旱地区可用的野外数据支持,我们假设EVI和蒸散量之间存在线性关系(Nagler et al.,2005;Guerschman et al.,2009;O'Grady et al.,2011)。有了这样的假设,我们能够估算仅使用当地降水且与长期降水平衡的植被覆盖的预期EVI值(Boer and Puigdefábregas,2003;Contreras et al.,2008)。在这种情况下,每年蒸散量(ET)接近MAP,我们的研究区域也达到这种情况的75百分位数阈值(Contreras et al.,2011)。在年度时间尺度上,我们定义了EVI异常的概念如下:

$$EVI_a = EVI_{ma} - EVI_{map} \qquad 式13-2$$

其中,EVI_{ma}是从卫星影像每个像素上计算的EVI的观测年平均值,EVI_{map}是式13-1中使用第75分位数阈值估算的EVI_{ref}。从功能的角度来看,EVI_a和EVI_{ma}两个指标在这里都被认为是初级生产力的替代性评价指标。

根据CRU CL 2.0数据集(New et al.,2002)中报告的长期平均月值计算了该区域的MAP图,该图在之前用CLIMWAT 2.0数据库(FAO,2006年数据)和当地气象站的数据进行了校正。最终将降水图重新采样到250米的空间分辨率,与卫星数据兼容(Contreras et al.,2011)。

从MODIS Collection 5(Solano et al.,2010)数据中提取的ENVMOD13Q1土地产品涵盖了2001年9月～2009年8月的9个水文年(每个水文年23景影像)。在处理之前,利用Timeat软件(Jönsson and Eklundh,2004)使用基于自适应Savitzk-Golay滤波器的局部多项式函数对250米空间分辨率的原始EVI数据进行滤波。结合了蓝色、红色和红外光谱带数据的EVI比NDVI更可取,因为大气干扰和土壤背景信号被更有效地去除,同时也因为其对高生物质情况更敏感(Huete et al.,2002)。

式13-1在125个符合低干扰率和缺乏人工或径流供水要求的参考站点进行了参数化(Contreras et al.,2008,2011)。根据EVI-MAP散点图,本文使用三个分位数阈值来提出地下水依赖类别的阈值梯度。首先,如前所述,我们使用平均年度 EVI 与 MAP 函数的 75 分位数回归作为保守值来生成 EVI_{map} 值。观测到的低于 EVI_{map} 的 EVI 值(即负的绿度异常值)被假定为不依赖于地下水资源。对于高于 EVI_{map} 的值,分别使用第 75 分位数阈值、第 90 分位数阈值和第 99 分位数阈值建立低、中和高依赖程度三个潜在类别的梯度。在本研究中,任意选择先前的分位数阈值,以评估结果依赖水平与从绿度时间分析提取的物候度量间的潜在一致性。

(三) 绿度时间和指标

选择了四个研究系统(泰尔特卡林地、罗萨里奥和瓜纳卡奇湿地及门多萨绿洲灌溉作物)的代表性像素样本,用于反演与初级生产力、季节性和物候植被特征相关的度量或物候指标参数(图 13-2)。在泰尔特卡保护区,对 78 个像素进行了采样,以覆盖由绿度异常(无依赖度、低、中、高)确定的所有潜在地下水依赖程度(无依赖度、低、中、高),但最终将中等和高依赖度的像素分组在一起,以获得类别之间更可靠的比较。

在湿地系统中,仅从具有潜在高度依赖地下水的像素中(瓜纳卡奇,$n=40$;罗萨里奥,$n=10$)提取植被指标。从 2000 年 9~2009 年 8 月的平均季节轨迹(年内变异性)和研究涵盖的九个水文年的年度轨迹(年际变异性)中,在年内和年际时间尺度上提取了与植被初级生产力、季节性和物候相关的度量。在本研究中,我们提取了与以下特征相关的 EVI 指标:(1)植被生产力:EVI_{ma} 和 EVI_{gs};(2)植被季节性:EVI_{min}、EVI_{max} 和 EVI_{nrange};(3)植被物候:LGS 和 T_{max}(详见图 13-2)。

(四) 地下水对植被动态的影响:概念模型

从功能角度来看,靠近地表的地下水位预计会以多种方式影响 EVI 动态的年内(季节)和年间(多年)变化。以下假设指导了我们的分析(表 13-1)。

在年内尺度上,我们假设随着生态系统对地下水的依赖程度的增加,植被的年平均 EVI(EVI_{ma})、生长季累积的平均 EVI(EVI_{gs})、最大绿度值(EVI_{max})和最小绿度值(EVI_{min})将增加。

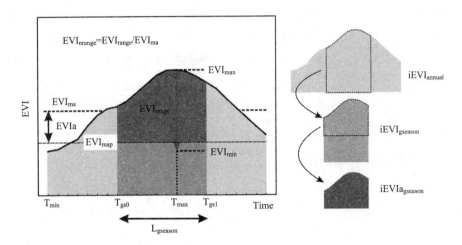

图 13-2

注：从增强植被指数（EVI）轨迹反演植被指标，以确定依赖地下水的生态系统并量化其对地下水的依赖。从年度（9～8 月）和平均长期季节轨迹中反演到所有的指标。指标与生产力性状有关：EVI_{ma} = 年平均 EVI；EVI_{gs} = 生长季节累积的平均 EVI；季节性性状：EVI_{max} 和 EVI_{min} = 最大和最小 EVI 值；EVI_{range} = EVI 的年振幅（$EVI_{max} - EVI_{min}$），EVI_{nrange} = 归一化年振幅（EVI_{range}/EVI_{ma}）；物候性状：T_{max} 和 T_{min} = 达到最大和最小 EVI 值的时间；T_{gs0} 和 T_{gs1} = 生长季节开始和结束的时间，L_{gseasy} = 生长季节长度。生产力特征是由年尺度（$iEVI_{annual}$）和生长季节尺度（$iEVI_{gs}$）的绿度综合值估算的。EVI_{map} 是根据当地平均年降水量预期的参考 EVI 值，需要计算年（$EVIa_{ma}$）和生长季节（$EVIa_{gs}$）的绿度异常。

表 13-1　植被性状趋势

植被性状	绿度指标	年内规模	年际变化
生产力	EVI_{ma}	↑	↓
	EVI_{gs}	↑	↓
季节性	EVI_{max}	↑	↓
	EVI_{min}	↓	↓
	EVI_{nrange}	↓	↓
物候学	L_{gs}	↑	↓
	T_{max}	↑	↓

注：随着地下水依赖程度的增加，陆地 GDEs 中预期的卫星指标测量的趋势（箭头）。指标与生产力性状有关：EVI_{ma} = 年平均 EVI；EVI_{gs} = 生长季节累积的平均 EVI；季节性特征：EVI_{max} 和 EVI_{min} = 最大和最小 EVI 值；EVI_{range} = EVI 的年振幅（$EVI_{max} - EVI_{min}$），EVI_{nrange} = 归一化年振幅（EVI_{range}/EVI_{ma}）；物候性状：T_{max} = 达到最大值的时间。

由于浅层地下水位代表了生态系统的常年水源，我们还假设，随着地下水资源的增加，任何依赖地下水的植被的生产力季节轨迹（EVI_{nrange}）变化较小。由于绿度轨迹的可变性较小（EVI_{nrange}较小），需要较长的时间才能达到年总生产力的50%，这里定义为生长季节长度（L_{gs}）。在年际尺度上，我们预计 GDEs 中描述生产力、季节性和物候特征的所有植被指标的变异性应低于非 GDEs，并应随着对地下水的依赖增加而降低。本文提出的评价生态系统对地下水依赖的概念规则矩阵是在假定植被对地下水的获取保持相对恒定，而地下水位不会发生大的变化的情况下设计的。然后，这种生态系统的地下水位或水文状况的变化预计将伴随其绿度动态和物候模式而改变。

三、结果与讨论

（一）MAP-EVI 区域功能

根据该地区描述的 MAP-EVI 函数（图 13-3），正 EVI 异常覆盖了 26 000 平方千米（约占总面积的 30%），其中 36% 分布在该地区灌溉绿洲（表 13-2；图 13-1）。高正异常约占自然生态系统、牧场上总正异常的 24%，但占灌溉绿洲总面积的 95%，说明灌溉对该地区农业发展的重要作用。

（二）泰尔特卡站点地下水依赖梯度的增强植被指数动态

在位于泰尔特卡自然保护区的牧豆树属林地中，EVI 的年内轨迹（图 13-4）和年际间轨迹（图 13-5）几乎是同步的，随着 EVI 异常的增加，年度和生长期值高于对照点（没有正绿度异常的地点）（图 13-6(a)）。随着 EVI 异常增加，季节性值 EVI_{max} 和 EVI_{min} 值较高，但季节性变化（EVI_{nrange}）较低，生产力指标的趋势也得到了证实（图 13-6(b)）。然而，在物候指标，即生长季节长度（L_{gs}）和最大值达到的时间（T_{max}）上，没有发现显著的趋势，尽管随着 EVI 异常增加，两个站点的平均值高于对照站点（图 13-6(c)）。

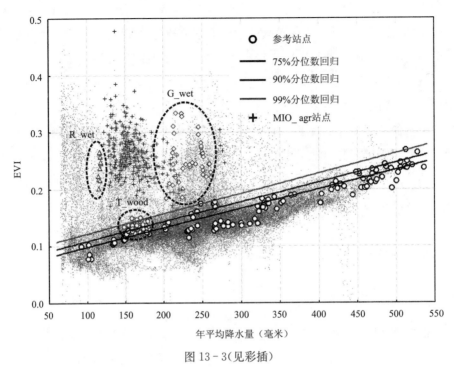

图 13-3（见彩插）

注：研究区年平均降水量——植被增强指数分位数回归函数。在每个控制区域选择的像素样本（棕色符号）被虚线包围。根据在 125 个参考位点（黑白圆圈）测量的 MAP-EVI 值计算函数。用 75、90 和 99 分位回归对应的 EVI 阈值，系统分别划分为除本地降水外对水输入的低、中和高依赖型。

表 13-2 植被增强指数的负向和正向异常

土地覆盖类型	负异常	正异常				合计
		低	中	高	小计	
自然生态系统或牧场	61.526	6.416	6.222	3.846	16.483	78.009
灌溉绿洲	0.265	0.139	0.334	8.796	9.269	9.534
合计	61.791	6.555	6.556	12.642	25.752	87.543

注：研究区域内 EVI 负异常和正异常的总面积（平方千米）。正异常和负异常是根据从第 75 分位数回归函数估计的 EVI_{map} 值计算的。如果 EVI 分别高于第 75、90 和 99 分位数阈值，则代表除降水外对水输入有任何依赖程度的景观部分的正异常被分为低、中和高依赖水平。

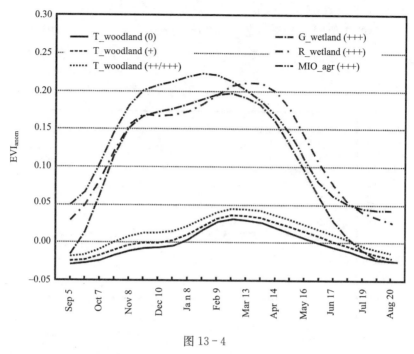

图 13-4

注：代表站点的平均季节增强植被指数异常轨迹。提取了除当地降水外对水输入依赖程度不同的像素样本组的轨迹(0:不依赖；+:低依赖；++:中依赖；+++:高依赖)；T_woodland＝泰尔特卡的开放牧豆树属林地；G_wetland＝瓜纳卡奇湿地；R_wetland＝罗萨里奥湿地；MIO_agr＝门多萨灌溉绿洲(农业生态系统)。

由于我们的概念模型预测的模式和趋势在泰尔特卡林地上是匹配的，因此绿度异常似乎可以很好地替代表示林地生态系统对地下水资源的依赖。除了所记录的指标和趋势外，在11月～12月的春季晚期，在EVI异常为中度偏高的站点，绿度轨迹的增长率也高于对照地点(图13-4)。这种"早期上升"使得各地点间的EVI差异在这个季节期间最高，因为对于地下水湿生植物，能量限制(低温)开始消失，但降雨输入对于促进非地下水湿生植物的植被生长来说仍然很少。

在整个研究期间，在对照点和站点EVI异常值中度偏高的站点发现EVI年值存在显著差异(图13-5)。虽然观察到EVI的微弱绝对差异，但差异在数值很低的EVI正异常位点和对照位点之间不显著。随着EVI异常增加，生长季节 EVI(EVI_{gs})、最小 EVI(EVI_{min})和年内变化(EVI_{nrange})的年际变化(以每个水文年的年值间的变异系数衡量)降低(表13-3)。其余的 EVI 指标没有显示出任何明显的趋势,尽管具有中等偏高 EVI 异常的站点的值略低于对照站点(表13-3)。

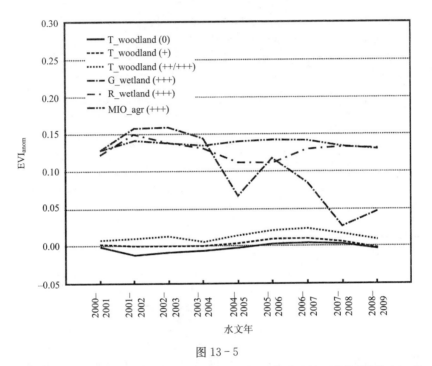

图 13-5

注:年际增强植被指数(EVI)在除当地降水外对水输入依赖程度不同的系统的异常轨迹(0:无依赖;+:低依赖;++:中依赖;+++:高依赖)。T_woodland=泰尔特卡开放的牧豆树属林地;G_wetland=瓜纳卡奇湿地;R_wetland=罗萨里奥湿地;MIO_agr=门多萨灌溉绿洲(农业生态系统)。

(三) 水生林地、湿地和灌溉作物的比较

门多萨绿洲的湿地和灌溉区的 EVI 轨迹与开阔林地牧豆树属的 EVI 轨迹明显不同(图13-4)。瓜纳卡奇和罗萨里奥湿地和灌溉地的年生产力(EVI_{ma})分别是泰尔特卡林地的8.7、9.4和12.4倍(图13-6(a))。根据其绿度异常,开阔林地牧豆树属的年平均蒸散速率达到约185毫米/年,其中估计约25毫米/年由浅层地下水储备提供(Contreras et al.,2011)。

表 13-3 绿度度量的变异系数

站点	生产力		季节性			物候学	
	EVI_{ma}	EVI_{gs}	EVI_{max}	EVI_{min}	EVI_{nrange}	L_{gs}	T_{max}
T_wood	6.02	0.89	11.88	6.36	25.17	5.96	18.42
(0)	(0.78)	(0.21)	(2.01)	(0.85)	(3.58)	(1.39)	(5.05)
T_wood	5.61	0.69	10.67	5.86	22.56	5.05	15.67
(+)	(0.79)	(0.31)	(1.85)	(1.29)	(3.62)	(1.08)	(4.30)
T_wood	6.00	0.50	10.72	5.55	22.12	5.34	18.97
(++/+++)	(1.23)	(0.25)	(3.52)	(1.40)	(5.52)	(1.08)	(5.50)
G_wet	26.52	1.36	23.31	30.49	20.43	9.46	30.65
(+++)	(15.53)	(1.25)	(12.90)	(19.64)	(12.59)	(4.40)	(12.15)
R_wet	9.62	0.51	11.75	8.20	14.49	5.80	22.65
(+++)	(3.55)	(0.41)	(5.68)	(2.12)	(5.53)	(1.44)	(7.57)
MIO_agr	7.54	1.08	9.85	12.61	16.13	7.56	26.45
(+++)	(5.31)	(0.72)	(4.79)	(6.15)	(6.51)	(2.84)	(10.47)

注：数值根据 2000 年 9 月～2009 年 8 月（9 个水文年）计算的年度指标得出。变异系数的标准偏差（在每个生态系统类型观察到的空间变异性）显示在括号之中。

这些结果与根据独立同位素和水化学证据计算的独立估算一致（Jobbágy et al.，2011）。湿地的补充水消耗量甚至高于开阔林地，除了降水量输入外，其速率可达到 450～500 毫米/年。没有关于地下水供应对瓜纳卡奇湿地和罗萨里奥湿地平均生产力的相对贡献的准确数据，但对不同季节平均 EVI 轨迹的观察表明了两种生态功能模式：瓜纳卡奇湿地的植被与罗萨里奥湿地的植被比较，具有较高的年际变化（EVI_{nrange}）、较低的最小 EVI 值（EVI_{min}；图 13-6(b)）和较短的生长季节（L_{gs}；图 13-6(c)）。虽然在两个湿地系统之间没有发现最大绿度的显著差异（图 13-6(b)），但瓜纳卡奇湿地系统达到最大绿度的时间比罗萨里奥湿地系统早大约 16 天（图 13-6(c)）。与瓜纳卡奇湿地相比，罗萨里奥湿地的所有 EVI 指标的年际变化较低，这表明罗萨里奥系统的生态功能比瓜纳卡奇系统更依赖地下水资源。瓜纳卡奇系统的年际绿度动态（图 13-5）证实了这一事实，其中平均年 EVI 值的突然上升和下降表明，对圣胡安河提供的水量的依赖性更高，因此，对上游圣胡安绿洲灌溉的取水作用也随之升高。

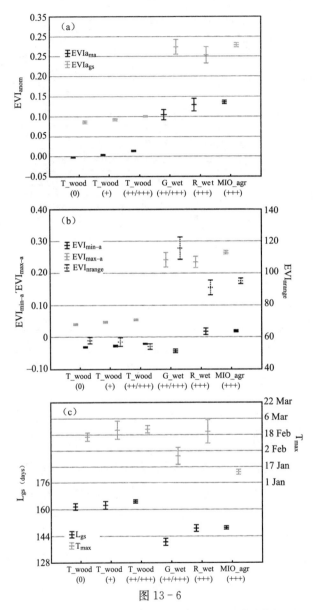

图 13-6

注：箱须图展示了以下指标的平均和置信区间（95%水平）(a)生产力指标（EVI_{ma} 和 EVI_{gs}）、(b)季节性指标（EVI_{min}、EVI_{max}、EVI_{nrange}）和(c)物候指标（L_{gs}、T_{max}）。在泰尔特卡牧豆树属林地（T_wood）、瓜纳卡奇和罗萨里奥湿地（G_wet、R_wet）和门多萨灌溉绿洲（MIO_agr）计算了指标。除当地降水外，每个系统的样本都是根据对水输入的不同依赖程度选择的（0：无依赖；+：低依赖；++/+++：中和高依赖）。出于比较的目的，所有平均指标都是根据其相应的 EVI 异常来计算的。

在门多萨灌溉绿洲观测到的季节绿度动态特征是其与其他生态系统类型缺乏耦合。灌溉绿洲的生产力指标(图 13 - 6(a))和季节性(图 13 - 6(b))与在两个湿地系统中观察到的值相似。然而,门多萨绿洲 EVI 年内平均轨迹(图 13 - 4)表明,最大植被活动较早,生长季节活动较高。灌溉绿洲和天然湿地系统的季节 EVI 轨迹之间存在相位差,表明了水资源的竞争过程。这一事实早先在圣胡安绿洲和瓜纳卡奇湿地之间得到了强调,门多萨绿洲和罗萨里奥系统也受到了同样预期。在瓜纳卡奇-圣胡安案例中,由于湿地对地下水资源的依赖有限,取水对湿地生产力的影响更加清晰。在罗萨里奥-门多萨案例中,罗萨里奥湿地似乎更多地依赖地下水资源,预计农业发展对湿地生产力的影响在干燥年比潮湿或平均降水量较大的水文年要明显。确定最干旱时期差异的更详细研究将有助于确定全球可持续发展及灌溉发展对其生态功能和生态系统服务提供情况的影响。

四、结论

GDEs 提供了人类福祉依赖生态系统服务的一个突出例子。在本研究中,我们证明了年度绿度异常概念(Contreras *et al.*,2011)对于确定景观系统的有用性,这些景观系统的植被活动除了降水外还依赖于异常高的水输入:即河岸生态系统、湿地(见第十七章)、地下水湿生植物和灌溉绿洲。特别是在低降水量地区,供应服务(如生物质或水资源可用性)、调节服务(如维持生命周期、栖息地和基因库,或当地气候调节)和文化服务(与独特景观的智力、精神或娱乐互动)都集中在这类生态系统中,生态系统与地下水动态密切相关。从卫星数据估算的年度绿度异常被证明是一种简单而可靠的测量方法,可在野外数据有限的广大区域对依赖水流流入的系统进行绘制,并对其需水量进行初步估算。关于此类生态系统对地下水依赖的更多信息,可通过对其 EVI 的年内和年间轨迹进行补充分析,并通过提取与此类生态系统碳收益的生产率、季节性和物候相关的指标来获得(见第九章)。

在本研究中,我们揭示了生长季节的平均绿度、年最小绿度和年际变率(归一化范围)是如何在地下水湿生植物和湿地高于非地下水湿生植物和湿地的。因此,我们建议使用这种指标来量化和绘制生态系统对地下水资源的依赖程度,以及生态系统服务对地下水动态的依赖程度。

基于植被绿度异常的空间分析和选择在具备干旱降雨条件的时间段跟踪其物候条件的卫星方法,提供了对地下水资源依赖的自然生态系统的初步表征。这两种方法都是非常有用的,是作为建立 GDEs 功能概念模型、量化其需水量、评估其生态保护和农业发展选择之间可能出现的生态系统服务权衡的第一步。

致谢

本文作者孔特雷拉斯感谢西班牙科学和创新部对胡安·德拉西尔瓦博士后研究金(JCI-2009-04927)给予的资助支持。感谢巴勃罗.E·维拉格拉、埃里卡·塞斯卡和三位匿名审稿者提供的技术数据方面的帮助和他们富有洞察力的评论。

参 考 文 献

Abraham, E., H. F. del Valle, F. Roig, et al. 2009. Overview of the geography of the Monte Desert biome (Argentina). *Journal of Arid Environments* 73:144–153.

Alcaraz-Segura, D., J. M. Paruelo, and J. Cabello. 2006. Identification of current ecosystem functional types in the Iberian Peninsula. *Global Ecology and Biogeography* 15:200–212.

Aranibar, J. N., P. E. Villagra, M. L. Gómez, et al. 2011. Nitrate dynamics in the soil and unconfined aquifer in arid groundwater coupled ecosystems of the Monte desert, Argentina. *Journal of Geophysical Research* 116:G04015.

Barron, O., I. Emelyanova, T. G. Van Niel, D. Pollock, and G. Hodgson. 2012a. Mapping groundwater-dependent ecosystems using remote sensing measures of vegetation and moisture dynamics. *Hydrological Processes*. doi:10.1002/hyp.9609.

Barron, O., R. Silberstein, R. Ali, et al. 2012b. Climate change effects on water-dependent ecosystems in south-western Australia. *Journal of Hydrology* 434–435:95–109.

Bergkamp, G., and C. Katharine. 2006. Groundwater and ecosystem services: Towards their sustainable use. *Proceedings of the International Symposium on Groundwater Sustainability*. Alicante, Spain: Instituto Geológico y Minero de España, 24–27 January, 177–193.

Bertrand, G., N. Goldscheider, J. M. Gobat, and D. Hunkeler. 2012. Review: From multi-scale conceptualization to a classification system for inland groundwater-dependent ecosystems. *Hydrogeology Journal* 20:5–25.

Boer, M. M., and J. Puigdefábregas. 2003. Predicting potential vegetation index values as a reference for the assessment and monitoring of dryland condition. *International Journal of Remote Sensing* 24:1135–1141.

Bradley, B. A., and J. F. Mustard. 2008. Comparison of phenology trends by land cover class: A case study in the Great Basin, USA. *Global Change Biology* 14:334–346.

Brauman, K. A., G. C. Daily, T. K. Duarte, and H. A. Mooney. 2007. The nature and value of ecosystem services: An overview highlighting hydrologic services. *Annual Review of Environment and Resources* 32:67–98.

Chen, J. S., L. Li, J. Y. Wang, et al. 2004. Groundwater maintains dune landscape. *Nature* 432:459–460.

Contreras, S., M. M. Boer, F. J. Alcalá, et al. 2008. An ecohydrological modelling approach for assessing long-term recharge rates in semiarid karstic landscapes. *Journal of Hydrology* 351:42–57.

Contreras, S., E. G. Jobbágy, P. E. Villagra, M. D. Nosetto, and J. Puigdefábregas. 2011. Remote sensing estimates of supplementary water consumption by arid ecosystems of central Argentina. *Journal of Hydrology* 397:10–22.

de Groot, R. S., M. A. Wilson, and R. M. J. Boumans. 2002. A typology for the classification, description and valuation of ecosystem functions, goods and services. *Ecological Economics* 41:393–408.

Devitt, D. A., L. F. Fenstermaker, M. H. Young, B. Conrad, M. Baghzouz, and B. M. Bird. 2011. Evapotranspiration of mixed shrub communities in phreatophytic zones of the Great Basin region of Nevada (USA). *Ecohydrology* 4:807–822.

Do, F. C., A. Rocheteau, A. L. Diagne, V. Goudiaby, A. Granier, and J. P. Homme. 2008. Stable annual pattern of water use by *Acacia tortilis* in Sahelian Africa. *Tree Physiology* 28:95–104.

Eamus, D. 2009. *Identifying groundwater dependent ecosystems: A guide for land and water managers.* Sydney: Land & Water Australia.

Eamus, D., R. Froend, R. Loomes, G. Hose, and B. R. Murray. 2006. A functional methodology for determining the groundwater regime needed to maintain the health of groundwater-dependent vegetation. *Australian Journal of Botany* 54:97–114.

Eamus, D., C. M. O. Macinnis-Ng, G. C. Hose, M. J. B. Zeppel, D. T. Taylor, and B. R. Murray. 2005. Ecosystem services: An ecophysiological examination. *Australian Journal of Botany* 53:1–19.

Eamus, D., A. P. O'Grady, and L. Hutley. 2000. Dry season conditions determine wet season water use in the wet-dry tropical savannas of northern Australia. *Tree Physiology* 20:1219–1226.

Ellery, W. N., and T. S. McCarthy. 1998. Environmental change over two decades since dredging and excavation of the lower Boro River, Okavango Delta, Botswana. *Journal of Biogeography* 25:361–378.

Ellis, T. W., and T. J. Hatton. 2008. Relating leaf area index of natural eucalypt vegetation to climate variables in southern Australia. *Agricultural Water Management* 95:743–747.

FAO (Food and Agriculture Organization). 2006. *ClimWat 2.0 for CropWat.* Rome: FAO, United Nations.

Fernández, N., J. M. Paruelo, and M. Delibes. 2010. Ecosystem functioning of protected and altered Mediterranean environments: A remote sensing classification in Doñana, Spain. *Remote Sensing of Environment* 114:211–220.

Groom, P. K., R. H. Froend, and E. M. Mattiske. 2000. Impact of groundwater abstraction on a *Banksia* woodland, Swan Coastal Plain, Western Australia. *Ecological Management and Restoration* 1:117–124.

Guerschman, J. P., A. I. J. M. Van Dijk, G. Mattersdorf, et al. 2009. Scaling of potential evapotranspiration with MODIS data reproduces flux observations and catchment water balance observations across Australia. *Journal of Hydrology* 369:107–119.

Hatton, T. J., and R. Evans. 1997. *Dependence of ecosystems on groundwater and its significance to Australia.* Canberra: Land and Water Resources Research and Development Corporation.

Huete, A. K. Didan, T. Miura, E. P. Rodriguez, X. Gao, and L. G. Ferreira. 2002. Overview of the radiometric and biophysical performance of the MODIS vegetation indices. *Remote Sensing of Environment* 83:195–213.

Jobbágy, E. G., M. D. Nosetto, P. E. Villagra, and R. B. Jackson. 2011. Water subsidies from mountains to deserts: Their role in sustaining groundwater-fed oases in a sandy landscape. *Ecological Applications* 21:678–694.

Jobbágy, E. G., O. E. Sala, and J. M. Paruelo. 2002. Patterns and controls of primary production in the Patagonian steppe: A remote sensing approach. *Ecology* 83:307–319.

Jönsson, P., and L. Eklundh. 2004. TIMESAT—A program for analyzing time-series of satellite sensor data. *Computers and Geosciences* 30:833–845.

Llamas, M. R. 1988. Conflicts between wetland conservation and groundwater exploitation: Two case histories in Spain. *Environmental Geology and Water Sciences* 11:241–251.

MacKay, H. 2006. Protection and management of groundwater-dependent ecosystems: Emerging challenges and potential approaches for policy and management. *Australian Journal of Botany* 54:231–237.

Morisette, J. T., A. D. Richardson, A. K. Knapp, et al. 2009. Tracking the rhythm of the seasons in the face of global change: Phenological research in the 21st century. *Frontiers in Ecology and the Environment* 7:253–260.

Muñoz-Reinoso, J. C., and F. García-Novo. 2005. Multiscale control of vegetation patterns: The case of Doñana (SW Spain). *Landscape Ecology* 20:51–61.

Murray, B. R., G. C. Hose, D. Eamus, and D. Licari. 2006. Valuation of groundwater-dependent ecosystems: A functional methodology incorporating ecosystem services. *Australian Journal of Botany* 54:221–229.

Murray, B. R., M. J. B. Zeppel, G. C. Hose, and D. Eamus. 2003. Groundwater-dependent ecosystems in Australia: It's more than just water for rivers. *Ecological Management and Restoration* 4:110–113.

Murray-Hudson, M., P. Wolski, and S. Ringrose. 2006. Scenarios of the impact of local upstream changes in climate and water use on hydro-ecology in the Okavango Delta, Botswana. *Journal of Hydrology* 311:73–84.

Nagler, P. L., J. Cleverly, E. P. Glenn, D. Lampkin, A. R. Huete, and Z. Wan. 2005. Predicting riparian evapotranspiration from MODIS vegetation indices and meteorological data. *Remote Sensing of Environment* 94:17–30.

New, M., D. Lister, M. Hulme, and I. Makin. 2002. A high-resolution data set of surface climate over global land areas. *Climate Research* 21:1–25.

O'Grady, A. P., J. L. Carter, and J. Bruce. 2011. Can we predict groundwater discharge from terrestrial ecosystems using existing eco-hydrological concepts? *Hydrology and Earth System Sciences* 15:3731–3739.

Palmer, A. R., S. Fuentes, D. Taylor, et al. 2010. Towards a spatial understanding of water use of several land-cover classes: An examination of relationships amongst pre-drawn leaf water potential, vegetation water use, aridity and MODIS LAI. *Ecohydrology* 3:1–10.

Paruelo, J. M., H. E. Epstein, W. K. Lauenroth, and I. C. Burke. 1997. ANPP estimates from NDVI for the central grassland region of the United States. *Ecology* 78:953–958.

Paruelo, J. M., E. G. Jobbágy, and O. E. Sala. 2001. Current distribution of ecosystem functional types in temperate South America. *Ecosystems* 4:683–698.

Patten, D. T., L. Rouse, and J. C. Stromberg. 2008. Isolated spring wetlands in the Great Basin and Mojave Deserts, USA: Potential response of vegetation to groundwater withdrawal. *Environmental Management* 41:398–413.

Richardson, S., E. Irvine, R. Froend, P. Boon, S. Barber, and B. Bonneville. 2011. *Australian groundwater-dependent ecosystem toolbox. Part 1: Assessment framework.* Canberra: National Water Commission.

Ridolfi, L., P. Odorico, and F. Laio. 2007. Vegetation dynamics induced by phreatophyte-aquifer interactions. *Journal of Theoretical Biology* 248:301–310.

Running, S. W., and R. R. Nemani. 1988. Relating seasonal patterns of the AVHRR vegetation index to simulated photosynthesis and transpiration of forests in different climates. *Remote Sensing of Environment* 24:347–367.

Solano, R., K. Didan, A. Jacobson, and A. Huete. 2010. *MODIS vegetation indices (MOD13) C5—User's guide*. Tucson, AZ: The University of Arizona.

Specht, R. L. 1972. Water use by perennial evergreen plant communities in Australia and Papua New Guinea. *Australian Journal of Botany* 20:273–299.

Specht, R. L., and A. Specht. 1989. Canopy structure in eucalyptus-dominated communities in Australia along climatic gradients. *Oecologia Plantarum* 10:191–202.

Stromberg, J. C., R. Tiller, and B. Richter. 1996. Effects of groundwater decline on riparian vegetation of semiarid regions: The San Pedro, Arizona. *Ecological Applications* 6:113–131.

Villagra, P. E., G. E. Defossé, H. F. del Valle, et al. 2009. Land use and disturbance effects on the dynamics of natural ecosystems of the Monte Desert: Implication for their management. *Journal of Arid Environments* 73:202–211.

Villagra, P. E., R. Villalba, and J. A. Boninsegna. 2005. Structure and dynamics of *P. flexuosa* woodlands in two contrasting environments of the central Monte Desert. *Journal of Arid Environment* 60:187–199.

第十四章 地表土壤水分遥感监测:在生态系统过程和尺度效应中的应用

M.J. 保罗　M.P. 冈萨雷斯-杜戈　C. 阿吉拉尔　A. 安德鲁

一、引言

地表土壤水分监测在量化含水量变化、验证、校正水文模型,确定生态系统的极限状态以及评估不同生态系统服务等方面扮演着重要的角色。同时,土壤水分在能量和水分收支平衡中发挥着重要作用,它关系着土壤和植被的蒸散发。水分梯度决定了土壤中水分通量的方向和强度(Brutsaert,1982,2005;Jury and Horton,2004),以及植被、优势物种和根系生长深度在中长期的分布。后者通过蒸腾过程、土壤表层的遮盖、土壤结构和组分的变化、地表有效粗糙度的增加以及能量和水分平衡中涉及的一系列因素来影响局部的湿度状况。在平流和扩散传输条件下(Jury and Horton,2004),土壤水分和植被对养分和污染物在土壤中的吸附、释放以及迁移具有很大影响(Sposito,1989)。因此,土壤水分的变化不仅会影响植物和动物的水分-养分-污染物再生的供应服务,还会影响其他因素的监管和维护服务(如降雨产生的径流、洪水风险、土壤维护、地表和地下水质量等)以及态系统的文化服务,如植被和湿地对景观的影响,以及它们对生态系统内人类活动的影响。

土壤属于多孔介质,该特性增加了土壤中水分时空演化特征的复杂性。土壤水分的变化主要取决于孔隙度,而孔隙度与土壤结构、质地及某些组分(如有机质、碳酸盐等)的含量有关。在不同孔隙度的土壤中,可以发现绝对值相同的含水量,但其水流状态不同。在饱和条件下,土壤水分被认为是完全填满的土孔隙的连续体,在这个假设下,可以应用流体力学定律。但在非饱和条件下,孔隙中液态水、水蒸气和空气共存,在这种情况下,会出现多相传输。由于饱和、非饱和流的分类是模拟和量化土壤水分运动和交换的关键条件,所以,土壤水分主要以相对术语(指给定土壤中的水分范围)表示接近饱和(最大土壤水分)的情况。因此,水和能量平衡方程,以及养分、盐和污染物的传输方程,利用相对湿度作为尺度因子,可在土壤剖面应用中实现不同类型的土壤和状态之间的比较。

异质性是土壤的固有特征。均质土壤性是简化土壤模型建立的理想条件,土壤水分也不例外。如果要排除干扰因素,则需要在野外对土壤水分实地取样以进行实验设计,这在大尺度调查中通常是不可行的。然而,土壤水分是水文模型中的主要状态变量之一,在大中型研究中,它通常会影响野外测量的直接校准结果。

太空观测地球能力的不断增强,使得在大范围内应用和校准物理、化学和环境模型成为可能。与许多其他地表特征类似,地表土壤水分可以通过分析特定时期内的地表反射光谱进行局部追踪。遥感传感器具有不同频率和空间分辨率,可以提供大范围的地表反射。土壤水分监测可以有多种来源,主被动微波遥感应用最为广泛,包括欧洲航空局发射的土壤水分和海洋盐度(Soil Moisture and Ocean Salinity,SMOS)卫星(Dente et al.,2012)以及美国航空局发射的主被动土壤水分(Soil Moisture Active and Passive,SMAP)卫星。遥感观测结果需要与地面实地测量数据进行比较,以提高卫星传感器探测地表的能力。由于土壤本身具有复杂的特征,利用遥感多时空分辨率观测结合相关模型是目前获取土壤水分的主要手段。

数据处理对于植被、水分应力、土壤温度等多种环境变量的相互关联具有重要意义。调查指定区域的土壤水分时序图对水文模型的校正具有重要的价值,但其易受野外观测的时间和成本的限制。由于土壤的建模"环境"定义,即土壤水分在有效深度内统一,因此需要进一步研究尺度效应,以使水文模拟的结果与遥感数据相匹配,因为遥感提供了地表信息,而模型处理的是由若干厘米(10～100厘米)组成的先验定义的地表层。

二、土壤水分的遥感监测

受大气波动效应的影响,土地覆盖、地形和土壤性质的空间异质性,导致土壤水分在空间和时间上差异很大。因此,需要将这些差异量化并将其列入需考虑的因素,以确保水文模型的正确。

传统上,土壤水分测量是将探测设备直接插入地面,获得该点高质量的剖面数据;调查区域内的水分分布通过探测点数据的插值得到。然而,这种方法不足以描述大面积的土壤水分情况,因此这需要进一步评估土壤水分的空间变化。从这个意义上讲,相对于高频率的逐点测量,遥感是提供土壤水分空间分布估算的有效数据源(Engman and Gurney,1991;Laguardia and Niemeyer,2008)。

接收地面反射的电磁频谱的几个波段可用于监测土壤水分。在过去的几十年里,大多数尝试都是基于微波波谱范围,使用无源或有源系统(表 14-1)。乌拉比等人(Ulaby et al.,1981)和冯(Fung,1994)对微波遥感的基本理论和首次应用进行了彻底的修订。使用微波最重要的原因之一是其能够穿透云层,并在某种程度上穿透雨水。在地表,微波可以比光学传感器更深入地穿透植被和土壤。穿透程度取决于水分含量、植被密度和微波的波长,波长越长,穿透深度越深。使用微波的其他原因是微波独立于太阳光,可以昼夜观察,并提供可见光和红外区域(Visible and Infrared Regions,VNIR)之外的补充信息。在微波频率下,研究区域地表的几何形态和介电性能决定了其反射特性,而植被或土壤表层的分子共振反射范围则主要在红外区域。水和干土的介电常数存在着巨大反差(分别为 80 和<5)导致微波区的发射率反差,从而为遥感监测地表土壤水分提供了途径。自 20 世纪 70 年代以来,许多研究都使用了这种方法,例如,施穆格和杰克逊(Schmugge and Jackson,1994)、维格纳龙等人(Wigneron et al.,1998,2003)、施穆格等人(Schmugge et al.,2002)和朱格利亚等人(Juglea et al.,2010)。这种方法的限制是微波测量的深度仅为波长的 0.1~0.2 倍,并且随湿度降低而减小。

表 14-1 具有代表性的传感器的特性和应用示例

传感器	波段	主动、被动	空间分辨率	应用实例
AMSR-E	C 波段(6.9GHz), X 波段(10.7 GHz), 18.7 GHz、23.8 GHz、36.5GHz 和 89GHz	P	25~50 千米	Gruhier et al.,(2010) Liu et al.,(2011) Su et al.,(2013)
TRMM-TMI	10.7、19.4、21.3、37 和 85.5GHz	P	6-50 千米	Bindlish et al.,(2003) Gao et al.,(2003) Gruhier et al.,(2010)
SMOS	L 波段(1.4GHz)	P	50 千米	Kerr et al.,(2001) Su et al.,(2013)
RADARSAT 雷达卫星	C 波段(5.3GHz)	A	10~100 米	Baghdadi et al.,(2002) Zribi and Dechambre (2002)
ENVISAT-ASAR	C 波段(5.3GHz)	A	30 米	Zribi et al.,(2005) Manninen et al.,(2005)
JERS-1	L 波段(1.27GHz)	A	18 米	Narayanan and Hegde (1995) Kasischke et al.,(2011) Kasischke et al.,(2011)
ERS-1/ERS-2	C 波段(5.3GHz)	A	30 米	Wagner et al.,(1999a, 1999b) Moran et al.,(2001) Ceballos et al.,(2005) Parajka et al.,(2006)

  被动微波的空间分辨率较低,通常在几十千米左右,更适合大气和海洋应用。此外,陆地目标通常具有复杂的介电和几何特性。然而,近年来,国际社会已在努力开发具有多种能力(多频、双极化或偏振、多角度观测)的先进仪器,以改善区域和全球范围内土壤水分的定量观测。其中,包括搭载在 AQUA 卫星的地球观测系统先进的微波扫描辐射计(AMSR-E),2009 年欧洲航天局(European Space Agency,ESA)发射的 SMOS 卫星。SMOS 使用 L 波段辐射计,是获取土壤水分信息的最佳选择(Kerr et al.,2001)。2014 年,美国宇航局计划发

射SMAP卫星,主动和被动微波传感器结合前所未有的分辨率、灵敏度、覆盖面积和重复周期提供精度较高的土壤水分产品。

主动微波雷达系统可将空间分辨率提高到几十米,特别是使用人工合成的虚拟天线,即所谓的合成孔径雷达。然而,由于地表粗糙度和地形的影响(Lievens et al.,2011),雷达系统受到一定限制。一些研究者(Engman and Chauhan,1995)提出使用多时相信息来监测土壤水分的相对变化,并将给定时间段内静态变量的影响最小化。维格纳龙等人(Wigneron et al.,2003)提出了四种主要的算法,以从微波遥感中获取土壤水分,并考虑了植被数量、土壤温度、积雪(见第十五章)、地形和地表粗糙度等因素对观测结果的影响。

这些算法基于(1)土地覆盖分类;(2)辅助遥感分类;(3)双参数反演;(4)三参数反演(即同时从微波观测中获取土壤水分、植被光学厚度和有效地表温度)。方法(3)和(4)是基于频率、极化或入射角的多因素观测。作者讨论了不同算法的优缺点,强调了基于新的多重微波特征反演土壤水分的关键问题。

格吕耶等人(Gruhier et al.,2010)介绍了不同的有源和无源微波产品。一些正在进行的研究工作是通过综合多种有源和无源微波数据,生成时间序列的土壤水分数据(Fernández-Prieto et al.,2012;Liu et al.,2012)。

一些作者(Carlson,2007;Crow et al.,2008;Sobrino et al.,2012)已经讨论了利用光学和热红外波段的数据间接估算土壤水分。这种方法是将遥感观测资料融合到土壤—植被—大气迁移(SVAT)模型中,其基本思路是:地表辐射温度(TR,由传感器直接测量)和地表湍流通量对地表土壤水分变化很敏感。间接估算技术也可用于获得定性分类(专栏14-1)或通过其他植被指数评估土壤水分(专栏14-2)。

专栏14-1 利用陆地卫星5号TM数据通过光谱混合分析进行土壤状态分类

将航空和野外多光谱数据相结合,以确定西班牙南部加的斯(Cadiz)退化盐沼地区的土壤水分状态,从而可以量化该系统的恢复潜力(Polo et al.,2009)。同时,可以校准潮沟地区的洪水极限,由于潮沟的水位较低,不能进行直接测量和监测。通过对需要研究的重要特征(湿土、干土、绿色植被、干燥植被和阴影)进行初次分类,对研究区域内的每个像素进行光谱混合分析(Spectral Mixture Analysis,SMA)以评估其所含成分的重要性。SMA 是一种反卷积技术,该技术分离出每个像素中的单独成分,将光谱中各波段的像素信号视为其各单独成分的线性组合(Roberts et al.,1998)。对 2004~2007 年期间不同季节的 9 幅陆地卫星 5 号(TM)图像进行了分析,并对 2007 年与传感器时间同步进行了两次不同地区的实地调查,对 SMA 进行校准。

使用 GER 3700 辐射计在 30 个选定的控制点进行测量,这些控制点覆盖了之前根据潮汐周期和水动力分析确定的潜在洪水区域。

将 SMA 的结果重新分类为三个显著的分量(干土、湿土和植被),从而清楚地识别了潮滩延伸和植被发展的变化,植被部分的相关值为 0.91,干土部分的相关值为 0.71。此外,当与潮汐和气象资料结合时,每个数据结果都与影响地表土壤水分演化的不同因素(降雨、每天潮汐周期和季节潮汐周期等)密切相关。潮沟地区的洪水极限评估通过数字化一维水动力模型实现,并根据图像获取时间进行了校准。

专栏 14-2　天然植被状态作为土壤水分动态指标

在生态学研究中应用归一化差异植被指数,可以对绿色植被进行量化和制图。NDVI 是在大时空尺度上耦合气候和植被分布及状态的有效方法(Pettorelli et al.,2005)。由于植被的状态与水文变量(降水量、蒸腾量等)有关,因此 NDVI 在景观尺度上可作为这些因素的综合衡量标准(Wang et al.,2003;Groeneveld and Baugh,2007)。

阿吉拉尔等人(Aguilar et al.,2012)将美国东部一个岛屿在景观层面上树木和植被绿色深浅的变化与过去的降水量和水文信息联系起来,通过地下水透镜量化了降水量与水可用性季节变化的响应。NDVI 年度变化由 TM 数据与降水量变化数据联合计算得出。干旱年份(2007 年和 2008 年)的 NDVI 平均值较低(图 14-1 中 0.75~0.9),累积降水量和平均地下水位深

> 度之间存在线性关系，NDVI 最大值和 NDVI 平均值可以作为衡量植被总体和平均状态的指标。
>
> 生长季内不同水文条件下的月度变化揭示了绿色植被的不同状态；在干旱年份(2007)，地下水位深度下降，NDVI 等级直方图在观测中期变得比较窄。
>
> NDVI 值与该水文观测年度的累积降水量之间存在线性关系，r^2 接近 1。这表明植被对近期系统中可用水量减少的快速反应，随干旱年份上个月累积的降水量而变化。NDVI 和地下水之间的月度关系仅在干旱年份(如图 14-1 中的 2007 年)出现，当时降水量稀少，蒸散量高，而淡水透镜成为植物的主要水源。
>
> 这些结果利用经验数据证明了绿色植被对淡水透镜变化的反应间的重要反馈，如邵等人(Shao et al., 1995)的模型所示。由于 NDVI 依赖于水资源利用率，在使用该指数确定生产率时，必须考虑遥感数据的获取时间。

三、水文模型的遥感同化

历史上最早建立的水文模型是分析模型。这些模型解决了描述水流、沉积物和伴生物质的生成和运输的基本物理模式。理论上，这些模型中涉及的参数是可测量的。然而，涉及的众多参数及环境特征的异质性使得有必要使用观测数据对结果进行校准。测量过程中的误差、模型算法的应用范围与测量尺度之间的差异增加了结果的不确定性(Mertens, 2003)。

地理信息系统的发展、计算机处理速度的提高和存储容量的增长，以及通过遥感技术获得空间数据的手段，使半分布模型应运而生。这些模型将分别计算系统划分的每个空间单元的信息，并将每个单元的计算结果在时间和空间上进行叠加，汇总产生了流域尺度的输出结果(Beven, 1989, 2000)。通过这种方式，分析模型得到广泛应用，但分析结果的严谨性和准确性将取决于模型所用的空间分布输入参数的质量，以及全球气候变化的适宜性和(或)分析模型对所研究

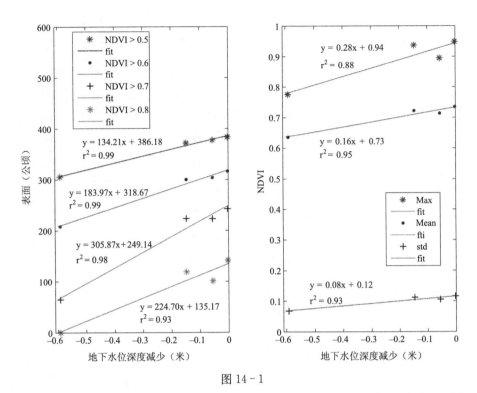

图 14-1

注:2007 年,随着每月地下水位深度下降(m/月),NDVI 值高于阈值和 NDVI>0.5 的像素样本统计值相应发生了线性变化。

的时空尺度的适用性(Gupta et al.,1986;Blöschl and Sivapalan,1995)。在基于自然形态的分布式水文模型中,描述单个过程的表达式的推导受到若干假设的约束,这些假设简化了过程,因此可以用数学方式表达。一般来说,这些模型中控制过程的方程是在受控条件下,在均质系统中以小尺度推导出的,但在实践中,这些方程可应用于更广泛的尺度范围:单元、子流域等以及不同的物理条件下。这些方程是从理论上推导出来的,用于空间和时间连续的数据,尽管其中一个像素可能代表流域中一个单元甚至一个区域。一些作者认为,在大多数物理模型中将小尺度物理过程作为单元进行拼接的方法是不可信的,因为以这种方式反演的参数仅提供了物理意义上失真的调整系数(Beven,1989)。因此,有必要考虑某些初始假设对待建模系统特性的影响。因此,遥感数据、模型和实地数据的有机组合所提供的潜力预示着未来的重大机遇(Fernández-Prieto et al.,

2012)。

数据同化是将遥感产品集成到水文模型中,作为一种将不确定观测值与不确定模型输出的融合手段,将遥感数据集成到水文模型中,以提高估算精度。

因此,数据同化不仅用于更新水文模型状态,还用于量化观测结果和水文模型误差(Moradkhani,2008)。其中水文模型状态完美结合了模型输出和观测结果。遥感数据可以作为输入数据融合到模型中(例如,降水量、云量、反射率、前期土壤水分条件等),作为状态变量(如土壤水分或雪水当量),或用于校正模型的输出。流域尺度的水文模型需要校正水文过程计算中涉及的参数。由于土壤水分数据经常丢失,校正过程通常使用流域内某些控制站记录的水流数据实现。然而,表层土壤水分估算可从遥感数据中得出,因此水文模型中土壤水分信息的加入可大大改善蒸发和径流预测(Brocca et al.,2010)。

目前,通过同化土壤水分观测获得的改进,其效果不佳可能与几个主要因素有关:首先是空间尺度不匹配,与模型估算相比,现场测量范围较小而卫星传感器的粒度较粗。其次,相关层深不同,遥感只能获取地表较薄的层面(2～5厘米)信息,该深度与通常降雨径流模型中模拟的土壤深度(1～2米)不匹配。最后,土壤水分数据的可用性有限,这是由近期当地土壤水分网络和具有水文应用所需的时间分辨率的卫星数据所决定的(Vereecken et al.,2008;Brocca et al.,2010)。

四、评估植被蒸散发在土壤水分胁迫的间接价值

土壤水分动态的主要推动力是土壤向大气的蒸发和蒸腾速率。实际上,在植被覆盖区分离这两种通量是一件复杂的事情,除非在小型且高度监控的地块上进行科学实地研究。因此,蒸散量被定义为两个水源的综合损失。相较于潜在最大的蒸散量速率(蒸散量速率相当于无限供水),当土壤水分低于其最大值(即饱和状态)时,植被和土壤基质影响了蒸散量的实际速率(虽然程度较轻)。因此,通常通过估算土壤水分状态和水分胁迫对植被的影响来分析与潜在蒸散量相关的实际蒸散量分布。

通过估算不同时间尺度的蒸散量演变来跟踪自然和农业种植区植被的状态,以进行植被的短期管理及土壤使用和水资源管理的中长期规划。在农业领

域,这种评估至关重要。

蒸散量是水文过程中的一个关键变量,已有多种评估方法。将遥感数据整合到蒸散量模型中,将这些模型的应用范围从点扩展到流域和区域尺度(见第十八章)。总之,将实地数据与这些分散数据一起使用会产生显著的尺度效应。

空间分布的遥感数据已在区域和大陆尺度上用于估算蒸散量和陆地表面水分、预测需水量、监测干旱和气候变化(Bastiaanssen *et al.*,2002;Chandrapala and Wimalasuriya,2003;Anderson *et al.*,2007),用于估算灌溉区域的耗水量和规划灌溉计划(Garatuza-Payan and Watts,2005;Rossi *et al.*,2010),分析灌溉和生产力绩效指标(Bastiaanssen *et al.*,1999;Akbari *et al.*,2007;Hamid *et al.*,2011)。水分应力通常被量化为模型中的实际蒸散量和估算量的差值(PET;Moran,2003)。

利用遥感数据监测蒸散量的两种不同方法已成功地应用于农业用水研究。在第一种方法中,地表能量平衡通过使用从热数据(8～14 微米)导出的辐射表面温度计算显热通量,并获得潜热通量作为能量平衡的残差(Moran *et al.*,1994;Kustas and Norman,1996;Gillies *et al.*,1997;Bastiaanssen *et al.*,1998)。这种方法必须处理 TR 和空气动力温度(To)间的差异。空气动力温度用于计算显热,特别是对于部分植被覆盖的表面(Kustas,1990)。为了解决这个问题,已建立了多种具有不同复杂程度和输入要求的方案。其中一些使用经验或半经验关系来用 TR 校准 To(Kustas *et al.*,1989;Lhomme *et al.*,1994;Chehbouni *et al.*,1996;Mahrt and Vickers,2004)。当用现场数据进行校正时,这些方法可提供准确的结果(Chavez *et al.*,2005)。另一种避免测定 To 的方案包括对表面温度进行内部校准(Bastiaanssen *et al.*,1998)。这种处理方式还减少了对辐射表面温度进行大气校正的需要,而校正辐射表面温度是一个可能引入额外误差的复杂过程。最佳模拟部分冠层覆盖影响的方法是双源方法(Shuttleworth and Wallace,1985;Norman *et al.*,1995;Kustas and Norman,1999)。该方法将不同地表通量分为地表和冠层部分。之前的研究(Timmermans *et al.*,2007;González-Dugo *et al.*,2009)显示了双源能量平衡模型相对于单源模型的优势。

此外,能量平衡应用得到的是图像采集时的瞬时通量估值;因此,有必要将该估值换算为全天数值(专栏 14-3)。

第二种方法使用从地表反射率导出的植被指数来估算作物系数(Kc)。这种方法依赖于作物生长速率和蒸腾速率间的密切对应关系(Tasumi et al.,2005;Tasumi and Allen,2007;Singh and Irmak,2009)。作物系数将给定作物的蒸散量与根据地面气象数据计算的参考植被表面的蒸散量联系起来。这种基于指数的方法直接获得每日数值,大多数高频卫星提供的光学输入数据足够详细,可以为该方法提供数据(Hunsaker et al.,2005a,2005b;Duchemin et al.,2006;Er-Raki et al.,2007,2010;González Dugo and Mateos,2008;Campos et al.,2010;Sánchez et al.,2010;Padilla et al.,2011)。

> **专栏 14-3 使用热卫星图像绘制蒸散图**
>
> 安德森等人(Anderson et al.,2011)的研究是利用热卫星图像绘制每日蒸散量图的一个例子。利用基于物理的大气-土地交换反演模型(ALEXI;Anderson et al.,1997,2007;Mecikalski et al.,1999)生成了多尺度蒸散量地图,该模型将双源地表模型与大气边界层模型和相关通量分解技术相耦合(DisALEXI;Norman et al.,2003)。这项研究描述了由此产生的信息如何用于美国、欧洲和非洲等地区的干旱监测、农业水资源和水文管理。亚历克西(ALEXI)模型应用于西班牙南部最大的流域(57 527平方千米)瓜达尔基维尔河流域(图14-2)。它有效支持了长期缺水地区的农业生产管理(约8 000平方千米的灌溉土地)。日常遥感估算可以为田间、灌区和流域尺度的水资源管理者提供准确信息,以不同的时间模式显示不同作物的实际用水量,以改进水资源管理决策。
>
> 作为瓜达尔基维尔(Guadalquivir)盆地作物灌溉耗水量监测的规划和操作工具,MINARET(灌溉农业蒸散量监控)提供了基于指数估算蒸散量的例子(González-Dugo et al.,2012)。
>
> 从2007年、2008年和2009年的一系列高分辨率卫星图像中获得植被指数,并可以从按天到按季度评估不同作物和各个地块的蒸散量,从而能够分析作物用水量。为了估算实际蒸散量,使用彭曼-蒙蒂斯(Penman-Monteith)组合方程计算参考蒸散量(Allen et al.,1998)。

第十四章 地表土壤水分遥感监测：在生态系统过程和尺度效应中的应用

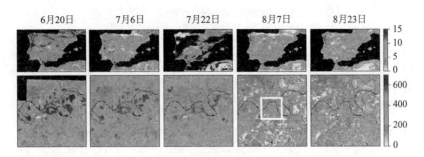

图 14-2（见彩插）

注：2009 年生长季 5 天内，西班牙南部瓜达尔奎维尔河沿岸灌溉农业区的 ALEXI 白天蒸散量（量化为 $MJ \cdot m^{-2} \cdot d^{-1}$ 中的潜热通量；第一行）和 DisALEXI 瞬时蒸散量（当地中午前不久；$W\,m^{-2}$；第二行）的图像。8 月 7 日 DisALEXI 地图上的白框指示 3 千米 MSG 像素大小。（资料来源：Anderson, M. C., et al., Hydrology and Earth System Sciences, 15, 223-239, 2011; http://www.hydrol-earth-syst-sci.net/15/223/2011/doi:10.5194/hess-15-223-2011 © Author(s) 2011. 该文献在 Creative Commons Attribution 3.0 许可下发布。）

为了验证蒸散量估算，需要进行蒸散量观测(Allen et al.,2011)。蒸散量观测系统包括蒸渗仪、闪烁计、波文比观测仪和涡动协方差方法。蒸渗仪通常用于提供校准和验证步骤的信息(Makkink,1957; Jensen,1974; Doorenbos and Pruitt,1977; Wright,1981,1982; Allen et al.,1989; Jensen et al.,1990)。波文比能量平衡(Bowen Ratio Energy Balance,BREB)是微气象方法(Bowen,1926)，其通过测量蒸发表面上方近表面层中的空气温度和蒸汽压力梯度来获取能量平衡方程。闪烁计是一种光学装置，用于测量空气折射率的微小波动。这些微小波动由温度、湿度和压力导致的密度变化引起。电流闪烁计测量显热通量，为了获得蒸散量，还需要测量净辐射(Rn)和土壤热通量(G)(见 Meijninger and De Bruin,2000; Meijninger et al.,2002; Hartogensis et al.,2003; de Bruin,2008)。

涡动协方差系统测量感热和潜热通量、动量通量、二氧化碳或取决于其配置的其他通量。这种方法基于温度和湿度波动及向上和向下湍流漩涡间的协方差。需要以 5～20 赫兹的频率高速测量温度、风速和湿度。该仪器相对脆弱和昂贵，需要定期维护，但方法高度可靠。

垂直（和水平）风分量通常使用声波风速计测量；温度测量使用超细导线热电偶测量，或者在超声波测量后根据湿度效应进行校正(Munger and Loescher,

2004)。湿度是用快速响应湿度计测量(Buck, 1976; Campbell and Tanner, 1985; Tanner, 1988; Burba and Anderson, 2008)。为了描述相同的涡流标度，必须在同一点进行测量，或至少在最靠近的区域进行测量。因此，由于仪器分离、不同的频率响应、坐标旋转和湿度计类型差异，需要对数据进行校正(Tanner et al.,1993; Villalobos, 1997; Aubinet et al., 2000; Horst, 2000; Massman, 2000, 2001; Paw et al., 2000; Twine et al., 2000; Rannik, 2001; Sakai et al., 2001; Wilson et al., 2002; Moncrieff et al., 2010; Mauder et al., 2013)。许多软件产品可用于处理和校正原始数据(EdiRE, Clement, 1999; TK3, Mauder and Foken, 2004; Eddy Soft, Kolle and Rebmann, 2007; ECPack, van Dijk et al., 2004)。

当湍流热通量的总和与可用能量 Rn-G(净辐射减去土壤热通量)相比时，湍流热通量往往被低估。平均闭合误差约为 20%～30%(Twine et al., 2000; Wilson et al., 2002; Foken, 2008; Hendricks Franssen et al., 2010)。可能的原因包括水平平流的影响、雨棚中的热量储存、通量发散、光合作用、净辐射(Rn)或热通量(G)的测量误差、传感器的频率响应、湍流通量的测量误差和仪器的分离。观测塔台场地必须满足以下条件：地势平坦，范围开阔，具有单一和长势均匀的植被。传感器放置的高度取决于植被的高度、采集的范围和仪器的频率响应。专栏 14-4 分别提供了全球和局部地区的两个应用示例。

> **专栏 14-4　通量网数据库的微气象塔站点说明**
>
> 为了满足科学界对二氧化碳、水蒸气和能量通量数据的需求，巴尔多奇等人开发了 FluxNet 全球通量网络数据库(Baldocchi et al., 2001)，其中包含 500 多个长期微气象塔站点。监测了不同的冠层覆盖，包括温带针叶林和阔叶林(落叶和常绿)、热带和北方森林、农作物、草原、灌木丛、湿地和苔原。观测塔由区域网络或单个项目维护。
>
> 诺曼等人(Norman et al., 1995)、库斯塔斯和诺曼(Kustas and Norman, 1999)的双源能量平衡表面模型正在应用于马丁·冈萨洛(Martin Gonzalo)流域瓜达尔基维尔河的一个子流域。为了校准模型并验证蒸散量估算，在典型的区域生态系统——德赫萨(dehesa)的树木稀疏的草原上建立了涡动协方差塔。原因在于该区域树木的平均高度(7 米)，塔高为 18 米，基

> 本平坦,且树冠覆盖均匀(图 14-3)。塔内设有 CSAT 三维声波风速计、KH20 氪湿度计、温度和相对湿度探头、净辐射计。在土壤中安装了两个土壤热流板,并安装了金属丝热电偶。在这种系统配置下,每小时测量四个能量通量。

五、结论

土壤水分是土壤中能量和水分平衡的最终结果,同时,其梯度和演化在不同的水循环过程中起着关键作用,迫使水分通过土壤运移;通过土壤蒸发和植物蒸腾控制水分损失;提供不同条件下的养分和污染物,以供生物吸收;通过土壤剖面传输,并作为土壤成分保留。

因此,土壤水分是提供生态系统服务的决定因素,首先它维持植被和土壤中微生物的充足条件,其次可以保证营养链的平衡。但它也会影响水资源在监管和维护服务的可用性,并对文化服务产生明显影响。土壤异质性使土壤水分在时间和空间上具有天然的高度可变性,这使得在地面观测系统中从中等范围到大范围的描述土壤水分状态的可行性较小。

遥感通过直接和间接信息提供了另一种可靠的高频度数据源。本章回顾了土壤水分信息获取方法,并引用了不同的应用示例。重点关注了需要考虑使用直接和间接来源的土壤水分空间分布所产生的显著尺度效应。此外,还对处理植被覆盖的应用作了进一步解释,这些应用通常是将实地和航空数据相结合进行研究的。植被覆盖是反映并影响土壤水分变化的最大规模的活性成分,也提供了更高尺度的供应服务。

详细信息可在本文引用的文献中找到,预计在未来几年内,这一主题相关方面将迅猛发展,特别是在传感器开发、数据分析和应用范围等方面。土壤水分的演变及其尺度变化在水文学中仍然具有挑战性,遥感技术已扩展了以往对中大尺度应用的期望。生态系统服务价值的评估将受益于高精度土壤水分的估算方法。此外,在当前气温上升趋势下,生态系统服务功能的演变将很大程度上取决

图 14-3(见彩插)

注:(a)设置在卡尔德尼亚(Martín Gonzalo 河流域)的涡流协方差塔;(b)"dehesa"景观;(c)土壤热通量面板系统和补充气象站。

第十四章　地表土壤水分遥感监测：在生态系统过程和尺度效应中的应用

于土壤水分状况的影响。遥感提供了土壤水分变化的区域评估，以及对相关生态系统供应、维护和文化服务的影响评估所需信息。

致谢

这项工作得到了塞勒斯(CERESS)项目(AGL2011-30498-02，Ministerio de Economía y Competitividad of Spain，co-funded FEDER)的支持。

参 考 文 献

Aguilar, C., J. C. Zinnert, M. J. Polo, and D. R. Young. 2012. NDVI as an indicator for changes in water availability to woody vegetation. *Ecological Indicators* 23:290–300.

Akbari, M., N. Toomanian, P. Droogers, W. Bastiaanssen, and A. Gieske. 2007. Monitoring irrigation performance in Esfahan, Iran, using NOAA satellite imagery. *Agricultural Water Management* 88:99–109.

Allen, R. G., M. E. Jensen, J. L. Wright, and R. D. Burman. 1989. Operational estimates of reference evapotranspiration. *Agronomy Journal* 81:650–662.

Allen, R. G., L. S. Pereira, T. A. Howell, and M. E. Jensen. 2011. Evapotranspiration information reporting: I. Factors governing measurement accuracy. *Agricultural Water Management* 98:899–920.

Allen, R. G., L. S. Pereira, D. Raes, and M. Smith. 1998. *Crop evapotranspiration. Guidelines for computing crop water requirements.* Irrigation and Drainage Paper No. 56. Rome: FAO.

Anderson, M. C., W. P. Kustas, J. M. Norman, et al. 2011. Mapping daily evapotranspiration at field to continental scales using geostationary and polar orbiting satellite imagery. *Hydrology and Earth System Sciences* 15:223–239.

Anderson, M. C., J. M. Norman, G. R. Diak, W. P. Kustas, and J. R. Mecikalski. 1997. A two-source time-integrated model for estimating surface fluxes using thermal infrared remote sensing. *Remote Sensing of Environment* 60:195–216.

Anderson, M. C., J. M. Norman, J. R. Mecikalski, J. A. Otkin, and W. P. Kustas. 2007. A climatological study of evapotranspiration and moisture stress across the continental United States based on thermal remote sensing: 2. Surface moisture climatology. *Journal of Geophysical Research: Atmospheres* 112:D11112.

Aubinet, M., A. Grelle, A. Ibrom, et al. 2000. Estimates of the annual net carbon and water exchange of forests: The EUROFLUX methodology. *Advances in Ecological Research* 30:113–175.

Baghdadi, N., C. King, A. Bourguignon, and A. Remond. 2002. Potential of ERS and RADARSAT data for surface roughness monitoring over bare agricultural fields: Application to catchments in Northern France. *International Journal of Remote Sensing* 23:3427–3442.

Baldocchi, D. D., E. Falge, L. Gu, et al. 2001. FLUXNET: A new tool to study the temporal and spatial variability of ecosystems-scale carbon dioxide, water vapor, and energy flux densities [Review]. *Bulletin of the American Meteorological Society* 82:2415–2434.

Bastiaanssen, W. G. M., M. D. Ahmad, and Y. Chemin. 2002. Satellite surveillance of evaporative depletion across the Indus 1 Basin. *Water Resources Research* 38:1273.

Bastiaanssen, W. G. M., M. Menenti, R. A. Feddes, and A. A. M. Holstlag. 1998. A remote sensing surface energy balance algorithm for land (SEBAL). 1. Formulation. *Journal of Hydrology* 212–213:198–212.

Bastiaanssen, W. G. M., S. Thiruvengadachari, R. Sakthivadivel, and D. J. Molden. 1999. Satellite remote sensing for estimating productivities of land and water. *Water Resources Development* 15:181–194.

Beven, K. J. 1989. Changing ideas in hydrology—The case of physically-based models. *Journal of Hydrology* 105:157–172.

Beven, K. J. 2000. On the future of distributed modelling in hydrology. *Hydrological Processes* 14:3183–3184.

Bindlish, R., T. J. Jackson, E. F. Wood, et al. 2003. Soil moisture estimates from TRMM Microwave Imager observations over the southern United States. *Remote Sensing of Environment* 85:507–515.

Blöschl, G., and M. Sivapalan. 1995. Scale issues in hydrological modeling—A review. *Hydrological Processes* 9:251–290.

Bowen, I. S. 1926. The ratio of heat losses by conduction and by evaporation from any water surface. *Physical Review* 27:779–787.

Brocca, L., F. Melone, T. Moramarco, et al. 2010. Improving runoff prediction through the assimilation of the ASCAT soil moisture product. *Hydrology and Earth System Science.* 14:1881–1893.

Brutsaert, W. 1982. *Evaporation into the atmosphere.* Dordrecht, The Netherlands: Kluwer Academic Publishers.

Brutsaert, W. 2005. *Hydrology. An introduction.* Cambridge, UK: Cambridge University Press.

Buck, A. 1976. The variable path Lyman-alpha hygrometer and its operating characteristics. *Bulletin of the American Meteorological Society* 51:1113–1118.

Burba, G., and D. Anderson. 2008. *Introduction to the eddy covariance method: General guidelines and conventional workflow.* Lincoln, NE: LiCor Corporation.

Campbell, G. S., and B. D. Tanner. 1985. A krypton hygrometer for measurement of atmospheric water vapor concentration. In *Moisture and humidity*, 609–612. Triangle Research Park, NC: Instrument Society of America.

Campos, I., C. M. U. Neale, A. Calera, C. Balbontin, and J. González-Piqueras. 2010. Assessing satellite-based basal crop coefficients for irrigated grapes (*Vitis vinifera L.*). *Agricultural Water Management* 97:1760–1768.

Carlson, T. 2007. An overview of the "Triangle Method" for estimating surface evapotranspiration and soil moisture from satellite imagery. *Sensors* 7:1612–1629.

Ceballos, A., K. Scipal, W. Wagner, and J. Martinez-Fernandez. 2005. Validation of ERS scatterometer-derived soil moisture data in the central part of the Duero Basin, Spain. *Hydrological Processes* 19:1549–1566.

Chandrapala, L., and M. Wimalasuriya. 2003. Satellite measurements supplemented with meteorological data to operationally estimate evaporation in Sri Lanka.

Agricultural Water Management 58:89–107.
Chavez, J. L., C. M. U. Neale, L. E. Hipps, J. H. Prueger, and W. P. Kustas. 2005. Comparing aircraft-based remotely sensed energy balance fluxes with eddy covariance tower data using heat flux source area functions. *Journal of Hydrometeorology* 6:923–940.
Chehbouni, A., D. Lo Seen, E. G. Njoku, and B. M. Monteney. 1996. Examination of difference between radiometric and aerodynamic surface temperature over sparsely vegetated surfaces. *Remote Sensing of Environment* 58:177–186.
Clement, R. 1999. EdiRE. Available from: http://www.geos.ed.ac.uk/abs/research/micromet/EdiRe.
Crow, W. T., W. P. Kustas, and J. H. Prueger. 2008. Monitoring root-zone soil moisture through the assimilation of a thermal remote sensing-based soil moisture proxy into a water balance model. *Remote Sensing of Environment* 112:1268–1281.
De Bruin, H. A. R. 2008. Theory and application of large aperture scintillometers. Short course notes. Scintec Corp.
Dente, L., Z. Su, and J. Wen. 2012. Validation of SMOS soil moisture products over the Maqu and Twente regions. *Sensors* 12:9965–9986.
Doorenbos, J., and W. O. Pruitt. 1977. Guidelines for predicting crop-water requirements. In *FAO Irrigation and drainage*. Paper No. 24. 2nd ed. Rome: FAO. 156 pp.
Duchemin, B., R. Hadria, S. Er-Raki, et al. 2006. Monitoring wheat phenology and irrigation in Central Morocco: On the use of relationships between evapotranspiration, crop coefficients, leaf area index and remotely-sensed vegetation indices. *Agricultural Water Management* 79:1–27.
Engman, E. T., and N. Chauhan. 1995. Status of microwave soil moisture measurements with remote sensing. *Remote Sensing of Environment* 51:189–198.
Engman, E. T., and R. J. Gurney. 1991. *Remote sensing in hydrology*. London: Chapman & Hall.
Er-Raki, S., A. Chehbouni, and B. Duchemin. 2010. Combining satellite remote sensing data with the FAO-56 dual approach for water use mapping in irrigated wheat fields of a semi-arid region. *Remote Sensing* 2:375–387.
Er-Raki, S., A. Chehbouni, N. Guemouria, B. Duchemin, J. Ezzahar, and R. Hadria. 2007. Combining FAO-56 model and ground-based remote sensing to estimate water consumptions of wheat crops in a semi-arid region. *Agricultural Water Management* 87:41–54.
Fernández-Prieto, D., P. van Oevelen, Z. Su, and W. Wagner. 2012. Advances in Earth observation for water science (Editorial). *Hydrology and Earth System Science* 16:543–549.
FLUXNET. Integrating worldwide CO_2, water and energy flux measurements. Available from: http://fluxnet.ornl.gov. Accessed July 7, 2012
Foken, T. 2008. The energy balance closure problem—An overview. *Ecological Applications* 18:1351–1367.
Fung, A. K. 1994. *Microwave scattering and emission models and their applications*. Norwood, MA: Artech House, Inc.
Gao, H., E. Wood, M. Drusch, M. McCabe, T. J. Jackson, and R. Bindlish. 2003. Using TRMM/TMI to retrieve soil moisture over southern United States from 1998 to 2002: Results and validation. *EOS Transactions* 84:618.

Garatuza-Payan, J., and C. J. Watts. 2005. The use of remote sensing for estimating ET of irrigated wheat and cotton in northwest Mexico. *Irrigation and Drainage Systems* 19:301–320.

Gillies, R. T., T. N. Carlson, J. Cui, W. P. Kustas, and K. S. Humes. 1997. A verification of the "triangle" method for obtaining surface soil water content and energy fluxes from remote measurements of the Normalized Difference Vegetation Index (NDVI) and surface radiant temperatures. *International Journal of Remote Sensing* 18:3145–3166.

González-Dugo, M. P., S. Escuin, L. Mateos, et al. 2012. Monitoring evapotranspiration of irrigated crops using crop coefficients derived from time series of satellite images. II. Application on basin scale. *Agricultural Water Management* 125:92–104.

González-Dugo, M. P., and L. Mateos. 2008. Spectral vegetation indices for benchmarking water productivity of irrigated cotton and sugarbeet crops. *Agricultural Water Management* 95:48–58.

González-Dugo, M. P., C. M. U. Neale, L. Mateos, et al. 2009. A comparison of operational remote-sensing-based models for estimating crop evapotranspiration. *Agricultural and Forest Meteorology* 149:1843–1853.

Groeneveld, D. P., and W. M. Baugh. 2007. Correcting satellite data to detect vegetation signal for eco-hydrologic analyses. *Journal of Hydrology* 344:135–145.

Gruhier, C., P. de Rosnay, S. Hasenauer, et al. 2010. Soil moisture active and passive microwave products: Intercomparison and evaluation over a Sahelian site. *Hydrology and Earth System Science* 14:141–156.

Gupta, V. K., I. Rodríguez-Iturbe, and E. F. Wood (eds.). 1986. *Scale problems in hydrology*. Dordrecht, The Netherlands: Reidel.

Hamid, S. H., A. A. Mohamed, and Y. A. Mohamed. 2011. Towards a performance-oriented management for large-scale irrigation systems: Case study, Rahad scheme, Sudan. *Irrigation and Drainage* 60:20–34.

Hartogensis, O. K., C. J. Watts, J. C. Rodriguez, and H. A. R. De Bruin. 2003. Derivation of an effective height for scintillometers: La Poza experiment in northwest Mexico. *Journal of Hydrometeorology* 4:915–928.

Hendricks Franssen, H. J., R. Stöckli, I. Lehner, E. Rotenberg, and S. I. Seneviratne. 2010. Energy balance closure of eddy-covariance data: A multisite analysis for European FLUXNET stations. *Agricultural and Forest Meteorology* 150:1553–1567.

Horst, T. W. 2000. On frequency response corrections for eddy covariance flux measurements. *Boundary-Layer Meteorology* 95:517–520.

Hunsaker, D. J., E. M. Barnes, T. R. Clarke, G. J. Fitzgerald, and P. J. Pinter. 2005a. Cotton irrigation scheduling using remotely sensed and FAO-56 basal crop coefficients. *Transactions of the ASAE* 48:1395–1407.

Hunsaker, D. J., P. J. Pinter, and B. A. Kimbal. 2005b. Wheat basal crop coefficients determined by normalized difference vegetation index. *Irrigation Science* 24:1–14.

Jensen, M. E. (ed.). 1974. *Consumptive use of water and irrigation water requirements*. New York: American Society of Civil Engineers.

Jensen, M. E., R. D. Burman, R. G. Allen (eds.). 1990. *Evapotranspiration and irrigation water requirements*. New York: American Society of Civil Engineers.

Juglea, S., Y. Kerr, A. Mialon, E. López-Baeza, D. Braithwaite, and K. Hsu. 2010. Soil moisture modelling of a SMOS pixel: Interest of using the PERSIANN database over the Valencia Anchor Station. *Hydrology and Earth System Science* 14:1509–1525.

Jury, W. A., and R. Horton. 2004. Chemical transport in soil. In *Soil physics*, 225–276. Hoboken, NJ: Wiley.

Kasischke, E. S., M. A. Tanase, L. L. Bourgeau-Chavez, and M. Borr. 2011. Soil moisture limitations on monitoring boreal forest regrowth using spaceborne L-band SAR data. *Remote Sensing of Environment* 115:227–232.

Kerr, Y., P. Waldteufel, J. P. Wigneron, J. M. Martinuzzi, J. Font, and M. Berger. 2001. Soil moisture retrieval from space. The Soil Moisture and Ocean Salinity (SMOS) mission. *IEEE Transaction on Geoscience and Remote Sensing* 39:1729–1735.

Kolle, O., and C. Rebmann. 2007. EddySoft—Documentation of a software package to acquire and process Eddy covariance data. Jena, Germany: Max Planck Institute for Biogeochemistry.

Kustas, W. P. 1990. Estimates of evapotranspiration with a one- and two-layer model of heat transfer over partial cover. *Journal of Applied Meteorology and Climatology* 29:704–715.

Kustas, W. P., B. J. Choudhury, M. S. Moran, et al. 1989. Determination of sensible heat flux over sparse canopy using thermal infrared data. *Agricultural and Forest Meteorology* 44:197–216.

Kustas, W. P., and J. M. Norman. 1996. Use of remote sensing for evapotranspiration monitoring over land surfaces. *Hydrological Sciences* 41:495–516.

Kustas, W. P., and J. M. Norman. 1999. Evaluation of soil and vegetation heat flux predictions using a simple two source model with radiometric temperatures for partial canopy cover. *Agricultural and Forest Meteorology* 94:13–29.

Laguardia, G., and S. Niemeyer. 2008. On the comparison between the LISFLOOD modelled and the ERS/SCAT derived soil moisture estimates. *Hydrology and Earth System Science* 12:1339–1351.

Lhomme, J. P., B. Monteny, and M. Amadou. 1994. Estimating sensible heat flux from radiometric temperature over sparse millet. *Agricultural and Forest Meteorology* 68:77–91.

Lievens, H., N. E. C. Verhoest, E. De Keyser, et al. 2011. Effective roughness modelling as a tool for soil moisture retrieval from C- and L-band SAR. *Hydrology and Earth System Science* 15:151–162.

Liu, Y. Y., W. A. Dorigo, R. M. Parinussa, et al. 2012. Trend-preserving blending of passive and active microwave soil moisture retrievals. *Remote Sensing of Environment* 123:280–297.

Liu, Y. Y., R. M. Parinussa, W. A. Dorigo, et al. 2011. Developing an improved soil moisture dataset by blending passive and active microwave satellite-based retrievals. *Hydrology and Earth System Science* 15:425–436.

Mahrt, L., and D. Vickers. 2004. Bulk formulation of the surface heat flux. *Boundary-Layer Meteorology* 110:357–379.

Makkink, G. P. 1957. Testing the Penman formula by means of lysimeters. *Journal of the Institution of Water Engineers* 11:277–288.

Manninen, T., P. Stenberg, M. Rautiainen, P. Voipio, and H. Smolander. 2005. Leaf area index estimation of boreal forest using ENVISAT ASAR. *IEEE Transactions on Geoscience and Remote Sensing* 43:2627–2635.

Massman, W. J. 2000. A simple method for estimating frequency response corrections for eddy covariance systems. *Agricultural and Forest Meteorology* 104:185–198.

Massman, W. J. 2001. Reply to comment by Rannik on: A simple method for estimating frequency response corrections for eddy covariance systems. *Agricultural and Forest Meteorology* 107:247–251.

Mauder, M., M. Cuntz, C. Drüe, et al. 2013. A strategy for quality and uncertainty assessment of long-term eddy-covariance measurements. *Agricultural and Forest Meteorology* 169:122–135.

Mauder, M., and T. Foken. 2004. *Documentation and instruction manual of the eddy covariance software package TK2*. Arbeitsergebnisse 26. Bayreuth, Germany: Universitaet Bayreuth, Abt. Mikrometeorologie.

Mecikalski, J. M., G. R. Diak, M. C. Anderson, and J. M. Norman. 1999. Estimating fluxes on continental scales using remotely sensed data in an atmosphere-land exchange model. *Journal of Applied Meteorology and Climatology* 38:1352–1369.

Meijninger, W. M. L., and H. A. R. De Bruin. 2000. The sensible heat fluxes over irrigated areas in western Turkey determined with a large aperture scintillometer. *Journal of Hydrology* 229:42–49.

Meijninger, W. M. L., O. K. Hartogensis, W. Kohsiek, J. C. B. Hoedjes, R. M. Zuurbier, and H. A. R. De Bruin. 2002. Determination of area-averaged sensible heat fluxes with a large aperture scintillometer over a heterogeneous surface—Flevoland field experiment. *Boundary-Layer Meteorology* 105:37–62.

Mertens, J. 2003. Parameter estimation strategies in unsaturated zone modelling. PhD diss., Leuven Catholic University.

Moncrieff, J. B., R. Clement, J. Finnigan, and T. Meyers. 2010. Averaging, detrending, and filtering of eddy covariance time series. In *Handbook of micrometeorology: A guide for surface flux measurement and analysis*, eds. X. Lee, W. Massman, and B. Law, 7–31. Dordrecht, The Netherlands: Kluwer Academic Publishers.

Moradkhani, H. 2008. Hydrologic remote sensing and land surface data assimilation. *Sensors* 8:2986–3004.

Moran, M. S. 2003. Thermal infrared measurement as an indicator of plant ecosystem health. In *Thermal remote sensing in land surface processes*, eds. D. A. Quattrochi and J. Luvall, 257–282. Philadelphia, PA: Taylor & Francis.

Moran, M. S., T. R. Clarke, Y. Inoue, and A. Vidal. 1994. Estimating cropwater deficit using the relation between surface-air temperature and spectral vegetation index. *Remote Sensing of Environment* 49:246–263.

Moran, M. S., D. C. Hymer, J. Qi, and Y. Kerr. 2001. Comparison of ERS-2 SAR and Landsat TM imagery for monitoring agricultural crop and soil conditions. *Remote Sensing of Environment* 79:243–252.

Munger, J. W., and H. W. Loescher. 2004. Guidelines for making eddy covariance flux measurements. Available from: http://public.ornl.gov/ameriflux/ measurement standards 4.doc. Accessed May 12, 2012.

Narayanan, R. M., and M. S. Hegde. 1995. Soil moisture inversion algorithms using ERS-1, JERS-1, and ALMAZ SAR data. Geoscience and Remote Sensing Symposium. IGARSS '95. *Quantitative Remote Sensing for Science and Applications* 1:504–506.

Norman, J. M., M. C. Anderson, W. P. Kustas, et al. 2003. Remote sensing of surface energy fluxes at 101-m pixel resolutions. *Water Resources Research* 39:1221.

Norman, J. M., W. P. Kustas, and K. S. Humes. 1995. A two-source approach for estimating soil and vegetation energy fluxes from observations of directional radiometric surface temperature. *Agricultural and Forest Meteorology* 77:263–293.

Padilla, F. L. M., M. P. González-Dugo, P. Gavilán, and J. Domínguez. 2011. Integration of vegetation indices into a water balance model to estimate evapotranspiration of wheat and corn. *Hydrology and Earth System Science* 15:1213–1225.

Parajka, J., V. Naeimi, G. Blöschl, W. Wagner, R. Merz, and K. Scipal. 2006. Assimilating scatterometer soil moisture data into conceptual hydrologic models at the regional scale. *Hydrology and Earth System Science* 10:353–368.

Paw, U. K. T., D. D. Baldocchi, T. P. Meyers, and K. B. Wilson. 2000. Correction of eddy covariance measurements incorporating both advective effects and density fluxes. *Boundary-Layer Meteorology* 97:487–511.

Pettorelli, N., J. O. Vik, A. Mysterud, J. M. Gaillard, C. J. Tucker, and N. C. Stenseth. 2005. Using the satellite-derived NDVI to assess ecological responses to environmental change. *Trends in Ecology and Evolution* 20:503–510.

Polo, M. J., J. Regodón, and M. P. González-Dugo. 2009. Tidal flood monitoring in marsh estuary areas from Landsat TM data. In *Remote sensing for agriculture, ecosystems, and hydrology XI*, eds. C. M. U. Neale and A. Maltese. Washington, DC: SPIE.

Rannik, U. 2001. A comment on the paper by W. J. Massman: A simple method for estimating frequency response corrections for eddy covariance systems. *Agricultural and Forest Meteorology* 107:241–245.

Roberts, D. A., G. T. Batista, J. L. G. Pereira, E. K. Waller, and B. W. Nelson. 1998. Change identification using multitemporal spectral mixture analysis: Applications in eastern Amazonia. In *Remote sensing change detection: Environmental monitoring methods and applications*, eds. R. S. Lunetta and C. D. Elvidge, 137–161. Ann Arbor, MI: Ann Arbor Press.

Rossi, S., A. Rampini, S. Bocchi, and M. Boschetti. 2010. Operational monitoring of daily crop water requirements at the regional scale with time series of satellite data. *Journal of Irrigation and Drainage Engineering* 136:225–231.

Rouse, J. W., Jr., R. H. Haas, J. A. Schell, and D. W. Deering. 1974. Monitoring vegetation systems in the Great Plains with ERTS. *Proceedings of the Third ERTS Symposium* 1:309–317.

Sakai, R. K., D. R. Fitzjarrald, and K. E. Moore. 2001. Importance of low-frequency contributions to eddy fluxes observed over rough surfaces. *Journal of Applied Meteorology and Climatology* 40:2178–2192.

Sánchez, N., J. Martínez-Fernández, A. Calera, E. Torres, and C. Pérez-Gutiérrez. 2010. Combining remote sensing and in situ soil moisture 1 data for the application and validation of a distributed water balance model (HIDROMORE). *Agricultural Water Management* 98:69–78.

Schmugge, T. J., and T. J. Jackson. 1994. Mapping surface soil moisture with microwave radiometers. *Meteorology and Atmospheric Physics* 54:213–223.

Schmugge, T. J., W. P. Kustas, J. C. Ritchie, T. J. Jackson, and A. Rango. 2002. Remote sensing in hydrology. *Advances in Water Resources* 25:1367–1385.

Shao, G., H. H. Shugart, and D. R. Young. 1995. Simulation of transpiration sensitivity to environmental changes for shrub (*Myrica cerifera*) thickets on a Virginia barrier island. *Ecological Modelling* 78:235–248.

Shuttelworth, W., and J. Wallace. 1985. Evaporation from sparse crops: An energy combination theory. *Quarterly Journal of the Royal Meteorological Society* 111:1143–1162.

Singh, R., and A. Irmak. 2009. Estimation of crop coefficients using satellite remote sensing. *Journal of Irrigation and Drainage Engineering* 135:597–608.

Sobrino J. A., B. Franch, C. Mattar, J. C. Jiménez-Muñoz, and C. Corbari. 2012. A method to estimate soil moisture from Airborne Hyperspectral Scanner (AHS) and ASTER data: Application to SEN2FLEX and SEN3EXP campaigns. *Remote Sensing of Environment* 117:415–428.

Sposito, G. 1989. *The chemistry of soils*. New York: Oxford University Press.

Su, C. H., D. Ryu, R. I. Young, A. W. Western, and W. Wagner. 2013. Inter-comparison of microwave satellite soil moisture retrievals over the Murrumbidgee basin, southeast Australia. *Remote Sensing of Environment* 134:1–11.

Su, Z., J. Wen, L. Dente, et al. 2011. The Tibetan Plateau observatory of plateau scale soil moisture and soil temperature (Tibet-Obs) for quantifying uncertainties in coarse resolution satellite and model products. *Hydrology and Earth System Science* 15:2303–2316.

Su, Z., J. Wen, and W. Wagner. 2010. Advances in land surface hydrological processes: field observations, modeling and data assimilation: Preface. *Hydrology and Earth System Science* 14:365–367.

Tanner, B. D. 1988. Use requirements for Bowen ratio and eddy correlation determination of evapotranspiration. In *Planning now for irrigation and drainage in the 21st century*, ed. D. R. Hay, 605–616. Lincoln, NE: ASCE.

Tanner, B. D., E. Swiatek, and J. P. Green. 1993. Density fluctuations and use of the krypton hygrometer in surface flux measurements. In *Management of irrigation and drainage systems: Integrated perspectives*, eds. R. G. Allen and C. M. U. Neale, 945–952. Lincoln, NE: ASCE.

Tasumi, M., and R. G. Allen. 2007. Satellite-based ET mapping to assess variation in ET with timing of crop development. *Agricultural Water Management* 88:54–62.

Tasumi, M., R. Trezza, R. G. Allen, and J. L. Wright. 2005. Operational aspects of satellite-based energy balance models for irrigated crops in the semi-arid U.S. *Irrigation and Drainage Systems* 19:355–376.

Timmermans, W. J., W. P. Kustas, M. C. Anderson, and A. N. French. 2007. An intercomparison of the surface energy balance algorithm for land (SEBAL) and the two source energy balance (TSEB) modeling schemes. *Remote Sensing of Environment* 108:284–369.

Twine, T. E., W. P. Kustas, J. M. Norman, et al. 2000. Correcting eddy-covariance flux underestimates over a grassland. *Agricultural and Forest Meteorology* 103:279–300.

Ulaby, F. T., R. K. Moore, and A. K. Fung. 1981. *Microwave remote sensing: Active and passive, Vol. I—Microwave remote sensing fundamentals and radiometry*. Reading, MA: Addison-Wesley, Advanced Book Program.

Van Dijk, A., A. F. Moene, and H. A. R. De Bruin. 2004. *The principles of surface flux physics: Theory, practice and description of the ECPACK library*. Internal Report 2004/1, Meteorology and Air Quality Group. Wageningen, The Netherlands: Wageningen University.

Vereecken, H., J. A. Huisman, H. Bogena, J. Vanderborght, J. A. Vrugt, and J. W. Hopmans. 2008. On the value of soil moisture measurements in vadose zone hydrology: A review. *Water Resources Research* 44:W00D06.

Villalobos, F. J. 1997. Correction of eddy covariance water vapor flux using additional measurements of temperature. *Agricultural and Forest Meteorology* 88:77–83.

Wagner, W., G. Lemoine, M. Borgeaud, and H. Rott. 1999a. A study of vegetation cover effects on ERS scatterometer data. *IEEE Transactions on Geoscience and Remote Sensing* 37:938–948.

Wagner, W., G. Lemoine, and H. Rott. 1999b. A method for estimating soil moisture from ERS scatterometer and soil data. *Remote Sensing of Environment* 70:191–207.

Wang, J., P. M. Rich, and K. P. Price. 2003. Temporal responses of NDVI to precipitation and temperature in the central Great Plains, USA. *International Journal of Remote Sensing* 24:2345–2364.

Wigneron, J. P., J. C. Calvet, T. Pellarin, A. A. V. D. Griend, M. Berger, and P. Ferrazzoli. 2003. Retrieving near-surface soil moisture from microwave radiometric observations: Current status and future plans. *Remote Sensing of Environment* 85:489–506.

Wigneron, J. P., T. Schmugge, A. Chanzy, J. C. Calvet, and Y. Kerr. 1998. Use of passive microwave remote sensing to monitor soil moisture. *Agronomie* 18:27–43.

Wilson, K. B., A. H. Goldstein, and F. Falge. 2002. Energy balance closure at FLUXNET sites. *Agricultural and Forest Meteorology* 113:223–243.

Wright, J. L. 1981. *Crop coefficients irrigation scheduling for water and energy conservation in the 80s*. St. Joseph, MO: American Society of Agricultural Engineers.

Wright, J. L. 1982. New evapotranspiration crop coefficients. *Journal of the Irrigation and Drainage Division ASCE* 108:57–74.

Zribi, M., N. Baghdadi, N. Holah, O. Fafin, and C. Guérin. 2005. Evaluation of a rough soil surface description with ASAR-ENVISAT radar data. *Remote Sensing of Environment* 95:67–76.

Zribi, M., and M. Dechambre. 2002. A new empirical model to inverse soil moisture and roughness using two radar configurations. *Proceedings of the IEEE International Geoscience and Remote Sensing Symposium and 24th Canadian Symposium on Remote Sensing*, Toronto, Canada: Institute of Electrical and Electronics Engineers. 2223–2225.

第十五章　积雪作为山区生态系统服务的关键要素：有效监测计划的设计思路

F.J. 博内特　A. 米拉尔斯　J. 埃雷罗

一、简介：从监测积雪覆盖到量化生态系统服务

积雪覆盖是在地面上堆积的雪层，它会经历一个分组和重新排列的过程，从而形成冰。在一定时期内，积雪保持着不同程度的变质作用。它由一个固体基质构成，其中含有一个可以储存气体和液体的中空部分。由于这种结构，积雪层具有一些物理特性，如密度、孔隙率或水力和热导率。雪的基质为冰冻的水。冰不能在温度高于0℃的环境下存在。在这些情况下，基质融化，成为孔隙的一部分。如果温度低于0℃，这种液态水可以再次变成冰（基质）。它也可以渗入地下或成为浅层径流的一部分。积雪是一个多孔系统，水的三种状态（气体、液体和固体）可能同时共存，是地球表面上的一个独特元素。而且，它非常奇特，因为它在空间和时间上都能经历巨大的波动。一场气旋事件可以将积雪的覆盖范围增加到1 000平方千米（Cohen and Rind, 1991）。

由于这些特征，积雪覆盖是塑造所在山区自然景观的关键因素。例如，它的储水能力就显示了它在水文循环中的作用。积雪的重量和热导率可以显示其在某些生态过程中的作用。根据国际生态系统服务通用分类（Haines-Young and Potschin, 2013），生态系统服务从根本上依赖于生命过程的产出，自然界的非生物产出基本不被视为生态系统服务。因此，不应将积雪视为能够提供生态系统服务的要素。尽管积雪为这些服务创造了一个非生物环境，但最终的服务却是由生态系统提供的。

第十五章 积雪作为山区生态系统服务的关键要素：有效监测计划的设计思路

上述推理可作为我们在本章中遵循的概念框架指引。(1)积雪覆盖具有多种结构和功能特征；(2)这些特征可以通过几种监测方案来观测；(3)观测的属性有助于山地生态系统服务。这最后一步归功于量化方法，从水文模型(见第十二章)到(例如)生态生理模型，这些模型能够利用雪的含水量信息模拟生物量的产生。图15-1显示了本章的概念模型。

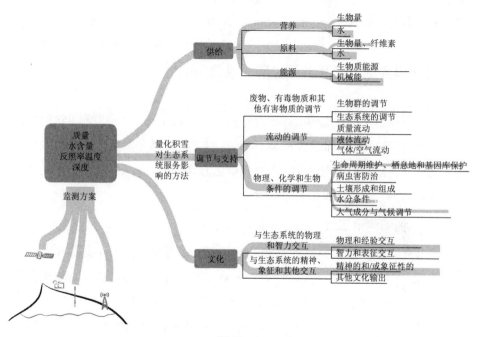

图 15-1

注：该图显示了我们在本章中所遵循的概念框架。积雪具有若干功能和结构特征（温度、水含量、深度等），可以通过监测方案进行评估。获得的信息可用于量化不同类型的生态系统服务（根据R. 海恩斯·杨和M. 波茨钦进行分类，2012年8月～12月CICES第4版磋商后编制的修订报告，欧洲经济协会框架合同编号EEA/IEA/09，2013）。

根据上述概念框架，本章的主要目标是：(1)阐述有助于实施监测计划以收集积雪覆盖信息的不同方法；(2)列出积雪覆盖提供的终端服务；(3)给出一个旨在量化终端服务的监测项目案例研究。

我们选择山区作为本章的空间对象。尽管雪覆盖了世界各地巨大的平坦区域，但我们选择了山区，因为它们是陆地系统中非常特殊的区域。它们浓缩了不同的气候类型、非常丰富的生物多样性,并在一个不大的区域内聚集了大量的人口。这些特征使它们成为一个小小的"地球系统",以此为对象分析全球变化对生物圈的影响具有特殊意义。它们可以被视为充当"煤矿中的金丝雀"的相关观测系统(Diaz et al.,2003),从而提醒我们注意影响地球的重要变化(译者注:金丝雀对甲烷和一氧化碳特别敏感,非常适合用于侦测瓦斯积聚危险。这里意指某种危险的指示信号)。此外,山区可以认为是保护生态系统服务和生物多样性成本低效益高的地区(Sanderson et al.,2002)。

二、设计和实施监测积雪覆盖所提供服务的方法

我们将描述一个理想的山区积雪监测计划。该计划在确定监测方法时将考虑到山区的特殊特征。如概念框架所示,这些方案能够定量描述积雪覆盖的结构和功能。这是对积雪提供的服务进行量化的第一步。拟议的监测计划还将考虑山区和监测过程中相关诸要素的时空层次。

(一) 现场测定法

本节包含现场收集信息的监测方法流程。基于这些方法,我们可以获得关于雪深、雪水当量的信息,也有可能获得有关雪层结构的信息。实施这些方法需要在现场安装传感器和其他设备,且需人工收集信息。这些方法对于收集雪的含水量等详细信息特别有用。因此,这些方法被用于量化积雪提供的供给和调节服务,有助于校准水文模型,校准结果可用于量化雪提供的调节和供给服务。

1. 横断面、雪地勘测和自动观测站

为了获得详细的雪量和含水量信息,有必要建立封闭式设施,在其内安装渗漏计、深度传感器(使用超声波)、压力传感器等仪器(Dingman,2002)。

这些高强度方法很难实施,尤其是在雪深非常不均匀的山区。创建横断面是获取山区雪水当量信息的一种更有效的方法。该方法更简单,需要在积雪覆盖上设计一组具有周期性采样点的横断面。使用金属圆筒提取雪柱,并测量其深度、重量和含水量(Chow,1964;Rallison,1981)。但这两种方法都难以在偏

远地区实施。测量的空间范围非常小,收集信息的周期非常长(经常测量横断面是不现实的)。为了避免这些问题,可以安装能够测量雪水当量的自动工作站。雪情遥测(SNOTEL;Rallison,1981;Watson et al.,2008)系统可能是全世界最常用的监测雪水当量的方案。该系统是美国农业部运营的一个网络,能够监测雪水当量、降水、温度和雪深。在美国西部有600个雪情遥测站。因为有了雪情遥测这样的系统,就有可能克服横断面和定时测雪的问题。

这些方法最主要的缺点之一是它们的空间范围有限。设计、实施和维护基于横断面或自动现场观测站的密集监测系统非常昂贵耗时。

大众参与(Silvertown,2009)有助于克服之前方法中空间范围受限的问题。一些公共行政部门,如加拿大政府,已经意识到可能有成千上万的人愿意收集有关冰雪覆盖的信息,他们的主要动机是希望能够用于监测全球变化。加拿大创建了一个名为 Icewatch(http://www.naturewatch.ca/english/icewatch)的计划,允许普通民众参与发现积雪和冰的变化方式和原因。该网络帮助科学家了解加拿大冰雪的冻融循环。

2. 倾斜摄影用于监测积雪覆盖

点采样方法对于获取积雪覆盖范围的信息没有用处。它们无法探测到不同区域显著的积雪变化(Hinkler et al.,2002)。标准数码相机可以帮助克服这个问题,允许从单个现场测量点监测积雪覆盖的范围(Hinkler et al.,2002;Laffly et al.,2011)。该技术结合了现场监测和遥感方法,尤其应用在小流域尺度具有重要意义(Parajka et al.,2012)。倾斜摄影照片以相对较低的成本实现了非常高的时空分辨率。标准数码相机连接到可以由电子计时器控制的自动触发器。

可通过太阳能电池板供应能源。信息存储在闪存卡中,在有信号的地方可以通过通用分组无线电服务(GPRS)自动发送。下一步是将获得的倾斜照片与数字高程模型(Digital Elevation Mode,DEM)进行配准,以获得可以测量雪面的图层。为了获得这样一个图层,有必要创建一个函数,将照片中的二维像素与数字高程模型中的三维点联系起来(Corripio,2004;Laffly et al.,2011)。还必须对图像中的地形和大气影响造成的反照率偏差进行纠正。最后,可以使用可见光波长或红色和红外波长(多光谱辐射)的组合来自动检测雪盖范围,通过计算归一化差异雪指数(NDSI；Hall et al.，1995；Dozier and Painter，2004)，将雪与其他白色地物，如云或岩石区分开。

(二) 卫星测定法

使用卫星监测积雪的历史由来已久(自 20 世纪 60 年代以来)(Dietz et al.，2012)。这是因为积雪在水文循环和气候系统中的重要性，以及它在为人类社会提供有用服务方面的作用。1960 年，第一台开始监测积雪覆盖的卫星是电视和红外观测卫星(TIROS-1)。从那时到现在，许多空间机构都设计了具有不同光谱通道和时空分辨率的传感器。本节简要回顾了使用遥感设备监测积雪的相关主要方法。大多数卫星都能测量出积雪覆盖的范围，其中的一些卫星可获得有关雪水当量和雪深的信息。这两个变量对于量化积雪提供的水供给服务非常有用。

这些技术的主要优点在于卫星从空中收集的信息。多亏了这些技术设备，才有可能在全球范围内监测积雪覆盖的情况。所有获得的图像都可按照类似的过程进行处理，并且可获得长时间序列的积雪覆盖范围显示。这些方法还包括校准传感器的技术，由于恶劣的空间环境，这些传感器的功能会退化(Wang et al.，2012)。有多种校准传感器的方法，包括基于星载设备(Green and Shimada，1997)、地面场地(使用地表的自然或人工场地)(Martiny et al.，2005)或飞行器(使用携带校准辐射计并同步测量辐射值的飞行器)等方法。

1. MODIS

中分辨率成像光谱仪传感器安装在两颗卫星上：Terra(1999 年 12 月发射)和 Aqua(2002 年 5 月发射)。

该传感器可捕获 36 个光谱波段(0.4～14.4 微米)的数据。这些波段允许创建有关土地利用、云量、气溶胶、海洋和陆地生物生产力、地表温度和积雪覆盖的相关地球物理信息。关于最后一个元素，MODIS 能够提供有关积雪覆盖范围和积雪覆盖比例(每像素的积雪百分比)的信息。积雪覆盖是通过一种称为 Snowmap 的算法获得的(Hall et al.，1995)。它使用可见光和红外线通道的反照率来获得归一化差异雪指数(NDSI)。MODIS 提供基于 Snowmap 算法生成的 MOD10A2 产品。该产品的周期为 8 天，空间分辨率为 500 米。在每幅

MOD10A2 图像中,如果在八天周期内的某一天出现过雪,则每个像素都将标为雪。因此,MOD10A2 显示了八天内积雪覆盖的最大范围。MODIS 还在一个名为 MOD10A1 的日常产品中提供积雪覆盖比例的信息。每个像素中雪覆盖的表面百分比是基于所罗门森和阿佩尔(Salomonson and Appel)在 2006 年开发的算法生成的。这两个产品对于校准和验证水文模型非常有用(Parajka and Blöschl,2008)。鲍威尔等人(Powell et al.,2011)使用 MODIS 雪覆盖面积作为每日水文模型的输入数据,该模型预测加利福尼亚河的河流流量。安德烈亚季斯和莱滕梅尔(2006)成功地将 MODIS 产品同化到美国水文模型的可变渗透能力(VIC)中。为满足一些特定目标,MODIS 产品已经过改进。蒂雷尔等人(Thirel et al.,2012)根据 MODIS 信息推出了一个实时积雪覆盖产品。该产品被用于验证水文模型的输出。

MODIS 产品最重要的优势之一是可以通过多种网络服务获取数据。美国国家航空航天局开发了多种程序,以支持轻松下载所有 MODIS 产品的原始数据。目前最先进的是 ECHO Reverb(http://reverb.echo.nasa.gov/reverb),它是一个强大的网络应用程序,允许在全球范围内下载 MODIS 雪覆盖产品。用户可以选择一个区域来获取所有现有图像。还可以订阅多个产品并通过文件传输协议(FTP)接收新图像,这为创建可自动下载、处理和显示基于 MODIS 的有关积雪信息的第三方应用程序成为可能。

2. Landsat

陆地卫星(Landsat)自 1972 年以来一直在运行。前三颗卫星有四个光谱波段。它们的空间分辨率为 79 米,每 18 天采集一次图像。陆地卫星 4~5(1982 年发射)和陆地卫星 7(1999 年发射)有七、八个光谱带,空间分辨率为 30 米,周期为 16 天。陆地卫星和 MODIS 提供了类似的积雪产品:积雪覆盖范围和积雪覆盖比例。然而,由于时间分辨率较差(16 天),陆地卫星不应用于监测。陆地卫星的产品可用于验证水文模型的输出(Herrero et al.,2011)。这两种产品都可使用类似于 MODIS 的算法获得(Rosenthal and Dozier,1996;Vikhamar and Solberg,2003)。遗憾的是,陆地卫星 5 已不再运行,陆地卫星 7 已出现严重问题(扫描线校正器故障)。"陆地卫星数据连续性任务"(LDCM)于 2013 年 2 月发射,被认为是下一代陆地卫星(Irons et al.,2012)。它的空间分辨率全色为 15

米,多光谱为30米,热红外波段为100米,地面轨道重复周期为16天,降交点为上午10点。

3. AVHRR

1978年,由美国国家海洋和大气管理局发射了第一台先进的高级甚高分辨率辐射计(AVHRR)传感器。目前,至少有两颗极地轨道卫星在轨道上运行。其空间分辨率为1 090米,并每天拍摄图像。AVHRR具有5个0.58~12.5微米的光谱带。该传感器提供的信息经常被用于监测全球的积雪覆盖情况。费尔南德斯和赵(2008)设计了一种算法,能够基于AVHRR影像生成积雪覆盖图。该算法使用AVHRR通道1和2的反射率,加上归一化差异植被指数、反照率和地表温度。最近,赫斯勒等人(Hüsler et al.,2012)开发了一种更精确的算法,用AVHRR绘制积雪覆盖图。虽然AVHRR为积雪覆盖监测提供了最长的时间序列,但它也有一些缺点。最重要的一点是,它的空间分辨率太粗,无法进行局部详细研究以量化积雪提供的服务。此外,其光谱分辨率仅支持生成积雪覆盖产品,却不能生成积雪覆盖比例图层。

(三) 机载传感器

虽然在几颗地球观测卫星上有各种各样的传感器,但它们的空间分辨率不够精确,无法监测雪的状态和含水量。为了提高收集积雪覆盖信息质量,机载传感器成为最佳选择。为了提高有关雪水当量、雪深和雪密度信息的质量,在过去的几十年中,已经开发了一些传感器。只要天气条件和预算允许,这些机载传感器几乎可以部署在任何地方。我们将介绍两种最重要的方法:微波和激光成像探测与测距(LIDAR)。还有其他一些相关技术,如高光谱数据(机载可见光、红外光谱仪或机载可见、红外光谱仪,Airborne Visible/Infrared Spectrometer,AVIRIS)和能够探测这种自然辐射的伽马传感器。

微波传感器是最令人关注的机载设备之一。微波可以穿透雪,可以收集雪的含水量(Chang et al.,1982)、雪的范围、深度和干湿状态(Sokol et al.,2003)等信息。根据微波辐射的来源,我们可以区分两种类型。(1)无源微波:积雪下方地面发出的微波强度取决于其结构特征(颗粒大小、密度、垂直分布异质性、雪的结晶状态和温度)。利用这些微波设备,可估计积雪中每单位面积的水质量,即雪水当量。然而,这种方法有几个缺点。其一受限于液态水的存在,其微波辐射掩盖了雪的信号。这使得微波传感器很难探测到此种类型的湿雪;(2)有源微波:这些传感器发射自己的辐射来记录地形的后向散射响应。这种后向散射取决于积雪特性(含水量、粒度大小等)。当有干雪时,地面的反应与裸露地面的反应相似。这使得这两种物质难以分离(Guneriussen,1997)。因此,有源微波对干雪覆盖无用。

机载激光雷达是另一种机载传感器,可以提供有关积雪深度和体积的信息。该技术(也称为激光测高)基于飞行平台上的设备发射的激光脉冲。光线从地面反射到传感器,从而可以建立该区域的高分辨率数字地形模型。要使用激光成像探测与测距方法监测积雪,有必要进行两次调查:第一次无积雪(获得一种基线数字高程模型);第二次有积雪,其积雪深度可被探测(Hopkinson et al.,2004)。然而,也可以使用发射脉冲并用光电二极管检测其反射的地面设备进行地面激光扫描(Prokop,2008)。

三、山区积雪提供的终端服务

(一) 供给服务

积雪覆盖提供的最明显的服务是其储水能力。由于其多孔性,积雪可以储存大量的水。这种储存容量取决于三个因素:每单位积雪的水量、积雪总量(深度和范围)及其在景观中的持久性。

因此,当积雪存在时,它是山区重要的水库。据估计,全球范围内的积雪覆盖面积超过 224 000 平方千米(Shiklomanov,2009)。这些区域可储存超过 2 400 万立方千米的淡水。这些水大部分来自河流并储存在土壤中,作为生态和人类系统的动力引擎。农业是山区积雪供水的最重要消耗者。根据奥基和金江村(2006年),农业每年需要消耗超过 12 000 立方千米的水。

融化的水也"携带"了一些由山脉中的海拔梯度产生的势能。人类利用这种能量来发电。一些发达国家(奥地利、瑞士)利用融水的水力发电站满足其 60%~70% 的电力需求(Lehner et al.,2005)。

(二) 调节和支持服务

雪的高反照率和低热导率是积雪覆盖影响当地和全球气候的主要因素(Rittger et al.,2012)。积雪可以改变低层大气和地表的辐射。这种物理效应对当地、区域和全球气候具有重要影响。积雪覆盖对气候有很大的影响(反之亦然),但这种影响很难量化。积雪对气候的影响可以用气候模型来研究(Vavrus,2007)。这些模型预测大部分地区的积雪覆盖会使对流层的空气明显降温。这种降温效应是 $2\times CO_2$ 情景中预期升温幅度的三分之一。

在过去的几十年里,积雪对气候的局部影响得到了广泛的研究。沃尔什研究(1984)表明,局部温度与积雪之间存在很强的关系。他的结论是,较冷的温度与高于正常水平的积雪有关。实际上,他的研究发现,积雪的存在可能是预测温度的好工具。区域效应也得到了很好的研究。全球环流模型有助于解开积雪与气候之间的复杂关系。一个很好的例子是喜马拉雅山积雪对南亚季风的影响。事实证明,冬季欧亚大陆积雪覆盖与次年夏天印度季风的降水量之间存在反向关系(Barnett et al.,1989;Yasunari et al.,1991;Douville and Royer,1996)。原因在于反照率:积雪覆盖的高反照减少了太阳辐射总量,而太阳辐射总量反过来又减少了大陆地表的加热。这种加热是季风地区暴雨形成的最重要驱动因素(Yasunari et al.,1991)。

在上一节中,我们注意到了积雪覆盖对水文循环的定量影响,但雪的重要性不仅仅在于数量。雪的固态使其在山区的水文循环中起着非常重要的缓冲作用。这意味着积雪覆盖可以保留水分,并根据当地温度以给定的速率将其排出。这种由逐渐融雪引起的排放延迟,增强了含水层的调节能力。当春季积雪逐渐融化时,含水层以一种最佳方式得到了补给。

融雪和含水层补给的结合解释了为什么大部分来自雪流域的水在水文年会显示出更恒定的流量(Manga,1999)。除了这些自然过程之外,人类还开发了一些试图调节水文循环的管理措施。从建造小型水坝储存水到使用灌溉渠道来分配水(Kamash,2012),几个世纪以来,人类一直试图调节山区的水文循环(Liu and Yamanaka,2012)。专栏 15-1 提供了一些人类行为的示例。

专栏 15-1　附属设施:调节地中海地区水文循环的灌溉渠道网络

　　高山积雪引发了一种特殊的水文状况,其水流量和水文过程较缓和。这一特点使人类聚落更容易利用水资源。在半干旱环境中,这些暂时保存在盆地最高部分的积雪资源代表了必要的储水量。在位于伊比利亚半岛东南部的瓜达尔费奥流域,内华达山脉以雪形式储存的水量占全年水资源的70%以上。在该地区,历史上对融雪的使用(可追溯到12世纪的复杂渠道或早期灌溉渠道系统)是地中海环境中看到的更复杂的水调节系统之一。在这里,融雪的再分配以两种方式发生:(1)以地表方式用于下游灌溉区,主要是梯田作物和高山牧场;(2)作为人工地下水补给,通过局部裂缝或硅镁层将水输送至含水层系统。后者是伊比利亚半岛上最早的水调节系统之一(Díaz-Marta,1989)。在这里,融雪的延迟是通过含水层的储放关系实现的。之后,这些水通过不同的来源返回下游,从而在旱季发挥更大的效用(Castillo and Fedeli,2002; Millares et al.,2009)。今天,由于美学和文化价值,出现了与这种人工排水网络相关的新植被和动物群。内华达山脉国家公园排水网络的价值促使公共管理人员在这些渠道的清查、保护和修复方面进行了大量投资。

　　积雪覆盖的支持作用很重要,因为它可以为各种生命形式创造栖息地。其中一些生物有助于创造其他生态系统服务。雪的物理和化学特性创造了适合微生物、动物和植物的栖息地(Jones et al.,2001)。

　　冰川和其他有永久冰雪的环境中,藏有由单细胞藻类、细菌和真菌组成的生态系统(Segawa et al.,2005)。这些生物体已经适应了极端的温度、pH值、辐射水平和缺乏营养的条件(Jones et al.,2001)。由于缺乏液态水,它们甚至会受到干旱的影响。然而,由于雪下环境的热状况,积雪覆盖可能是小动物和植物的合适栖息地。积雪覆盖在干燥、多风、寒冷的大气和下面潮湿、温暖的空气之间起着某种过渡地带的作用。无脊椎动物和小型哺乳动物利用这种热力维护来度过严酷的冬季。积雪覆盖的相关因子也对植物的生命产生了深远的影响。一些研究表明,山地植物的物候(Ostler et al.,1982)和分布模式(Kudo and Ito,1992)主要受融雪日期和积雪持续时间的控制。

> 积雪覆盖对景观结构也有重要影响。事实上,它被认为是影响高山地区植被格局的最重要变量(Jones et al.,2001)。积雪覆盖持续时间和积雪累积量是解释高山地区植物群落结构和组成的关键因素。此外,雪对植被功能有很深的影响。最大积雪量的变化可以解释美国内华达山脉地区森林绿色年际变化的50%以上(Trujillo et al.,2012)。某些积雪覆盖服务的状况影响到某些生态系统(森林)提供服务(初级生产)的能力。

(三) 文化服务

积雪覆盖提供的文化服务很容易说清楚,但由于涉及高度的主观性,最具挑战性的问题是如何量化这些服务。在本节中,我们将列出积雪覆盖提供的一些最重要的文化服务。滑雪项目可能是雪提供的最明显的文化服务。山脉和积雪覆盖吸引了全世界大量的游客。全世界有超过 3 500 个滑雪场,其中大部分位于北美和欧洲。保守估计,全球有超过 6 500 万滑雪者。因此,这项服务的经济影响非常重要,在 2000 年达到 6 210 亿美元(Hudson,2000)。这是影响一些国家当地经济的重要因素。例如,在奥地利,冰雪旅游约占国内生产总值的 18%(GDP;Amelung and Moreno,2009)。

积雪覆盖(和山区)提供的另一项重要服务是风景价值。山脉是世界上最受景仰的景观之一(Beza,2010),积雪覆盖是许多山脉的与生俱来的部分。尽管雪在人类对美的感知中的作用难以量化,但一些研究指出,它是山地景观中最重要的五个特征之一(Clay and Daniel, 2000)。

四、监测积雪覆盖、量化生态系统服务案例研究:内华达山脉生物圈保护区(西班牙)

内华达山脉是一座位于西班牙最南端的地中海山脉。它海拔 3 400 米,被认为是地中海地区最重要的生物多样性热点之一。这座山周围有 50 多个城市区域,总人口约为 50 万人。大部分人口依赖内华达山脉进行灌溉和提供饮用水。这意味着内华达山脉提供的服务对人类福祉非常重要。由于地中海气候的频繁干旱,积雪作为一个基本的蓄水池变得越发重要。内华达山脉拥有一个长期生态研究站点(http://www.ilternet.edu),该站点收集有关全球变化对生态系统影响的信息,并培养其应对此等影响的适应力。

该观测站首先完成的任务之一是建立一个综合系统,以监测不同时空尺度的积雪覆盖,模拟积雪的水文效应,并量化积雪为生态系统和人类社区提供的服务(图15-2),这个系统可以用多层方法来描述。

第一层是一套用于收集内华达山脉的降雪信息的监测流程,包括六种方案,结合了空间、时间和主题等不同研究尺度。其中三个监测方法用于现场收集信息:(1)杆;(2)配备数码相机和超声波传感器、用来测量积雪深度的自动气象站;(3)可用于量化雪水当量和其他物理性质的横断面。其他三种方法使用远程设备收集积雪信息:(1)能够下载、处理和分析2000年至今MODIS积雪产品图像的自动化系统;(2)基于定期处理陆地卫星图像并提取积雪覆盖范围的更精细监测方法;(3)从格拉纳达市(距离内华达山脉30千米)到内华达山脉的山丘,每天都会拍摄倾斜照片,目的是监测春季和初夏的融雪过程。

第一层获得的原始数据被传输到第二层,第二层的主要功能是通过归纳和分析信息来形成有用的知识。该分析层由一组能够模拟雪形成过程的复杂工具构成。我们开发了两种分析工具:(1)最强大的是 WiMMed (Polo et al.,2009)。它是一个分布式水文模型,可以模拟河流中的水流、雪水含量、积雪范围,甚至土壤湿度(见第十四章)。WiMMed 使用陆地卫星和倾斜照片来验证积雪动态。使用雪样带、极点和气象站提供的雪水当量和雪深来校准模型;(2)Linaria 是另一套分析程序,用于下载、处理和分析 MODIS 雪产品提供的所有信息。该系统依靠 Kepler(Altintas et al.,2004)软件来处理美国国家航空航天局提供的所有分级数据(HDF 格式)文件,并产生显示积雪状态的四个重要指标的图表和表格。这些指标包括降雪持续时间、降雪开始日期、积雪融化日期和每个水文年的融化周期数(Wang and Xie,2009)。这些信息通过网络门户对外提供访问(在 http://linaria.obsnev.es 免费注册)。

第三层包括一组方法,应通过使用上述原始和处理过的信息,对积雪提供的服务进行量化。为此,有必要创建混合生态水文模型,模拟不同类型生态系统服务(支持、调节、供应)的数量、质量和时间演变。这一层可以使用的一个很好的

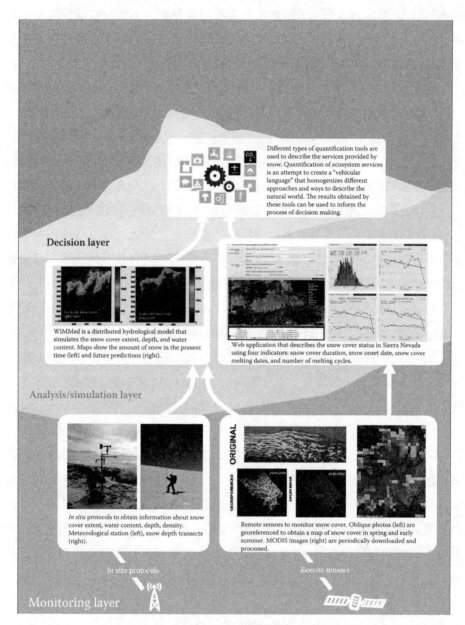

图 15-2(见彩插)

注:示意图显示了一个综合系统的结构,该系统能够使用不同的方法监测积雪,分析原始数据以获得积雪指标,并最终量化积雪提供的服务。该系统正在内华达山脉(西班牙)实施。

模型案例是"生态系统服务的综合评估和权衡（Tallis et al., 2011）"软件。该软件旨在量化生态系统服务，并在多种人类用途之间做出协调。

五、结论

积雪覆盖是塑造山地景观的一个非常重要的非生物元素。尽管不能将其视为生态系统服务提供者（因为它不是一个生态系统），但积雪与几个最终的生态系统服务有关。

遥感和其他技术对于监测积雪覆盖范围、含水量、深度等非常有用。设计良好的监测方案应该结合不同监测技术，从收集含水量信息的实地活动，到评估积雪范围的遥感信息。这对山区尤为重要，因为山区的地形决定了雪在地表的分布。

获取积雪覆盖结构信息只是生态系统服务量化过程中的第一步。有必要在积雪覆盖结构和山区生态系统提供的服务之间建立联系。积雪与几个最终的生态系统服务有关：水的生产可能是最明显的一个，但不是唯一的。另一项供给服务是发电能力。调节和支持服务也是非常重要的。雪的反照率在局部和区域范围内对气候调节有重要作用（见第十六章）。雪的固体状态使其在山区的水文循环中起着非常重要的缓冲作用。积雪可以截留水分，并根据当地温度以一定的速率排出。最后，积雪覆盖被认为是若干物种和生物群落的有利栖息地。一些植物、无脊椎动物，甚至哺乳动物都利用了雪下环境的热状态。雪还影响如物候等生态功能，以及一些植物物种的分布。

但真正需要解决的挑战是对与积雪覆盖有关的生态系统服务的量化。根据雪的上述特性，可以量化它们对生态系统服务的影响。这种量化可以通过使用水文和生态模型来解决。通过监测方法获得的原始数据可以用作这些分布式仿真模型的输入数据。

致谢

本章是在内华达山脉全球变化观测台（http://obsnev.es）的合作框架下编

写的。资助是由生物多样性基金会(西班牙政府)和来自格拉纳达大学的 CEI 生物组织提供。

参 考 文 献

Altintas, I., C. Berkley, E. Jaeger, M. Jones, B. Ludascher, and S. Mock. 2004. Kepler: An extensible system for design and execution of scientific workflows. *Proceedings of the 16th International Conference on Scientific and Statistical Database Management, Greece.* Vol. I. doi:10.1109/SSDM.2004.1311241.

Amelung, B., and A. Moreno. 2009. *Impacts of climate change in tourism in Europe. PESETA-Tourism Study, Luxembourg: Office for Official Publications of the European Communities.* doi:10.2791/33218.

Andreadis, K. M., and D. P. Lettenmaier. 2006. Assimilating remotely sensed snow observations into a macroscale hydrology model. *Advances in Water Resources* 29:872–886.

Barnett, T. P., L. Dümenil, U. Schlese, E. Roeckner, and M. Latif. 1989. The effect of Eurasian snow cover on regional and global climate variations. *Journal of the Atmospheric Sciences* 46:661–685.

Beza, B. B. 2010. The aesthetic value of a mountain landscape: A study of the Mt. Everest trek. *Landscape and Urban Planning* 97:306–317.

Castillo, A., and B. Fideli. 2002. Algunas pautas del comportamiento hidrogeológico de rocas duras afectadas por glaciarismo y periglaciarismo en Sierra Nevada (España). *Geogaceta* 32:189–191.

Chang, A. T. C., J. L. Foster, and D. K. Hall. 1982. Snow water equivalent estimation by microwave radiometry. *Cold Regions Science and Technology* 5:259–267.

Chow, V. T. 1964. *Handbook of applied hydrology.* New York: McGraw-Hill.

Clay, G. R., and T. C. Daniel. 2000. Scenic landscape assessment: The effects of land management jurisdiction on public perception of scenic beauty. *Landscape and Urban Planning* 49:1–13.

Cohen, J., and D. Rind. 1991. The effect of snow cover on the climate. *Journal of Climate* 4:689–706.

Corripio, J. G. 2004. Snow surface albedo estimation using terrestrial photography. *International Journal of Remote Sensing* 25:5705–5729.

Diaz, H. F., M. Grosjean, and L. Graumlich. 2003. Climate variability and change in high elevation regions: Past, present and future. *Climatic Change* 2001:1–4.

Díaz-Marta, M. 1989. Esquema histórico de la ingeniería y la gestión del agua en España. *Revista de obras públicas* 13:8–23.

Dietz, A. J., C. Kuenzer, U. Gessner, and S. Dech. 2012. Remote sensing of snow—A review of available methods. *International Journal of Remote Sensing* 33:4094–4137.

Dingman, L. 2002. *Physical hydrology.* Upper Saddle River, NJ: Prentice Hall.

Douville, H., and J. F. Royer. 1996. Sensitivity of the Asian summer monsoon to an anomalous Eurasian snow cover within the Météo-France GCM. *Climate Dynamics* 12:449–466.

Dozier, J., and T. H. Painter. 2004. Multispectral and hyperspectral remote sensing of alpine snow properties. *Annual Review of Earth and Planetary Sciences* 32:465–494.

Fernandes, R., and H. Zhao. 2008. Mapping daily snow cover extent over land surfaces using NOAA AVHRR imaginery. In *Remote sensing of land ice and snow*, 11–13. The Netherlands: IOS Press.

Green, R. O., and M. Shimada. 1997. On-orbit calibration of a multi-spectral satellite sensor using a high altitude airborne imaging spectrometer. *Advances in Space Research* 19:1387–1398.

Guneriussen, T. 1997. Backscattering properties of a wet snow cover derived from DEM corrected ERS-1 SAR data. *International Journal of Remote Sensing* 18:375–392.

Haines-Young, R., and M. Potschin. 2013. CICES V4.3–Revised report prepared following consultation on CICES Version 4, August–December 2012. EEA Framework Contract No EEA/IEA/09/003.

Hall, D. K., G. A. Riggs, and V. V. Salomonson. 1995. Development of methods for mapping global snow cover using moderate resolution imaging spectroradiometer data. *Remote Sensing of Environment* 54:127–140.

Herrero, J., M. J. Polo, and M. A. Losada. 2011. Snow evolution in Sierra Nevada (Spain) from an energy balance model validated with Landsat TM data. Proc. SPIE 8174, *Remote Sensing for Agriculture, Ecosystems, and Hydrology* XIII, 817403.

Hinkler, J., S. B. Pedersen, M. Rasch, and B. U. Hansen. 2002. Automatic snow cover monitoring at high temporal and spatial resolution, using images taken by a standard digital camera. *International Journal of Remote Sensing* 2:4669–4682.

Hopkinson, C., M. Sitar, L. Chasmer, and P. Treitz. 2004. Mapping snowpack depth beneath forest canopies using airborne LIDAR. *Photogrammetric Engineering & Remote Sensing* 70:323–330.

Hudson, S. 2000. *Snow business: A study of the international ski industry*. London: Cassell.

Hüsler, F., T. Jonas, S. Wunderle, and S. Albrecht. 2012. Validation of a modified snow cover retrieval algorithm from historical 1-km AVHRR data over the European Alps. *Remote Sensing of Environment* 121:497–515.

Irons, J. R., J. L. Dwyer, and J. A. Barsi. 2012. The next Landsat satellite: The Landsat Data Continuity Mission. *Remote Sensing of Environment* 122:11–21.

Jones, H. G., J. W. Pomeroy, D. A. Walker, and R. W. Hoham. (eds.). 2001. *Snow ecology: An interdisciplinary examination of snow-covered ecosystems*. Cambridge, UK: Cambridge University Press.

Kamash, Z. 2012. Irrigation technology, society and environment in the Roman near east. *Journal of Arid Environments* 86:65–74.

Kudo, G., and K. Ito. 1992. Plant distribution in relation to the length of the growing season in a snow-bed in the Taisetsu Mountains, northern Japan. *Vegetatio* 98:165–174.

Laffly, D., E. Bernard, M. Griselin, et al. 2011. High temporal resolution monitoring of snow cover using oblique view ground-based pictures. *Polar Record* 48:11–16.

Lehner, B., G. Czisch, and S. Vassolo. 2005. The impact of global change on the hydropower potential of Europe: A model-based analysis. *Energy Policy* 33:839–855.

Liu, Y., and T. Yamanaka. 2012. Tracing groundwater recharge sources in a mountain-plain transitional area using stable isotopes and hydrochemistry. *Journal of*

Hydrology 464–465:116–126.

Manga, M. 1999. On the timescales characterizing groundwater discharge at springs. *Journal of Hydrology* 219:56–69.

Martiny, N., R. Santer, and I. Smolskaia. 2005. Vicarious calibration of MERIS over dark waters in the near infrared. *Remote Sensing of Environment* 94:475–490.

Millares, A., M. J. Polo, and M. A. Losada. 2009. The hydrological response of baseflow in fractured mountain areas. *Hydrology and Earth System Science* 13:1261–1271.

Oki, T., and S. Kanae. 2006. Global hydrological cycles and world water resources. *Science* 313:1068–1072.

Ostler, W. K., K. T. Harper, K. T. McKnight, and D. C. Anderson. 1982. The effects of increasing snowpack on a subalpine meadow in the Uinta Mountains, Utah, USA. *Arctic and Alpine Research* 14:203–214.

Parajka, J., and G. Blöschl. 2008. The value of MODIS snow cover data in validating and calibrating conceptual hydrologic models. *Journal of Hydrology* 358:240–258.

Parajka, J., P. Haas, R. Kirnbauer, J. Jansa, and G. Blöschl. 2012. Potential of time-lapse photography of snow for hydrological purposes at the small catchment scale. *Hydrological Processes* 26:3327–3337.

Polo, M. J., J. Herrero, and C. Aguilar. 2009. WiMMed, a distributed physically-based watershed model (I): Description and validation. In *Environmental hydraulics: Theoretical, experimental and computational solutions, IWEH09*, ed. P. A. López-Jiménez, 225–228. London: CRC Press.

Powell, C., L. Blesius, J. Davis, and F. Schuetzenmeister. 2011. Using MODIS snow cover and precipitation data to model water runoff for the Mokelumne River Basin in the Sierra Nevada, California (2000–2009). *Global and Planetary Change* 77:77–84.

Prokop, A. 2008. Assessing the applicability of terrestrial laser scanning for spatial snow depth measurements. *Cold Regions Science and Technology* 54:155–163.

Rallison, R. E. 1981. Automated system for collecting snow and related hydrological data in mountains of the Western United States. *Hydrological Sciences* 26:83–89.

Rittger, K., T. H. Painter, and J. Dozier. 2012. Assessment of methods for mapping snow cover from MODIS. *Advances in Water Resources* 51:367–380.

Rosenthal, W., and J. Dozier. 1996. Automated mapping of montane snow cover at subpixel resolution from the Landsat Thematic Mapper. *Water Resources Research* 32:115–130.

Salomonson, V. V., and I. Appel. 2006. Development of the Aqua MODIS NDSI fractional snow cover algorithm and validation results. *IEEE Transactions on Geoscience and Remote Sensing* 44:1747–1756.

Sanderson, E. W., M. Jaiteh, M. A. Levy, K. H. Redford, A. V. Wannebo, and G. Woolmer. 2002. The human footprint and the last of the wild. *BioScience* 52:891–904.

Segawa, T., K. Miyamoto, and K. Ushida. 2005. Seasonal change in bacterial flora and biomass in mountain snow from the Tateyama Mountains, Japan, Analyzed by 16S rRNA gene sequencing and real-time PCR. *Applied and Environmental Microbiology* 71:123–130.

Shiklomanov, I. A. 2009. *The hydrological cycle*. Vol. I. St. Petersburg, Russia: EOLSS Publishers.

Silvertown, J. 2009. A new dawn for citizen science. *Trends in Ecology and Evolution* 24:467–471.

Sokol, J., T. J. Pultz, and A. E. Walker. 2003. Passive and active airborne microwave remote sensing of snow cover. *International Journal of Remote Sensing* 24:5327–5344.

Tallis, H. T., T. Ricketts, A. D. Guerry, et al. 2011. InVEST 2.4.3 user's guide: Integrated valuation of environmental services and tradeoffs. A modeling suite developed by the Natural Capital Project to support environmental decision-making. Stanford, CA: The Natural Capital Project.

Thirel, G., C. Notarnicola, M. Kalas, et al. 2012. Assessing the quality of a real-time snow cover area product for hydrological applications. *Remote Sensing of Environment* 127:271–287.

Trujillo, E., N. P. Molotch, M. L. Goulden, A. E. Kelly, and R. C. Bales. 2012. Elevation-dependent influence of snow accumulation on forest greening. *Nature Geoscience* 5:705–709.

Vavrus, S. 2007. The role of terrestrial snow cover in the climate system. *Climate Dynamics* 29:73–88.

Vikhamar, D., and R. Solberg. 2003. Subpixel mapping of snow cover in forests by optical remote sensing. *Remote Sensing of Environment* 84:69–82.

Walsh, J. E. 1984. Snow cover and atmospheric variability. *American Scientist* 72:50–57.

Wang, D., D. Morton, J. Masek, et al. 2012. Impact of sensor degradation on the MODIS NDVI time series. *Remote Sensing of Environment* 119:55–61.

Wang, X., and H. Xie. 2009. New methods for studying the spatiotemporal variation of snow cover based on combination products of MODIS Terra and Aqua. *Journal of Hydrology* 371:192–200.

Watson, F. G. R., T. N. Anderson, and W. B. Newman. 2008. Modeling spatial snow pack dynamics. *Terrestrial Ecology* 7961:18–23.

Yasunari, T., A. Kitoh, and T. Tokioka. 1991. Local and remote responses to excessive snow mass over Eurasia appearing in the northern spring and summer climate. A study with the MRI-GCM. *Journal of the Meteorological Society of Japan* 69:473–487.

第五部分

地表能量平衡相关生态系统服务

第十六章 气候调节服务的表征与监测

D. 阿尔卡拉斯-塞古拉   E. H. 博柏利   O. V. 穆勒   J. M. 帕雷洛

一、简介

(一) 气候调节生态系统服务

大气成分和气候调节组(Haines-Young and Potschin, 2013)相关的生态系统服务,与维持有利于健康、作物生产和其他人类活动等全球和地方性气候条件之间有着密切联系。在全球范围内,生态系统的生物地球化学过程通过向大气排放/吸收温室气体和气溶胶来影响气候。森林获得并储存二氧化碳,沼泽地、湖泊、稻田和牧场则释放甲烷。生态系统的生物物理属性,如反照率、潜热和感热,会影响当地或区域的温度、降水和其他气候因素(Pielke et al., 2002; Bonan, 2008; Oki et al., 2013)。例如,热带森林中入射太阳辐射中部分用于水的蒸散(即潜热通量),从而降低地表温度(Bonan, 2008)。此外,森林的蒸散有利于云层形成,成为当地气候的一部分,同时也有助于保持空气质量。在热带沙漠中,辐射使土壤加热,进而使空气升温(即感热通量)。

尽管生态系统通过生物地球化学(如温室气体交换)和生物物理学(如水和能源平衡)进行气候调节,但目前政策只重点关注生物地球化学方面的影响(如二氧化碳排放)。最近,安德森-蒂克西拉等人(Anderson-Teixeira et al., 2012)提出了气候调节价值(Climate Regulation Value, CRV)指数,该指数反映了影响生态系统-气候服务价值的生物地球化学和生物地球物理的生态系统特征。CRV 将生物物理效应转化为生物地球化学单元。因此,CRV 提供了满足当前全球政策和碳市场拓展所需考虑的一系列气候调节服务的可能性。CRV 的生物物理部分是根据生态系统的地表净辐射和潜热通量估算出来的,主要使用陆面模式进行模拟,例如生物圈集成模拟器(IBIS)(Foley et al.,1996;Kucharik et al.,2000)或诺亚陆面模型(Noah LSM)(Chen et al.,1996;Chen and Dudhia,2001;Ek et al.,2003)。一般而言,这些模拟主要涉及使用与植被相关的变量,如叶面积指数、气孔阻抗、根系深度、可见光和近红外的反射率和透射率、热量、水量、雪量等。本章提出了一种利用卫星信息和耦合气候区域模式来估算和监测这些变量的方法。

(二) 生态系统与气候间的响应和反馈

气候是生态系统结构和功能的主要区域驱动力,决定了系统中可用能量(热量和太阳辐射)以及水的节奏和数量(Stephenson,1990)。反之,生态系统也通过多种途径影响气候,主要是通过反照率、长波辐射、表面粗糙度、蒸散量、温室气体或气溶胶的变化来确定地表和大气间的能量、动量、水和化学平衡(Chapin et al.,2008)。因此,自然和人类对生态系统的影响可能改变生态系统与气候间反馈的一条或多条路径,最终给区域和全球气候带来影响。事实上,一些研究(Weaver and Avissar,2001;Pielke et al.,2002;Werth and Avissar,2002;Kalnay and Cai,2003)已得出结论,土地利用变化对气候变化的贡献率约占全球变化总量的10%,但土地利用变化在区域上的相对贡献则更为显著,甚至超过温室气体排放的贡献。已知的案例表明,土地利用变化可能会改变区域气候,例如罗马时期地中海盆地的干旱化(Reale and Dirmeyer,2000;Reale and Shukla,2000),或毁林后亚马孙水文气象的变化(Gedney and Valdes,2000;Roy and Avissar,2002)。

气候条件的年际变化会给植被结构和功能特性带来显著影响(Brando et al.,2010;Zhao and Running,2010),进而可能影响区域气候。例如,干旱造成的植被密度下降导致反照率的增加,进一步可能引起对流抬升运动和水汽平流运动的减少(Bonan,2008)。虫灾爆发(Maness et al.,2013)和过度放牧会进一步降低植被密度,从而加剧这些影响,进而增强反照率引起的对流抬升运动的减少。与此同时,南美洲广大地区正遭受着人为改造土地覆盖和作物系统管理等人类实践活动所带来的影响,这些变化可能会影响生态系统与气候间的反馈(Foley et al.,2003)。其中,为了推动农业和畜牧业发展而进行森林砍伐与土地开垦是最重要的影响因素(Foley et al.,2007;Volante et al.,2012)。土地开垦会产生以下影响:①反照率的增加,导致向生态系统(及随后向大气层)的能量传递的减少;②蒸腾作用的减少,导致从土壤和地表含水层向大气输送水分的减少;③二氧化碳的净释放,导致大气吸热能力的增强。反之,南美洲其他大面积的土地利用变化,如草地造林(Nosetto et al.,2005),会导致:(1)反照率的降低,进而导致向生态系统能量传递的增强;(2)蒸散量的增加,使得从土壤和地表含水层向大气输送水分的增加;(3)表面粗糙度的增加(Beltranán-Przekurat et al.,2012)。然而,生态系统与气候间反馈的其他例子来自免耕农业的广泛实践或旱地灌溉农业的大范围推广(De Oliveira et al.,2009),这导致蒸散量的增加并降低了反照率。

生态系统与气候间的反馈和响应既是模拟气候系统中的陆地与大气相互作用中面临的核心问题(Mahmood et al.,2010),也是其他生物和环境等诸多课题面临的核心问题(Oki et al.,2013)。如前文所述,生态系统与大气间的相互作用和反馈,取决于底层土地覆盖的物理性质。因此,生态系统结构和功能的变化将对这些物理性质产生影响,从而影响地表的辐射平衡及动量、热量、水分和其他气体/气溶胶物质间的交换。

质量和能量的交换至少可通过两种机制来改变:一是通过人类活动改变地表条件(土地覆盖和土地利用变化);二是通过气候的自然变化来影响生态系统的健康、性能和生物物理属性,例如区域气候的低频调节,旱季与雨季的交替等。

正如气候影响生态系统一样,反之,生态系统影响气候也很常见。发生土地覆盖变化区域的大小、地理位置和斑块分布可以决定其对当地、区域乃至全球气候的影响程度(Marland et al.,2003;Pielke et al.,2007)。土地覆盖小范围的变化(例如10千米左右)会导致局部降水模式和强度的变化(Pielke et al.,2007),但在频繁发生大雷暴的热带地区,其造成的影响将会扩大,甚至可能升级到全球范围(Pielke,2001;Werth and Avissar,2002)。在美国,通过开展观测和数值研究,表明了土地覆盖变化和灌溉技术的推广会给区域气候和植被带来影响(Stohlgren et al.,1998;Baidya Roy et al.,2003;Diffenbaugh,2009)。

(三) 生态系统与大气间相互作用建模

生态系统与气候间的反馈是气候系统中陆地与大气相互作用建模所需要探

讨的核心问题。但是将陆地与大气相互作用和反馈纳入当前区域和全球环流模型并非易事。目前存在一些方法是通过使用土地覆盖图来估算生物物理性质图谱(West et al.,2011)。这种估算依赖于特定的植物功能性状(如具有常绿、落叶或一年生生命形式)与不同生态系统功能性质间的关系(Smith et al.,1997)。然而,对于这些功能性状是如何决定生态系统和生物地球化学循环的升尺度效应的相关机理尚不完全清楚(Lavorel et al.,2007)。事实上,现有一些研究已经表明,植物功能类型分类在预测生态系统功能方面是并不可靠的(Wright et al.,2006;Bret-Harte et al.,2008)。首先,植物功能性状和特性通常只能通过有限的观察来确定;其次,尽管在现实世界中它们在植物功能类型内或在植物功能类型之间是存在差异的,但是在设置模型时通常假设其在植物功能类型或土地覆盖范围内是恒定的(例如,即使只在少数地点进行了观测,并且观测结果可能有所不同,但是会存在针叶林在世界任何地方都具有相同且恒定特性的假设)(Reich and Oleksyn,2004;Wright et al.,2004,2005,2006;Reich et al.,2006)。同时,这些土地覆盖图很难每年更新,而植被的一些结构特征(如叶片寿命等)对环境变化并不敏感(McNaughton et al.,1989)。总的来说,对于植被特性的简化表示很有可能导致响应延迟,并降低模型描述中包括土地使用变化、火灾、洪水(见第十七章)、干旱和虫灾爆发等在内的快速应变能力。因此,非常有必要改进陆面模式中植被空间和年际动态变化方式,以便进一步说明土地利用/覆盖变化对环流模式的影响。

由于包含碳循环的动态植被模型考虑到了植被变化,并顾及地表过程有了更进一步的假设,在生态系统与大气相互作用的领域有了显著改进。然而,诸多陆面模型并没有考虑生态系统的概念,这是当前用于业务预报的中等复杂度模型(如诺亚 LSM 模型)普遍存在的情况,这些模型具有恒定或静态植被类别,并带有查找表,以便标识其相应生物物理属性的值或年度循环周期。更复杂的模型是采用土地覆盖分类,根据植物的功能类型组成来识别地表斑块。植物功能类型是指具有相似功能特征的种群,例如叶片寿命、代谢途径或固氮等(Smith et al.,1997)。在现实中,被假定恒定不变的土地覆盖类型可能会经历重大变化。例如,典型植被类型在丰水期的生物物理属性与在干旱时期应该是存在巨大差异的。同样的情况也发生在处于强降雨的异常时期,会形成诸多池塘或洪水。但需要注意的是,对于一个规定了每年地表性质年内周期变化都相同的模型而言,所有上述情形下,在陆地与大气相互作用、辐射收支及地表水、能量和碳循环等方面却都别无二致。

描述生态系统尺度上生物群和大气之间能量和物质交换的植被功能属性(Valentini et al.,1999;Virginia et al.,2001)可能有助于满足这些需求,因为相较于结构变化,它们对环境变化的响应更快(McNaughton et al.,1989)。此外,许多生态系统功能性质使用基于卫星遥感数据生成的光谱指数进行监测会相对更容易。归一化差异植被指数是最广泛用于描述和监测生态系统功能的光谱指数,该指数与被绿色植被吸收的入射有效光合辐射的比例密切相关(Sellers et al.,1986;Tucker and Sellers,1986),因此与初级生产(Monteith,1972)密切相关。初级生产是生态系统功能最具综合性的指标(Virginia et al.,2001)。归一化差异植被指数已被广泛、可靠地用于描述植被动态的年内和年际变化(Alcaraz-Segura et al.,2009),监测生态系统结构和功能的变化(Pettorelli et al.,2005),探测植被生长和物候的长期趋势(Kathuroju et al.,2007),为初级生产模型提供输入(Cao et al.,2004),并为全球碳平衡建模提供参考(Potter et al.,2005)。

衍生于归一化植被指数的植被拦截辐射的功能描述因子,也被用于生态系统功能类型(EFTs)分类,这些生态系统功能类型表征了生态系统功能的空间异质性(Paruelo et al.,2001)。EFT被定义为以共同的方式与大气交换质量和能量的地表斑块,显示出对环境因素的协调和特定响应(Soriano and Paruelo,1992;Valentini et al.,1999;Paruelo et al.,2001)。因此,相较于与更传统的基于植物功能类型的自下而上的土地覆盖类型方法,EFT可被认为是一种自上而下的方法,从更高分类层级对生态系统进行分类。阿尔卡拉斯-塞古拉等人(Alcaraz-Segura et al.,2006)修改了帕鲁埃洛等人(Paruelo et al.,2001)基于类之间的固定界限来定义EFT的方法,提供了一种通过时间维度来监测和比较生态系统运行状态的可能性。

本章讨论了EFT的年际变化及其相应的生物物理属性。然后,提出了一种使用具有相应特性的时变EFT代替区域模式中传统常规的土地覆盖类型的方法。我们将分析重点放在南美洲南部,因为它受到厄尔尼诺南部涛动现象引发气候条件年际变化的强烈影响,并且在过去几十年中地表覆盖遭受翻天覆地的变化,这为利用卫星信息评估气候变化和土地变化对植被生物物理属性的影响提供了特殊机会。为此,我们首先利用 NDVI 动态监测和比较的三个指标制作了 1982~1999 年的年度 EFT 地图,然后,基于美国地质调查局(USGS)土地覆盖类型的诺亚 LSM 参数化,获得了每种 EFT 的生物物理属性。最后,我们通过案例研究,说明使用 EFT 能更真实地表示下边界条件对区域气候模拟的影响。

二、生态系统服务类型的识别

(一) 卫星数据记录

EFT 的识别是基于土地长期数据记录团队所制作的 NDVI 数据集(Pedelty et al., 2007)。NDVI 由可见的红色和近红外波长的反射率计算得到(Tucker and Sellers, 1986),即: $NDVI = (NIR - R)/(NIR + R)$, NIR 和 R 分别代表近红外和红色区域的光谱反射率。激光目标指示器与测距仪(LTDR)是来源于先进超高分辨率辐射计档案文件的最新 NDVI 数据集。LTDR 是美国宇航局资助的一个 REASoN 项目(研究、教育和应用解决方案网络项目),旨在从先进超高分辨率辐射计、中分辨率成像光谱仪和可见光/红外成像辐射计(VIIRS)传感器中产生一致且长期的数据集。LTDR 项目通过机载导航雷达项目 AVHRR Land II(PAL-II)中确定的预处理改进及 MODIS 预处理步骤中使用的大气和双向反射分布函数(Bidirectional Reflectant Distribution Function, BRDF)校正(http://ltd.nascom.nasa.gov)(Pedelty et al., 2007),对 1981 年至今的全球覆盖(GAC)档案进行再处理。第 2 版 LTDR 数据集由 1981~1999 年期间的每日全球影像组成,空间分辨率为 $0.05° \times 0.05°$(赤道处约 25 平方千米)。图像已针对传感器退化、水和臭氧蒸汽吸收、瑞利散射和传感器间差异进行了校正。在本研究中,我们计算了 15 天的最大值合成值(Holben, 1986),以尽量减少日常图像中由于云层覆盖、云层阴影和气溶胶污染产生的干扰(尽管剩余干扰可能仍会残留在时间序列中; Nagol et al., 2009)。我们还使用质量评估信息来淘汰低质量 NDVI 值(即,无效的 NDVI 通道、大太阳天顶角、存在太阳耀斑、云层阴影、多云或部分多云)(更多细节见 Alcaraz-Segura et al., 2010, 2013)。

(二) 生态系统服务类型的定义

首先,我们将 LTDR 质量评估标记信息中掩码为水(在所有年份的每组 15 天中超过 80%)的所有像素分类为水体。按照阿尔卡拉斯-塞古拉等人(Alcaraz-Segura et al.,2006)引入的方法,将剩余像素分类为 EFT,其中,EFT 采用固定的类间界限加以识别,这样可进行年际比较。为此,我们使用了三种 NDVI 季节动态度量指标,来反映 NDVI 时间序列的主要变化(Paruelo et al.,2001;Alcaraz-Segura et al.,2009):NDVI 年均值(NDVI-m)、季节变异系数(CV_{seas},即年内标准差除以年平均值)和 NDVI 最大值出现的日期(DMAX)。NDVI-m 是初级生产力的线性估计值,CV_{seas} 是季节性指标(即生长季和非生长季碳收益的年内变差),DMAX 是生长季节的物候指标。

然后,将每个 NDVI 度量的取值范围划分为四个固定区间,给出了 $4 \times 4 \times 4 = 64$ 个 EFT 的潜在数目。就 DMAX 而言,四个区间与温带生态系统中的四季一致。就 NDVI-m 和 CV_{sea} 而言,三个固定限值由每个变量获得的 18 个直方图中的第一、第二和第三四分位数的 18 年中位数计算而得。我们根据帕鲁埃洛等人(Paruelo et al.,2001)提出的术语,依据两个字母和一个数字(三个字符)为每个 EFT 分配代码。代码的第一个字母(大写)对应 NDVI-m 水平,范围从"A"到"D",表示 NDVI-m 由低至高。第二个字母(小写)显示了季节性 CV,CV 从高到低由"a"到"d"范围表示。数字表示 NDVI 最大值出现的季节(见表 16-1)。EFT 的定义和编码仅基于对生态系统功能的描述,并允许对图例直接进行生态逻辑解释。最后,我们按照前面的过程确定每年的固定限值,由此我们可比较 EFT 分布在不同年份间的变化情况。为了绘制 1982~1999 年期间 EFT 平均分布图,我们计算了与每个像素 EFT 直方图的 18 年中位数相对应的 EFT。

表 16-1 LTDR/NDVI 衍生性状的范围

	功能代码	下限	上限
NDVI-m	A	0.000 1	0.347 6
	b	0.347 6	0.518 6
	c	0.518 6	0.583 4
	d	0.583 4	0.889 1

续表

	功能代码	下限	上限
CVseas	d	0.000 0	0.138 8
	c	0.138 8	0.181 3
	b	0.181 3	0.230 7
	A	0.230 7	3.897 4
DMAX	1	春天	—
	2	夏天	
	3	秋天	
	4	冬天	

注：1982～1999年南美洲生态系统功能类型定义中使用的范围特征：NDVI年平均值（NDVI-m）、季节变异系数NDVI（CVsea）和最大NDVI出现的日期（DMAX）。大写字母对应NDVI-m水平，范围从"A"到"D"（表示NDVI-m由低至高）。小写字母显示季节变异系数，由CVsea的"a"到"d"范围表示从高到低。数字表示NDVI最大值出现的季节。

在研究区域中，几乎所有可能的NDVI-m、CVseas和DMAX组合都得到确定（图16-1）。

（三）南美洲南部的生态系统服务类型

图16-1（d）所示的EFT图简明扼要地显示了1982～1999年期间南美洲温带生态系统运行的空间格局。EFT地图整体反映了三个生态系统运行综合指标的空间分异特征：生产力（NDVI-m，图16-1（a））、季节性（CVseas，图16-1（b））和物候（DMAX，图16-1（c））。研究区域内几乎所有可能NDVI-m、CVseas和DMAX的组合都被确定。一般而言，大多数生态系统在秋季和夏季显示出NDVI最大值（图16-1（c））。夏季极大值的EFT具有中低生产力和高季节性，而秋季和春季极大值的EFT具有生产力和季节性的大部分可能组合。冬季NDVI最大值的EFT表现出非常低或非常高的生产力，季节变异值很低。EFT的定义和编码使得人们可以根据与生产力（NDVI-m）、季节性（CVsea）和物候（DMAX）相关的三个NDVI指标对图例进行生态解释。最大的NDVI-m（D）出现在阿尔托帕拉纳大西洋森林和亚马孙西南部潮湿森林（图16-1（a））。本章中的所有空间参考都基于世界野生动物基金会（WWF）的世界生态区域地

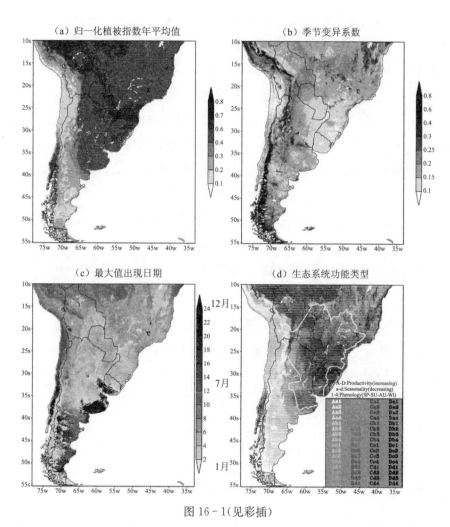

图 16-1（见彩插）

注：基于 LTDR/NDVI 动态的南美洲生态系统功能类型分布及相对扩展。该图显示（a）NDVI 年平均值（NDVI-m）、（b）季节变异系数 NDVI（CVseas）、（c）NDVI 最大值出现的日期（DMAX）和（d）生态系统功能类型（EFT）与 1982～1999 年 17 年间的中位数相对应。EFT 命名代码的解释见表 20-1。

图（Olson et al.，2001）。最低的 NDVI-m（A）发生在阿塔卡马-塞丘拉沙漠、中部安第斯普纳高原的干燥地区和巴塔哥尼亚草原的部分中部地区（图 16-1（a））。整个研究区域的 CVseas 相对较低（c,d）（图 16-1（b））。最大的季节性（a）发生在（1）南部安第斯山脉（瓦尔迪维亚温带森林的最高部分）和（2）北部安第斯山脉(玻利维亚山地干林和中部安第斯山脉湿草原)的最高海拔(图16-1(b))。阿根廷中部和西北部(横跨埃斯皮纳尔和干旱的查科地区)、巴西东部(横跨卡廷加和大西洋干燥森林)的农业用地也存在高季节性(a,b)(图16-1(b))。在(1)最干燥的生态区(如阿塔卡马沙漠和巴塔哥尼亚草原)和(2)非常潮湿的生态区(如阿尔托帕拉纳大西洋森林和亚马孙河西南部的潮湿森林)观察到最低的季节性(d)(图16-1(b))。生长季节的物候指标DMAX显示,温带南美洲大部分地区夏季(2)和秋季(3)NDVI最大值(图16-1(c))。春季(1)极大值仅出现在湿润的潘帕斯和乌拉圭大草原的东南部。冬季(4)极大值很少,主要在智利最北部的马托拉尔和亚马孙河西南部的潮湿森林(图16-1(c))。

三、生态系统服务类型的生物物理属性

(一) 生态系统服务类型陆面参数化

为了获得每个EFT的地表属性,首先,我们使用1992年USGS的1千米像素分辨率的全球土地覆盖图(见http://edc2.usgs.gov/1KM/1kmhomepage.php;Eidenshink and Fauneen,1994)和诺亚陆面模式物理性质表,制作了15幅1992年地表参数地图。然后,将之前15幅基于USGS数据绘制的地图在空间上与1992年的EFT分类相叠加,以计算每个EFT的各地表属性的空间均值。EFT的定义能够捕捉同一土地覆盖类型内生态系统功能的差异。例如,通过初级生产力变动幅度的差异,将密植灌木丛与开阔灌木丛或灌溉农田区分开来。为了揭示EFT方法在地表属性中引入的空间分异性,我们计算了基于USGS数据的属性图和基于EFT的属性图间的相对差异,计算方式为:[(USGS-EFT)/USGS](图16-2)。

(二) 基于美国地质调查局数据与生态系统服务类型的生物物理属性间的对比

所有地表属性的大尺度区域模式在USGS和EFT衍生的图谱之间是相似的(图16-2)。空间自相关是一种生态变量的内在特性(Legendre,1993),它在由EFT导出的图谱中要比在USGS所导出的图谱中稍强(EFT和USGS图谱

中所有属性的全局莫兰指数均值分别为 0.82 和 0.72，P 值<0.05，n=335，534）。这主要是因为相较于基于 USGS 数据的属性图易发生突变，基于 EFT 的属性图变化更为平缓。这两种方法在某些属性上的空间差异比其他属性更大，如：图 16-2(a)、(b)、(e)、(f)、(i) 和(j)，最小(Z_{0m})和最大(Z_{0x})表面粗糙度平均绝对差值分别为 69% 和 53%；图 16-2(c)、(d)、(g)、(h)、(k) 和(l)，最小和最大反照率的平均绝对差值分别为 16% 和 17%。由于 EFT 性质是以空间平均值计算的，其副作用是 EFT 方法倾向于减小生物物理属性的取值范围。例如，由于空间平滑，EFT 方法增加了接近零的变量值，如无植被和几乎无植被区域的最小和最大表面粗糙度（图 16-2(i) 和(j)）。同样导致了在此类相同的地区（未显示）的绿色植被覆盖度、叶面积指数和根系深度的低值增加较平缓，在植被密集的亚热带地区，最大和最小反照率增加也较平缓（图 16-2(k) 和(l)）。另一方面，EFT 方法倾向于降低非常高的值，如植被密集地区的最小和最大表面粗糙度（如图 16-2(i) 和(j)），及植被覆盖稀疏地区的最小和最大反照率（图 16-2(k) 和(l)）。大多数地区的平均差异低于 30%，但有部分特定区域重复集中了大多数属性（未显示）上的较大差异（大于 70%），如巴塔哥尼亚南部草原的灌丛、安第斯山脉南部和中部的干旱和半干旱地区、瓦尔迪维亚温带森林和塞拉多生态保护区的农业区。在埃斯皮纳尔和 Low Monte 生态区及西南亚马孙潮湿森林间的过渡带中存在平均差异（生态区名称遵循奥尔森等人的说法（Olson et al.，2001））。

（三）植被特性的年际变化

我们还根据 1982~1999 年 EFT 分布的年际变化和前述每个 EFT 的陆面参数化结果，描述了地表性质的年际变化（图 16-3）。计算每个地表性质的变异系数（标准差除以平均值；未显示）和四分位距除以中位数[（第三四分位数-第一四分位数)/中位数×100]，作为年际差异的相对指标。

有些地表性质在整个研究区表现出比其他更大的年际变化（图 16-3）。最小和最大表面粗糙度、气孔阻抗和最小叶面积指数的年际变化较大（分别为 34%、28%、27% 和 23%）（图 16-3(a) 至(d)）。最大和最小辐射率和辐射应力（低于 6%）的年际变化较小（图 16-3(e) 至(g)）。根系深度、最小和最大背景反照率、绿色植被覆盖度和最大叶面积指数表现出中等变异性（未显示）。平均而

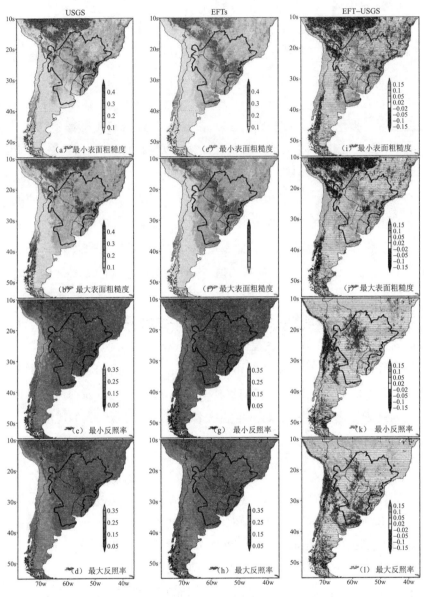

图 16-2(见彩插)

注:选定的生物物理属性:最小表面粗糙度(Z0m)、最大表面粗糙度(Z0x)、最小反照率(ALBm)和最大反照率(ALBx),根据 1982~1999 年 EFT 中值图(左列,(a)~(d))估算;从 USGS 地表覆盖区域(中位列,e~h)及其相对差异[(USGS-EFT)/USGS](右列,i~l)中获取。

第十六章 气候调节服务的表征与监测

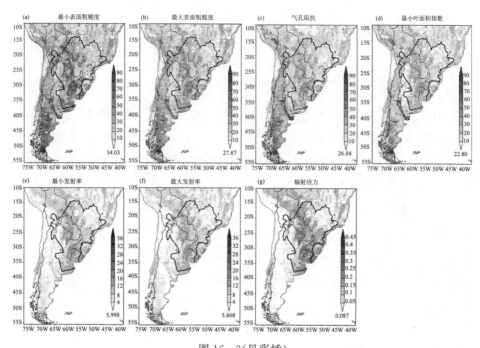

图 16-3（见彩插）

注：基于 1982~1999 年 EFT 分布的年际变化，南美选定生物物理属性的年际变化（IQR/M）。IQR/M =（第三四分位数-第一四分位数）/中位数 × 100。(a)最小表面粗糙度（Z0m）；(b)最大表面粗糙度（Z0x）；(c)气孔阻抗；(d)最小叶面积指数（LAIM）；(e)最小辐射率（Em）；(f)最大辐射率（Ex）；(g)辐射应力（Rad 应力）。每个组底部的数字表示整个区域（空间）的平均年际变化。

言，整个研究区域在所有生物物理属性上的年际变异系数相对较低（13%）（未显示）。然而，一些地区在所有性质上都反复表现出很高的年际变率。巴塔哥尼亚草原半干旱区与南部、西部和东北部湿润区间的边界即交错带年际变化最大（大于 60%）。在(1)巴塔哥尼亚草原和科迪勒拉山系的安第斯山脉的半干旱地区；(2)玻利维亚东南部至乌拉圭的西北—东南横剖面；以及(3)巴西大西洋高原（巴西东部的一个山脉，在圣弗朗西斯科河和大西洋海岸间的南北方向）观察到较高的年际变化（在 30% 和 60% 之间）。

四、区域模拟中的气候调节服务

(一) 区域气候模式中生物物理属性变化的核算分析

为了研究生态系统生物物理属性的年际变化对区域气候的影响，以 2008 年南美拉普拉塔盆地干旱事件为例，将常规固定土地覆盖类型的模型模拟结果与使用 EFT 作为下边界条件的模型模拟结果进行了比较。该事件的细节将在下一节中给出。

使用与诺亚 LSM 耦合的天气研究和预测（Weather Research and Forecasting，WRF）模型进行长时模拟。WRF 模型是一个中尺度天气预报系统，旨在满足大气研究和业务预报的需要。它为区域和全球应用提供米到数千千米尺度的服务。WRF 模式通过求解完全可压缩的非静力方程来模拟大气的行为。诺亚 LSM 具有四个土壤层，自上而下厚度为 10、30、60 和 100 厘米（总深度为 2 米），并且包括根区、植群、每月植被覆盖度和土壤质地等表示。诺亚 LSM 能够解决表面能和水平衡，为边界层提供地表状况条件，如水分和热量通量。

该耦合模型在南美洲南部，即跨越 40°S 和 12°S、77°W 和 42°W 的范围内使用。该区域包括太平洋沿岸的安第斯山脉和巴西海岸的巴西高原等重要地形地貌特征。与低地和平原构成了完整的地形。在水文方面，主要特征是拉普拉塔流域的存在，它是南美洲流量第二大河，对该地区发展具有巨大经济价值。模型可配置为 18 千米×18 千米的空间分辨率，大气垂直层次有 28 层，土壤深度有 4 层，时间步长为 90s。使用国家环境预测中心或者国家大气研究中心（National Centers for Environmental Prediction/National Center for Atmospheric Research，NCEP/NCAR）的再分析项目（NNRP）数据集（Kalnay and Cai，2003）作为初始条件和 6 小时侧向边界条件。

耦合模型中土地覆盖类型的定义是诺亚 LSM 的一部分。具体而言，诺亚 LSM 使用的地图中每个象元表示一种土地覆盖类型。然后，通过查找表为每种土地覆盖类型赋予了相关的 15 个生物物理属性值，包括：绿色植被覆盖度、根系深度、气孔阻抗、辐射应力函数中使用的参数、蒸汽压力差函数中使用的参数、100% 积雪覆盖的阈值、水当量、积雪深度、深雪最大反照率上限、全年最小和最大叶面积指数、全年最小和最大背景反照率及全年最小和最大背景粗糙度长度。该模型提供了 USGS 和国际地球生物圈计划（IGBP）两种土地覆盖分类方法。根据 1992/1993 年 AVHRR(Eidenshink and Faunden, 1994)资料, USGS 有 27 个类别; 根据 2001 年 MOD12Q1 产品(Friedl et al., 2010)信息, IGBP 有 20 种土地覆盖类型。

为了评估新的生物物理属性数据集的性能及其对地表状况的影响, 将使用常规 USGS 土地覆盖图的模型模拟结果与使用 2008 年 EFT 图的模拟结果进行比较, 其相关属性作为下边界条件。2008 年的 EFT 图来源于 2001~2009 年期间的 MODIS MOD13C1 产品, 阿尔卡拉斯-塞古拉等人(Alcaraz-Segura et al., 2013)提供了全部细节。该过程与前面解释的 1982~1999 年期间使用 LTDR NDVI 的过程(在十六章第二节第三部分)相同。

对这两组进行五次模拟, 每次模拟都选择在连续的几天时间中进行, 以生成两个被标识为 WRF-USGS 和 WRF-EFT 的集合。WRF-USGS 集合使用传统的 USGS 土地覆盖类别来表示植被。WRF-EFT 集合使用了 2008 年的 EFT 图及其相关性质。本实验旨在帮助理解使用现实植被数据的可能优势及模型对这种信息的敏感性。该假说认为, 由于极端事件或人类活动引起的土地利用变化和植被变化改变了可能影响行星边界层热通量的植被性质, 从而可能导致上方大气状态变化。结果和相关讨论将在下一节中介绍, 穆勒等人(Müller et al., 2013)在 2013 年对此进行详细说明。

(二) 干旱事件案例研究

2008 年影响南美洲东南部的严重干旱主要是由拉尼娜事件和温暖的热带北大西洋共同造成的。此外, 伴随着干旱的发展, 由土地覆盖变化驱动的地表过程可能对热通量产生局部影响, 热通量进而可能对低层大气产生作用。图 16-4 显示了 2008 年期间降水和 NDVI 异常在空间分布和时间演变方面的相似性。降水距平图(图 16-4(b))和 NDVI 距平图(图 16-4(e))在乌拉圭和阿根廷东北部呈现类似的负值模式。在阿根廷、巴拉圭和巴西三方边界周围, 茂密的亚热带森林具有承受降水不足而不减损植被的能力。图 16-4(c)和(f)显示了所选定区域中干旱的演变(63°W~55°W, 38°S~28°S), 该区域干旱主要由作物和牧场主导。从 2007 年 6 月开始, 降水时间序列显示了干旱的持续存在, 两年来降

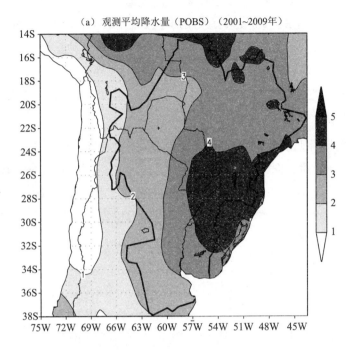

(a) 观测平均降水量（POBS）（2001~2009年）

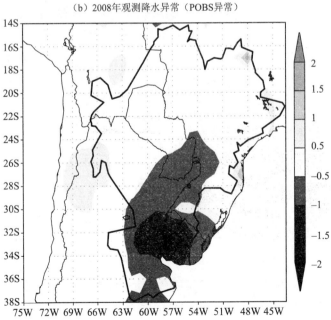

(b) 2008年观测降水异常（POBS异常）

(c) 2007~2009年平均降水量时间序列（毫米/天）

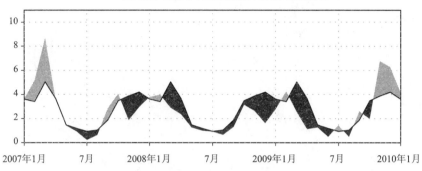

(d) 2001~2009年归一化植被指数黏度平均值

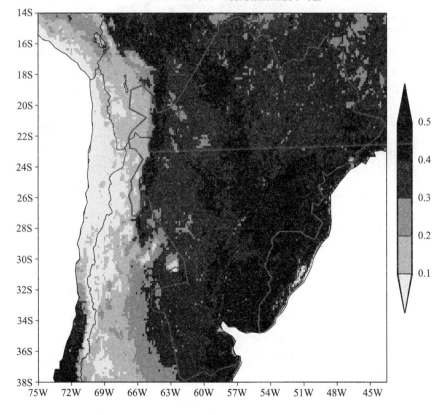

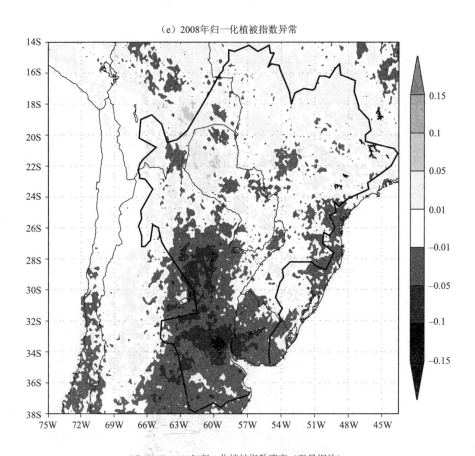

(e) 2008年归一化植被指数异常

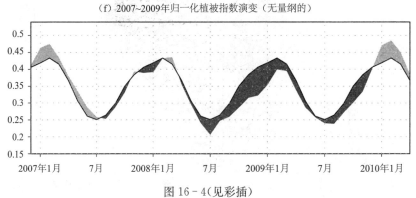

(f) 2007~2009年归一化植被指数演变(无量纲的)

图 16-4(见彩插)

注:(a)观测平均降水量(POBS,2001~2009年)和(b)2008年观测降水异常(POBS异常);(c)2007~2009年平均降水量的时间序列(黑线表示平均年周期;黑线上的绿色面积表示降水过多,黑线下的棕色面积表示降水不足);(d)~(f)为(a)~(c),仅针对NDVI。

水距平几乎呈连续的负趋势。NDVI 的时间演变似乎与降水量减少存在一定的滞后关系,滞后约 1 个月。低于正常值的数值从 2007 年 8 月开始,到 2009 年 11 月结束。只有 2007 年 11 月和 2008 年 3 月的两个特例,显示为弱正向异常。距平图和时间序列直观地显示了降雨不足给植被带来的直接影响,特别是在无人工灌溉的作物和牧场地区。正如我们将在下文进一步解释的那样,植被可能根据土地覆盖类型及其当前状态而对大气有不同反馈。

如图 16-4 所示,干旱等极端事件对某些支持生态系统服务具有直接影响,例如年度初级生产力的减少(如图 16-4(e)和(f)中的负异常,NDVI-m 的减少)和碳收益的季节变异性的增加(未显示,CVseas 的增加)。2008 年的干旱也最终影响了地表生物物理属性。针对 EFT 和 USGS 的差异,在图 16-4 中选择同一干旱区域即(63°W~55°W,38°S~28°S)绘制时间序列,我们观察到,在干旱期间 EFTs 数据集的绿色植被覆盖度(−13.2%)、最大和最小叶面积指数(−13.9%和−20.7%)显著低于 USGS 数据集。相比之下,最大和最小反照率(分别为 0.7%和 9.0%)和气孔阻抗(105.8%)更大(Müller et al.,2013)。

图 16-5 比较了 WRF-USGS 和 WRF-EFT 集合的模拟降雨结果。图 16-5(a)中 WRF-USGS 的空间分布呈现东北区域为高值,向西南方向递减。距平图(图 16-5(b))显示了 WRF-USGS 集合对巴西、巴拉圭和玻利维亚南部等区域湿度偏差正值(绿色)、区域东南部的负距平(棕色)的数量误差。这意味着,该模型模拟了干旱区大部分地区低于观测值的降水(即,模型夸大了干旱),但该区域北部地区的模型偏向较大值。

为了分析 WRF-EFT 集合的性能,图 16-5(c)绘制了两个集合间的差异图,穆勒等人(Müller et al.,2013)于 2013 年在其研究成果第五节中做了进一步分析。当使用 EFT 作为下边界条件时,区域 1 和 2 显示出湿度偏差减少(图 16-5(b)中的棕色阴影),但区域 3 上显示干燥偏差减少(图 16-5(b)中区域 3 中的绿色阴影)。两个集合都倾向于高估区域 1 的降水,区域 1 没有受到干旱的影响(WRF-EFT 集合未显示)。然而,WRF-EFT 集合减少了湿度偏差,因为它可以从负值(区域 1 中的棕色阴影)图形推断出来。当使用 EFT 时,模型误差降低约 19%。偏差减小与主要由表面粗糙度降低引起的水分平衡所有组成部分的普遍减少有关,表面粗糙度的降低致使近表面湍流度降低,进而加强了边

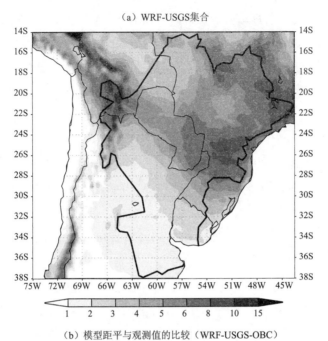

(a) WRF-USGS集合

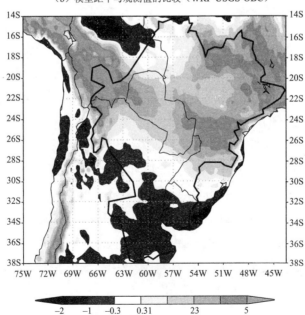

(b) 模型距平与观测值的比较（WRF-USGS-OBC）

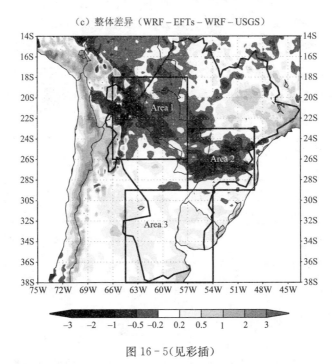

图 16-5（见彩插）

注：干旱期间的时均降水量：(a)WRF-USGS 集合，(b)模型距平与观测值的比较；干旱期间的时均降水量：(c)WRF-USGS 和 WRF-EFT 的差异。红线标注三个选定区域，称为区域1（顶部矩形，正常降水）、区域2（中部矩形，干旱）和区域3（底部矩形，干旱）。

界层的稳定性。此外，与 USGS 相比，EFT（未显示）引入了更大的气孔阻抗，导致蒸散量的减少。在区域 2 中，虽然两个集合都不能描述干旱区的北部区域（WRF-EFT 集合未显示），但 EFT 将 USGS 的湿度偏差降低了约 7%。近年来，该地区经历了农作物取代天然林的巨大变动(Izquierdo et al., 2008)。由于 EFT 解释说明了年际变化（见第十六章第三节），因此其特性代表了粗糙度长度大幅减少，同时伴随着植被覆盖度和叶面积指数（未显示）减少等年际变化。

植被破坏与减少致使蒸散量降低，进而使得降水减少（土壤水分和径流也减少了）。最后，WRF-EFT 集合略微增加了干旱核心区域（区域3）上的降水，从而校正了 WRF-USGS 集合的过度干旱偏差。在这一地区，占主导地位的作物和牧场暴露在干旱条件下，丧失在叶子上蓄水的能力。随后降雨直接渗入土层，使土壤直接蒸发量增加，有利于增加降雨。

综上所述，采用常规 USGS 地表覆盖模式虽然能够反映降水的空间分布和时间演变，但有放大正、负距平的倾向。然而，使用 EFT 年度植被图在很大程度上有助于通过反映生态系统生物物理属性的现实条件，来减少若干区域在这方面的误差，并使生态系统生物物理属性与气候服务之间有更好的联系。

五、讨论和结论

气候调节服务的生物物理部分是根据生态系统的地表净辐射和潜热通量估算的，并利用陆面模式进行模拟。一般而言，此等模拟涉及以下生态系统生物物理属性的使用，如叶面积指数、气孔阻抗、根系深度、可见光和近红外的反照率和透射率、热、水和雪容量等。在此，我们提供了一种基于卫星地图来估算和监测这些变量的方法。我们还提供了一个干旱案例的气候模式模拟结果，表明使用该方法可减少降水偏差。

我们的方法基于 EFT 定义，它源于 NDVI 衍生出的三个简单指标，即生产力、季节性和碳收益的物候特征(图 16-1)，为将植被生物物理属性的年际变化纳入地表和气候模拟建模开辟了一条更便捷的途径。目前，使用中等复杂程度的陆面方案的气候模式没有考虑到植被生物物理属性或土地覆盖类型的年际变化(Oki et al., 2013)。我们的研究表明，在整个研究区域某些生物物理属性在年际间是如何变化，以及这些年际变化在几个区域中是如何凸显出来的(图 16-3)。我们的 EFT 分类完全基于 NDVI 季节动态的三个指标，作为初级生产力的替代，由此被用作整体生态系统运行的综合指标(Virginia et al., 2001)。由于 EFT 可逐年定义，因此可以用来反映植被功能实际特征的时变地表特性，而不仅仅是固定时间的植被类型。从这个意义上讲，时变 EFT 的使用反映了在人类活动驱动下，土地利用和管理变化产生的影响。此外，特定年份的 NDVI 动态不仅反映了植被对特定年份环境条件的响应，而且表现了系统对过去年份气候条件和干扰效应的记忆(Wiegand et al., 2004)。

本文所提出的方法还可以进一步改进，在对基于 NDVI 的各个指标进行直方图计算之前，通过为每年设置一个决策规则来对非植被像元(水体、雪和纯粹沙漠)进行分类。例如，可使用 LTDR 或 MODIS 质量评估特征信息来识别水体和被积雪覆盖的像元,沙漠也可归类为具有极低 NDVI 值(例如,全年 NDVI 值总是低于 0.15)的像元。最后,剩下的像元可以按照这里所描述的分类到 EFT 中。如引言所述,EFT 的定义取决于所包含的变量。遥感可提供额外的功能变量来改进对生态系统的描述(Nemani and Running,1997)。蒸散量(见第十八章)、短波反照率(见第十七章)和地表温度(见第十九章)是与生态系统的水和能量交换密切相关的三个变量,根据碳收益的季节动态,可对本文提出的变量进行补充描述(Piñeiro et al.,2002;Garbulsky and Paruelo,2004;Fernández et al.,2010)。使用卫星信息的一个主要优势是信息的近实时分发和全球可用性。在不久的将来,我们将推出基于全球植被指数(GVI)的 EFT 地图业务化产品,该指数来源于国家环境卫星、数据和信息服务(NESDIS)网站的气象业务卫星(MetOp)AVHRR 全球 1 千米数据。

致谢

美国国家科学基金会(Grant GEO-0452325)下属的美国间全球变化研究所(IAI、CRN Ⅱ 2031 和 2094)、马里兰大学、美国国家航空航天局拨款的 NNX08AE50G、布宜诺斯艾利斯大学的 Proyecto Estratégico、阿根廷科学技术研究委员会(CONICET)、FONCYT、FEDER 基金、安达卢西亚政府(GLOCHARID 项目和 SEGALERT P09-RNM-5048)、西班牙国家公园管理局(066/2007 项目)、西班牙科学和创新部(CGL2010-22314 项目,国家 I+D+I 2010 计划)及阿尔梅里亚大学 Zonas Áridas 研究小组,提供了资金支持。LTDR 数据来自 LTDR 团队网站。

参考文献

Alcaraz-Segura, D., J. Cabello, and J. Paruelo. 2009. Baseline characterization of major Iberian vegetation types based on the NDVI dynamics. *Plant Ecology* 202:13–29.

Alcaraz-Segura, D., E. Liras, S. Tabik, J. M. Paruelo, and J. Cabello. 2010. Evaluating the consistency of the 1982–1999 NDVI trends in the Iberian Peninsula across four time-series derived from the AVHRR sensor: LTDR, GIMMS, FASIR, and PAL-II. *Sensors* 10:1291–1314.

Alcaraz-Segura, D., J. Paruelo, and J. Cabello. 2006. Identification of current ecosystem functional types in the Iberian Peninsula. *Global Ecology and Biogeography* 15:200–212.

Alcaraz-Segura, D., J. Paruelo, H. Epstein, and J. Cabello. 2013. Environmental and human controls of ecosystem functional diversity in temperate South America. *Remote Sensing* 5:127–154.

Anderson-Teixeira, K. J., P. K. Snyder, T. E. Twine, S. V. Cuadra, M. H. Costa, and E. H. DeLucia. 2012. Climate-regulation services of natural and agricultural ecoregions of the Americas. *Nature Climate Change* 2:177–181.

Baidya Roy, S., G. C. Hurtt, C. P. Weaver, and S. W. Pacala. 2003. Impact of historical land cover change on the July climate of the United States. *Journal of Geophysical Research: Atmospheres* 108: ACL 11-1–ACL 11-14.

Beltrán-Przekurat, A., R. A. Pielke Sr., J. L. Eastman, and M. B. Coughenour. 2012. Modelling the effects of land-use/land-cover changes on the near-surface atmosphere in southern South America. *International Journal of Climatology* 32:1206–1225.

Bonan, G. B. (2008). *Ecological climatology: concepts and applications*. Second Edition. Cambridge University Press.

Bonan, G. B. 2008. Forests and climate change: Forcings, feedbacks, and the climate benefits of forests. *Science* 320:1444–1449.

Brando, P. M., S. J. Goetz, A. Baccini, D. C. Nepstad, P. S. A. Beck, and M. C. Christman. 2010. Seasonal and interannual variability of climate and vegetation indices across the Amazon. *Proceedings of the National Academy of Sciences of the United States of America* 107:14685.

Bret-Harte, M. S., M. C. Mack, G. R. Goldsmith, et al. 2008. Plant functional types do not predict biomass responses to removal and fertilization in Alaskan tussock tundra. *Journal of Ecology* 96:713–726.

Cao, M., S. D. Prince, J. Small, and S. J. Goetz. 2004. Remotely sensed interannual variations and trends in terrestrial net primary productivity 1981–2000. *Ecosystems* 7:233–242.

Chapin, F. S., J. T. Randerson, A. D. McGuire, J. A. Foley, and C. B. Field. 2008. Changing feedbacks in the climate-biosphere system. *Frontiers in Ecology and the Environment* 6:313–320.

Chen, F., and J. Dudhia. 2001. Coupling an advanced land surface-hydrology model with the Penn State-NCAR MM5 modeling system. Part I: Model implementation and sensitivity. *Monthly Weather Review* 129:569–585.

Chen, F., K. Mitchell, J. Schaake, et al. 1996. Modeling of land surface evaporation by four schemes and comparison with FIFE observations. *Journal of Geophysical Research: Atmospheres* 101:7251–7268.

De Oliveira, A. S., R. Trezza, E. Holzapfel, I. Lorite, and V. P. S. Paz. 2009. Irrigation water management in Latin America. *Chilean Journal of Agricultural Research* 69:7–16.

Diffenbaugh, N. S. 2009. Influence of modern land cover on the climate of the United States. *Climate Dynamics* 33:945–958.

Eidenshink, J. C., and J. L. Faundeen. 1994. The 1 km AVHRR global land data set: First stages in implementation. *International Journal of Remote Sensing* 15:3443–3462.

Ek, M. B., K. E. Mitchell, Y. Lin, et al. 2003. Implementation of Noah land surface model advances in the National Centers for Environmental Prediction operational mesoscale Eta model. *Journal of Geophysical Research: Atmospheres* 108:GCP 12-1–GCP 12-16.

Fernández, N., J. Paruelo, and M. Delibes. 2010. Ecosystem functioning of protected and altered Mediterranean environments: A remote sensing classification in Doñana, Spain. *Remote Sensing of Environment* 114:211–220.

Foley, J. A., G. P. Asner, M. H. Costa, et al. 2007. Amazonia revealed: Forest degradation and loss of ecosystem goods and services in the Amazon Basin. *Frontiers in Ecology and the Environment* 5:25–32.

Foley, J. A., M. H. Costa, C. Delire, N. Ramankutty, and P. Snyder. 2003. Green surprise? How terrestrial ecosystems could affect Earth's climate. *Frontiers in Ecology and the Environment* 1:38–44.

Foley, J. A., I. C. Prentice, N. Ramankutty, et al. 1996. An integrated biosphere model of land surface processes, terrestrial carbon balance, and vegetation dynamics. *Global Biogeochemical Cycles* 10:603–628.

Friedl, M. A., D. Sulla-Menashe, B. Tan, et al. 2010. MODIS Collection 5 global land cover: Algorithm refinements and characterization of new datasets. *Remote Sensing of Environment* 114:168–182.

Garbulsky, M. F., and J. M. Paruelo. 2004. Remote sensing of protected areas to derive baseline vegetation functioning characteristics. *Journal of Vegetation Science* 15:711–720.

Gedney, N., and P. J. Valdes. 2000. The effect of Amazonian deforestation on the northern hemisphere circulation and climate. *Geophysical Research Letters* 27:3053–3056.

Haines-Young, R., and M. Potschin. 2013. *Common International Classification of Ecosystem Services (CICES): Consultation on Version 4.* Nottingham, UK: University of Nottingham.

Holben, B. N. 1986. Characteristics of maximum-value composite images from temporal AVHRR data. *International Journal of Remote Sensing* 7:1417–1434.

Izquierdo, A. E., C. D. De Angelo, and T. M. Aide. 2008. Thirty years of human demography and land-use change in the Atlantic forest of Misiones, Argentina: An evaluation of the forest transition model. *Ecology and Society* 13:3.

Kalnay, E., and M. Cai. 2003. Impact of urbanization and land-use change on climate. *Nature* 423:528–531.

Kathuroju, N., M. A. White, J. Symanzik, M. D. Schwartz, J. A. Powell, and R. R. Nemani. 2007. On the use of the Advanced Very High Resolution Radiometer for development of prognostic land surface phenology models. *Ecological Modelling* 201:144–156.

Kucharik, C. J., J. A. Foley, C. Delire, et al. 2000. Testing the performance of a dynamic global ecosystem model: Water balance, carbon balance, and vegetation structure. *Global Biogeochemical Cycles* 14:795–825.

Lavorel, S., S. Díaz, J. Cornelissen, et al. 2007. Plant functional types: Are we getting any closer to the Holy Grail? In *Terrestrial ecosystems in a changing world*, eds. J. G. Canadell, D. E. Pataki, and L. F. Pitelka, 149–164. New York: Springer.

Legendre, P. 1993. Spatial autocorrelation: trouble or new paradigm? *Ecology* 74:1659–1673.

Mahmood, R., R. A. Pielke, K. G. Hubbard, et al. 2010. Impacts of land use/land lover change on climate and future research priorities. *Bulletin of the American Meteorological Society* 91:37–46.

Maness, H., P. J. Kushner, and I. Fung. 2013. Summertime climate response to mountain pine beetle disturbance in British Columbia. *Nature Geoscience* 6:65–70.

Marland, G., R. A. Pielke Sr., M. Apps, et al. 2003. The climatic impacts of land surface change and carbon management, and the implications for climate-change mitigation policy. *Climate Policy* 3:149–157.

McNaughton, S. J., M. Oesterheld, D. A. Frank, and K. J. Williams. 1989. Ecosystem-level patterns of primary productivity and herbivory in terrestrial habitats. *Nature* 341:142–144.

Monteith, J. 1972. Solar radiation and productivity in tropical ecosystems. *Journal of Applied Ecology* 9:747–766.

Müller, O. V., E. H. Berbery, and D. Alcaraz-Segura, M. B. Ek. 2013. Regional model simulations of the 2008 drought in Southern South America using a consistent set of land surface properties. *Journal of Climate*. (In revision.)

Nagol, J. R., E. F. Vermote, and S. D. Prince. 2009. Effects of atmospheric variation on AVHRR NDVI data. *Remote Sensing of Environment* 113:392–397.

Nemani, R., and S. Running. 1997. Land cover characterization using multitemporal red, near-IR, and thermal-IR data from NOAA/AVHRR. *Ecological Applications* 7:79–90.

Nosetto, M. D., E. G. Jobbágy, and J. M. Paruelo. 2005. Land-use change and water losses: The case of grassland afforestation across a soil textural gradient in central Argentina. *Global Change Biology* 11:1101–1117.

Oki, T., E. M. Blyth, E. H. Berbery, and D. Alcaraz-Segura. 2013. Land cover and land use changes and their impacts on hydroclimate, ecosystems and society. In *Climate science for serving society: Research, modeling and prediction priorities*, eds. G. R. Asrar and J. W. Hurrel. Dordrecht, The Netherlands: Springer Science+Business Media.

Olson, D. M., E. Dinerstein, E. D. Wikramanayake, et al. 2001. Terrestrial ecoregions of the worlds: A new map of life on Earth. *BioScience* 51:933–938.

Paruelo, J. M., E. G. Jobbagy, and O. E. Sala. 2001. Current distribution of ecosystem functional types in temperate South America. *Ecosystems* 4:683–698.

Pedelty, J., S. Devadiga, E. Masuoka, et al. 2007. Generating a long-term land data record from the AVHRR and MODIS instruments. *IGARSS 2007 IEEE International Geoscience and Remote Sensing Symposium* 1021–1025.

Pettorelli, N., J. O. Vik, A. Mysterud, J. M. Gaillard, C. J. Tucker, and N. C. Stenseth. 2005. Using the satellite-derived NDVI to assess ecological responses to environmental change. *Trends in Ecology & Evolution* 20:503–510.

Pielke, R. A., Sr. 2001. Influence of the spatial distribution of vegetation and soils on the prediction of cumulus convective rainfall. *Reviews of Geophysics* 39:151–177.

Pielke, R. A., Sr., J. Adegoke, A. BeltráN-Przekurat, et al. 2007. An overview of regional land-use and land-cover impacts on rainfall. *Tellus B* 59:587–601.

Pielke, R. A., Sr., G. Marland, R. A. Betts, et al. 2002. The influence of land-use change and landscape dynamics on the climate system: Relevance to climate-change policy beyond the radiative effect of greenhouse gases. *Philosophical Transactions of the Royal Society A: Mathematical, Physical and Engineering Sciences* 360:1705–1719.

Piñeiro, G., D. Alcaraz, J. Paruelo, et al. 2002. A functional classification of natural and human-modified areas of 'Cabo de Gata,' Spain, based on Landsat TM data. *Presented at the 29th International Symposium of Remote Sensing of Environment*, Buenos Aires, Argentina, 8–12.

Potter, C., S. Klooster, P. Tan, M. Steinbach, V. Kumar, and V. Genovese. 2005. Variability in terrestrial carbon sinks over two decades. Part III: South America, Africa, and Asia. *Earth Interactions* 9:1–15.

Reale, O., and P. Dirmeyer. 2000. Modeling the effects of vegetation on Mediterranean climate during the Roman Classical Period: Part I: Climate history and model sensitivity. *Global and Planetary Change* 25:163–184.

Reale, O., and J. Shukla. 2000. Modeling the effects of vegetation on Mediterranean climate during the Roman Classical Period: Part II. Model simulation. *Global and Planetary Change* 25:185–214.

Reich, P. B., and J. Oleksyn. 2004. Global patterns of plant leaf N and P in relation to temperature and latitude. *Proceedings of the National Academy of Sciences of the United States of America* 101:11001.

Reich, P. B., M. G. Tjoelker, J. L. Machado, and J. Oleksyn. 2006. Universal scaling of respiratory metabolism, size and nitrogen in plants. *Nature* 439:457–461.

Roy, S. B., and R. Avissar. 2002. Impact of land use/land cover change on regional hydrometeorology in Amazonia. *Journal of Geophysical Research* 107:8037.

Sellers, P. J., Y. Mintz, Y. C. Sud, and A. Dalcher. 1986. A simple biosphere model (SIB) for the use within general circulation models. *Journal of the Atmospheric Sciences* 43:505–531.

Smith, T. M., H. H. Shugart, and F. I. Woodward. (eds.). 1997. *Plant functional types: Their relevance to ecosystem properties and global change*. Cambridge, UK: Cambridge University Press.

Soriano, A., and J. M. Paruelo. 1992. Biozones: Vegetation units defined by functional characters identifiable with the aid of satellite sensor images. *Global Ecology & Biogeography Letters* 2:82–89.

Stephenson, N. L. 1990. Climatic control of vegetation distribution—The role of the water-balance. *American Naturalist* 135:649–670.

Stohlgren, T. J., T. N. Chase, R. A. Pielke, T. G. F. Kittel, and J. S. Baron. 1998. Evidence that local land use practices influence regional climate, vegetation, and stream flow patterns in adjacent natural areas. *Global Change Biology* 4:495–504.

Tucker, C. J., and P. J. Sellers. 1986. Satellite remote-sensing of primary production. *International Journal of Remote Sensing* 7:1395–1416.

Valentini, R., D. D. Baldocchi, J. D. Tenhunen, and P. Kabat. 1999. Ecological controls on land-surface atmospheric interactions. In *Integrating hydrology, ecosystem dynamics and biogeochemistry in complex landscapes*, eds. D. Tenhunen and P. Kabat, 105–116. Berlin: John Wiley & Sons.

Virginia, R. A., D. H. Wall, and S. A. Levin. 2001. Principles of ecosystem function. In *Encyclopedia of biodiversity*, ed. S. Levin, 345–352. San Diego, CA: Academic Press.

Volante, J. N., D. Alcaraz-Segura, M. J. Mosciaro, E. F. Viglizzo, and J. M. Paruelo. 2012. Ecosystem functional changes associated with land clearing in NW Argentina. *Ecosystem Services and Land-Use Policy* 154:12–22.

Weaver, C. P., and R. Avissar. 2001. Atmospheric disturbances caused by human modification of the landscape. *Bulletin of the American Meteorological Society* 82:269–281.

Werth, D., and R. Avissar. 2002. The local and global effects of Amazon deforestation. *Journal of Geophysical Research* 107:8087.

West, P. C., G. T. Narisma, C. C. Barford, C. J. Kucharik, and J. A. Foley. 2011. An alternative approach for quantifying climate regulation by ecosystems. *Frontiers in Ecology and the Environment* 9:126–133.

Wiegand, T., H. A. Snyman, K. Kellner, and J. M. Paruelo. 2004. Do grasslands have a memory: Modeling phytomass production of a semiarid South African grassland. *Ecosystems* 7:243–258.

Wright, I. J., P. B. Reich, J. H. C. Cornelissen, et al. 2005. Assessing the generality of global leaf trait relationships. *New Phytologist* 166:485–496.

Wright, I. J., P. B. Reich, M. Westoby, et al. 2004. The worldwide leaf economics spectrum. *Nature* 428:821–827.

Wright, J. P., S. Naeem, A. Hector, et al. 2006. Conventional functional classification schemes underestimate the relationship with ecosystem functioning. *Ecology Letters* 9:111–120.

Zhao, M., and S. W. Running. 2010. Drought-induced reduction in global terrestrial net primary production from 2000 through 2009. *Science* 329:940–943.

第十七章 与能量平衡相关的生态系统服务：以湿地反射能量为例

C.M. 迪贝拉 M.E. 贝格特

一、引言

生态系统在向人类提供商品和服务方面发挥着重要作用。它对人类直接或间接的益处包括供给服务（如食物、水、燃料、纤维和遗传资源）、支持服务（如生态系统初级生产、土壤形成、氧气生成）、文化发展服务（如娱乐、认知发展、反思和精神充实）和调节服务（如空气净化、水净化、人类疾病调控、侵蚀防治和气候调节）（MA，2005）。近年来，生态系统服务的研究、量化和制图越来越受到科学界的关注。

在生态系统层面，有若干调节流向大气的水、质量和能量通量的过程（如Baldocchi and Wilson，2001；Noe et al.，2011）。如：光合作用，影响大气中二氧化碳的水平；蒸散量，控制土壤和植物向大气释放的潜热和水；气溶胶的产生，改变大气的辐射加热；太阳辐射的反射（反照率），调整可用能量，从而改变陆地表面温度（House et al.，2005；Smith et al.，2011）。

地表反照率在能量平衡和气候调节中起着重要作用，因为它决定了进入生态系统的能量数量。正如阿尔卡拉斯-塞古拉（见第十六章）所解释的那样，由于土地利用变化，反照率的变化会改变能量传递到生态系统的总量，继而改变传递到大气的总量。例如，热带和亚热带旱地的荒漠化导致反照率增加，蒸散量减少，从而导致区域降雨量的减少（House et al.，2005）（见第十八章）。在季节性积雪覆盖的森林中，由于反照率的增加和蒸散量的减少，土地开垦分别导致了区域雪季降温和夏季变暖。基于不同季节和地理位置，通过地表反照率的生物地球物理过程提供生态系统气候调节服务(House et al.,2005)。

地表反照率定义为在太阳光谱范围内,入射漫射与所有方向上反射的直接辐射的比值(Pinty and Verstraete,1992),它包含了 0.3 微米和 3 微米范围内的电磁光谱带上的反射辐射。然而,该范围可进一步分为"可见反照率"(0.3~0.7 微米)和"近红外反照率"(0.7~3 微米)(Bsaibes et al.,2009)。

为了测量任何一块土地上的反照率,可使用一种特定的仪器:日射强度计。该仪器是一个具有 180°视野的辐射宽带传感器。反照率是反射太阳辐射(日射强度计向下方向)和入射太阳辐射(日射强度计向上方向)的比率。尽管安装了一系列现场测量网络(如 Amerifolux、AsiaFlux、Fluxnet、CarboEurope),由于空间和时间变化幅度大,反照率制图成为一项困难的任务。

在这种背景下,通过卫星观测进行地表反照率制图成为一种有利方法。在 20 世纪 80 年代,在开展地球辐射收支实验(Earth Radiation Budget Experiment,ERBE)时发射了两个仪器,专门用于测量宽带地表反照率。这些装置被组装在两个卫星平台上:地球辐射收支卫星(Earth Radiation Budget Satellite,ERBS)和国家海洋和大气管理局卫星。从 1984~1999 年,这些仪器生成了 2.5°空间分辨率的月度产品(S-9 和 S-10 产品),从而提升了地表反照率的全球观测水平(Hatzianastassiou et al.,2004)。

如今,诸多现有的卫星平台获得的光谱数据可以用于近似估算地表反照率(表 17-1),从而有助于生态系统能量平衡的研究(表 17-2)。

表 17-1 不同卫星传感器的反照率计算

传感器/平台	波段(微米)	空间分辨率	参考文献
ASTER/TERRA	0.52~0.6 0.63~0.69 0.78~0.86 1.6~1.7 2.15~2.18 2.18~2.22 2.23~2.28 2.29~2.36 2.36~2.43	15 米(可见光和近红外)/ 30 米(红外)/90 米	Nasipuri et al.,2006

续表

传感器/平台	波段(微米)	空间分辨率	参考文献
AVHRR/NOAA	0.57~0.71 0.72~1.01	1.25千米/5千米	Csiszar and Gutman, 1999; Key et al., 2001
MISR/TERRA	0.42~0.45 0.54~0.55 0.66~0.67 0.85~0.87	275米	Diner et al., 1998
MODIS/TERRA	0.62~0.67 0.84~0.87 0.46~0.48 0.54~0.56 1.23~1.25 1.63~1.65 2.11~2.15	1千米	Wang and Zender, 2010
PODER/ADEOS	0.43~0.46 0.66~0.68 0.74~0.79 0.84~0.88	6千米×7千米	Deschamps et al., 1994
VEGETATION/SPOT	0.43~0.47 0.61~0.68 0.78~0.89 1.58~1.75	1千米(NADIR)	Stroppiana et al., 2002
LANDSAT/ETM+	0.45~0.51 0.63~0.69 0.75~0.9 1.55~1.75 2.09~2.35	30米	Shuai et al., 2011

表17-2 与能量平衡和反照率生态系统服务相关的遥感研究

参考文献	应用
Consoli et al., 2006; Mariotto and Gutschick, 2010	蒸散估算
Mueller et al., 2011; Posselt et al., 2012	气候监测与分析
Alton, 2009	陆地表面碳、水和能量交换的模拟
Beck et al., 2011	森林火灾影响

续表

参考文献	应用
Bright et al.,2012	生物能源气候影响
Kuusinen et al.,2012	北方森林季节变化
Puzachenko et al.,2011	生物圈热力学参数估算
Allen et al.,2013	土地覆被动态
Maes et al.,2011	生态系统演替与能量耗散
Yang et al.,2012	灌溉土地的水管理
Zwart et al.,2010	水分生产率
Ju et al.,2010	土壤含水量监测

中分辨率成像光谱仪提供的产品是用于估算全球尺度反照率最受欢迎的遥感产品之一(Lucht et al.,2000)。它是基于梁等人(Liang et al.,1999)提出的窄波段到宽波段转换系数。在计算之前,可通过双向反射分布函数模型对 MODIS 数据进行大气和角度校正。

自 2000 年 3 月以来,通过整合每日连续多角度观测值(在精确周期内),形成一平方千米空间分辨率(MOD43;http://modis.gsfc.nasa.gov/data/dataprod)、每 16 天提供一次的产品(Lucht et al.,2000;Schaaf et al.,2002)。梁(Liang,2000)随后改进了将方向/半球反射率(针对每个太阳天顶角)和双半球反射率转换为反照率的算法。总短波反照率(α,0.25～2.5 微米)转换由式 17-1 给出:

$$\alpha = 0.160\alpha_1 + 0.291\alpha_2 + 0.243\alpha_3 + 0.116\alpha_4 + 0.112\alpha_5 + 0.081\alpha_7 - 0.0015$$

式 17-1

其中,α_n 表示 MODIS 第 n(1-7)波段的反照率。

人工或天然湿地是商品和服务的重要来源。由于生态系统的开放性和公共性,与湿地利用和保护相关的政治和决策多次被低估(Brander et al.,2006)。湿地利用评估已成为一个非常重要的课题,并引领科学界开始关注服务评价。在本文的例子中,关注的焦点是湿地条件对能量平衡的影响,特别是对地表反照率的影响。

在湿地生态系统中,水会改变植被冠层的光谱行为(Beget and Di Bella, 2007);反照率估算将取决于水位和水面上下的叶面积指数(Beget and Di Bella, 2007;Sumner et al.,2011;Beget et al.,2013)。尽管地表反照率或者是在它作为一个整体提供的商品和服务方面具有重要意义,但很少有研究考虑到由于湿地被其他土地覆被所取代(例如森林、牧场或水稻作物)或生长条件变化(水位或叶面积指数)而产生的反照率变化。为了回答这些问题,使用辐射传输模型(SAILHFlood;Beget et al.,2013)来模拟多种条件下(例如土地利用类型、叶面积指数和水位)的冠层反射率。本文假设模拟反射率作为计算光谱反射率的中间变量(Liang,2000),将有助于评估(在第一种方法中)土地利用和覆盖的变化对区域尺度能量评估的所有影响。

二、湿地反射能量

湿地是最具生产力的陆地生态系统之一,在所有大陆和各种气候条件下覆盖约700万~900万平方千米(Mitsch and Gosselink,2000)。不管是自然还是人工湿地(如水稻种植园),都是人类商品和服务的重要来源。这些生态系统除了提供可供人类使用的物品外,还提供基础服务,如蓄水层的排放和补给、防洪、风暴控制、水质的改善、生物多样性和物种进化物质的来源、交通、娱乐和旅游以及气候调节(Lambert,2003)。

特别是在气候调节方面,水在湿地反射能量(地表反照率)中起着重要作用。就反射而言,它的环境是高度可变的。毫无疑问,在覆盖土地上改变光谱行为的这些因素中,水起着主要作用,因为它的反射特性与浊度、沉积物类型或藻类以及每个系统中水位产生的变化存在关联。土地覆盖、裸土比例和叶面积指数(淹没和露出)是强烈影响地表光谱响应的其他因素。假如考虑在湿地中这些特征可以无限地组合,而且响应通常是这些因素之间多重相互作用的结果,那么这些变化对湿地光谱响应的影响就很难了解,更不用说理解了。此外,由于水位波动及被农作物取代,湿地也在不断地变化(Kingsford, 2000;Bunn and Arthington,2002;Schottler et al.,2013)。

反照率产品MOD43在计算不同生态系统和不同土地利用条件下的此类变

量时能提供很好的近似解(Disney et al.,2004;Stroeve et al.,2005;Román et al.,2010;Wang and Zender,2010;Cescatti et al.,2012)。然而,现存文献很少有介绍土地覆盖/利用变化对湿地反照率的影响,而关于水位变化对生态系统的影响的文献则更少。若考虑到大多数卫星产品空间分辨率为1平方千米,这项研究将变得更加困难。从这个层面出发,评估这类系统的光谱行为,从而防止地表反照率的可能变化是非常有意义的。在本研究中,模拟不同土地覆盖、水位和叶面积指数的反射能量,从而外推出这类变化,以估算地表反照率。反射率模型为研究陆地覆盖区的光谱特性提供了可能性,这在实验条件下很难进行(如Jacquemoud et al.,2009)。光谱反射率可集成在任何宽带传感器中,例如MODIS(Liang,2000)。通过比较不同土地覆盖的模拟反射率,可直接研究土地利用/覆盖变化或水位差异对光谱反照率的影响。

(一)冠层反射率模拟模型

辐射传输模型是研究土地覆被特征对光谱行为影响的非常有用的工具(如Jacquemoud et al.,2000,2009;Weiss et al.,2000)。考虑到叶片的光学特性,这类模型通过描述色散/消光辐射通量来计算冠层反射率。

以湿地生态系统为例,模型应该考虑水,因为与空气不同,水吸收和散射由树叶传递和反射的能量。另一方面,由于两种介质的折射率不同,水-空气界面会干扰能量通量,改变其方向(Beget et al.,2010)。辐射传输模型 SAILHFlood 由贝格特等人(Beget et al.,2010,2013)开发,用于模拟部分淹没的树冠反射。SAILHFlood 模型在 SAILH(任意倾斜叶片散射)模型(Verhoef,1984,1985)中加入了一个浸没的植被层,其传播介质为水(图17-1)。SAILHFlood 模型根据与照明和观测几何、叶片和土壤的光学特性及冠层特征相关的变量计算部分淹没冠层的反射率(表17-3)。

SAILHFlood 模型为 SAILH 模型引入了三个新的输入。这些输入是淹没叶面积指数、露出叶面积指数和水位(表17-3)。在实验室条件下使用人工淹没冠层测试了 SAILHFlood 模型(Beget et al.,2013)。试验模型包括三个光照天顶角、九个观测天顶角、五层叶面积指数和三个水位。模拟反射率与实测反射率间的均方根误差(RMSE)为0.0355。试验结果表明模型性能良好。

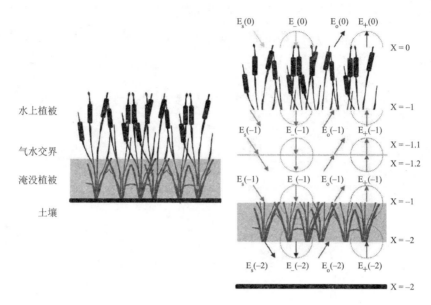

图 17-1(见彩插)

注:SAILHFlood 模型示意图。X 表示向上的垂直尺寸。箭头表示通量(直接通量由箭头表示,扩散通量由箭头和半球共同表示)。ES 代表太阳直接辐照度;E_- 和 E_+ 分别是向下和向上扩散辐照度;Eo 是 π 乘以观察者方向辐照度。

(二) 模拟案例研究

在 0.01 微米光谱分辨率范围内,使用 SAILHFlood 模型对三种土地覆盖(湿地、水稻作物和森林)和不同叶面积指数(1~12)和不同水位(0~1.4 米)的光谱反射率进行了在 0.4~2.5 微米之间的模拟(表 17-4)。选择林地覆盖的原因是,一些地区是在湿地地区开展森林种植(Baigún et al.,2008)。水稻因其自身的洪水特性而成为人工湿地,但与普通芦苇相比,其所在水体有季节特征和水位变化。对于 90 次模拟中的每一次,光谱反射率都被平均到 MODIS 光谱波段中(表 17-1 和图 17-4)。

选择普通芦苇(拉丁学名:Phragmites australis[Cav.]Trin. ex Steud)来代表一块湿地的光谱特性。芦苇广泛分布于世界各地(Clevering and Lissner, 1999),并在叶面积指数上具有较大的可塑性(范围在 2.2~12.2 之间)(Mal and Narine,2004)。对于反射率模拟,伯贝等人(Burba et al.,1999)及玛尔和纳林

(Mal and Narine,2004)的文献中采用芦苇的冠层特征。将六个叶面积指数以及八个水位结合起来运行该模型(表17-4)。冠层垂直分布被视为与其他草类相似(Zheng et al.,2007)。

表 17-3　SAILHFlood 模型的输入变量

SAILHFlood 输入变量	符号	单位/范围
水位	h	米
出水叶面积指数	LAIem	—
淹没叶面积指数	LAIsum	—
平均叶倾角	θl	度
热点	hot	—
叶片反射率	refl	0～1
叶片透光率	tran	0～1
太阳天顶角	θs	度
观察天顶角	θo	度
方位相对角度	ψ	度
漫射辐射系数	skyl	0～1
土壤双向反射	ros	0～1
土壤双半球反射	rdd	0～1
土壤定向半球反射	rsd	0～1
土壤半球定向反射	rdo	0～1

资料来源:修改自 Beget, M. E. et al., Ecol Model, 257, 25-35, 2013。

注:$\theta s=8°,\theta o=30°,\psi=0°,skyl=0$。

在假设水位淹没后没有变化的条件下,叶面积指数值的垂直分布使得每个水位处的出水和淹没的叶面积指数是可计算的。热点参数和平均叶倾角在所有模拟中都是固定的,参照巴库等人(Bacour et al.,2002)的研究进行设置。

以农田为例,以一个灌溉水稻种植园作为案例研究。大部分水稻产量是在低地灌溉系统中生产的(来自国际水稻研究所的数据是:世界产量的75％)。灌溉意味着在田间施用5～25厘米的水。使用六种叶面积指数值和六种水位值的组合进行模拟(表17-4),水稻作物的叶面积指数值可达到7(Casanova et al.,1998;Stropiana et al.,2006)。郑等人(Zheng et al.,2007)利用水稻冠层结构三维模型对汕优63的冠层垂直分布和平均叶倾角进行了模拟。所采用的热点参数与穆林等人(Moulin et al.,2002)对小麦作物采用的参数相似。

表17-4 SAILHFlood模拟中使用的输入参数值

土地覆盖	物种	叶面积指数	水位(米)	(θ_l 度)	Hot 值	模拟次数
湿地(芦苇)	芦苇	LAI=[2,4,6,8,10,12]	h=[0、0.2、0.4、0.6、0.8、1、1.2、1.4]	70	0.2	48
农田(水稻)	水稻	LAI=[2,3,4,5,6,7]	h=[0、0.1、0.2、0.3、0.4、0.5]	70	0.1	36
人工林(杨树)	加拿大杨树	LAI=[1,2,3,4,5,6]	无水	56.5	0.005	6

以加拿大杨树(Populus canadensis)的反射冠层模拟为例研究了人工林。在该案例中,对不同的叶面积指数水平进行了不同的模拟,叶面积指数范围在表示幼林的1到表示成熟林的7之间(Meroni et al.,2004)。热点参数和平均叶倾角在所有模拟中被设定为固定值,并且参考梅罗尼等人(Meroni et al.,2004)的研究进行取值。由于杨树种植在高地,模拟时不考虑水。

对于每个物种,土壤和叶片的光学特性在模拟时保持不变。从LOPEX93(叶片光学特性实验93)数据库中以0.005微米的分辨率(Hosgood et al.,1994)获取所有物种的叶片光学特性(反射率和透射率)。光谱分辨率为0.001微米且采用三次样条平滑。使用手持式光谱辐射计(0.4~2.5微米,Analytical Spectral Devices© FieldSpecPro,ASD)测量粉质黏土上的土壤光谱反射率,并假定满足朗伯反射。

(三)模拟反射率

以0.001微米的光谱分辨率,在0.4~2.4微米范围内模拟反射率。对于具有相同叶面积指数(4)的冠层,森林呈现出更明显的红色边缘(Gitelson et al.,

1996),近红外部分的反射率更高,短波红外吸收峰比水稻更明显,甚至超过湿地(图17-2)。相比之下,湿地在近红外波段的反射率值最低,在可见光和短波红外波段的吸收率最低(图17-2)。当湿地和水稻冠层被水淹没时(h=0.4m)时,反射率从0.6微米下降。根据贝格特等人(Beget et al.,2013)的结果,随着水位的增加,被淹没冠层的反射率的下降幅度更大。

湿地和水稻种植园在7个MODIS波谱带中的大部分显示出相似的光谱特性(图17-3),不同模拟间的易变性更大。这种响应的发生主要是受到水位、低叶面积指数(Beget and Di Bella,2007)或高比例裸土(Oguro et al.,2003)的影响(图17-3)。水引入了易变性,因为(如前所述)反射率随着水位的增加而降低。

除了蓝色带(459~479纳米),人工林与湿地和水稻人工林几乎在所有光谱带上都有区别。然而,突出的波段是中红外波段(1 628~1 652纳米和2 005~2 155纳米)(分别为图17-3(f)和(g))。在红外范围内强调了水对冠层反射率的影响(图17-3(d)和(g))(Beget et al.,2013)。尽管预计到叶面积指数会影响(在更大程度上)红外反射率(Jacquemoud et al.,2009),但无论模拟的土地覆盖之间的叶面积指数差异如何大(高达50%),湿地、水稻种植园和林地的近红外波段反射率变化都是相似的(图17-3(d))。

反照率是根据梁(Liang,2000)的算法(式17-1)对湿地、稻田和森林的模拟情况进行估算的(图17-4)。在这种情况下,森林的反照率值为0.22,分别比稻田和湿地高17%和59%。尽管人工林总体上表现出相对于其他覆盖物(如草地和灌木)的低反照率值(Houldcroft et al.,2009;Cescatti et al.,2012),但正如已知所述,其表现出比水体相对更高的反照率值(Houldcroft et al.,2009)。这就是为什么在湿地和水稻作物中存在地表水降低反照率的原因。变差(评估为最大和最小反照率间的差异)在湿地和水稻种植园之间非常高且相似(分别介于0.185和0.179之间)。

希布斯等人(Bsaibes et al.,2009)发现,当计算不同土地覆盖(水稻、小麦、玉米、裸土和草地)的反照率时,水稻表现出更多的困难,主要是因为水和植被的组合系统的复杂性。在人工林的情况下,反照率的变差仅约为0.1(图17-4)。

当分析不同叶面积指数的反照率模拟结果时,发现湿地和水稻种植园的反

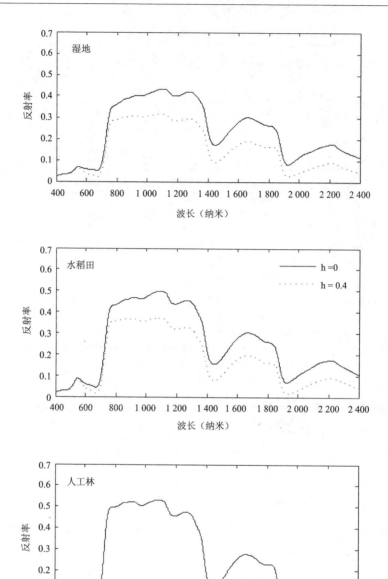

图 17-2

注：湿地、水稻田和人工林高光谱波段的模拟反射。所有例子的叶面积指数值都是4。实线表示没有水的例子，虚线表示淹没的例子，h=0.4米。

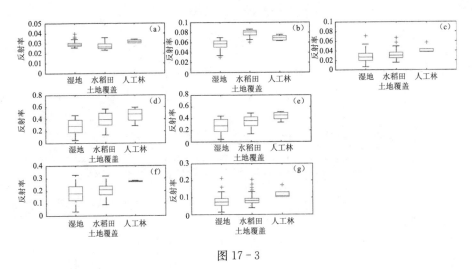

图 17-3

注:研究土地覆盖的每个 MODIS 波段模拟反射率箱线图(nweld=48,nrice=36,nforest=6)。(a)波段 459~479 纳米;(b)波段 545~565 纳米;(c)波段 620~670 纳米;(d)波段 841~876 纳米;(e)波段 1 230~1 250 纳米;(f)波段 1 628~1 652 纳米;(g)波段 2 005~2 155 纳米。每个波段的反射轴尺度不同。在每个箱体中,中心标记是中位数,上下边缘是第 25 和 75 百分位,箱须线延伸到极值点,离群值单独绘制为"+"符号。离群值是远离箱体顶部或底部的四分位范围的 1.5 倍以上的值。

照率呈正相关(图 17-5(a))。这类反照率值的差异仅在叶面积指数低值和高值之间显著(湿地的叶面积指数低值和高值分别为 0.058 和 0.183;水稻的叶面积指数的低值和高值分别为 0.106 和 0.233)。劳蒂艾宁等人(Rautiainen et al.,2011)没有发现叶面积指数变化对所计算的森林反照率有任何可测量影响。在这种情况下,叶面积指数和反照率间的关系与所进行的模拟有关。除叶面积指数外,模型中使用的输入变量对于所有模拟都是恒定的。然而,在自然界中很难找到此等情况,因为冠层特征(输入模型变量)本身是相关且协变。模拟结果表明,在叶面积指数增加的瞬间,近红外波段的反射率显著增加,而其他波段的反射率没有显著变化。

因此,与预期相反,反照率随叶面积指数的影响而增加。在湿地中,无论叶面积指数的值如何,水位都会导致反照率的变差很大(图 17-5(a))。变差似乎不会随着叶面积指数值的升高而降低,可能是因为高水位和淹没/露出水面的叶面积指数的比例变化很大。在湿地和水稻种植园中发现的变差差异来自于这样一个事实:在湿地模拟中,水的范围在 0~1.4 米之间,而在水稻种植园中,水的

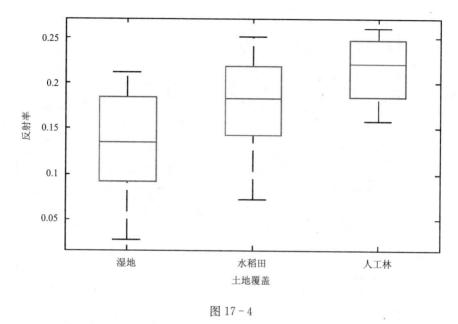

图 17-4

注：研究土地覆盖的反照率箱线图（nwelet＝48，nrice＝36，nforest＝6）。在每个框箱体中，中心标记是中位数，上下边缘是第 25 和 75 百分位，箱须线延伸到极值点，离群值是远离箱体顶部或底部的四分位范围的 1.5 倍以上的值。

范围在 0～0.5 米之间。

随着湿地和稻田水位的升高，反照率显著降低（图 17-5(b)）。因为水中的反照率值较低，因此这一结果是在预期当中（Rautiainen et al., 2011）。在湿地的情况下，三个测量水位的反照率值差异显著（$P<0.5$）（0、0.6 和 1.4 米的水位对应的 P 值分别为 0.207、0.147 和 0.071）。水稻种植园仅在极端水位值（0 和 0.5 米）上表现出显著差异。在这两种情况下，由于叶面积指数（淹没水中和露出水面）的影响，水位增加了变差。当系统中有水时，叶面积指数对高于水位的冠层反射率的贡献（即露出水面的叶面积指数）非常重要（Beget et al., 2013）。同样，总叶面积指数的微小差异导致了露出水面的叶面积指数的巨大差异；这反过来又导致了冠层反射率的巨大差异。特别是在水稻的情形中，我们注意到除了露出水面的和淹没水中的叶面积指数外，裸土对综合值的变差有重要影响（图 17-5(b)）。

除了来自反照率的即时值变化（图 17-4 和图 17-5）外，我们还预计由于叶

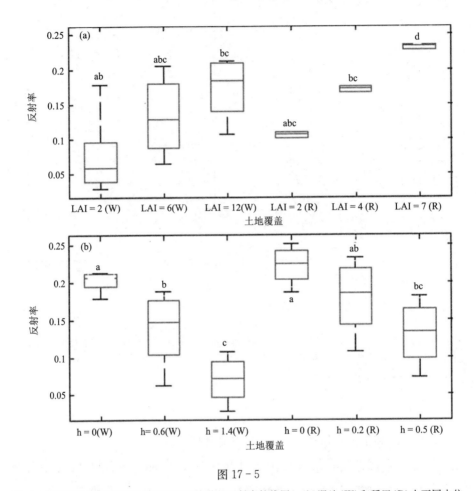

图 17 - 5

注：(a)湿地(W)和稻田(R)中不同叶面积指数的反射率箱线图。(b)湿地(W)和稻田(R)中不同水位(h)的综合反射率(MODIS 波段带 1~7 模拟反射率的整合)箱线图。在每个框箱体中，中心标记是中位数，上下边缘是第 25 和 75 百分位，箱须线延伸到极值点。用 T-Student 的 t 检验比较均值间的差异(nwowlands=48, nrice=36, p=0.05)。具有相同字母的均值没有显著差异(Tukey 显著性差异标准)。

面积指数的变化或不同水位，以及通过用新作物替换地表自然覆盖层所产生的变化，反照率会发生非常重要的时态演变。此等变化可能来自土地利用/土地覆盖变化或严重的干旱和洪水。为了回答这些问题，我们分析了生长季节模拟中反照率值的演变(图 17-6)。在 7 月，普通芦苇湿地(生长速度较慢)在高水位中保持着重要的绿色覆盖和衰老物质的逐步积累(Mal and Narine, 2004)。然而，在水稻的情形下，土壤是休耕的，并且大部分是低湿度的。在此条件下，水稻

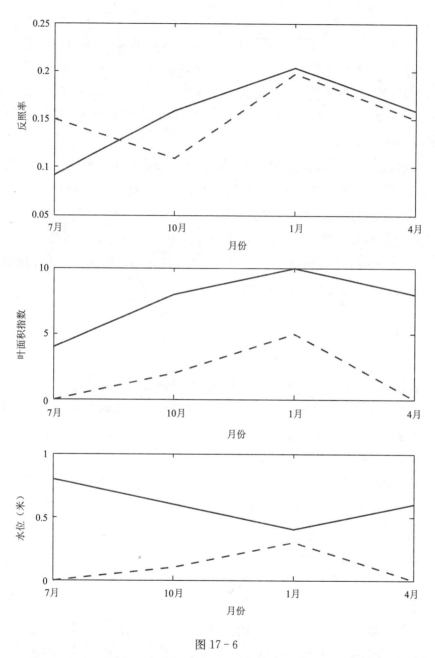

图 17-6

注：湿地状况（实线）和水稻状况（虚线）时间演变的反照率、叶面积指数和水位（单位：米）。7 月（冬季）、10 月（春季）、1 月（夏季）和 4 月（秋季）模拟了南半球的时间演化。

休耕地反照率比湿地高 40%(分别为 0.150 和 0.091)。10 月,即水稻播种后的一个月,叶面积指数值较低(叶面积指数=2),反照率极低(0.109),甚至低于湿地的反照率(0.159)。在 1 月(夏初),即花前和花期,湿地的叶面积指数是最大的,反照率值在两种情况下非常相似(0.230 和 0.194)。最后,在非常接近水稻收获和湿地物种生长最高点的 4 月(秋季),反照率值再次略高于新作物(水稻为 0.150,湿地为 0.159)。

一般而言,森林在捕获辐射方面非常有效(House et al.,2005)。然而,与森林相比,湿地生态系统通过非常低的反照率值能更有效地捕获辐射。这一事实改变了地表能量平衡,即较低的反照率值意味着增加了向生态系统和之后向大气的能量传递,从而改变了当地气候(见第十六章)。将辐射传输模型(如 SAIL-HFlood)集成到生态系统服务模型中,将使科学家能够以基础的方式评估能量平衡的反照率组成及各成分间的权衡。

三、结论

我们使用 SAILHFlood 模型模拟冠层(部分淹没水中和未淹没)的反射率。基于这些案例,我们能够间接研究土地利用变化将如何影响地表反照率。研究还考虑了稻田或人工林对湿地的替代作用,以及叶面积指数和不同水位等土地覆盖特征的变化。

反射率模拟模型可成为一种用作评估土地覆盖/土地利用变化引起的反射率变化的有益工具。光谱反照率可通过宽带转换、大气与角度校正后的窄谱带估算。

洪水会降低所有湿地和水稻的反照率。反照率的降低将取决于叶面积指数和水位。土地利用变化,如湿地替代或洪水水位下降,将增加地表反照率,从而改变地表能量平衡。

辐射传输模型(SAILHFlood)的使用使我们能够研究在实际土地覆盖中难以复制的情况,这不仅是因为湿地系统和水稻种植园中水的复杂性,还因为与时空易变性相关的复杂性。

辐射传输模型使评估不同土地利用间的取舍成为可能。例如,土地利用变化将捕获更多的二氧化碳(见第十六章),但也将使更多的地表变暖(较低的反照率)。在湿地的情况下,较低的反照率值并不总是与较高的二氧化碳捕获相关联,因为辐射也被水吸收。辐射传输模型的使用成为量化与生物地球物理过程相关的生态系统服务的有价值的近似法,如反照率。这些模型可通过对反射能力测定中涉及的相关因素的理解来补充卫星反照率评估。SAILHFlood 模型特别包括了洪水的情况;因此,它能够评估能量平衡,并在此基础上量化湿地生态系统的气候调节服务。

致谢

本研究由国际研究发展中心(IDRC-Canada,106601-001 项目)和阿根廷气候与水研究所 INTA(AERN4-AERN4642 项目)资助。

参 考 文 献

Allen, T. R., Y. Wang, and T. W. Crawford. 2013. Remote sensing of land cover dynamics. In *Treatise on geomorphology*, ed. J. F. Shroder, 80–102. San Diego, CA: Academic Press.

Alton, P. 2009. A simple retrieval of ground albedo and vegetation absorptance from MODIS satellite data for parameterisation of global land-surface models. *Agricultural and Forest Meteorology* 149:1769–1775.

Bacour, C., S. Jacquemoud, M. Leroy, et al. 2002. Reliability of the estimation of vegetation characteristics by inversion of three canopy reflectance models on airborne POLDER data. *Agronomie* 22:555–565.

Baigún, C. R. M., A. Puig, P. G. Minotti, et al. 2008. Resource use in the Parana River Delta (Argentina): Moving away from an ecohydrological approach? *Ecohydrology and Hydrobiology* 8:245–262.

Baldocchi, D. D., and K. B. Wilson. 2001. Modeling CO_2 and water vapor exchange of a temperate broadleaved forest across hourly to decadal time scales. *Ecological Modelling* 142:155–184.

Beck, P. S. A., S. J. Goetz, M. C. Mack, et al. 2011. The impacts and implications of an intensifying fire regime on Alaskan boreal forest composition and albedo. *Global Change Biology* 17:2853–2866.

Beget, M. E., V. A. Bettachini, C. M. Di Bella, and F. Baret. 2013. SAILHFlood: A radiative transfer model for flooded vegetation. *Ecological Modelling* 257:25–35.

Beget, M. E., and C. M. Di Bella. 2007. Flooding: The effect of water depth on the spectral response of grass canopies. *Journal of Hydrology* 335:285–294.

Beget, M. E., C. M. Di Bella, F. Baret, and J. F. Hanocq. 2010. Modeling reflectance of partially submerged canopies. *III International Symposium on Recent Advances in Quantitative Remote Sensing*. Valencia, Spain.

Brander, L. M., R. J. Florax, and J. E. Vermaat. 2006. The empirics of wetland valuation: A comprehensive summary and a meta-analysis of the literature. *Environmental and Resource Economics* 33:223–250.

Bright, R. M., F. Cherubini, and A. H. Strømman. 2012. Climate impacts of bioenergy: Inclusion of carbon cycle and albedo dynamics in life cycle impact assessment. *Environmental Impact Assessment Review* 37:2–11.

Bsaibes, A., D. Courault, F. Baret, et al. 2009. Albedo and LAI estimates from FORMOSAT-2 data for crop monitoring. *Remote Sensing of Environment* 113:716–729.

Bunn, S. E., and A. H. Arthington. 2002. Basic principles and ecological consequences of altered flow regimes for aquatic biodiversity. *Environmental Management* 30:492–507.

Burba, G. G., S. B. Verma, and J. Kim. 1999. Surface energy fluxes of *Phragmites australis* in a prairie wetland. *Agricultural and Forest Meteorology* 94:31–51.

Casanova, D., G. F. Epema, and J. Goudriaan. 1998. Monitoring rice reflectance at field level for estimating biomass and LAI. *Field Crops Research* 55:83–92.

Cescatti, A., B. Marcolla, S. K. Santhana Vannan, et al. 2012. Intercomparison of MODIS albedo retrievals and in situ measurements across the global FLUXNET network. *Remote Sensing of Environment* 121:323–334.

Clevering, O., and J. Lissner. 1999. Taxonomy, chromosome numbers, clonal diversity and population dynamics of *Phragmites australis*. *Aquatic Botany* 64:185–208.

Consoli, S., G. D'Urso, and A. Toscano. 2006. Remote sensing to estimate ET-fluxes and the performance of an irrigation district in southern Italy. *Agricultural Water Management* 81:295–314.

Csiszar, I., and G. Gutman. 1999. Mapping global land surface albedo from NOAA AVHRR. *Journal of Geophysical Research—Atmospheres* 104:6215–6228.

Deschamps, P. Y., F. Brion, M. Leroy, et al. 1994. The POLDER mission: Instrument characteristics and scientific objectives. *IEEE Transactions on Geoscience and Remote Sensing* 32:598–615.

Diner, D. J., J. C. Beckert, T. H. Reilly, et al. 1998. Multi-angle Imaging Spectroradiometer (MISR) instrument description and experiment overview. *IEEE Transactions on Geoscience and Remote Sensing* 36:1072–1087.

Disney, M., P. Lewis, G. Thackrah, T. Quaife, and M. Barnsley. 2004. Comparison of MODIS broadband albedo over an agricultural site with ground measurements and values derived from Earth observation data at a range of spatial scales. *International Journal of Remote Sensing* 25:5297–5317.

Gitelson, A. A., M. N. Merzlyak, and H. K. Lichtenthaler. 1996. Detection of red edge position and chlorophyll content by reflectance measurements near 700 nm. *Journal of Plant Physiology* 148:501–508.

Hatzianastassiou, N., C. Matsoukas, D. Hatzidimitriou, C. Pavlakis, M. Drakakis, and I. Vardavas. 2004. Ten year radiation budget of the Earth: 1984–93. *International Journal of Climatology* 24:1785–1802.

Hosgood, B., S. Jacquemoud, G. Andreoli, J. Verdebout, A. Pedrini, and G. Schmuck. 1994. Leaf Optical Properties EXperiment 93 (LOPEX93). *Report EUR 16095 EN*. Ispra, Italy: European Commission, Joint Research Centre, Institute for Remote Sensing Applications.

Houldcroft, C. J., W. M. F. Grey, M. Barnsley, et al. 2009. New vegetation albedo parameters and global fields of soil background albedo derived from MODIS for use in a climate model. *Journal of Hydrometeorogy* 10:183–198.

House, J., V. Brovkin, R. Betts, et al. 2005. Climate and air quality. In *Millennium ecosystem assessment. Ecosystems and human well-being: Current states and trends*, eds. P. Kabat, S. Nishioka, 355–390. Washington, DC: Island Press.

Jacquemoud, S., C. Bacour, H. Poilvé, and J. P. Frangi. 2000. Comparison of four radiative transfer models to simulate plant canopies reflectance—Direct and inverse mode. *Remote Sensing of Environment* 74:471–481.

Jacquemoud, S., W. Verhoef, F. Baret, et al. 2009. PROSPECT+SAIL models: A review of use for vegetation characterization. *Remote Sensing of Environment* 113:S56–S66.

Ju, W., P. Gao, J. Wang, Y. Zhou, and X. Zhang. 2010. Combining an ecological model with remote sensing and GIS techniques to monitor soil water content of croplands with a monsoon climate. *Agricultural Water Management* 97:1221–1231.

Key, J. R., X. Wang, J. C. Stoeve, and C. Fowler. 2001. Estimating the cloudy-sky albedo of sea ice and snow from space. *Journal of Geophysical Research—Atmospheres* 106:12489–12497.

Kingsford, R. T. 2000. Ecological impacts of dams, water diversions and river management on floodplain wetlands in Australia. *Austral Ecology* 25:109–127.

Kuusinen, N., P. Kolari, J. Levula, A. Porcar-Castell, P. Stenberg, and F. Berninger. 2012. Seasonal variation in boreal pine forest albedo and effects of canopy snow on forest reflectance. *Agricultural and Forest Meteorology* 164:53–60.

Lambert, A. 2003. *Economic valuation of wetlands: An important component of wetland management strategies at the river basin scale*. Ramsar: Gland.

Liang, S. 2000. Narrowband to broadband conversions of land surface albedo I: Algorithms. *Remote Sensing of Environment* 76:213–238.

Liang, S., A. H. Strahler, and C. W. Walthall. 1999. Retrieval of land surface albedo from satellite observations: A simulation study. *Journal of Applied Meteorology* 38:712–725.

Lucht, W., C. B. Schaaf, and A. H. Strahler. 2000. An algorithm for the retrieval of albedo from space using semiempirical BRDF models. *IEEE Transactions on Geoscience and Remote Sensing* 38:977–998.

MA (Millennium Ecosystem Assessment). 2005. *Ecosystems and human well-being: Current state and trends*. Washington, DC: Island Press.

Maes, W. H., T. Pashuysen, A. Trabucco, F. Veroustraete, and B. Muys. 2011. Does energy dissipation increase with ecosystem succession? Testing the ecosystem energy theory combining theoretical simulations and thermal remote sensing observations. *Ecological Modelling* 222:3917–3941.

Mal, T. K., and L. Narine. 2004. The biology of Canadian weeds. 129. Phragmites australis (Cav.) Trin. ex Steud. *Canadian Journal of Plant Science* 84:365–396.

Mariotto, I., and V. P. Gutschick. 2010. Non-Lambertian Corrected Albedo and Vegetation Index for estimating land evapotranspiration in a heterogeneous semi-arid landscape. *Remote Sensing* 2:926–938.

Meroni, M., R. Colombo, and C. Panigada. 2004. Inversion of a radiative transfer model with hyperspectral observations for LAI mapping in poplar plantations. *Remote Sensing of Environment* 92:195–206.

Mitsch, W. J., and J. G. Gosselink. 2000. *Wetlands*. 3rd ed. New York: John Wiley & Sons.

Moulin, S., L. Kergoa, P. Cayrol, G. Dedieu, and L. Prévot. 2002. Calibration of a coupled canopy functioning and SVAT model in the ReSeDA experiment. Towards the assimilation of SPOT/HRV observations into the model. *Agronomy* 22:681–686.

Mueller, R., J. Trentmann, C. Träger-Chatterjee, R. Posselt, and R. Stöckli. 2011. The role of the effective cloud albedo for climate monitoring and analysis. *Remote Sensing* 3:2305–2320.

Nasipuri, P., T. J. Majumdar, and D. S. Mitra. 2006. Study of high-resolution thermal inertia over western India oil fields using ASTER data. *Acta Astronautica* 58:270–278.

Noe, S. M., V. Kimmel, K. Hüve, et al. 2011. Ecosystem-scale biosphere–atmosphere interactions of a hemiboreal mixed forest stand at Järvselja, Estonia. *Forest Ecology and Management* 262:71–81.

Oguro, Y., Y. Suga, S. Takeuchi, H. Ogawa, and K. Tsuchiya. 2003. Monitoring of a rice field using Landsat-5 TM and Landsat-7 ETM+ data. *Advanced Space Research* 32:2223–2228.

Pinty, B., and M. Verstraete. 1992. On the design and validation of surface bidirectional reflectance and albedo model. *Remote Sensing of Environment* 41:155–167.

Posselt, R., R. W. Mueller, R. Stöckli, and J. Trentmann. 2012. Remote sensing of solar surface radiation for climate monitoring—The CM-SAF retrieval in international comparison. *Remote Sensing of Environment* 118:186–198.

Puzachenko, Y. G., R. B. Sandlersky, and A. Svirejeva-Hopkins. 2011. Estimation of thermodynamic parameters of the biosphere, based on remote sensing. *Ecological Modelling* 222:2913–2923.

Rautiainen, M., P. Stenberg, M. Mottus, and T. Manninen. 2011. Radiative transfer simulations link boreal forest structure and shortwave albedo. *Boreal Environment Research* 16:91–100.

Román, M. O., C. B. Schaaf, P. Lewis, et al. 2010. Assessing the coupling between surface albedo derived from MODIS and the fraction of diffuse skylight over spatially-characterized landscapes. *Remote Sensing of Environment* 114:738–760.

Schaaf, C. B., F. Gao, A. H. Strahler, et al. 2002. First operational BRDF, albedo nadir reflectance products from MODIS. *Remote Sensing of Environment* 83:135–148.

Schottler, S. P., J. Ulrich, P. Belmont, et al. 2013. Twentieth century agricultural drainage creates more erosive rivers. *Hydrological Processes*. doi:10.1002/hyp.9738.

Shuai, Y., J. G. Masek, F. Gao, and C. B. Schaaf. 2011. An algorithm for the retrieval of 30-m snow-free albedo from Landsat surface reflectance and MODIS BRDF. *Remote Sensing of Environment* 115:2204–2216.

Smith, P., H. Black, C. Evans, et al. 2011. Regulating services. In *UK National Ecosystem Assessment. Understanding nature's value to society*. Technical Report, 535–596. Cambridge, UK: UNEP-WCMC.

Stroeve, J., J. E. Box, F. Gao, S. Liang, A. Nolin, and C. Schaaf. 2005. Accuracy assessment of the MODIS 16-day albedo product for snow: Comparisons with Greenland in situ measurements. *Remote Sensing of Environment* 94:46–60.

Stroppiana, D., M. Boschetti, R. Confalonieri, S. Bocchi, and P. A. Brivio. 2006. Evaluation of LAI-2000 for leaf area index monitoring in paddy rice. *Field Crops Research* 99:167–170.

Stroppiana, D., S. Pinnock, J. M. C. Pereira, and J. M. Grégoire. 2002. Radiometric analysis of SPOT-VEGETATION images for burnt area detection in northern Australia. *Remote Sensing of Environment* 82:21–37.

Sumner, D. M., Q. Wu, and C. S. Pathak. 2011. Variability of albedo and utility of the MODIS albedo product in forested wetlands. *Wetlands* 31:229–237.

Verhoef, W. 1984. Light scattering by leaf layers with application to canopy reflectance modeling: The SAIL model. *Remote Sensing of Environment* 16:125–141.

Verhoef, W. 1985. Earth observation modeling based on layer scattering matrices. *Remote Sensing of Environment* 17:165–178.

Wang, X., and C. S. Zender. 2010. MODIS snow albedo bias at high solar zenith angles relative to theory and to in situ observations in Greenland. *Remote Sensing of Environment* 114:563–575.

Weiss, M., F. Baret, R. Myneni, A. Pragnère, and Y. Knyazikhin. 2000. Investigation of a model inversion technique for the estimation of crop characteristics from spectral and directional reflectance data. *Agronomie* 20:3–22.

Yang, Y., S. Shang, and L. Jiang. 2012. Remote sensing temporal and spatial patterns of evapotranspiration and the responses to water management in a large irrigation district of north China. *Agricultural and Forest Meteorology* 164:112–122.

Zheng, B., L. Shi, Y. Ma, Q. Deng, B. Li, B. and Y. Guo. 2007. Canopy architecture quantification and spatial direct light interception modeling of hybrid rice. *5th Workshop Functional Structural Plant Models Proceedings*, Napier, New Zealand, November 4–9, 37.1–37.3.

Zwart, S. J., W. G. M. Bastiaanssen, C. de Fraiture, and D. J. Molden. 2010. WATPRO: A remote sensing based model for mapping water productivity of wheat. *Agricultural Water Management* 97:1628–1636.

第十八章 能量平衡和蒸散：一种评价生态系统服务的遥感方法

V. A. 马尔切西尼　J. P. 古尔施曼　J. A. 索布里诺

一、生态系统服务、能量平衡和蒸散

在过去的40年里，人类已经改变了生态系统功能和生态系统为人类提供基本服务的能力，例如水资源供给和局部与区域的天气控制（MA，2005）。"生态系统服务"一词是指生态系统中支持和维持人类生活的所有过程（Daily，1997；Kremen and Ostfeld，2005；Kinzig et al.，2011）。自然生态系统通过调节天气和减缓气候变化来维持生物多样性，在维持人类生存方面发挥着不可替代的作用。蒸散是调节局部与区域气候的生态系统功能的综合衡量指标（Carlson et al.，1995；Anderson and Kustas，2008；Jung et al.，2010）。

蒸散被定义为水从植被、土壤和水体释放到大气中的过程（Allen et al.，1998）。蒸散通过调节植物冠层的水汽和热量，来调节水分动力与地下和地表过程的相互作用。蒸散通常表示为单位时间内的水的体积（如毫米每天），当考虑在特定温度下蒸发水所需的能量时，也可表示为能量单位（如瓦每平方米）。影响蒸散的主要变量是太阳辐射、空气湿度、风速、气温和土壤湿度（Carlson et al.，1990；Allen et al.，2006b；Jung et al.，2010）。局部土壤质地和盐度等因素也会对蒸散造成影响（Fernandez-Illescas et al.，2001；El-Nahry and Hammad，2009）。植被是影响蒸散的主要生物因子，植被的功能群和物种组成调节着蒸散的大小（Nosetto et al.，2005；Miao et al.，2009）。植被覆盖和植被结构的变化会显著影响能量的反射和传输，从而影响蒸散和水循环，对气候和生态系统服产生重要影响(Chapin et al.,2002;Chen et al.,2009)。

由于从土壤和植被蒸发水分的过程需要能量,事实上全球超过一半的地球表面所用能量是用于蒸发水分的(Jung et al.,2010),可以通过考虑能量平衡来估算和量化蒸散。简单来说,就是地表能量平衡方程(将在下一节中详细描述)假定地面上可用能量是到达地面的能量总量减去通过传导转移至土壤表面的能量通量。这些能量可分为两种湍流通量:感热通量和潜热通量。感热是指来自传导和对流过程的热通量,会导致温度发生变化(Jackson et al.,1977)。潜热与从地表(例如植被覆盖)蒸发水所需的能量有关。潜热和感热之间的比例关系(蒸散率)决定了水汽和热量间的平衡。因此,利用能量平衡来估算蒸散率对许多水文和生物地球化学过程的评估具有深远的意义(Wang et al.,2012)。

根据空间尺度的不同,还可以使用其他方法来估算蒸散:在不同空间尺度($10^0 \sim 10^3$平方米),最常见的方法包括蒸发皿法(Brutsaert and Parlange,1998;Blight,2002)、蒸渗仪法(Paruelo et al.,1991)和液流传感器(Burgess et al.,2001;Doody and Benyon,2011)。这些方法的最大局限之一是需要大量的重复测量来反映土壤的空间异质性。另一个问题是,将在不同空间尺度上获得的数据扩展到整个冠层可能会产生误导结果,这是由于尺度的变化可能导致非线性生态系统响应(Wang et al.,2012)。中等空间尺度的蒸散($10^4 \sim 10^6$平方米)最佳方式是通过通量塔获得:在平台上安装各种传感器,主要用于估算能源、碳和水动力(Baldocchi,2003;Cleugh et al.,2007)(见第二章)。通量塔的使用在过去的几年中已变得普遍,这是因为它们测量蒸散的尺度适用于获得卫星观测的地面真实数据。尽管这些技术提供了较好的时间分辨率,但仍然受限于扩展大范围的覆盖所需的设备数量及获取区域的精度不足。

尽管人们努力提高碳、水和能源在不同尺度上的估算精度,但以地面观测为补充的遥感技术仍然是覆盖大尺度(全球10^6平方米)和整合长时间序列数据的最佳方法,这是由于一些卫星平台和遥感产品已存在40多年。例如,陆地卫星(Landsat)卫星数据自1970年代初就可获得,甚高分辨率扫描辐射计(AVHRR)自1980年以来一直在收集信息。通过遥感估算蒸散主要基于两种方法:(1)通过光谱指数(如归一化植被指数)将蒸散与植被覆盖和绿度联系起来(NDVI;Caparrini et al.,2003;Mu et al.,2007;Contreras et al.,2011;Yebra et al. ,2013);(2)建立蒸散与地表温度之间的联系(Norman et al. ,1995;Su,2002;Sobrino et al. ,2005;Kalma et al. ,2008)。其中一些模型采用能量平衡方程和可利用能量在感热通量和潜热通量之间的比例关系来与蒸散和能量平衡建立联系。

二、建立蒸散和能量平衡与生态系统服务的联系

地球上的生命是由能量控制的,通过能量和水的结合开展光合作用等过程来支持生态系统的生命。如前所述,到达地球表面的能量主要是入射太阳辐射总量减去存量辐射总量的结果。简单地说,净辐射(Rn)可被认为是三个主要项的总和:

$$R_n = G + H + LET \qquad 式18-1$$

其中,Rn 为净辐射,G 为地面热通量,LET 为潜热通量,H 为感热通量。

式中所有项都用时间和面积的能量单位表示,如瓦秒每平方米(Jackson et al. ,1977;Hurtado and Sobrino,2000)。

这个简化的平衡方程没有考虑到其他次要的能量成分,如土壤和植被中储存的热量或作为代谢过程结果的通量。此外,该方程只考虑垂直通量,而不考虑平流过程导致的水平方向流动的能量(Allen et al. ,2006b)。从所有项中,潜热通量和感热通量都是与生态系统服务功能密切相关的项,因为蒸散是由潜热估算而来的,它代表由植被与水循环之间的直接联系。潜热通量与感热通量的比例关系,即蒸散率(λET),决定了地表水汽与热量的平衡,可作为地表覆盖变化的指标,具有特定的生态意义。土地利用的变化,如用作物替代森林或景观破碎化,以及因城市扩张而砍伐森林,都会对潜热通量与感热通量的再分配产生重大影响,从而影响蒸散和水平衡(Anderson et al. ,2008)。蒸散、降水、径流和深层排水的变化对提供气候和土壤稳定、调节局部降雨量、影响养分动态和地下水回灌率的生态系统服务有着深远的影响。蒸散还占到陆地水输入的50%以上,特别是在干旱和半干旱地区(Wilcox,2002;Reynolds et al. ,2004)。

三、利用遥感估算蒸散及其与生态系统服务的关系

近年来,随着技术的进步,在与变化相同的尺度上开展生态系统服务人类影响评估更加便利(Kerr and Ostrovsky,2003)。星载遥感传感器,是用来评估大面积区域变化影响为数不多的方法之一。许多用于评估地表和大气之间通量交换的方法均将感热通量获取为能量平衡方程的残差。这些模型的缺点是,很多需要局部自适应,且依赖于空间尺度,为获得"纯像素"需要的较大范围。然而,结合不同卫星平台在不同时空尺度上提供的产品可以弥补这些缺陷。

例如,地表能量平衡系统(SEBS;Su,2002)结合了反照率、地表温度(LST)、植被覆盖、空气湿度、风速和向下太阳辐射等变量来估算蒸散。该方法的优点是,大多数的输入变量可直接从遥感数据中获得。例如,当植被覆盖数据不可用时,归一化植被指数可被用作替代数据(Su,2002)。

一种用于估算感热、潜热和土壤热通量(G)的开拓性模型是由诺曼等(Norman et al.,1995)开发的两源能量平衡模型(TSEB)。该两层模型主要基于热遥感数据,需要输入净辐射、方向角视角、空气温度、风速,以及叶面积指数、植被高度或叶片尺寸等植被参数。该模型采用植被覆盖度来估算土壤温度和冠层温度。植被和土壤成分的分离是该模型的一个重要属性,这是由于它可以解释亚像素水平的变异性(Timmermans et al.,2011)。随后,安德森等(Anderson et al.,2008)通过结合冠层阻力光能利用效率模型改进了原模型对水分损失的预测,该模型考虑了气孔导度和气孔关闭的影响,这是由于水汽压亏缺导致的结果。

另一个用于估算蒸散的遥感模型是罗林克等(Roerink et al.,2000)开发的简化地表能量平衡指数(Simplifed Surface Energy Balance Index,S-SEBI)模型。该方法结合了在干湿条件下的反射率和表面温度来确定蒸发率,并从中获得感热和潜热。模型依据的原理如下:在水分充分保障条件下(例如灌溉作物),低反照率值决定了稳定的LST,这是由于系统中的可用能量被用于蒸发水。如果反照率值上升,LST增加到一个被称为"蒸发受控"的限定点,此时蒸散和潜热最大,感热为零($Rn-G=H=0=LET_{max}$)。当土壤水分保障率降低,蒸散值

接近于零时,系统中的所有可用能量都被用于加热土壤,因此地表温度增加。超过反射率阈值,地表温度达到一个称为"辐射受控"的点,此时蒸散为零,热通量最大($LET=0, H_{max}=Rn-G$)。给定这两个温度点($LET=0$ 和 $LET=max$),可将蒸散按比例估算为这两个温度线间值。潜热通量 LET 计算为蒸发率乘以净可利用能与土壤热通量之差。净有效能是通过考虑入射太阳辐射、大气透射率和地面反照率来计算的,而土壤热通量是通过表面反射率、表面温度和归一化植被指数或其他植被指数(如改良土壤调整植被指数 MSAVI 或叶面积指数 LAI)之间的关系根据经验估算的。该模型已在西班牙西南部(Sobrino et al., 2005)和意大利(Roerink et al., 2000)及阿根廷中部(Marchesini et al., 2012)等地区得到验证。

蒂默曼斯等(Timmermans et al., 2011)在德国东北部的农业区对简化地表能量平衡指数模型(S-SEBI)和两源能量平衡模型进行了比较。作者没有发现辐射通量(如净辐射或土壤热通量)上有大的差异,但他们发现湍流通量(H, LET)有着显著的差异,特别是在高粗糙度地区例如暴露在干燥条件下的高大作物的地区,这表明这两种模型在潮湿和均匀地区的一致性更好。

地表能量平衡系统、简化地表能量平衡指数模型等相关方法使用遥感来求解能量平衡方程项,而其他方法则将植被指数(和其他衍生指数)与蒸散进行更多的经验联系。中分辨率成像光谱仪获得的叶面积指数产品被用作表示冠层水平气孔导度,并结合潜在蒸散的 Penman-Monteith 模型来估算实际蒸散(Cleugh et al., 2007)。基于同样的方法,慕等(Mu et al., 2007, 2011)将该模型应用于全球数据,获得了全球尺度 8 天蒸散估算值。格尔施曼等(Guerschman et al., 2009)同样利用来源于 MODIS 传感器的增强植被指数(EVI)将潜在蒸散(ET_p)换算为实际蒸散(ET_a)。潜在蒸散是使用 Priestley-Taylor 公式从气象学数据中导出的。除了量化植被覆盖的增强植被指数外,还使用了全球植被水分指数(GVMI)。当植被覆盖(因而 EVI)较低时,全球植被水分指数允许区分开阔水面和裸露土壤,从而提供了一种更好的开阔水面蒸发的表示方式。该方法使用澳大利亚 7 个站点的通量塔测量数据进行校准,并使用约 200 个来源于集水区的蒸散(作为降雨和水流间的长期差异)进行评估。图 18-1 显示了由格尔施曼等(Guerschman et al., 2009)的方法推导出的横跨澳大利亚大陆的长期水平衡。通过在大区域内动态估算蒸散，并将其与降雨插值数据相结合，就能识别出降雨以外的蒸发水分来源区域(图 18-1，红色)。这些区域对应着依赖地下水的生态系统、灌溉作物、湿地以及它们的组合。这些地区往往是粮食、纤维、木材生产和生物多样性维护等生态系统产品和服务的重要来源地。

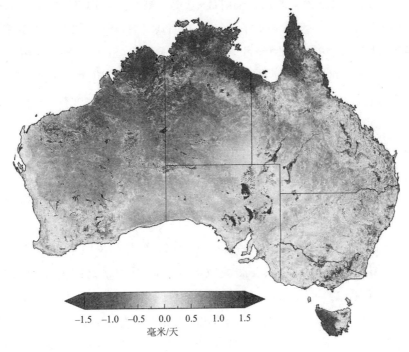

图 18-1(见彩插)

注：2000~2006 年澳大利亚平均降水量与平均月实际蒸散之差。(改编自 Guerschman, J.P., et al., 2009.)

四、蒸散、反照率、地表温度及其与生态系统服务的关系

植被结构简化(如叶面积指数由高到低)、物种丰富度下降或土壤覆盖减少导致的生态系统服务功能的恶化，可以通过增加所反映的能量比例来改变能量收支，从而影响地表温度和反照率(Dirmeyer and Shukla,1994)。

地表温度是最能综合大气现象和发生在地表的过程之间关系的变量。安德

森和库斯塔斯(Anderson and Kustas,2008)将地表温度描述为一种温度计,在这种温度计中,地表温度的增加可以作为异常情况(如缺水压力)的指示器。地表温度可由地球观测平台上的传感器测量的热红外数据进行反演。这些传感器包括:Landsat 平台上的专题制图仪(TM)或增强型专题绘图仪(ETM+);Terra 平台上的先进星载热发射和反射辐射计,它提供高空间分辨率(约 100 米)的数据;美国国家海洋和大气管理局平台上的 AVHRR;Terra 上的 MODIS;第二代气象卫星(MSG)上的旋转增强可见光和红外成像器(SEVIRI),提供低空间分辨率(1~5 千米)的数据。地表温度的估算有着广泛的应用需求,例如研究自然生态系统、农业和城市规划(例如城市热岛生态学)、减缓气候变化、林业、气象预测和水文学等方面。在新墨西哥州乔纳达地区,利用 Landsat 影像进行的研究显示,在地面植被覆盖较低的地区,灌木的侵入使地表温度和辐射增加了 5°C (Ritchie et al.,2000)。

热红外数据也被用于利用云温度区分云与非雨云,进而从空间和时间上评估降水趋势(Feidas et al.,2009)。

另一个可用于反映生态系统特性变化的综合变量是反照率,即从一个表面反射的太阳辐射通量与同一表面截取的太阳辐射通量之间的比值(Liang,2000)。在干旱和半干旱地区,反照率的增加导致地表吸收的能量的损失。土地开荒和裸土覆盖增加通过改变表面粗糙度,来提高地表反照率,并减少蒸散和局部降雨(Dirmeyer and Shukla,1994)(见第十七章)。据报道,由于灌木的扩散,奇瓦瓦沙漠的反照率和辐射率发生变化,对近地表气候产生了影响(Beltrán-Przekurat et al.,2008)。"辐射胁迫"一词在很大程度上被用于表示可能的全球变暖效应,即辐射平衡的变化,包括对流层顶水平面的反照率变化(Forster and Ramaswamy,2007)。豪斯帕诺西安等(Houspanossian et al.,2012)利用 MODIS 反照率产品 MCD43A3,评估了阿根廷半干旱地区由木本植物向以作物为主导的生态系统的转变是如何改变反照率和地表温度的。他们发现,作物的日地表温度为 2.5°C,反照率比森林高 50%。在亚马孙森林反照率从 0.09 增至 0.3 的地区,观察到了放晴后降水减少(Dirmeyer and Shukla,1994)。反照率的增加和潜热的增加意味着系统含水量的减少和地表温度的增加。然而,在中高纬度地区,一种"违反直觉"的冷却效应得到证实,森林覆盖率的减少会增加积雪的暴露和反照率,从而导致在昼夜循环中白天和夜间对大气的能量损失更大(Lee et al.,2011)。

通过遥感估算反照率可使用宽波段或窄波段观测来进行;然而,若已知大气层顶部的大气表面条件,则可获得精确的反照率值。由于具有更好的分辨率来表征地表和大气层面上的非均匀性,窄波段多光谱传感器在估算反照率方面更有效(Liang,2000)。从窄波段观测获得的反照率值需要进行角度和大气校正。目前,已有许多算法提出将宽波段反照率值转换为窄波段反照率值(Brest and Goward,1987;Liang,2000)。一般而言,根据气溶胶的光学深度,反照率可以被估算为完全直接照明(黑天反照率)和完全漫射照明(白天反照率)之间的外推值。要准确估算反照率,需要计算双向反射分布函数。该函数考虑从表面散射到半球上方任何方向表面的辐射。因此,双向反射分布函数被用于将具有可变太阳视图的几何形状的观测,标准化为一个通用的标准几何形状的观测。

由于地表结构会影响双向反射分布函数,该函数也可作为观测地表生物物理特征的信息来源。双向反射分布函数的估算需要在不同的观测几何条件下获取可见光近红外(VNIR)和短波近红外(SWIR)数据。

当可见光近红外和短波近红外数据不可用时,可以使用诸如桑德斯(Saunders,1990)提出的简单方法。该简单方法通过使用可见光(RED)和近红外(NIR)反射通道来获得反照率,如:

$$\text{Albedo}(\alpha) = 0.5(\rho \text{RED}) + \rho \text{NIR} \qquad 式 18-2$$

其中,ρRED 和 ρNIR 分别是可见光和近红外的反射率(Saunders,1990)。反照率的估算被认为是主要的辐射不确定因素之一。许多用于预测该变量的模型均存在大于 15% 的误差(Liang,2000)。

本节最后要强调的一点是地表温度、反照率和蒸散之间的联系。这三个变量通过瞬时净辐射通量(RNI)联系在一起,瞬时净辐射通量是能量平衡方程(式18-1)中的参数,用于估算感热通量和蒸散分数并由此推导蒸散量:

$$R_{ni} = (1-\alpha)Rc\lambda \downarrow + \varepsilon Rg\lambda \downarrow - \varepsilon\sigma T4S \qquad 式 18-3$$

其中,α 为反照率,$Rc\lambda \downarrow$ 为入射短波辐射(瓦每平方米),$\varepsilon Rg\lambda \downarrow$ 为入射长波辐射(瓦每平方米),σ 为 Stefan Boltzmann 常数,ε 为表面发射率,TS 是表面温度(K)(Sobrino et al.,2004,2005)。

目前,利用热红外(与地表温度相关)和反照率数据来估算蒸散变化的文献非常丰富(Carlson et al.,1995;Anderson and Kustas,2008;Kalma et al.,2008),在水文学、城市规划(见第十九章)、天气预报、农业和自然资源管理方面具有潜在的应用价值。

五、案例研究:阿根廷中部大规模毁林对干性森林的影响——景观尺度下蒸散、地表温度和反照率的变化

森林砍伐是人类在 21 世纪面临的主要环境问题之一(Malhi et al.,2008)。土地覆盖变化和植被格局的变化(如从木本植被为主导向草地/作物变化)会影响蒸散的大小。森林更替可增加生态系统服务的供给,例如粮食生产、木材和住房,但同时,森林砍伐可能对缓解干旱和洪水等生态系统服务调节功能产生负面影响(Carpenter and Folke,2006)。

澳大利亚的旱地盐碱化是最著名的例子之一,其中当地森林被农田(从高到低蒸散)取代,导致地下水位升高,造成土地盐碱化和数千公顷的肥沃地损失(Barrett-Lennard,2003;Lambers,2003)。相反的现象在阿根廷的潘帕斯草原被观察到,当地的草原上因树木种植使得地下水位和溪流减少了近 50%(Nosetto et al.,2012)。

南美洲是世界上原始森林比例最高、森林生物质碳储量最大的地区(>1 000 亿吨碳),同时也是森林砍伐率最高的地区。根据联合国粮食及农业组织(FAO,2010)的数据,南美洲的原始森林损失比例仍然非常高,每年有近 400 万公顷的森林损失(Hamilton,2008)。这些森林提供了广泛的生态系统服务:水土保持、荒漠化控制和减缓气候变化。在工业和家庭用水的供给中,森林集水区也有着重要比例。森林与能量平衡和水循环是通过调节过程而联系在一起,这些过程包括局部降水、土壤和冠层直接蒸发、蒸腾、土壤渗透(茎流和根孔)、雾水收集和辐射反射等(Dirmeyer and Shukla,1994;Miles et al.,2006;Hansen et al.,2010)。

利用 SEBAL 模型对玻利维亚森林进行的蒸散估算表明,超过 50% 的年降水量依赖于由蒸腾作用和湿叶蒸发产生的原位蒸散。这些地区的大规模森林砍

伐意味着局部总的水投入的减少,影响森林再生并导致极端温度升高等过程。塞勒和摩恩尼(Seiler and Moene,2011)在亚马孙森林中也得到了类似的结果,森林砍伐使蒸散减少了22%。

本节描述了干旱林地向草地发生的变化是如何减少了蒸散,并提高地表温度和反照率,利用卫星影像和基于能量平衡方程的模型被用于评估这一过程对阿根廷中部某干性森林的影响。通过卫星图像,我们能够对干扰发生之前、期间和之后的同一地点进行比较。

在阿根廷中部的圣路易斯省,灌木和乔木是森林顶层的优势种(Aguilera et al.,2003;Marchesini et al.,2012)。在过去的几十年里,农业扩张力度加大,特别是在大豆成为主要作物的地区。作为以大豆为代表的农业扩张的结果,集约化的牧场生产被推向半干旱地区,在历史上这些地区被用于自由牧场和选择性采伐(Gasparri and Grau,2009)。一种最常见的去除灌木林植物的方法是切碎辊,这种技术使用由推土机移动的重金属圆筒,有选择性地挑选乔木和灌木,最终通过切碎和碾碎将它们清除(Blanco et al.,2005)。这种做法的结果是,在乔木和灌木被清除后最终增加了几倍的草料覆盖,牛可以很容易地获得牧草。

大规模毁灭木本植被可能会影响生态系统通过蒸散整体利用水分的能力。森林的一些物理特征,如冠层高度和深层根系分布,会影响水分流失(Kelliher et al.,1993;Moore and Heilman,2011)。空气湍流是影响蒸散的最主要变量之一,在高冠层系统中,由于较高的粗糙度,空气湍流通常会升高,这有利于大气水分需求与冠层水分流失之间的强耦合。

2006～2009年期间,为了增加牧草产量,在阿根廷中部的一个大型干性林基地,有选择地清理了8块面积从60～300公顷的土地。其余未受干扰的干性林对照样地用于比较蒸散、地表温度和反照率。利用来自Landsat/TM5平台的遥感数据分析受干扰和未受干扰的样地,包括从2004年1月～2009年12月每年至少包括4幅影像。

使用简化的地表能量平衡模型(Roerink et al.,2000)估算蒸散,模型细节已在前一节中解释。为了比较和验证卫星影像估算蒸散的结果,使用了一种基于气温(从当地气象站获得的)的间接和简单的方法(Blaney and Criddle,1950)。虽然Blaney-Criddle方法由于基于标准条件而被认为是不准确的,但它可作为一个标准参数来比较在扰动样地和未扰动样地之间观测到的蒸散的大小和趋势。

对干性森林的砍伐导致蒸散减少30%,地表温度的增加高达4℃。在清除树木和灌木后,反照率也增加了60%以上。未扰动样地和扰动样地的差异在扰动后立即出现,并且在夏季和雨季的差异更大。清除的区域蒸散分布也表现出较大的空间变异性(图18-2)。根系和地上生物量的减少、叶面指数的变化、物种间的互补和正生态位相互作用是导致蒸散显著下降的主要原因。尽管由于能源收入减少(通过反照率增加),森林砍伐可对地球表面产生冷却效应(Lee,

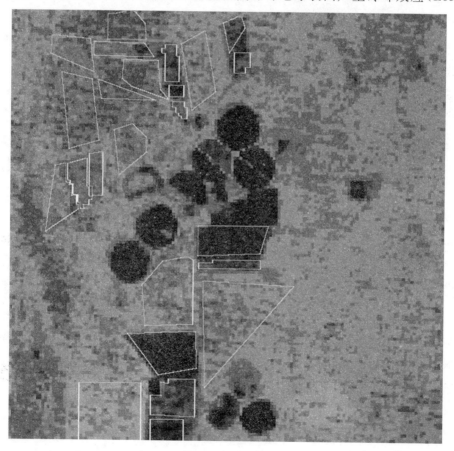

图18-2(见彩插)

注:2008年10月,砍伐树木后,利用Landsat估算的蒸散(毫米 day-1)。黑色圆圈代表被清除用于农业的区域。红色的多边形代表树木被移除的区域;蓝绿色区域对应的是未受干扰的林地。

2010),但本研究中在开阔区域观察到的变暖效应似乎更多是潜热和感热比率变化的结果(Davin et al.,2007)。森林似乎在更新地表和消除潜热方面比开阔草地表现得更好。

本节展示了遥感应用于研究由于植被覆盖变化而引起的水和能量平衡变化的一个案例。在生态系统服务方面,本研究表明,即使在森林砍伐后草地覆盖和生物量显著增加的情况下,阿根廷中部干性森林的木本植被对水分动态也有很强的控制作用。乔木和灌木生物量的减少也通过增加地表温度和地面释放的长波辐射的比例影响了局部气候。

由于水是调节半干旱地区大多数生物过程的关键因素,这种大规模的森林砍伐对蒸散的影响可能具有深远的实际意义。

六、结论

蒸散是一个与生态系统服务密切相关的术语:作为水循环的一部分,蒸散决定了热量和水蒸气的平衡,对气候减缓、控制降水、土壤平衡、养分通量和地下水补给具有重要影响。

由于蒸散与植被关系密切,植被覆盖率和叶面积指数的变化直接影响蒸散的大小,从而改变生态系统的服务功能。

在过去的一个世纪里,一些地区剧烈的土地利用变化影响了蒸散格局,从而影响了生态系统水分蒸发量。数千年来,生态系统通过蒸散的整体水分利用平衡被破坏,导致土地急剧退化,其中包括局部降水的减少和大规模的盐碱化等。这些大尺度的环境问题以及生态系统服务的损失只能利用适当的时间和空间分辨率的技术进行评估。卫星平台上的对地观测传感器仍然是从区域到全球尺度上实现这一目标的最佳方法。

本章提出的通过遥感估算蒸散的模型使用不同的输入变量——植被或物理参数,如反照率和地表温度。同时,利用能量平衡方程预测潜热通量和感热通量的分配,得到蒸散量。利用不同时间和空间尺度的卫星影像,如 Landsat 或 MODIS,来评估阿根廷干性森林在大规模砍伐前后的蒸散值或评估整个澳大利亚实际蒸散值,并且识别生态系统服务供给的相关区域。

在过去几十年里,与遥感、能量平衡和蒸散估算相关的前沿知识加速发展,使得包括普通公众在内的广大利益相关者都能够获得卫星产品和信息。尽管还需要进一步的改进,但新的模型和应用正在迅速出现,为跨时间的多尺度分析提供了可能性,并为生态系统服务研究创造了新的机会。在健康、环境和教育科学等基础和应用知识领域实施这种高科技信息,可以帮助我们作出更好的决策,从而更好地促进环境和人类福祉的平衡。

致谢

作者感谢 F. 泰斯特和两位匿名审稿人,他们对初稿给予了很好的建议,感谢 J. 斯特拉施诺伊的编辑工作。V. A. 马尔切西尼感谢瓦伦西亚大学影像处理实验室在其遥感实习期间的款待。

参 考 文 献

Aguilera, M. O., D. F. Steinaker, and M. R. Demaría. 2003. Runoff and soil loss in undisturbed and roller-seeded shrublands on semiarid Argentina. *Journal of Range Management* 56:227–233.

Allen, R. G., L. S. Pereira, D. Raes, and M. Smith. 1998. *Evapotranspiración del cultivo: guías para la determinación de los requerimientos de agua de los cultivos*. [Crop evapotranspiration - Guidelines for computing crop water requirements]. Irrigation and drainage paper 56, Rome: FAO.

Allen, R. G., W. O. Pruitt, J. L. Wright, et al. 2006b. A recommendation on standardized surface resistance for hourly calculation of reference ETo by the FAO56 Penman–Monteith method. *Agricultural Water Management* 81:1–22.

Anderson, M., and W. Kustas. 2008. Thermal remote sensing of drought and evapotranspiration. *EOS Transactions American Geophysical Union* 89:233–234.

Anderson, M. C., J. M. Norman, W. P. Kustas, R. Houborg, P. J. Starks, and N. Agam. 2008. A thermal-based remote sensing technique for routine mapping of land-surface carbon, water and energy fluxes from field to regional scales. *Remote Sensing of Environment* 112:4227–4241.

Baldocchi, D. D. 2003. Assessing the eddy covariance technique for evaluating carbon dioxide exchange rates of ecosystems: Past, present and future. *Global Change Biology* 9:479–492.

Barrett-Lennard, E. G. 2003. The interaction between waterlogging and salinity in higher plants: Causes, consequences and implications. *Plant and Soil* 253:35–54.

Beltrán-Przekurat, A., R. A. Pielke Sr., D. P. C. Peters, K. A. Snyder, and A. Rango. 2008. Modeling the effects of historical vegetation change on near-surface atmosphere in the northern Chihuahuan Desert. *Journal of Arid Environments* 72:1897–1910.

Blanco, L. J., C. A. Ferrando, F. N. Biurrum, et al. 2005. Vegetation response to roller chopping and buffelgrass seeding in Argentina. *Rangeland Ecology and Management* 58:219–224.

Blaney, H. F., and W. D. Criddle. 1950. Determining water requirements in irrigated areas from climatological and irrigation data. U.S. Department of Agriculture, SCS-TP 96, 44 pp. Washington, D.C.

Blight, G. E. 2002. Measuring evaporation from soil surfaces for environmental and geotechnical purposes. *Water SA* 28:381–394.

Brest, C. L., and S. Goward. 1987. Deriving surface albedo measurements from narrowband satellite data. *International Journal of Remote Sensing* 8:351–367.

Brutsaert, W., and M. B. Parlange. 1998. Hydrologic cycle explains the evaporation paradox. *Nature* 396:30.

Burgess, S. S. O., M. A. Adams, N. C. Turner, et al. 2001. An improved heat pulse method to measure low and reverse rates of sap flow in woody plants. *Tree Physiology* 21:589–598.

Caparrini, F., F. Castelli, and D. Entekhabi. 2003. Mapping of land-atmosphere heat fluxes and surface parameters with remote sensing data. *Boundary-Layer Meteorology* 107:605–633.

Carlson, T. N., W. J. Capehart, and R. R. Gillies. 1995. A new look at the simplified method for remote sensing of daily evapotranspiration. *Remote Sensing of Environment* 54:161–167.

Carlson, T. N., E. M. Perry, and T. T. Shmugge. 1990. Remote estimation of soil moisture availability and fractional vegetation cover for agricultural fields. *Agricultural and Forest Meteorology* 52:45–69.

Carpenter, S. R., and C. Folke. 2006. Ecology for transformation. *Trends in Ecology and Evolution* 21:309–315.

Chapin, F. S., P. Matson, and H. A. Mooney. 2002. Terrestrial water and energy balance. In *Principles of terrestrial ecosystem ecology*, eds. F. S. Chapin, P. Matson, and H. A. Mooney, 71–96. New York: Springer-Verlag.

Chen, S., J. Chen, L. Guanghui, et al. 2009. Energy balance and partition in Inner Mongolia steppe ecosystems with different land use types. *Agricultural and Forest Meteorology* 149:1800–1809.

Cleugh, H., R. Leuning, Q. Mu, and S. Running. 2007. Regional evaporation estimates from flux tower and MODIS satellite data. *Remote Sensing of Environment* 106:285–304.

Contreras, S., E. G. Jobbágy, P. E. Villagra, M. D. Nosetto, and J. Puigdefábregas. 2011. Remote sensing estimates of supplementary water consumption by arid ecosystems of central Argentina. *Journal of Hydrology* 397:10–22.

Daily, G. 1997. *Nature's services: Societal dependence on natural ecosystems*. Washington, DC: Island Press.

Davin, E. L., N. de Noblet-Ducoudré, and P. Friedlingstein. 2007. Impact of land cover change on surface climate: Relevance of the radiative forcing concept. *Geophysical Research Letters* 34:13702.

Dirmeyer, P. A., and J. Shukla. 1994. Albedo as a modulator of climate response to tropical deforestation. *Journal of Geophysical Research* 99:20863–20877.

Doody, T. M., and R. G. Benyon. 2011. Direct measurement of groundwater uptake through tree roots in a cave. *Ecohydrology* 4:644–649.

El-Nahry, A. H., and A. Y. Hammad. 2009. Assessment of salinity effects and vegetation stress, west of Suez Canal, Egypt, using remote sensing techniques. *Journal of Applied Sciences Research* 5:316–322.

FAO. 2010. Global Forest Resources Assessment. FAO Forestry Paper 163. Rome: FAO

Feidas, H., G. Kokolatos, A. Negri, M. Manyin, N. Chrysoulakis, and Y. Kamarianakis. 2009. Validation of an infrared-based satellite algorithm to estimate accumulated rainfall over the Mediterranean basin. *Theoretical and Applied Climatology* 95:91–109.

Fernandez-Illescas, C. P., A. Porporato, F. Laio, and I. Rodriguez-Iturbe. 2001. The ecohydrological role of soil texture in a water-limited ecosystem. *Water Resources Research* 37:2863–2872.

Forster, P., V. Ramaswamy, P. Artaxo, et al. 2007. Changes in atmospheric constituents and in radiative forcing. In Climate change 2007: The physical science basis. Contribution of Working Group I to the Fourth Assessment Report of the Intergovernmental Panel on Climate Change [Solomon, S., D. Qin, M. Manning, Z. Chen, M. Marquis, K.B. Averyt, M. Tignor and H.L. Miller (eds.)]. Cambridge University Press, Cambridge, United Kingdom and New York, NY, USA.

Gasparri, N. I., and R. Grau. 2009. Deforestation and fragmentation of Chaco dry forest in NW Argentina (1972–2007). *Forest Ecology and Management* 258:913–921.

Guerschman, J. P., A. I. J. M. Van Dijk, G. Mattersdorf, et al. 2009. Scaling of potential evapotranspiration with MODIS data reproduces flux observations and catchment water balance observations across Australia. *Journal of Hydrology* 369:107–119.

Hamilton, L. S. 2008. *Forests and water: A thematic study prepared in the framework of the Global Forest Resources Assessment 2005*. Rome: FAO.

Hansen, M. C., S. V. Stehman, and P. V. Potapov. 2010. Quantification of global gross forest cover loss. *Proceedings of the National Academy of Sciences of the United States of America* 107:8650–8655.

Houspanossian, J., M. Nosetto, and E. G. Jobbágy. 2012. Radiation budget changes with dry forest clearing in temperate Argentina. *Global Change Biology* 19:1211–1222.

Hurtado, E., and J. A. Sobrino. 2000. Daily net radiation estimated from air temperature and NOAA-AVHRR data: A case study for the Iberian Peninsula. *International Journal of Remote Sensing* 22:1521–1533.

Jackson, R. D., R. J. Reginato, and S. B. Idso. 1977. Wheat canopy temperature: A practical tool for evaluating water requirements. *Water Resources Research* 13:651–656.

Jung, M., M. Reichstein, P. Ciais, et al. 2010. Recent decline in the global land evapotranspiration trend due to limited moisture supply. *Nature* 467:951–954.

Kalma, J., T. McVicar, and M. McCabe. 2008. Estimating land surface evaporation: A review of methods using remotely sensed surface temperature data. *Surveys in Geophysics* 29:421–469.

Kelliher, F. M., R. Leuning, and E. D. Schulze. 1993. Evaporation and canopy characteristics of coniferous forests and grasslands. *Oecologia* 95 153–163.

Kerr, J. T., and M. Ostrovsky. 2003. From space to species: Ecological applications for remote sensing. *Trends in Ecology and Evolution* 18:299–305.

Kinzig, A. P., C. Perrings, F. S. Chapin, et al. 2011. Paying for ecosystem services—Promise and peril. *Science* 334:603–604.

Kremen, C., and R. Ostfeld. 2005. A call to ecologists: Measuring, analysing and managing ecosystem services. *Frontiers in Ecology and the Environment* 3:540–548.

Lambers, H. 2003. Dryland salinity: A key environmental issue in southern Australia. *Plant and Soil* 257:5–7.

Lee, X. 2010. Forest and climate: A warming paradox. *Science* 328:1479.

Lee, X., M. L. Goulden, D. Y. Hollinger, et al. 2011. Observed increase in local cooling effect of deforestation at higher latitudes. *Nature* 479:384–387.

Liang, S. 2000. Narrowband to broadband conversions of land surface albedo: I. Algorithms. *Remote Sensing of Environment* 76:213–238.

MA (Millennium Ecosystem Assessment). 2005. *Ecosystems and human well-being: desertification synthesis*. Washington, DC: Island Press.

Malhi, Y., J. T. Roberts, R. A. Betts, T. J. Killeen, W. Li, and C. A. Nobre. 2008. Climate change, deforestation, and the fate of the Amazon. *Science* 319:169–172.

Marchesini, V. A., R. J. Fernández, and E. G. Jobbágy. 2012. Salt leaching leads to drier soils in disturbed semiarid woodlands of central Argentina. *Oecologia* 171:1003–1012.

Miao, H., S. Chen, J. Chen, et al. 2009. Cultivation and grazing altered evapotranspiration and dynamics in Inner Mongolia steppes. *Agricultural and Forest Meteorology* 149:1810–1819.

Miles, L., A. Newton, R. DeFries, et al. 2006. A global overview of the conservation status of tropical dry forest. *Journal of Biogeography* 33:491–505.

Moore, G. W., and J. L. Heilman. 2011. Proposed principles governing how vegetation changes affect transpiration. *Ecohydrology* 4:351–358.

Mu, Q., F. A. Heinsch, M. Zhao, and S. W. Running. 2007. Development of a global evapotranspiration algorithm based on MODIS and global meteorology data. *Remote Sensing of Environment* 111:519–536.

Mu, Q., M. Zhao, and S. W. Running. 2011. Improvements to a MODIS global terrestrial evapotranspiration algorithm. *Remote Sensing of Environment* 115:1781–1800.

Norman, J. M., W. P. Kustas, and K. S. Humes. 1995. Source approach for estimating soil and vegetation energy fluxes in observations of directional radiometric surface temperature. *Agricultural and Forest Meteorology* 77:263–293.

Nosetto, M. D., E. G. Jobbágy, A. B. Brizuela, and R. B. Jackson. 2012. The hydrologic consequences of land cover change in central Argentina. *Agriculture, Ecosystems and Environment* 154:2–11.

Nosetto, M. D., E. G. Jobbágy, and J. M. Paruelo. 2005. Land-use change and water losses: The case of grassland afforestation across a soil textural gradient in central Argentina. *Global Change Biology* 11:1101–1117.

Paruelo, J. M., M. R. Aguiar, and R. A. Golluscio. 1991. Evaporation estimates in arid environments: An evaluation of some methods for the Patagonian steppe. *Agricultural and Forest Meteorology*. 55:127–132.

Reynolds, J. F., P. R. Kemp, K. Ogle, and R. J. Fernández. 2004. Precipitation pulses, soil water and plant responses: Modifying the 'pulse-reserve' paradigm for deserts of North America. *Oecologia* 141:194–210.

Ritchie, J. C., T. J. Schmugge, A. Rango, and F. R. Schiebe. 2000. Remote sensing applications for monitoring semiarid grasslands at the Sevilleta LTER, New Mexico, 1969–1971. In *IEEE 2000 International Geoscience and Remote Sensing Symposium Proceedings*. Honolulu, HI: International Association of Hydrological Sciences.

Roerink, G. J., Z. Su, and M. Menenti. 2000. S-SEBI: A simple remote sensing algorithm to estimate the surface energy balance. *Physics and Chemistry of the Earth Part B* 25:147–157.

Saunders, R. W. 1990. The determination of broad band surface albedo from AVHRR visible and near-infrared radiances. *International Journal of Remote Sensing* 11:49–67.

Seiler, C., and A. F. Moene. 2011. Estimating actual evapotranspiration from satellite and meteorological data in central Bolivia. *Earth Interactions* 15:1–24.

Sobrino, J. A., M. Gómez, J. C. Jiménez-Muñoz, A. Olioso, and G. Chehbouni. 2005. A simple algorithm to estimate evapotranspiration from DAIS data: Application to the DAISEX campaigns. *Journal of Hydrology* 315:117–125.

Sobrino, J. A., J. C. Jiménez, and L. Paolini. 2004. Land surface temperature retrieval from Landsat TM 5. *Remote Sensing of Environment* 90:434–440.

Su, Z. 2002. The Surface Energy Balance System (SEBS) for estimation of turbulent heat fluxes. *Hydrology and Earth System Sciences* 6:85–99.

Timmermans, W. J., J. C. Jimenez-Munoz, V. Hidalgo, et al. 2011. Estimation of the spatially distributed surface energy budget for AgriSAR 2006, Part I: Remote sensing model intercomparison. *IEEE Journal of Selected Topics in Applied Earth Observations and Remote Sensing* 4:465–481.

Wang, L., P. D'Odorico, J. P. Evans, et al. 2012. Dryland ecohydrology and climate change: Critical issues and technical advances. *Hydrology and Earth System Sciences* 9:4777–4825.

Wilcox, B. P. 2002. Shrub control and streamflow on rangelands: A process based viewpoint. *Journal of Range Management* 55:318–326.

Yebra, M., A. Van Dijk, R. Leuning, A. Huete, and J. P. Guerschman. 2013. Evaluation of optical remote sensing to estimate actual evapotranspiration and canopy conductance. *Remote Sensing of Environment* 129:250–261.

第十九章 城市热岛效应

J. A. 索布里诺　R. 奥尔特拉-卡里奥　G. 索里亚

一、引言

根据人口资料局（Population Reference Bureau，一个专注于进行人口数据分析的私营非营利组织，其总部设立在美国华盛顿特区——译者注）编纂的世界人口数据表（PRB，2011）可知，2011年全世界范围内居住在城市地区的总人口占世界总人口的比例达到51%。尽管各国定义城市的标准有所不同，有的将拥有100个及以上居民住宅的视为城市，有的只将居住在首都和省会城市的人口视为城市人口，但有一点可以明确的是：世界的城市化进程正在推进。生态系统通过调节地表能量和水平衡的生物物理过程来调节气候（West et al.，2011）（见第十六章）。

因此，如果用具有不同热特性的人工表面代替自然表面就会对生态系统产生相应的影响：自然生态系统将被转变为城市或人类主导的生态系统。因此，城市的发展会对当地气候的变化造成很大的影响。如奥凯（Oke，1987）文章所述，在某些情况下，这些变化是因特定人类用途改善大气环境而引起的，比如为防止种植蔬菜被风吹雨打，需要为植物提供温室的温暖环境。其他情况下，土地覆盖的变化或污染物对大气的直接污染也会带来变化，但这些变化是无意中发生的。无论如何，以上两类情况都与城市化有关。建造房屋是为了给居民提供舒适的环境，同时也改变了外界的风热环境。

城市热岛（Urban Heat Island，UHI）现象是局部气候变化的一个例子（Oke，1981）。与非城市化环境相比，这一现象的特点是城市区域的升温发热。

因此，UHI 被定义为城市内的空气温度（Air Temperature, AT）与其周围环境的空气温度间的差异（式19-1）（Morris et al., 2001）。UHI 的测定可作为生态系统变化的一个指标。事实上，城市生态系统通常比其他生态系统温度更高，而这会对当地动物群产生影响，改变动物群落的物种组成（Grimm et al., 2008），也会对植物群落产生同样影响。例如，城市地区的影响可改变植被物候，比如导致初花期提前（Roetzer et al., 2000）。

UHI 可以在城市大气的不同层和各种表面被定义。在城市冠层（Urban Canopy Layer, UCL），即从地面向上延伸到近似平均建筑物高度的城市大气层，以及位于 UCL 上方的城市边界层（Urban Boundary Layer, UBL）上可以确定大气热岛。

$$UHI = AT_{Urban} - AT_{Rural} \qquad 式19-1$$

奥凯等人（Oke, 1982; Oke et al., 1991）的研究提出了 UHI 一些成因的解释，沃格特（Voogt, 2002）对这些研究进行了完善，下文将对其进行阐述。UHI 的其中一个成因是城市的表面几何特征。由于城市峡谷的几何特征，城市区域有效表面积增加，太阳辐射被多次反射捕获，从而导致变暖。此外，密集的建筑物降低了天空可视指数，减少了辐射热损失。同时，建筑物还可作为遮蔽物，减少地表和近地表空气的对流热损失。其他影响因素包括表面热性能和相关条件。城市材料具有更高的热容量和更大的热导纳，比天然材料有更好的蓄热能力。此外，城市里的人造材料隔热防水，因此减少了水分蒸发，并将更多的能量传导至可使空气变热的显热（而非潜热）中。我们还应该考虑人为热量，即建筑物、车辆使用的城市能源以及人类自身释放出的热量。

造成 UHI 的最后一个原因是城市温室效应。受污染、潮湿和温暖的城市大气向下、向城市表面发射更多的热辐射。城市热岛效应可以为城市带来正面和负面影响，具体取决于纬度、气候区域和一年中的时间。

例如，能量消耗和 UHI 间的关系是被明确定义的。气候寒冷的城市可以通过空间供暖来节能。相反，炎热气候下的城市将面临额外的空调制冷成本，而这可能导致正反馈过程的形成，因为使用传统空调系统会加剧 UHI 效应，甚至可能加剧气候变化本身（WHO, 2003）。

UHI 最重要的影响之一是其对人类健康的影响。高温与人类健康问题有关,如夜间暴露在高温下可能会增加失眠发作的几率。人类疾病和UHI的联系十分紧密,世界卫生组织(World Health Organization,WHO)也在报告中有所强调(WHO,2003)。WHO将这一现象归类为土地利用在地方和局部层面上对气候带来的影响,并将热应力和空气污染定义为对气候变化和健康问题具有重大影响的因素。热应力和空气污染两个问题与城市生态系统有关,二者紧密相连,因为炎热的天气可能会加剧有害的光化学烟雾的产生。因此,在热敏感地区,最脆弱的人群是城市人口。但是,城市温度的升高也有可能会产生积极的影响。例如,马滕斯(Martens,1998)结合针对20个城市气候条件变化做出的预测,研究了温度变化对死亡率的影响。该研究发现,对于大多数城市来说,气候变化可能会导致死亡率的下降,原因是冬季死亡人数减少,主要是老年人因心血管问题死亡的人数减少。

  关于UHI在当前全球气候情景中的作用,可能会出现一个合理的问题。政府间气候变化专门委员会(Intergovernmental Panel on Climate Change,IPCC)的第四次评估报告(IPCC,2007)提到了自然变化或人类活动而导致的气候变化随时间推移的全球影响(文献中也称为全球变暖)。报告指出,由于气候过程和反馈相关的时间尺度,即使温室气体(Green House Gas,GHG)的排放量减少到足以使温室气体浓度稳定下来,人类变暖仍将持续几个世纪。城市地区是温室气体排放最多的地区;然而,UHI本身并不对全球气候变化负责。正如上文指出,UHI是局部气候的改变。尽管由于全球气候变化,预计城市气候会将变暖,但热岛的强度不太可能增加,因为据估计,城市和农村地区间的温度梯度将会保持在与今天相似的水平(Voogt,2002)。然而,奥凯(Oke,1997)等部分研究人员,将UHI视作全球变暖的模拟,认为发生在城市地区的气候变化可用于类比研究全球气候变化可能带来的改变。

  经过上述信息梳理和讨论,城市生态系统的研究很显然是至关重要的。然而,量化城市化进程带来的种种影响并不容易,因为通常我们没有汇集某地城市化之前各项测量结果的数据库。相反,比较来自城市中心和周围地区农村站点的数据却是很常见的做法。这样的城乡比较最多也只是得出城市变化的近似值。此外,城市内部的气象网络并不总是如人们预期的那样完整,并且城市内的气象站点并不总是在空间上均匀分布。因此,一些面积较大的区域可能仍然没有覆盖。从这个意义上说,遥感数据是研究城市环境的有力工具,解决了以往的难题。事实上,一些研究已经在强调热红外遥感数据对研究城市环境和为居民提供更友好、更舒适的生活环境的重要性(Johnson et al., 2009)。

当用遥感数据监测 UHI 时,我们必须考虑城市地表热岛(Surface Urban Heat Island, SUHI; Voogt and Oke, 2003),因为所研究的参数不再是空气温度,而是地表温度(land surface temperature, LST;式 19-2)。

$$SUHI = LST_{Urban} - LST_{Rural} \qquad 式19-2$$

二、城市热岛效应评估

举例来说,有数个研究开展了实验活动旨在研究城市气候,且大多数情况下,这些研究量化了不同城市区域的 UHI。实现这一研究目的的策略之一便是使用固定气象或移动监测样带以记录空气温度、湿度和风速测量值(Oke, 1973; Voogt and Oke, 1998; Fernández et al., 2004; Jasche and Rezende, 2007; Masson et al., 2008)。许多其他研究已经应用了遥感技术,使用卫星影像(Voogt and Oke, 2003; Hartz et al., 2006; Li et al., 2009; Stathopoulou et al., 2009)或空中传感器影像(Voogt and Oke, 1997; Lagouarde et al., 2004; Masson et al., 2008)来描述城市表面的变化。在本节中,我们将介绍欧洲航天局(ESA)在其对地观测计划框架内发起的两个实验活动。第一个是 2008 年在马德里进行的双用途欧洲安全红外实验(Dual-use European Security IR Experiment, DESIREX);第二个是 2009 年在雅典开展的 Thermopolis 实验活动。两个实验都收集了地面和大气数据及星载和机载影像。因此,可对 SUHI 和 UHI 两种效应进行评估,并在二者间建立相关性。航空影像由西班牙国家航空研究所(Instituto Nacional de Técnica Aeroespacial, INTA)的机载高光谱扫描仪(Airborne Hyperspectral Scanner, AHS)记录。

(一) DESIREX 实验活动:马德里案例研究

AHS 传感器是机载成像 80 个波段辐射计,在热红外(Thermal In Frared, TIR)范围内有 10 个波段(有效波长为 8.18、8.66、9.15、9.60、10.07、10.59、11.18、11.78、12.35 和 12.93 微米)。在 DESIREX 实验活动期间(Sobrino et

$al.$,2013),在6月和7月的不同日子进行了AHS飞行观测。实验获得了30张AHS影像,并将其分为从西北到东南(现称为P01飞段)和从南到北(现称为P02飞段)两种不同的模式。两个飞段过境时都横跨市中心,二者覆盖面积相同,均延伸约17平方千米。表19-1给出了每次飞行的具体说明。实验又对AHS影像进行了大气校正,并应用温度与发射率分离算法(Temperature and Emissivity Separation,TES)反演LST(Gillespie $et~al.$,1998)。

表19-1 DESIREX 2008运动中的机载高光谱扫描仪飞行观测数据

DOY	开始时间(UTC)	飞段编号
177	11:11/11:27	P01/P02
177	22:15/22:31	P01/P02
178	04:12/04:26	P01/P02
180	11:32/11:53	P01/P02
180	21:29/21:44	P01/P02
183	11:21/11:44	P01/P02
183	21:59/22:12	P01/P02
184	04:09/04:26	P01/P02
186	11:16/11:32	P01/P02
186	21:59/22:14	P01/P02

注:空间分辨率为4米的飞行;DOY=一年中的日期。

图19-1显示的为UHI和SUHI效应。对于SUHI现象,该实验计算了AHS的两次过境时的数值。式19-2中的LST_{Urban}被认为是AHS影像中城市区域的平均LST,而LST_{Rural}是农村区域的平均LST,农村区域是总面积减去城市区域。实验根据城市轮廓和社区边界划定了城乡界限。尽管每次过境时覆盖的城乡区域有所不同,但两次搜集影像中展示的变化趋势是相似的。大气温度数据由位于建筑物上方UBL中的固定观测点搜集记录,观测点使用3米高的桅杆固定住。此外,一个农村观测点被安置在城外,位于一片覆盖着绿草的地面上方。此后,通过计算城市观测点和农村观测点分别记录的平均大气温度间的差值,得出UHI效应(式19-1)。

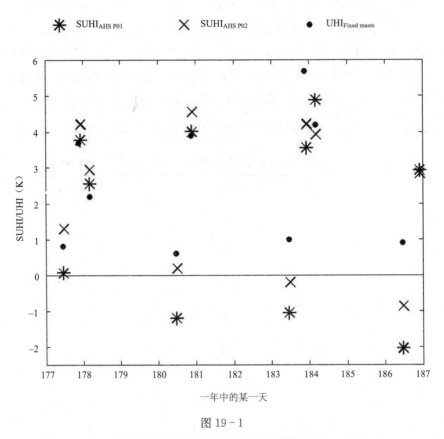

图 19-1

注:马德里市 DESIREX 实验期间测得的城市地表热岛效应和城市热岛效应。

在图 19-1 中需要强调的一个相关问题是,在同时捕获到的 UHI 和 SUHI 值之间存在明显的差异。对于中午获得的测量数据,UHI 和 SUHI 之间的差异最大;而在夜间,我们发现两种效应的数值呈现出合理的一致性。

因此,在夜间使用遥感影像观测可得出地表和冠层层面的热岛值。本多尔和萨罗尼(Ben-Dor and Saaroni,1997)在比较特拉维夫市夜间飞行观测时获得的辐射温度和空气温度时也取得了同样的结果。

中午时,由于乡村地表比城市地表温度更高,因此测得的 SUHI 值较低(有时为负值)。由于城市对热量的吸收速度较为缓慢,以及高层建筑带来的大阴影区域(Voogt,2002),午时观测到的负热岛值呈现为集中在一些城市市中心区域的特征,比如马德里。需注意的是,对于 UHI 效应而言,午时记录的数据中未曾有负值出现,但测量出的温度值约为 1 开尔文。

因此,在马德里,午时热岛效应的减退在大气数据中也有所体现,在上海等其他城市的数据中也验证了这一结果(Liu et al.,2007)。

尽管在 DESIREX 实验活动期间的 SUHI 最高值(5 开尔文)在 UHI 达到最高值(6 开尔文)数小时后才出现,但二者之间的差值落在 1 开尔文以内。奥凯(Oke,1973)预测的最高 UHI 期待值为 9 开尔文,比这次实验中得出的最高 UHI 值高出 3 开尔文。

综上所述,夜间的 UHI 强度最大,白天的 UHI 效应消失,换句话说城市比农村环境更凉爽,这与阿恩菲尔德(Arnfield,2003)在回顾 20 年的城市气候研究中观察到的变化趋势一致。

(二) Thermopolis 实验活动:雅典案例研究

在雅典城市上空开展搭载热红外遥感传感器的飞机航测实验,实验目的是为了研究 2009 年 6 月 18 日～2009 年 7 月 24 日期间城市的热岛现象,实验中使用了 AHS 仪器。图 19-2 描绘了在对影像进行大气校正和从 TES 反演 LST 之后获得的 SUHI 值。与 DESIREX 实验活动一样,在 Thermopolis 实验中,白天检测到冷岛(UHI 值为负)效应,而正 UHI 效应则发生在夜间。

三、适用于分析城市热岛效应的遥感传感器参数

过去文献中有几项研究报告了使用不同星载平台测量的 SUHI 效应,例如,高级星载热发射和反射辐射仪(Hartz et al.,2006;Lu and Weng,2006)、改进型甚高分辨率辐射计(Streutker,2003)、中分辨率成像光谱仪(Pu et al.,2006)和陆地卫星(Landsat;Pu et al.,2004;2007;Rajasekar and Weng,2009)。然而,现有的卫星遥感能力(重访时间、空间分辨率和光谱分辨率)限制了其成为有效监测 SUHI 效应的最佳方式。

为达成这一目标,本节主要目标是对卫星的空间分辨率和过境时间等参数提出建议,明确卫星成为有效监测 SUHI 效应所应具备的最佳参数特性。为了阐明上述问题,本节参考了来自 DESIREX 实验活动的影像(Sobrino et al.,2012a)。

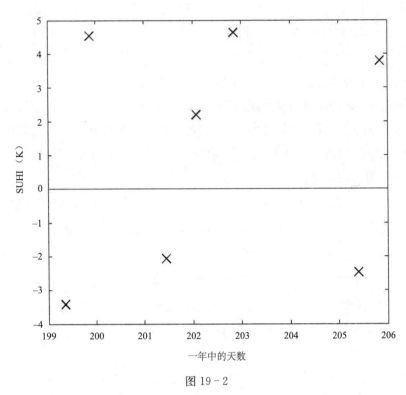

图 19-2

注：雅典市 Thermopolis 实验期间测得的城市地表热岛效应。

（一）空间分辨率

为了分析空间分辨率对有效监测城市热岛现象的影响，本文对 AHS 影像进行了重采样，通过对所有与输出像素有关的像素值求平均值，对经几何校正的传感器辐亮度影像进行了聚合处理，这一操作是使用©ENVI 应用程序的像素重采样工具完成的，该工具不考虑相邻像素之间产生的任何影响，过程中修正了大气影响并反演了 LST 值。

图 19-3 显示了聚合生成的 LST 影像。当空间分辨率降低时，影像中存在明显的信息缺失。这一现象在 1 000 米分辨率的聚合影像中尤为明显，在这些影像中不同区域的热结构已无法区分。这表明（在 1 000 米的空间分辨率下）现有卫星在观测大城市 SUHI 效应的热结构方面的能力有限；当然，这一局限对

第十九章 城市热岛效应

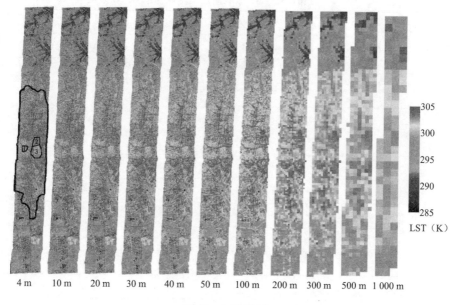

图 19 - 3（见彩插）

注：从机载高光谱扫描仪 4 米分辨率影像中获得的陆地表面温度以及在 10、20、30、40、50、100、200、300、500 和 1 000 米分辨率进行上述的聚合处理所获得的影像。影像左侧的最外层黑色线条圈出的区域为马德里市的边界。内部黑色线条对应文中提及的三个城区。

于小城市来说更关键。上述发现在表 19 - 2 中得到了定量的证实，表 19 - 2 显示了城市区域（图 19 - 3 中 4 米分辨率影像上绘制的极限范围内的区域）在不同集合 LST 的标准偏差。我们可观察到 3.6 开尔文的剧烈变化（从原始 4 米分辨率影像的 4.4 开尔文到重采样到 1 000 米分辨率影像的 0.8 开尔文）。表 19 - 2 还展示了针对每个聚合影像测量得到的 SUHI 效应。可以观察到，SUHI 值几乎不随分辨率的降低而降低，4 米空间分辨率下的 SUHI 值与 1 000 米空间分辨率下的 SUHI 值仅相差 0.3 开尔文。因此，为找到合适的空间分辨率，我们还需进一步探讨。

通常，文献中的 SUHI 值往往被定义为代表城市平均表面温度与周围农村地区温度之差的平均值。这是因为所使用的遥感影像空间分辨率低。然而，在同一城市内的不同区域的热舒适性可能有较大的差异（Toy *et al*., 2007）。为了研究这一影响，本文选择了三个具有代表性的地区（如图 19 - 3）。第一个是市

表 19-2 城市区域 LST 标准差及聚类影像的 SUHI 效应

空间分辨率(米)	σ(开尔文)	SUHI(开尔文)
4	4.4	4.56
10	3.3	4.56
20	2.8	4.55
30	2.5	4.54
40	2.4	4.54
50	2.2	4.53
100	1.7	4.49
200	1.4	4.49
300	1.2	4.51
500	1.0	4.40
1 000	0.8	4.25

注：LST=地表温度；SUHI=城市地表热岛。

中心，其特点是狭窄的街道和顶部有红砖的小建筑。第二个区域的特点是宽阔的街道和高层建筑物，屋顶上覆盖着隔热材料。第三个几乎被一座花园所覆盖，诸如街道或小花园等热结构，在 10 米和 50 米的空间分辨率下可以清楚地观测到。但在 100 米分辨率的影像中，街道和花园等的像元混杂在一起，但我们仍可以辨识出区域内的一些结构。在 500 米和 1 000 米的空间分辨率下，这些区域同质性高，而之间的异质性消失了。作为城市内部热效应变化的参数，本章选择的是最大 SUHI 值(Maximum SUHI，SUHIM)，它是根据每个区域内的最大 LST 与农村地区 LST 间的差值计算获得的。当把 SUHIM 表示为空间分辨率的函数时(图 19-4)，可观察到一些趋势：(1)各地区 SUHIM 值均高于 10 开尔文，并且从 10~40 米空间分辨率的 LST 影像可观察到地区之间 SUHIM 值的差异；(2)50 米的空间分辨率(各地区的 SUHIM 值接近 10K)可被视为临界空间分辨率；(3)低于 50 米的分辨率影像中呈现出较低的 SUHIM 值(从 100 米的 8 开尔文到 1 000 米的 4 开尔文)。

(二) 过境时间

在卫星过境的那一刻，必须满足一定的条件才能获得适当的观测用以监测

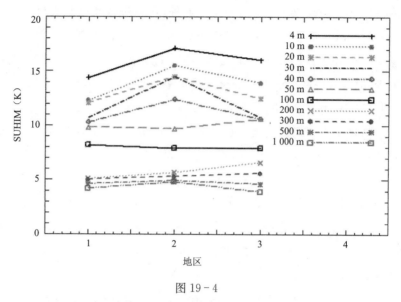

图 19-4

注:不同空间分辨率下三个地域的最大 SUHI 值(Maximum SUHI,SUHIM)。

SUHI 现象。观测几何条件对记录数据的影响应尽可能小,这样更容易比较不同日期记录的数值。在这方面,我们已观察到,在夜间,由于城市场景的 3D 结构,传感器所观测到的不同表面引起的城市表面各向异性降低。当分析 AHS 过境 P01 和 P02 共同覆盖的区域时,这一趋势得到了证明,因为每次飞行观测到的几何结构不同;即使观测同一区域,卫星捕获到的也是不同的表面。一方面,当卫星在白天进行观测时,每次过境获得的 LST 直方图均有不同。另一方面,在夜间,当没有光线直接照射表面时,两个直方图几乎相同。要施加的另一个条件是 UHI 值和 SUHI 值不应该有太大差别,或至少 UHI 值应该能简单而准确地从 SUHI 值中获得。造成这一情况的原因是大气效应能够直接影响居民的舒适度,尤其是在 UCL 中。图 19-1 中,从大气和地表两个层面分析了城市热岛,两种现象在夜间具有相似的值。因此,夜间正在成为卫星过境的最佳选择。可以基于在实验活动期间对固定点观测所记录的数据进行更深入的研究。

为了分析卫星过境时间对于 SUHI 估计的影响,我们对上节所述的固定点同时测得的每小时空气温度和表面温度逐一进行了比较。

若分析 LST 减去 AT 的每小时的演变,我们会发现对于所有站点,夜间

LST 和 AT 的差值均接近于 0 开尔文。此外,在住房密度较低的观测点观测到了最小温差。所有观测点均在 20:00 和 6:00 UTC 时段出现较低值。

图 19-5 给出了 LST 和 AT 温度的相关性分析,其中展示了相关系数每小时的演变,以及假设两个温度之间存在线性相关时所得到的误差。

绘制影像使用的是时间平均值,例如,横坐标轴上时间等于 4 的点,图中绘制的对应值是实验期间每天从 4:00 UTC 到 4:55 UTC 记录数据的平均值。根据图例,M1 对应的是城市外的放置在绿色的草地表面上观测点。M2、M3、M4 和 M5 观测点被放置在建筑物顶部的人造表面上。M2 位于房屋密度较低的区域,M3 和 M5 位于一个中等密度住宅区域,M4 位于在高密度住宅区域。白天与夜间的相关系数较低,误差较大。位于人工表面上方的观测杆,其相关系数在 21:00 UTC 到 5:00 UTC 时段更高(图 19-5(a))。此外,日出前后所有观测杆的均方根误差(Root Mean Square Error,RMSE)约为 1 开尔文(图 19-5(b))。因此,很明显,若要估算得出的 SUHI 值能够与气温数据得出的 UHI 值进行比较,日出之前的过境时间是观测的最佳选择。

四、反演城市地区地表发射率的不同程序的评估

包括 LST 在内的热遥感器的测量参数过于依赖地表特征,特别是发射率和几何形状(Voogt and Oke,2003)。实际上,1% 的发射率估计误差可导致 LST 出现高达 0.78K 的误差(Van de Griend and Owe,1993)。城市地表发射率(land surface emissivity,LSE)的正确反演至关重要,这是因为城市表面是非常不均匀的,且反演出的 LST 是研究 UHI 等城市现象的关键。

地表发射率是地表将积累的能量转化为辐射能时固有效率的度量。引入发射率项来表征以具有相同表面温度的黑体为基准的表面发射(式 19-3)

$$\varepsilon = L(T)/B(T) \qquad \text{式 19-3}$$

式中 L(T) 是温度 T 下表面的辐射亮度,B(T)(式 19-4)是表面温度下黑体的辐射亮度:

$$B(T) = \frac{c_1}{\lambda^5 \left[\exp\left(\frac{c_2}{\lambda T}\right) - 1\right]} \qquad \text{式 19-4}$$

第十九章 城市热岛效应

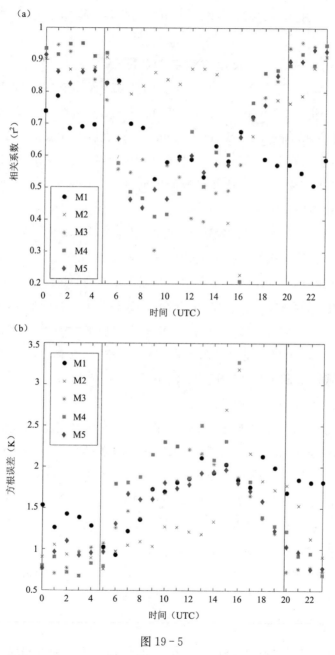

图 19-5

注：图(a)为相关系数的每日的演变，图(b)为每个观测点(M1、M2、M3、M4 和 M5)获得的每组大气温度和陆地表面温度间的线性拟合的估算误差。

其中，$c_1 = 1.191 \times 10^{-8}$ W/(m² sr cm⁻⁴)，$c_2 = 1.439$ cm K，λ 为波长，单位为厘米。

在本节中，三种 LSE 反演方法在 DESIREX 实验的框架下得到了应用 (Oltra-Carrió et al.，2012)。三种方法分别是 NDVI 阈值法（NDVI Threshold Method，NDVI™）(Sobrino et al.，2008a)、TES(Gillespie et al.，1998) 和 TISI (Temperature Independent Spectral Indices)算法(Becker and Li，1990)。所选择的每个算法都使用一种分离方法来解决 LST 和 LSE 的组合效应问题。首先，NDVI 利用的是来源于植被指数的辅助信息。我们可在文献中找到几个 NDVI 应用的例子，甚至有些文献将其用于城市表面的测算(Rigo et al.，2006；Stathopoulou et al.，2007)，当把使用 NDVI 方法反演得到的发射率与公开参考值进行比较时，发现相关系数（r^2）为 0.79，RMSE 为 0.010。其次，TES 使用的是热红外遥感波段光谱对比度。该算法依赖于观测到的热红外遥感发射率范围与其最小值间的经验关系。TES 是为 ASTER 设计的。基于数值模拟，TES 大约可在均匀区域弥补大约 ±1.5 开尔文的温度和大约 ±0.015 的发射率 (Gillespie et al.，1998)。因此，在几乎所有使用 ASTER 影像的文献中，TES 算法都被应用于 LSE 和 LST 的反演。其中，蒂安科等(Tiangco et al.，2008)和尼科尔等(Nichol et al.，2009)使用该算法分别开展了两项关于城市地区的研究。最后，TISI 算法使用中红外表面反射率作为解决发射率/温度分离问题的关键。在达什等(Dash et al.，2005)的研究中，TISI 算法被应用于 AVHRR 数据，而在珀蒂科兰和维莫特(Petitcolin and Vermote，2002)的研究中，TISI 算法被应用于非洲南部的 MODIS 数据，其中提取的发射率和表面温度分别在 ±0.01 和 ±1 开尔文范围内。以上三种算法均未考虑城市的三维结构，因此我们不会分析相邻表面对某一特定材料辐射平衡的贡献。

最后，在使用 LSE 图反演 LST 时通过应用劈窗（Split Window，SW）算法 (Sobrino et al.，2008b)，检验了上述方法的实际精度。

（一）地表发射率比较

为了在城市地区测试 LSE 反演方法，我们选择了三种不同的城市表面：城市内部一条大型道路上的沥青、沥青瓦（以下简称瓦）和地砖。后两个是马德里最常用的屋顶材料。本研究将 TES 算法应用于多波段辐射计记录的实地数据，测量了沥青和瓦的实地发射率。地砖光谱数据提取自 ASTER 光谱库(http://speclib.jpl.nasa.gov);样品制备和样品测量的细节在鲍德里奇等的研究(Sobrino et al.,2009)中有记录。

注意,本研究中不涉及金属表面,因为金属表面不遵循 TES 算法应用所需的最小发射率和光谱对比度间的经验关系(Sobrino et al.,2012b)。表 19-3 展示了 AHS 值和 LSE 方法得出的地面真实估计的对比统计结果。表中还列出了经典的统计参数,如 AHS EST 与地面真实 EST(偏差-平均值)的均值差和标准偏差(σ)。此外,表中给出了与经典参数相比受异常值影响较小的稳健的统计参数,比如 AHS EST 与地面真实 EST 之差的中位数(偏差-中值)和绝对中位差(median absolute deviation, MAD)。同时,表中还给出了每个验证过程的 RMSE。

表 19-3 反演的地表发射率产品验证实验的统计值

	NDVI	TES	TISI
r^2	0.250	0.640	0.730
偏差-平均值	0.024	0.015	−0.008
σ	0.051	0.036	0.029
偏差-中值	0.009	−0.003	−0.007
MAD	0.033	0.025	0.023
RMSE	0.056	0.039	0.030

注:TES=温度与发射率分离算法;TISI=温度无关光谱指数算法;MAD=绝对中位差;RMSE=均方根误差。

从我们的分析中可以得到一些信息。LSE 地面真值从 9.154 微米处(瓦)的 0.775 变到 8.695 微米处(沥青)的 0.977。然而,上述三种算法反演的发射率范围并不那么宽。对于 TISI,这一范围为 0.817~0.974,对于 TES,该范围为 0.881~0.969。NDVI 的范围最窄,为 0.939~0.972。此外,NDVI 和 TES 不能正确地再现 LSE 的低值。事实上,TES 给出了大于 0.93 的良好值。另外还要注意,NDVI 算法并不能区分不同的城市材料。NDVI 的各统计指标(表 19-3)显示两个反演值间的相关性较低,即 r^2 为 0.25,RMSE 为 0.06。这一结果是预料之中,因为该方法是基于植被指数进行计算,并不是表征人造材料的最佳方法。

对于 TES 算法和 TISI 算法,相关性(r^2)分别提高至 0.64 和 0.73。对于 TES,偏差-平均值和偏差-中值均在可接受的范围内,分别为 0.015 和 −0.003。对于 TISI,偏差-平均值和偏差-中值分别为 −0.008 和 −0.007。尽管如此,本研究中使用这两种算法获得的 RMSE 值高于先前研究中对上述三种算法的估计精度。

(二) 地表温度比较

本研究通过利用 SW 算法获得了 AHS 的每次飞行的 LST 图(Sobrino et al.,2008b)。

表 19-4 中展示的是 LST 验证的统计值。本研究分为两组。第一,考虑到所有夜间影像的 4 个验证目标,包括 2 种自然表面(绿草、裸土)和 2 种人造表面(城市内 2 种不同屋顶),总共有 18 个验证值。第二,只考虑人造表面和所有夜间影像,总共有 8 个验证值。用 NDVI 发射率反演算法获得的 LST 具有全部数据中的最高分散性(MAD 值和 σ 值最高),以及两组研究中的最大偏差和 RMSE 值(即分别考虑 18 个验证值和 8 个验证值时)。当研究自然和人造表面时,用 TES 和 TISI 发射率反演的 LST 表现出相似的趋势,MAD 值相等,σ 值也十分接近。然而,对于城市人工表面,利用 TES 发射率反演的 LST 比 TISI 算法反演的 LST 偏差更高,但数据的离散度更低,并且大气 LST 和实地 LST 之间也具有更高的线性相关性。从 RMSE 来看,我们可观察到当分析所有表面时,TES 和 TISI 呈现相似的值(分别为 1.7 开尔文和 1.6 开尔文)。若仅考虑人造表面,两种算法的 RMSE 是相同的(2 开尔文)。然而,在这两种情况下,TES 的 RMSE 比 TISI 更符合其理论精度。

需注意的是,与其他两种方法相比,TISI 算法需要更多的光谱信息(热红外波段和中红外区域中的至少一个波段)和更多时间的信息(一个夜间加一个白天的影像)。

但 TES 算法的应用只需五个 TIR 波段和一幅影像。此外,该算法还具有可同时反演 LST 与 LSE 的优点。总而言之,我们的研究结果表明,TES 发射率反演算法似乎是在城市区域反演 LST 的最佳算法,且不需要传感器具备高时间分辨率。

表 19-4 利用劈窗算法从机载高光谱扫描仪影像中反演地表温度—验证实验统计值

	18个验证值:绿草、裸土、2个人造表面			8个验证值:2个人造表面		
	NDVI	TES	TISI	NDVI	TES	TISI
r^2	0.86	0.92	0.90	0.83	0.88	0.67
偏差-平均值(开尔文)	-1.50	-1.00	-0.70	-2.10	-1.60	-0.58
σ(开尔文)	1.70	1.30	1.40	1.90	1.20	1.90
偏差-中位数(开尔文)	-1.20	-1.10	-0.50	-1.90	-1.60	0.29
MAD(开尔文)	1.40	1.10	1.10	1.70	0.90	1.40
RMSE(开尔文)	2.30	1.70	1.60	2.90	2.00	2.00

注：地表温度使用NDVI、TES和TISI算法计算。TES=温度与发射率分离算法；TISI=温度无关光谱指数算法；MAD=绝对中位差；RMSE=均方根误差。

五、结论

城市的发展会使得自然地表被人工表面所代替,而且农村生态系统将转变为城市生态系统。这一过程中热条件和风场会发生改变,这些变化将对当地气候有直接影响。UHI便是环境变化的一个例子。在本研究中,我们使用真实数据来引入热岛的概念。2008年的DESIREX实验活动为收集地面测量数据以及验证从城市地区感兴趣的遥感数据中反演生化物理参数的不同算法提供了极好的机会。SUHI和UHI效应从LST影像和AT数据测算得出。结果显示两种效应具有相似的趋势,夜间值较高,中午值较低。由此本研究得出三个主要结论:(1)午时地面出现一个负热岛,且UBL出现了1开尔文左右的低效应;(2)SUHI和UHI的最大值相差1开尔文,二者分别为5开尔文和6开尔文;然而,最高UHI值是在SUHI最大值出现之前的几个小时记录的;(3)最高UHI值和最高SUHI值间的差异在夜间达到最小,在白天则达到最大,这表明在夜间使用遥感影像可以得出更准确的表面和边界层面的热岛强度值。

此外,本研究还展示了星载传感器在大城市进行区域级 SUHI 监测所应必须具备的参数特性(在空间分辨率和卫星过境时间方面的特性)。结果表明,在区域层面上,若想要适当地估算 SUHI 效应需要高于 50 米的空间分辨率。低于 50 米的空间分辨率会低估效应的大小,且无法区分城市内部的不同区域。需注意的是,更高空间分辨率可提供更大信息量,但也伴随着较低重访时间频率和较低的底点条带。此外,本研究还表明,日出前的过境时间是估算 SUHI 效应的最佳时间,此时获得的结果可与从气温数据获得的 UHI 效应进行比较。目前,还没有一个星载热传感器同时满足上述时空条件,因此在规划未来空间任务时必须参考上述结论。

最后,我们比较了使用 AHS 传感器估算城市区域 LSE 的三种算法(NDVI、TES 和 TISI)。结果表明,NDVI 并不适用,因为该算法的误差最大,并且不能区分不同类型的人工表面,而 TES 和 TISI 算法均可以区分。至于 TES 和 TISI,两种算法分别得出 0.04 和 0.03 的 RMSE 值。当比较 LST 估算结果时,我们仍观察到 TISI 算法和 TES 算法与 NDVI 算法相比,得到的结果与实地值的一致性更高。TISI 算法得出的 RMSE 比仅观察人造表面时的精度高 1 开尔文。对于 TES 算法而言,该差异仅为 0.5 开尔文,这一差别对应着观测城市地区的多项应用需求,比如 UHI 效应的监测。此外,与其他两种方法相比,TISI 算法需要更多的光谱信息(TIR 波段和 MIR 区域中的至少一个波段)和温度信息(一个夜间加一个白天的影像),而 TES 算法的应用只需五个 TIR 波段和一幅影像。此外,TES 算法还具有可同时反演 LST 与 LSE 的优点。总之,结果表明,TES 发射率反演算法,在不要求具有高时间分辨率的情况下,能够最好地反演城市区域内 LST。

致谢

本文作者感谢欧洲空间局(DESIREX 2008,21717/08/I-LG 项目和 Thermopolis 2009 实验活动项目,22693/09/I-EC 项目),以及所有参加这两个实验活动的个人和团体,并感谢西班牙科学与创新部(CEOS-SPAIN,AYA2011-

29334-C02-01 项目)的财政支持。在本研究中,R. 奥尔特拉-卡里奥获得了瓦伦西亚大学的 V Segles 资助。

参 考 文 献

Arnfield, A. J. 2003. Two decades of urban climate research: A review of turbulence, exchanges of energy and water, and the urban heat island. *International Journal of Climatology* 23:1–26.

Baldridge, A. M., S. J. Hook, C. I. Grove, and G. Rivera. 2009. The ASTER spectral library version 2.0. *Remote Sensing of Environment* 113:711–715.

Becker, F., and Z. L. Li. 1990. Temperature-independent spectral indices in thermal infrared bands. *Remote Sensing of Environment* 32:17–33.

Ben-Dor, E., and H. Saaroni. 1997. Airborne video thermal radiometry as a tool for monitoring microscale structures of the urban heat island. *International Journal of Remote Sensing* 18:3039–3053.

Dash, P., F. M. Göttsche, F. S. Olesen, and H. Fischer. 2005. Separating surface emissivity and temperature using two-channel spectral indices and emissivity composites and comparison with a vegetation fraction method. *Remote Sensing of Environment* 96:1–17.

Fernández, F., J. P. Montávez, J. F. González-Rouco, and F. Valero. 2004. Relación entre la estructura espacial de la isla térmica y la morfología urbana de Madrid [Relationship between the spatial distribution of the urban heat island and the urban pattern of Madrid]. *El Clima entre el Mar y la Montaña* 4:641–650.

Gillespie, A., S. Rokugawa, T. Matsunaga, et al. 1998. A temperature and emissivity separation algorithm for Advanced Spaceborne Thermal Emission and Reflection Radiometer (ASTER) images. *IEEE Transactions on Geoscience and Remote Sensing* 36:1113–1126.

Grimm, N. B., S. H. Faeth, N. E. Golubiewski, et al. 2008. Global change and the ecology of cities. *Science* 319:756–760.

Hartz, D. A., L. Prashad, B. C. Hedquist, J. Golden, and A. J. Brazel. 2006. Linking satellite images and hand-held infrared thermography to observed neighborhood climate conditions. *Remote Sensing of Environment* 104:190–200.

IPCC (Intergovernmental Panel on Climate Change). 2007. Fourth assessment report: Climate change. Geneva, Switzerland: Intergovernmental Panel on Climate Change.

Jasche, A., and T. Rezende. 2007. Detection of the urban heat-island effect from a surface mobile platform. *Revista de Teledetección* 27:59–70.

Johnson, D. P., J. S. Wilson, and G. C. Luber. 2009. Socioeconomic indicators of heat-related health risk supplemented with remotely sensed data. *International Journal of Health Geographics* 8:57.

Lagouarde, J. P., P. Moreau, M. Irvine, et al. 2004. Airborne experimental measurements of the angular variations in surface temperature over urban areas: Case study of Marseille (France). *Remote Sensing of Environment* 93:443–462.

Li, J. J., X. R. Wang, X. J. Wang, W. C. Ma, and H. Zhang. 2009. Remote sensing evaluation of urban heat island and its spatial pattern of the Shanghai metropolitan area, China. *Ecological Complexity* 6:413–420.

Liu, W., C. Ji, J. Zhong, X. Jiang, and Z. Zheng. 2007. Temporal characteristics of the Beijing urban heat island. *Theoretical and Applied Climatology* 87:213–221.

Lu, D., and Q. Weng. 2006. Spectral mixture analysis of ASTER images for examining the relationship between urban thermal features and biophysical descriptors in Indianapolis, Indiana, USA. *Remote Sensing of Environment* 104:157–167.

Martens, W. J. M. 1998. Climate change, thermal stress and mortality changes. *Social Science and Medicine* 46:331–344.

Masson, V., L. Gomes, G. Pigeon, et al. 2008. The Canopy and Aerosol Particles Interactions in Toulouse Urban Layer (CAPITOUL) experiment. *Meteorology and Atmospheric Physics* 102:135–157.

Morris, M., E. P. McClain, and N. Plummer. 2001. Quantification of the influences of wind and cloud on the nocturnal urban heat island of a large city. *Journal of Applied Meteorology* 40:169–182.

Najjar, G., P. P. Kastendeuch, M. P. Stoll, et al. 2004. Le projet Reclus. Télédétection, rayonnement et bilan d'énergie en climatologie urbaine à Strasbourg. *La Météorologie* 46:44–50.

Nichol, J. E., W. Y. Fung, K. S. Lam, and M. S. Wong. 2009. Urban heat island diagnosis using ASTER satellite images and "in situ" air temperature. *Atmospheric Research* 94:276–284.

Oke, T. R. 1973. City size and the urban heat island. *Atmospheric Environment* 7:769–779.

Oke, T. R. 1981. Canyon geometry and the nocturnal urban heat island: Comparison of scale model and field observations. *Journal of Climatology* 1:237–254.

Oke, T. R. 1982. The energetic basis of the urban heat-island. *Quarterly Journal of the Royal Meteorological Society* 108:1–24.

Oke, T. R. 1987. *Boundary layer climates*. London: Routledge.

Oke, T. R. 1997. Urban climates and global environmental change. In *Applied climatology: Principles and practice*, eds. R. D. Thompson and A. Perry, 273–287. London: Routledge.

Oke, T. R., G. T. Johnson, D. G. Steyn, and I. D. Watson. 1991. Simulation of surface urban heat islands under ideal conditions at night. Part 2: Diagnosis of causation. *Boundary-Layer Meteorology* 56:339–358.

Oltra-Carrió, R., J. A. Sobrino, B. Franchand, and F. Nerry. 2012. Land surface emissivity retrieval from airborne sensor over urban areas. *Remote Sensing of Environment* 123:298–305.

Petitcolin, F., and E. Vermote. 2002. Land surface reflectance, emissivity and temperature from MODIS middle and thermal infrared data. *Remote Sensing of Environment* 83:112–134.

PRB (Population Reference Bureau). 2011. *2011 world population data sheet*. Technical Report. Washington, DC: Population Reference Bureau.

Pu, R., P. Gong, R. Michishita, and T. Sasagawa. 2006. Assessment of multi-resolution and multi-sensor data for urban surface temperature retrieval. *Remote Sensing of Environment* 104:211–225.

Rajasekar, U., and Q. H. Weng. 2009. Spatio-temporal modelling and analysis of urban heat islands by using Landsat TM and ETM plus imagery. *International Journal of Remote Sensing* 30:3531–3548.

Rigo, G., E. Parlow, and D. Oesch. 2006. Validation of satellite observed thermal emission with in-situ measurements over an urban surface. *Remote Sensing of Environment* 104:201–210.

Roetzer, T., M. Wittenzeller, H. Haeckel, and J. Nekovar. 2000. Phenology in central Europe—Differences and trends of spring phenophases in urban and rural areas. *International Journal of Biometeorology* 44:60–66.

Sobrino, J. A., J. C. Jimenez-Munoz, G. Soria, et al. 2008a. Land surface emissivity retrieval from different VNIR and TIR sensors. *IEEE Transactions on Geoscience and Remote Sensing* 46:316–327.

Sobrino, J. A., J. C. Jiménez-Muñoz, G. Sòria, et al. 2008b. Thermal remote sensing in the framework of the SEN2FLEX project: Field measurements, airborne data and applications. *International Journal of Remote Sensing* 29:4961–4991.

Sobrino, J. A., R. Oltra-Carrió, J. C. Jiménez-Muñoz, et al. 2012b. Emissivity mapping over urban areas using a classification-based approach: Application to the Dual-use European Security IR Experiment (DESIREX). *International Journal of Applied Earth Observation and Geoinformation* 18:141–147.

Sobrino, J. A., R. Oltra-Carrió, G. Sòria, R. Bianchi, and M. Paganini. 2012a. Impact of spatial resolution and satellite overpass time on evaluation of the surface urban heat island effects. *Remote Sensing of Environment* 117:50–56.

Sobrino, J. A., R. Oltra-Carrió, G. Sòria, et al. 2013. Evaluation of the surface urban heat island effect in the city of Madrid by thermal remote sensing. *International Journal of Remote Sensing* 34:3177–3192.

Stathopoulou, M., C. Cartalis, and M. Petrakis. 2007. Integrating CORINE land cover data and Landsat TM for surface emissivity definition: Application to the urban area of Athens, Greece. *International Journal of Remote Sensing* 28:3291–3304.

Stathopoulou, M., A. Synnefa, C. Caralis, et al. 2009. A surface heat island study of Athens using high-resolution satellite imagery and measurements of the optical and thermal properties of commonly used building and paving materials. *International Journal of Sustainable Energy* 28:59–76.

Streutker, D. R. 2003. Satellite-measured growth of the urban heat island of Houston, Texas. *Remote Sensing of Environment* 85:282–289.

Tiangco, M., A. M. F. Lagmay, and J. Argete. 2008. ASTER-based study of the nighttime urban heat island effect in metro Manila. *International Journal of Remote Sensing* 29:2799–2818.

Toy, S., S. Yilmaz, and H. Yilmaz. 2007. Determination of bioclimatic comfort in three different land uses in the city of Erzurum, Turkey. *Building and Environment* 42:1315–1318.

Van de Griend, A. A., and M. Owe. 1993. On the relationship between thermal emissivity and the normalized difference vegetation index for natural surfaces. *International Journal of Remote Sensing* 14:1119–1131.

Voogt, J. A. 2002. Urban heat island. In *Encyclopedia of global environmental change*, ed. T. Munn. Chichester: Wiley.

Voogt, J. A., and T. R. Oke. 1997. Complete urban surface temperatures. *Journal of Applied Meteorology* 36:1117–1132.

Voogt, J. A., and T. R. Oke. 1998. Radiometric temperatures of urban canyon walls obtained from vehicle traverses. *Theoretical and Applied Climatology* 60:199–217.

Voogt, J. A., and T. R. Oke. 2003. Thermal remote sensing of urban climates. *Remote Sensing of Environment* 86:370–384.

Weng, Q. H., D. S. Lu, and J. Schubring. 2004. Estimation of land surface temperature-vegetation abundance relationship for urban heat island studies. *Remote Sensing of Environment* 89:467–483.

West, P. C., G. T. Narisma, C. C. Barford, C. J. Kucharik, and J. A. Foley. 2011. An alternative approach for quantifying climate regulation by ecosystems. *Frontiers in Ecology and the Environment* 9:126–133.

WHO (World Health Organization). 2003. *Climate change and human health: Risk and responses*. Geneva, Switzerland: World Health Organization.

Yuan, F., and M. E. Bauer. 2007. Comparison of impervious surface area and normalized difference vegetation index as indicators of surface urban heat island effects in Landsat imagery. *Remote Sensing of Environment* 106:375–386.

第六部分

生态系统服务的其他维度

第二十章 生态系统服务评估的多维方法

A. J. 卡斯特罗・马丁内斯 M. 加西亚-略伦特 B. 马丁-洛佩斯
I. 帕洛莫 I. 伊涅斯塔-阿兰迪亚

一、生态系统服务评估对多维度跨学科框架的需求

如今,学术界越来越关注可持续发展科学,即研究人与自然关系的科学(MA,2005;Perrings,2007;Perrings et al.,2011),尤其关注将生态系统服务的概念应用于环境保护和管理(Seppelt et al.,2011;Burkhard et al.,2012a)。过去20年,全世界的科学家、管理者和决策者都愈加注重生态系统服务。该概念结合自然科学及社会科学两个视角,体现了社会运行对生态系统的依赖(Bastian et al.,2012a)。多项国际倡议,如千年生态系统评估、生态系统经济学和生物多样性研究(The Economics of Ecosystems and Biodiversity,TEEB)及生物多样性和生态系统服务政府间平台(IPBES)(Carpenter et al.,2009;de Groot et al.,2010;Seppelt et al.,2011;Burkhard et al.,2012a)都已制定跨学科框架,以面向生态系统造福社会的不同价值维度,从而使生态系统服务概念具有可操作性。

尽管已经取得一些进展,但将生态系统服务的概念整合到行动框架的过程仍存在诸多挑战(de Groot et al.,2010)。泽佩尔特等人(2011)最近研究了生态系统服务领域,以提供指导,增强概念的应用和结果的可信度。尽管越来越多的出版物从不同的视角提出了创新和互补的看法,但所用研究方法和技术的恰当性却越来越不确定,因此研究的可比性和适用性受限(Seppelt et al.,2012)。因

此,目前的主要挑战之一,是制定一个综合框架来整合生态系统服务的多个维度(即生物物理、社会文化和经济层面)(Lamarque et al.,2011;Chan et al.,2012)。

诸多作者已注意到确定生态系统提供服务的能力(供应端)及其社会需求(需求端)的重要性,强调生态系统服务不仅受生态系统自身特性的影响,而且受社会需求的影响(Paetzold et al.,2010;Syrbe and Walz,2012;Burkhard et al.,2012b)。在供应端,生态系统和生物多样性提供服务的能力目前严重退化。同时,随着人口增长和生活标准的提高,社会对某些生态系统服务的需求正在迅速增加(Liu et al.,2010)。布克哈德等人(2012a)将供应端定义为特定区域在给定时间段内提供特定生态系统服务的能力,而需求方定义为特定区域在给定时间段内所有生态系统服务之消耗、使用或价值的总和。

就该方面而言,过去20年,遥感方法和技术主要用于量化与绘制生态系统服务的供应端,尤其是绘制供应服务(如木材或粮食生产)与调节服务(如空气质量、气候、极端事件、废物处理、侵蚀和土壤肥力)的分布图(Ayanu et al.,2012)。遥感信息最常被用来推算生物物理变量(如生物质;见第五章),而后者又被进一步用来测量特定生态系统服务(如碳储存)。阿亚努等人(2012)最近回顾了用于量化生态系统供应与调节服务的相关遥感系统、传感器和方法。结果表明,在通过地球观测技术对生态系统服务进行量化的过程中使用回归模型(通过将遥感信息与有限数量的现场观测联系起来),或使用土地利用、土地覆盖分类(见第十章),可进一步用其测定生态系统服务。

本研究中,我们提出了生态系统服务评估的概念框架(图20-1),从生态系统服务评估的多个维度,从供给侧和需求侧进行了探索,将服务提供单元(Service-Providing Units,SPUs)和生态系统服务受益方(ESBs)的概念结合起来。在此框架下,我们试图探索有哪些可用于评估生态系统服务供应端和需求端的遥感方法与技术。

考虑到生态系统服务评估的多维性,从供给侧和需求侧进行了探索。在此框架内,我们探讨了目前可用于评估生态系统服务供给和需求的遥感方法和技术。

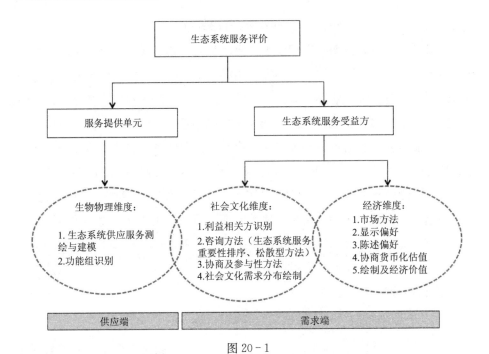

图 20-1

注：概念框架显示了服务提供单位和生态系统服务受益方之间的联系，以及如何通过生态系统服务评估中的不同维度对其进行探索。

二、生态系统服务供应端

（一）服务提供单元

学界已经提出若干概念以实现对生态系统服务提供过程进行操作化定义。其中之一则是服务提供单元概念。此概念能将物种种群与服务提供结合起来，主要是在小规模上(Luck et al., 2003)实现，但也可以扩展到更大的规模。服务提供单元在这里被定义为特定物种个体的组合及其为受益方提供生态系统服务所对应的特征。该概念也被进一步推广到其他层次的组织和分析规模。事实上，卢克等人(Luck et al., 2009)建议将社会生态景观单元看作服务提供单元，其中景观提供服务的能力将与其在社会生态背景下的结构、功能和社会属性相关。人们已经逐渐形成这样的共识，即在所有生物多样性组成部分中，该组成部分的功能多样性主要确保了生态系统服务的提供，特别是调节服务(Díaz et al.,

2006；Cardinale et al.，2012；Alcaraz et al.，2013）。

  遥感可以有效地帮助我们理解服务提供单元及其在区域和全球生态系统服务评估中的应用。光谱信息在绘制提供相同生态系统服务的土地单元时会起到辅助作用。生态系统功能类型的概念就是如此。既然植物物种可被分作不同的植物功能类型，那么生态系统也可有不同的功能类型（Paruelo et al.，2001）。生态系统功能类型代表的生态系统群在生物群和物理环境之间的物质与能量交换方面具有相同的功能特征（Paruelo et al.，2001；Alcaraz-Segura et al.，2006）。换句话说，既然同一生态系统功能类型中的土地单元以同样的方式与大气交换质量和能量，因此可以认为，这些土地单元即是具有同等能力的景观单元，其提供的生态系统服务源自分类中涉及的生态过程（如碳收益）。从这个意义上讲，进一步研究生态系统功能类型提供生态系统服务的能力可能会成为生态系统服务对地观测的重大突破。

（二）生态系统服务评估中的生物物理指标

  一些生物物理指标已被用于评估生态系统服务。指标类型则取决于案例搜集到的可用信息及愿意用于生态系统服务评估的资源量。一些研究对最常见的指标进行了分组（Maes et al.，2011；Burkhard et al.，2012b）。表20-1列出了一些最常用于评估生态系统服务供给的指标，以及一些可用于生态系统服务需求的指标。在"遥感研究"列，表格标出使用或可能使用了遥感技术来推导生态系统服务供需两侧指标的研究。

  对生态系统服务的时空异质性的了解亟待深入，自然保护效果也急需加强（Polasky et al.，2008；Nelson et al.，2009）。在此情况下，使用遥感技术可有效获取生态系统服务供应端的生物物理指标。通过直接或间接监测并结合生态系统模型，遥感方法可用于生态系统服务制图（Feng et al.，2010）。直接监测的生态系统服务，如物种栖息地（见第三章）、碳固定（见第六章）、水供应（见第十二章）和气候调节（见第十七章和第二十章）需要植被和水方面的信息。间接监测的服务，例如，基于土壤的服务，则需要使用替代信息，如土壤状况或冠层反射率。最后，通过遥感空间数据的输入，可在生态系统模型中监测和调控洪水调节或土壤侵蚀等服务。表20-2显示了一些与生态系统服务相关的遥感产品（Feng et al.，2010）。

第二十章 生态系统服务评估的多维方法

表 20-1 生态系统服务供需评估中的生物物理指标列表

生态系统服务	供应端生物物理指标	遥感研究	需求端社会指标	遥感研究
供应服务				
食物供应	作物产量	Doraiswamy et al., 2003	作物消耗量	
食物供应	放牧量		牲畜消耗量	Stephen et al., 2013
食物供应	鱼群数量	Chassot et al., 2010	捕获或消耗的鱼或生物质数量	Stuart et al., 2006
生物材料	木材储量	Clementel et al., 2012	木材消耗量	
水供应	可用水量（即降水量减去蒸散量）	Bahadur, 2011	耗水量	
医药资源	可用于提取天然药物的物种数		使用天然化合物的药物数量	
遗传库	某地区/地表的作物及牲畜品种数	Heller et al., 2012	某地区使用的作物及牲畜品种数量	
原材料	自然/地表存在的原材料	Mengzhi, 2009	提取、使用或购买的原材料数量	
调节服务				
空气质量调节	大气净化能力（污染物清除量）		避免的空气污染相关疾病	
水质调节	水域生态系统去除的氮磷营养物生物质		避免的水污染或不健康用水相关疾病	
气候调节	固存的碳（或甲烷）总量	Myeong et al., 2006	避免产生的气候难民及遭到气候变化影响之人数	Lobitz et al., 2000
极端天气事件减缓	抑制极端事件（洪水、风暴、雪崩）的自然因素或生物的种数量		避免产生的洪水、风暴、雪崩损害	

续表

生态系统服务	供应端生物物理指标	遥感研究	需求端社会指标	遥感研究
侵蚀防护	土壤侵蚀率、侵蚀相关变量(坡度、降雨量、植被覆盖率等)	Vrieling,2006	水库泥沙溢出量	
授粉	野生授粉者的丰度和物种丰富度	Schulp and Alkemade, 2011	授粉对作物生产或生物多样性维护的增益	
文化服务				
审美享受	视域范围、自然风光			
休闲旅游	可供旅游的自然区域数目	Nichol and Wong,2005	景区线路使用人数	
休闲旅游(休闲狩猎)	动物种群大小、物种更新率	Ropert-Cudert and Wilson,2005	参观人数	
知识与经验(科学知识)	具有科学价值的景观特征或物种丰富度	Mertes,2002	捕猎动物数量	
			从事环境资产研究的人数	

资料来源:de Groot, R. S., et al., 2010; Maes, J., et al.,2011; Burkhard, B., et al., 2012。

表 20-2 遥感数据量化生态系统服务

指标	生态系统服务	光谱指数或技术	传感器类型	参考文献
不同用途的土地/土地覆盖的碳储存/封存	气候调节；空气质量调节	绘制土地覆盖分布的高级甚高分辨率辐射计	LANDSAT	Konarska et al.，2002
动植物丰富度及物种丰富度	生物多样性保护	归一化差异植被指数——景观异质性；激光雷达——冠层结构	MODIS	Gould，2000；Balvanera et al.，2006；Carlson et al.，2007
生态系统碳收益的碳吸收	气候调节	归一化差异植被指数，净初级生产力；植被增强指数；总初级生产力；叶面积指数	MODIS	Gianelle and Vescovo，2007；Olofsson et al.，2008；Gianelle et al.，2009
水循环（绿、蓝水）	水温调节；侵蚀防护；防洪	归一化植被指数——地表参数；——表面参数	SWIM WEPP LASCAM SEB	Krysanova et al.，2007；Liu and Li，2008；Minacapilli et al.，2009；Williams et al.，2010
土壤类型属性	气候调节；土壤肥力	归一化差异水体指数——裸土比例；归一化差异植被指数——土壤颜色指数——土壤饱和度；土壤饱和度；激光雷达——土壤粗糙度测量	LANDSAT MODIS	Lobell et al.，2009；Kheir et al.，2010

注：本表提供了最常见的指标，光谱指数或遥感技术，所用传感器类型。LANDSAT 为陆地卫星（http：//landsat.gsfc.nasa.gov）；MODIS 为中分辨率成像光谱仪（http：//modis.gsfc.nasa.gov）；SEB 为地表能量平衡；SWIM 为土壤和水综合模型；WEPP 为水蚀预测项目模型；LASCAM 为大规模流域模型。

(三) 空间分布测量:早期及目前工具

生态系统服务地图似乎是将生态系统服务概念纳入生态系统管理的最有效方法之一(Balvanera et al.,2001;Daily and Matson,2008)。该方法可确定具有高保护价值的地区(Chan et al.,2006;Naidoo and Rickets,2006)或确定生态系统服务供应和需求,可能有助于实现生态系统服务的可持续利用(Kroll et al.,2012)。从最初的生态系统服务地图(Eade and Moran,1996;Costanza et al.,1997)到当前的生态系统服务模型或工具箱(例如 InVEST 或 ARIES;见专栏 20-1),我们见证了绘制生态系统服务地图的不同目的及方法的增加。如今,生态系统服务分布测量已经有了可支持景观管理的新工具(见专栏 20-1)(Jackson et al.,2013)。绘图工作开始侧重于供应端(Burkhard et al.,2012b),但参与性方法越来越多地将需求端也囊括在内(Bryan et al.,2011;Palomo et al.,2012)。

专栏 20-1 生态系统服务测绘的主要工具

InVEST:生态系统服务评估与权衡

InVEST 是"自然资本项目"设计的一系列用于绘制生态系统服务图的工具,旨在为自然资源管理相关决策提供信息,并通过估算当前景观或未来情景下生态系统服务的数量和价值,为评估生态系统服务间的权衡提供有效工具。InVEST 模型是空间直观模型,使用地图作为信息源,评估结果也以地图形式呈现。InVEST 既可生成生物物理结果(如储存碳吨数),也可生成具有经济意义的结果(如固存碳当前的净现值)。InVEST 还可绘制若干种生态系统服务图,包括:(1)土地和水域:生物多样性、碳、水力、水净化、水库淤积、管理木材生产、作物授粉;(2)海洋和海岸:波浪能、沿海脆弱性、海域水产养殖、美学价值、重叠分析(渔业娱乐、生境风险评估)。该模型不同的工具作为 ArcMap 的拓展运行,可从"自然资本项目"网站(http:www.naturalcapitalproject.org/InvEST.html)获得。用户可通过相关论坛交流和交流经验。

土地用途、土地覆盖：遥感信息

土地用途、土地覆盖信息已经被广泛用作生态系统服务量化和制图的替代指标，不同的土地利用、土地覆盖类型可指向不同的生态系统价值。遥感则为土地用途、土地覆盖分类提供了有用数据。分类方法则是基于统计学分析得出不同类别，其精度取决于所设置的训练区域。因此，最终的土地用途、土地覆盖分类取决于可用遥感数据的特性。因此，生态系统服务量化的准确性取决于土地用途、土地覆盖分类的准确性（Ayanu et al.，2012）。

ARIES：生态系统服务人工智能

ARIES是一种基于网络的技术，由众多机构（包括佛蒙特大学）开发，支持进行快速生态系统服务评价与评估。

该技术提供了一个智能建模平台，能够基于用户指定的系列模型组合复杂的生态系统服务模型。它可以绘制生态系统服务（源）、人类受益者（用户）及任何可能耗尽服务流（汇）的生物物理特征的位置和数量。因为用ARIES建模语言编写的诸多模型均基于概率贝叶斯方法，能够对数据输入和结果输出的不确定性进行明确表达，并且在数据稀缺时模型也能够运行，这是确定性模型无法实现的。在版本1.0 beta版本中，生态系统服务地图覆盖世界七个地区，并包括以下生态系统服务：供水、自给型渔业、碳、洪水和沉积物调节、海岸保护、美学价值和娱乐价值。所有信息均可在项目网站（http://ariesonline.org）获得。

POLYSCAPE：新型工具

POLYSCAPE是一个地理信息系统框架，旨在探索生态系统服务间的空间显性协同和权衡，以支持景观管理。POLYSCAPE目前包括探索土地覆盖变化对洪水风险、栖息地连通性、侵蚀和相关沉积物向受体输送、碳封存和农业生产力的影响的算法。

遥感为生态系统服务模型或常用于模拟生态系统服务的工具箱提供了宝贵的输入数据。生态系统服务模型在待量化的生态系统服务与遥感参数之间提供了明确联系。例如,"环生态系统服务评估与权衡"工具箱利用可从遥感数据中获得的土地覆盖、蒸散、降水和地形的空间明确信息,量化和绘制生态系统服务图。专栏 20-1 提供了遥感数据在 InVEST 模型内表征和绘制土地利用和土地覆盖特征方面的适用性信息。

三、生态系统服务需求端

(一) 生态系统服务受益方

目前,大多数生态系统服务研究并未明确包括不同生态系统服务受益方的偏好和价值(Menzel and Teng 2010; Seppelt et al.,2011)。

然而,不同利益相关方就哪些生态系统服务对其福祉最为重要方面具有不同的优先权(McMichael et al.,2003; Díaz et al.,2011),因此,这也应当被纳入生态系统服务评估中(Egoh et al.,2007)。目前的普遍做法是,将生态系统服务与其感知价值剥离开来。这意味着可以定义这些服务,但不包括从中受益的人给出的价值。这种方法不符合生态系统服务的定义。生态系统服务包括生态系统对人类福祉的直接或间接贡献;因此,将人类偏好的重要性纳入评估中是必要的(de Groot et al.,2010; EME, 2011)。

最近的地中海保护区研究确定了评估中应包括的三类利益相关者概况:旅游人口、管理者与环境专业人士、与生态系统服务有不同关系的当地人口或居民(Martín-López et al.,2007; Castro et al.,2011; García-Llorente et al.,2011a, 2011b)。当地人主要受益于生态系统带来的与农业养殖活动相关的供应服务和与地方归属感及文化遗产相关的文化服务。专业人士关注调解服务,以及与管理与知识系统和保护区相关的文化服务(如环境教育、科学知识或自然旅游)。游客通常偏爱与娱乐活动相关的文化服务和与都市人口亲近自然的需求相关的审美价值。理解此等观点的多样性有助于分析潜在的社会冲突和理解生态系统服务的权衡。一方面,识别不同的生态系统服务受益方并归纳其特征是其参与设计或促进环境管理政策改善过程的第一步(Baker and Landers, 2004; Reed,

2008；Palomo et al.，2011)。另一方面，纳入不同的生态系统服务受益方概况可促进不同知识来源的结合，即实验性(当地生态知识)和经验性(技术或科学知识)两种(García-Llorente et al.，2011b)。

(二) 社会文化价值评判

在服务评估方面使用社会文化视角的研究数量非常有限，所使用的技术也没有像在经济评估中那样正式化(这里解释)。然而，作为一种评估文化服务、增加非物质利益(Chan et al.，2006，2012)及反映服务价值多样性的手段，该视角正日益受到关注。生态系统服务的社会文化评估使用非经济方法来分析人类对生态系统服务的需求、使用、享受和价值偏好，其中在道德、伦理、历史或社会层面起着重要作用。要了解人类的偏好、对自然的态度和行为意图，除经济视角外，还需要分析心理、历史和伦理因素(Spash et al.，2009)。

在文化评估中，我们认识到生态系统及其生物多样性提供了与非使用价值相关的服务，如保护生物多样性的满意度、当地身份或当地生态知识。当我们想要探索物种和生态系统的内在价值时，在经济维度中使用的效用函数并没有涵盖所有的人类动机(Chan et al.，2006)。正如库马尔和库玛(2008)所述，有些问题超越了经济选择的特定逻辑，因此也超越了人类福祉的不同维度，如社会关系、健康、安全、选择和行动自由(MA，2003)。此等问题无法用经济维度的分析来解决(Wegner and Pascual，2011)。

可以利用涉及直接和间接咨询方法的定性和定量技术，探讨生物多样性和生态系统服务所具有的特殊社会文化价值。直接咨询方法包括探索个人认知和集体偏好方法的技术。一方面，分析个人对生态系统服务重要性或使用情况的感知的技术，通过调查和使用量表(如李克特量表)，对首选的生态系统服务进行排名或评级。在排名技术中，受访者通常从特定生态系统中现有的服务组中决定最重要的生态系统服务(如 Castro et al.，2011)。在评级过程中，受访者对每项服务进行独立评级(Agbenyega et al.，2009)，通常使用某种视觉辅助工具(Calvet-Mir et al.，2012)。另一方面，分析集体偏好，如基于话语的分析(Wilson and Howarth，2002)，是基于这样一个假设，即公共产品(如大多数生态系统服务)的价值评判应该来自一个包含社会公平问题的自由公开的公共辩论过程，而非个人看法的集合。在这套技术中，一小群个体(通常超过2人，但不超过 20 人)就生态系统服务价值进行辩论并达成共识。

在间接协商过程中,受访者需要列出描述特定生态系统的概念或术语;随后,研究人员将此概念与该生态系统服务类型联系起来(Quetier et al.,2010)。此外,此等表达的观点可与传播媒介(书籍、文章、法律、保护项目、网络等)中明确包含的观点、想法和语言相结合。这可使用内容分析技术来实现,该技术是评估生态系统服务受益方对生物多样性和生态系统服务重要性的看法和价值评判的合适工具(Xenarios and Tziritis, 2007; Webb and Raffaelli, 2008)。Q-methodology 混合研究法也有望成为确定生态系统服务价值的一种方法。该方法侧重于理解生态系统服务受益方对环境问题的看法和感受(Sandbrook et al.,2011),并有可能根据受益方概况探索生态系统服务优先级和权衡。其他间接方法包括:首先得出社会对景观的偏好(鉴于景观往往是生态系统服务提供能力的外部体现),并使用德尔菲法对生态系统服务进行后续归纳(García-Llorente et al.,2012)。

(三) 经济价值评判

过去几十年,学术和政治领域都优先考虑经济层面问题(Gómez-Baggethun et al.,2010)。20 世纪 90 年代,随着德·格鲁特(1992)、戴利(1997)或科斯坦萨等人(1997)研究作品的出版,经济价值评判开始了迈向主流的第一步。其中将全球 16 个生物群的 17 个生态系统服务价值估算为每年 33 万亿美元。(与此同时,全球国内生产总值每年约为 18 万亿美元。)因此,按照此估计,自然"生产"是人类"生产"的 1.8 倍。从那时起,生态系统服务经济学的主导地位可以从科学论文数量的增加看出(Loomis et al.,2000; Kontogianni et al.,2010),以及不同的全球项目,比如 TEEB,它量化了生物多样性损失和生态系统服务退化的成本(TEEB,2010)。这一方法被广泛地应用于评估不同生态系统,如森林(Croitoru, 2007; Zandersen and Tol, 2009)、湿地(Woodward and Wui, 2001; Brander et al.,2006)、海洋生态系统(Turpie et al.,2003; Ressurreição et al.,2011)或物种(Losey and Vaughan, 2006; García-Llorente et al.,2011b)提供的生态系统服务。

下文将给出一些对该维度优势的解释。首先,考虑到生物多样性持续降低(Burkhard et al.,2012),除了承认传统保护主义者支持的生物多样性内在价值外,人们呼吁寻找支持生物多样性保护的更有力的论据。有人认为,由于生物多样性内在价值不能测量,在决策过程中往往被忽视(Bateman et al.,2002)。根据这一观点,由于生物多样性所提供的"看不见"的生态系统服务缺乏基准价格(而不是价值),人们越来越热衷于让生态系统服务发挥的正外部性效应可以"看见"。

大量的调节服务(如授粉、侵蚀防护或水文调节)和文化服务(如审美价值、当地生态知识或物种保护带来的满意度)都是如此(Rodríguez et al.,2005;Gee and Burkhard,2010;Vejre et al.,2010)。决策者对行动成本尤为敏感,大多数土地使用政策背后的经济研究都采用成本效益分析来衡量政策措施的效益与成本(Balmford et al.,2011)。这一论点认为即关于生态系统和生物多样性对人类福祉的贡献在商业意义上的信息越完整,决策的成功程度就越高(de Groot 2006)。

生态系统服务的经济学的确有助于证明那些没有市场价格的生态系统服务的重要性,例如害虫控制和授粉对农业生产的贡献(Gallai et al.,2009),了解此类概念之后能更好地决定应把哪些保护战略放在优先位置(Martín-López et al.,2007;García-Llorente et al.,2011b),或分析不同生物多样性和生态系统管理选项间的权衡或协同效应。

世界各地的一些案例研究表明,维护不同自然生态系统,不仅要具有生态价值,而且要具有经济价值,而非将其转化为密集的用途上。研究表明,喀麦隆热带森林或加拿大湿地生态系统的重要性(Balmford et al.,2002)、湿地可持续管理的经济重要性(Birol et al.,2006)、珊瑚礁管理策略(Hicks et al.,2009)或地中海保护区提供的生态系统服务(Martín-López et al.,2011)都是如此。从这个意义上讲,考虑到全球变化的背景和主流经济思维,对生态系统服务的经济价值评判创造了科学家和决策者之间务实的共同语言,也创造了围绕如何明确人类对生态系统及其生物多样性的依赖的公共讨论论坛。

环境经济学将生态系统服务看作正外部性,可通过总经济价值(Total Economic Value,TEV)框架来衡量。不同服务类别具有不同类型的价值,这些价值可以被聚合和分离进行分析(Pearce and Turner,1990)。总经济价值由使用价值和非使用价值组成(图 20-2)。使用价值与我们从生态系统中所获得的直接或间接贡献有关；非使用价值与维护生物多样性及与其使用价值无关的生态系统服务的道德或伦理考虑有关。同时，使用价值由直接使用价值、间接使用价值和期权价值组成。直接使用价值通常在市场上有所体现，是人类直接使用生态系统及其生物多样性的结果，无论这是消耗性或采掘性（如木材或淡水）或非消耗性或非采掘性（如自然旅游）。同时，间接使用价值通常不会体现在传统市场中，而是源自生态过程和调节服务（如水生植物的净水作用或红树林生态系统在侵蚀防护或减缓方面的作用）。最后，期权价值与未来维持生态系统服务流的重要性相关，因此与任何生态系统服务类别都有关。非使用价值可包括与保护生态系统及其生物多样性的满意度相关的存在价值，即使我们不会享受或使用这些价值；也就是说，那些代表文化服务或慈善价值观，与了解后代将获得生态系统服务（遗赠价值）的满意度，以及知道其他人可以获得生态系统服务（利他价值）的满足感相关。

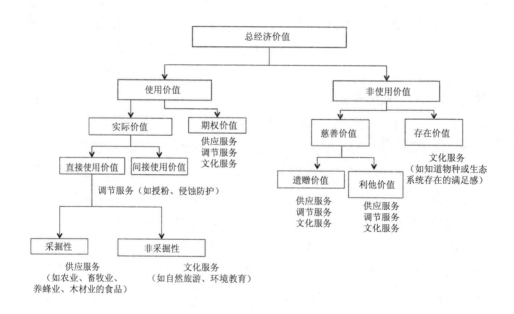

图 20-2

注：表示总经济价值、价值类型及其相关的生态系统服务类别。

因此，期权价值、遗赠价值和利他价值可与所有生态系统服务类别相关（Bateman et al.，2002；Martín-López et al.，2009；TEEB 2010 for more detail）。

为了评估每种价值类型，环境经济学中设计了诸多方法。这些方法可分为三类：直接市场、显示偏好和陈述偏好。目前的挑战是了解哪种评估方法最适合不同的生物多样性价值类型及其生态系统服务，与此同时考虑背景和特定的政策目的（de Groot et al.，2010）。在回顾之前的工作时，我们就如何确定哪种评估方法最适合评估不同的评估类型和服务类别提供了一些指导原则。

直接市场使用价格反映价值，然后使用来自实际市场的数据来估算直接使用价值。这些数据包括：(1)市面上的供应服务价格，如从农业或森林服务中获得的商品（如木材或非木材林产品）；(2)生产函数，用于估算不具市场价格的特定生态系统服务对在市场上销售的另一种服务的提供有多大贡献（例如授粉对养蜂或农业生产的贡献）；(3)成本法，成本法通过估算避免成本或重置成本（TEEB 2010），估算生态系统服务贡献需要通过人工市场重建时将产生的费用。生产函数法和成本法通常用于间接估算调节服务的价值（如通过保护海岸线免受风暴和洪水的影响可保护多少湿地）（图 20-3）。

显示偏好通过观察与服务相关的替代市场来估算给定无市场定价服务的价值。包括两种主要方法：(1)旅行成本（可估算特定自然区域内的自然旅游等娱乐服务），其基础是到达特定区域的成本至少应等于所获得的效用（Shrestha et al.，2002；Martín-López et al.，2009）；(2)特征价格，其中市场商品（通常为房产）被描述为拥有若干属性，包括环境属性（如大小、周边，窗外的风景）。随后，通过估算该房产的需求函数，我们可以推断环境属性价值的变化（如窗外视野好、不好）（Lansford and Jones，1995；Geoghegan et al.，1997）。旅游成本和特征价格主要用于估算与娱乐活动相关的文化服务（如自然旅游、休闲捕鱼、休闲狩猎、景观审美价值）的间接使用价值。然而，特征价格也可用于对一些调节服务价值进行评估，例如，在房地产价格中估算空气质量服务的隐含价格（图 20-3）。

最后，陈述偏好方法则在调查中创建假想市场，用于计算与使用价值和非使

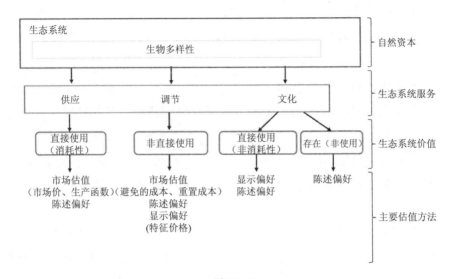

图 20-3

注：自然资本、生态系统服务类别、相应经济价值及最常用的估值方法的图形表示。

用价值相关的生态系统服务价值，并可应用于所有生态系统服务类别（图 20-3）。三种主要的方法是：(1) 条件评估法。直接询问人们在假想市场中，愿意（或接受）为给定生态系统服务的数量或质量的变化支付多少费用，从而得出公众的偏好（Mitchell and Carson, 1989）。该技巧是最广泛使用的技巧之一（Jorgensen et al., 2001; Gürlük, 2006; García-Llorente et al., 2011a）；(2) 选择建模或联合分析法（选择实验、选择排名和选择评级的可能性不同）。通过要求受访者从一系列选择中选择他们的首选项来得出公众偏好，每个选择集合都根据与生态系统服务或不同环境计划相关的不同属性和水平进行描述（Hanley et al., 2003; Westerberg et al., 2010; Zander and Straton, 2010）；(3) 最后，审慎的货币法是一种混合方法，其在小群体中应用陈述偏好法，以促进参与过程（Zografos and Howarth, 2008; Kenter et al., 2011）。关于不同评估方法的概述，见贝特曼等人（2002）、徐等人（Xu et al., 2004）以及特纳等人（Turner et al., 2010）的文献。

以上所有估值法都存在与信息和方法误设、战略应对、公平问题、不熟悉性或排序效应相关的优缺点，因此都需改进（Carson et al., 2001; Barkmann et al.,2008；Schläpfer,2008；Turner et al.,2010)。然而,其中一些缺点也与经济分析框架本身固有的局限性有关。这类方法基于新古典主义经济学,以功利主义框架为前提。该框架假设社会中的每名个人都有理性偏好,并试图最大化他们的利润、优势或利益,而社会利益是个人利益的总和(Dequech,2007；García-Llorente et al.,2011a)。此外,由于道德、伦理或心理动机的影响,生物多样性及其生态系统服务的评估与其他商品的评估不同(Hanley and Milne,1996)。此类价值不能也不应完全转化为经济意义上的价值,必须使用社会文化分析等其他工具进行补充或处理(TEEB,2010)。然而,从生态价值的视角看,经济评估途径的主导意味着存在将复杂问题简单化的风险,容易使人忽略生态系统过程在提供生态系统服务方面的重要性(Norgaard,2010；Sagoff,2011),并且,支持创建市场来将某些生态系统服务商品化,会对正在解决的问题产生反作用(Gómez-Baggethun and Ruiz-Pérez,2011)。这再次导致了将经济评估与社会和生物物理评估相结合或使用替代方案的必要性。

四、讨论与延伸:走向混合方法和新概念

该领域的最新发展包括结合非货币化和货币化法或多维法的混合方法。其中之一是应用绘图工具。一方面,这可改进许多经济活动的设计和描述(Troy and Wilson,2006；Balmford et al.,2011),例如特征价格法、旅行成本法或选择建模法(Geoghegan et al.,1997；Brouwer et al.,2010),其中地理信息系统允许在经济活动中整合环境数据和生态复杂性。此外,绘图工具允许我们在分析生态系统中的生态系统服务可持续性的同时,还考虑到此等服务的社会价值和需求(Sherrouse et al.,2011；Kroll et al.,2012)。生成的地图提供了易于解释的信息,有助于分析不同获得空间直观经济数据的管理选项(Bateman and Jones,2003；Goldstein et al.,2012)。

陈述偏好法的其他例子有选择建模,但其中不应包括货币化或成本属性(Sayadi et al.,2005),也不包括传统的支付意愿以外的支付形式,例如愿不愿意牺牲个人时间进行生物多样性保护或生态系统服务维护(Higuera et al.,2013)。这一选择得到了不同生态系统服务受益人概况的支持;此等方式避免了

公平问题,也避免了将货币价值分配给那些一般认为无法用金钱衡量的事物(García-Llorente et al.,2011a)。正如考林等人(2008)所言,"因为货币是最常用的可互换商品,用单一的货币化术语对服务进行评估可能会发出这样一个错误信息:人为制造的供应商可以轻易取代某些服务"。因此,生态系统服务的人类的价值分析应从经济(结合货币化和非货币文化法)和社会文化的两个视角进行。

当前工具箱(如 InVEST 或 ARIES)中可用的生态系统服务的空间直观绘图基于密集的数据需求,而数据调查的高成本限制了其在本地范围的应用(Ayanu et al.,2012)。

相比之下,遥感以相对较低的成本为量化和绘制生态系统服务图提供数据,并为监测提供了频繁和标准化观测的可能性(如区域或全球范围内碳储量的量化和绘制)。目前,生态系统服务科学的大部分知识缺口都涉及空间直观的量化与制图。因此,遥感有可能解决生态系统服务空间量化的基本问题。这主要与生态系统服务的供应与管理有关,例如碳储存或固存(Konarska et al.,2002)或侵蚀防护(Krysanova et al.,2007)。然而,如今,生态系统服务评估的社会文化和经济层面也可利用遥感技术这种成本效益高的方式来收集服务供应的空间分布信息。我们只发现洛比茨等人(2000)的一项研究使用了供需两端的遥感指标来间接衡量传染病的影响。

此外,该领域的进展引入了新的概念,如服务受益区域(指特定区域生态系统服务需求)和服务连接区域(指连接生态系统服务供需的区域)(Syrbe and Walz,2012)。从这个意义上讲,服务提供单元和服务受益地区的空间直观量化,以及将服务提供者与生态系统服务受益人联系起来的重要性可被看作是为生物多样性保护制定更好评估方案的机会(Luck et al.,2003,2009),并有机会促进更有力和更充分的评估应用。

五、结论

生态系统服务评估在包括自然、社会和经济学界在内的各界学者中都变得愈加重要。然而,尽管学术上取得了进展,但将生态系统服务概念整合到一个可用于决策的操作框架中仍存在诸多挑战。本章根据当今使用的方法和技术的跨学科性质，回顾了对生态系统管理服务的评估现状。我们特别关注生态系统服务的多维性（即生物物理、社会文化和经济），从供应端和需求端对其进行评估。

关于生态系统服务制图，本章提供了一个清单，其中包括一些评估生态系统服务供应的常用指标（如测定淡水供应使用降水量减去蒸散量），我们还注意到一些可用于评估生态系统服务需求（如每个地表和时间段消耗的水）的指标。

本章也列出了生态系统服务绘制的主要现有工具箱（如由"自然资本项目"设计的 InVEST 工具箱）。本章还确定了利用最常见的遥感指标，绘制从地区到全球范围的服务地图，例如归一化差异植被指数，用于在多时间和空间尺度上量化净初级生产力。

遥感数据的使用主要用于生态系统供应服务供应端的定量和制图。通常，此等研究只是描述生态过程，如净初级生产力，而没有将其与潜在效益（如粮食生产或气候调节）联系起来。我们的研究揭示了使用生态服务标准化术语的必要性，该术语体系不仅关注生态过程或功能，而且关注公众认为有益的后续生态系统服务。我们的研究还表明，在生态系统服务需求侧评估中，遥感数据运用仍然不足。这可能是由于难以从遥感数据中跟踪社会文化方面的信息，如社会偏好或感知。因此，需要进一步的研究来提供指导，以帮助将遥感信息、定量和绘制生态系统服务于供应端、社会感知利益（即需求端）。

尽管使用社会文化评估的研究数量有限，但本文描述了一些分析人类偏好的方法，包括非货币化方法。本文分析了经济维度之所以在学术和政治领域中都占主导的一些原因，并在总经济价值框架内提供了一些指导性原则，即如何确定哪种评估方法更适合衡量不同的价值类型和服务类别。总而言之，本文建议从生物物理、社会文化和货币化及非货币化法相结合的经济维度形成生态系统服务地图。

致谢

笔者在此向三位匿名评论者对本章提出了十分宝贵的意见致以感谢。本项目由安达卢西亚政府环境部 GLOCHARID 项目和生物多样性基金会通过西班牙千年生态系统评估项目(http://www.ecomilenio.es/)提供支持。同样感谢安达卢西亚全球变革评估与监控中心(CAESCG)安东尼奥 J. 卡斯特罗的支持。

请支持 M. 加西亚·略伦特的 BESAFE 项目(生物多样性和生态系统服务:关于未来环境的不同论据; www.besafe-project.net),该项目由欧盟委员会第七框架计划(合约编号 282743)资助。

参 考 文 献

Agbenyega, O., P. J. Burgess, M. Cook, and J. Morris. 2009. Application of an ecosystem function framework to perceptions of community woodlands. *Land Use Policy* 26:551–557.

Alcaraz-Segura, D., J. M. Paruelo, and J. Cabello. 2006. Identification of current ecosystem functional types in the Iberian Peninsula. *Global Ecology and Biogeography* 15:200–212.

Alcaraz-Segura, D., J. M. Paruelo, H. E. Epstein, and J. Cabello. 2013. Environmental and human controls of ecosystem functional diversity in temperate South America. *Remote Sensing* 5:127–154.

Ayanu, Y. Z., C. Conrad, T. Nauss, et al. 2012. Quantifying and mapping ecosystem services supplies and demands: A review of remote sensing applications. *Environmental Science and Technology* 46:8529–8541.

Bahadur, K. C. 2011. Assessing strategic water availability using remote sensing, GIS and a spatial water budget model: Case study of the Upper Ing Basin, Thailand. *Hydrological Sciences Journal* 56:994–1014.

Baker, J. P., and D. H. Landers. 2004. Alternative-futures analysis for the Willamette River Basin, Oregon. *Ecological Application* 14:311–312.

Balmford, A., J. Birch, R. Bradbury, et al. 2011. *Measuring and monitoring ecosystem services at the site scale*. Cambridge, UK: Cambridge Conservation Initiative and BirdLife International.

Balmford, A., A. Bruner, P. Cooper, et al. 2002. Economic reasons for conserving wild nature. *Science* 297:950–953.

Balvanera, P., G. C. Daily, P. R. Ehrlich, et al. 2001. Conserving biodiversity and ecosystem services. *Science* 291:2047.

Balvanera, P., A. B. Pfisterer, N. Buchmann, et al. 2006. Quantifying the evidence for biodiversity effects on ecosystem functioning and services. *Ecology Letters* 9:1146–1156.

Barkmann, J., K. Glenk, A. Keil, et al. 2008. Confronting unfamiliarity with ecosystem functions: The case for an ecosystem service approach to environmental valuation with stated preference methods. *Ecological Economics* 65:48–62.

Bastian, O., D. Haase, and K. Grunewald. 2012. Ecosystem properties, potentials and services—The EPPS conceptual framework and an urban application example. *Ecological Indicators* 21:7–16.

Bateman, I. J., and A. P. Jones. 2003. Contrasting conventional with multi-level modelling approaches to meta-analysis: An illustration using UK woodland recreation values. *Land Economics* 2:235–258.

Bateman, I. J., A. P. Jones, A. A. Lovett, I. Lake, and B. H. Day. 2002. Applying geographical information systems (GIS) to environmental and resource economics. *Environmental and Resource Economics* 1–2:219–269.

Birol, E., K. Karousakis, and P. Koundouri. 2006. Using economic methods to inform water resource management policies: A survey and critical appraisal of available methods and an application. *Science of the Total Environment* 365:105–122.

Brander, L., R. Florax, and J. E. Vermaat. 2006. The empirics of wetland valuation: A comprehensive summary and a meta-analysis of the literature. *Environmental and Resource Economics* 33:223–250.

Brouwer, R., T. Dekker, J. Rolfe, and J. Windle. 2010. Choice certainty and consistency in repeated choice experiments. *Environmental and Resource Economics* 46:93–109.

Bryan, B. A., C. M. Raymond, N. D. Crossman, and D. King. 2011. Comparing spatially explicit ecological and social values for natural areas to identify effective conservation strategies. *Conservation Biology* 25:172–181.

Burkhard, B., R. de Groot, R. Costanza, R. Seppelt, S. E. Jorgensen, and M. Potschin. 2012b. Solutions for sustaining natural capital and ecosystem services. *Ecological Indicators* 21:1–6.

Burkhard, B., F. Kroll, S. Nedkov, and F. Müller. 2012a. Mapping ecosystem service supply, demand and budgets. *Ecological Indicators* 21:17–29.

Calvet-Mir, L., E. Gómez-Baggethun, and V. Reyes-García. 2012. Beyond food production: Ecosystem services provided by home gardens. A case study in Vall Fosca, Catalan Pyrenees, northeastern Spain. *Ecological Economics* 74:153–160.

Cardinale, B. J., J. E. Duffy, A. Gonzalez, et al. 2012. Biodiversity loss and its impact on humanity. *Nature* 486:59–67.

Carlson, K. M., G. P. Asner, R. F. Hughes, et al. 2007. Hyperspectral remote sensing of canopy biodiversity in Hawaiian lowland rainforests. *Ecosystems* 4:536–549.

Carpenter, S. R., H. A. Mooney, J. Agard, et al. 2009. Science for managing ecosystem services? Beyond the Millennium Ecosystem Assessment. *Proceedings of the National Academy of Sciences of the United States of America* 106:1305–1312.

Carson, D., A. Gilmore, C. Perry, and K. Gronhaug. 2001. *Qualitative marketing research*. London: Sage.

Castro, A. J., B. Martín-López, D. García-Llorente, P. A. Aguilera, E. López, and J. Cabello. 2011. Social preferences regarding the delivery of ecosystem services in a semiarid Mediterranean region. *Journal of Arid Environments* 75:1201–1208.

Chan, K. M. A., T. Satterfield, and J. Goldstein. 2012. Rethinking ecosystem services to better address and navigate cultural values. *Ecological Economics* 74:8–18.

Chan, K. M. A., M. R. Shaw, D. R. Cameron, E. C. Underwood, and G. C. Daily. 2006. Conservation planning for ES. *PLoS Biology* 4:2138–2152.

Chassot, E., S. Bonhommeau, N. K. Dulvy, et al. 2010. Global marine primary production constrains fisheries catches. *Ecological Letters* 13:495–505.

Chee, Y. E. 2004. An ecological perspective on the valuation of ecosystem services. *Biological Conservation* 120:549–565.

Clementel, F., G. Colle, C. Farruggia, et al. 2012. Estimating forest timber volume by means of "low cost" LiDAR data. *Italian Journal of Remote Sensing* 1:125–140.

Costanza, R., R. d'Arge, R. de Groot, et al. 1997. The value of the world's ecosystem services and natural capital. *Nature* 387:253–260.

Cowling, R. M., B. Egoh, A. T. Knight, et al. 2008. An operational model for mainstreaming ES for implementation. *Proceedings of the National Academy of Sciences of the United States of America* 105:9483–9488.

Croitoru, L. 2007. How much are Mediterranean forests worth? *Forest Policy and Economics* 9:536–545.

Daily, G. 1997. *Nature's services: Societal dependence on natural ecosystems*. Washington, DC: Island Press.

Daily, G. C., and P. Matson. 2008. Ecosystem services: From theory to implementation. *Proceedings of the National Academy of Sciences of the United States of America* 105:9455–9456.

de Groot, R. 2006. Function-analysis and valuation as a tool to assess land use conflicts in planning for sustainable, multi-functional landscapes. *Landscape and Urban Planning* 75:175–186.

de Groot, R. S. 1992. *Functions of nature: Evaluation of nature in environmental planning, management and decision making*. Groningen, The Netherlands: Wolters-Noordhoff B. V.

de Groot, R. S., R. Alkemade, L. Braat, L. Hein, and L. Willemen. 2010. Challenges in integrating the concept of ecosystem services and values in landscape planning, management and decision making. *Ecological Complexity* 7:260–272.

Dequech, D. 2007. Neoclassical, mainstream, orthodox, and heterodox economics. *Journal of Post Keynesian Economics* 30:279–302.

Díaz, S., J. Fargione, F. S. Chapin III, and D. Tilman. 2006. Biodiversity loss threatens human well-being. *PLoS Biology* 4:1300–1305.

Díaz, S., F. Quétier, D. M. Cáceres, et al. 2011. Linking functional diversity and social actor strategies in a framework for interdisciplinary analysis of nature's benefits to society. *Proceedings of the National Academy of Sciences of the United States of America* 3:895–902.

Doraiswamy, P. C., S. Moulin, and P. W. Cook. 2003. Crop yield assessment from remote sensing. *Photogrammetric Engineering and Remote Sensing* 69:665–674.

Eade, J. D. O., and D. Moran. 1996. Spatial economic valuation: Benefits transfer using geographical information systems. *Journal of Environmental Management*

48:97–110.

Egoh, B., M. Rouget, B. Reyers, et al. 2007. Integrating ecosystem services into conservation assessments: A review. *Ecological Economics* 63:714–721.

EME (Spanish Millennium Ecosystem Assessment). 2011. La Evaluación de los Ecosistemas del Milenio de España. Síntesis de resultados. Fundación Biodiversidad. Ministerio de Medio Ambiente, y Medio Rural y Marino, Spain.

Feng, X., F. Bojie, and Y. Yang. 2010. Remote sensing of ecosystem services: An opportunity for spatially explicit assessment. *Chinese Geographical Science* 20:522.

Gallai, N., J. M. Salles, J. Settele, and B. E. Vaissière. 2009. Economic valuation of the vulnerability of world agriculture confronted with pollinator decline. *Ecological Economics* 68:810–821.

García-Llorente, M., B. Martín-López, S. Díaz, and C. Montes. 2011b. Can ecosystem properties be fully translated into service values? An economic valuation of aquatic plant services. *Ecological Applications* 21:3083–3103.

García-Llorente, M., B. Martín-López, and C. Montes. 2011a. Exploring the motivations of protesters in contingent valuation: Insights for conservation policies. *Environmental Science and Policy* 14:76–88.

García-Llorente, M., B. Martín-López, P. A. L. D. Nunes, A. J. Castro, and C. Montes. 2012. A choice experiment study for land use scenarios in semi-arid watersheds environments. *Journal of Arid Environments* 87:219–230.

Gee, K., and B. Burkhard. 2010. Cultural ecosystem services in the context of offshore wind farming: A case study from the west coast of Schleswig-Holstein. *Ecological Complexity* 7:349–358.

Geoghegan, J., L. A. Wainger, and N. E. Bockstael. 1997. Spatial landscape indices in a hedonic framework and ecological economics analysis using GIS. *Ecological Economics* 23:251–264.

Gianelle, D., and L. Vescovo. 2007. Determination of green herbage ratio in grasslands using spectral reflectance. Methods and ground measurements. *International Journal of Remote Sensing* 5:931–942.

Gianelle, D., L. Vescovo, B. Marcolla, et al. 2009. Ecosystem carbon fluxes and canopy spectral reflectance of a mountain meadow. *International Journal of Remote Sensing* 2:435–449.

Goldstein, J. H., G. Caldarone, T. K. Duarte, et al. 2012. Integrating ecosystem-service tradeoffs into land-use decisions. *Proceedings of the National Academy of Sciences of the United States of America* 1:6.

Gómez-Baggethun, E., R. de Groot, P. L. Lomas, and C. Montes. 2010. The history of ecosystem services in economic theory and practice: From early notions to markets and payment schemes. *Ecological Economics* 69:1209–1218.

Gómez-Baggethun, E., and M. Ruiz-Pérez. 2011. Economic valuation and the commodification of ecosystem services. *Progress in Physical Geography* 5:613–628.

Gould, W. 2000. Remote sensing of vegetation, plant species richness and regional biodiversity hotspots. *Ecological Applications* 10:1861–1870.

Gürlük, S. 2006. The estimation of ecosystem services' value in the region of Misi rural development project: Results from a contingent valuation survey. *Forest Policy and Economics* 9:209–218.

Hanley, N., and J. Milne. 1996. Ethical beliefs and behaviour in contingent valuation surveys. *Journal of Environmental Planning and Management* 2:255–272.

Hanley, N., F. Schläpfer, and J. Spurgeon. 2003. Aggregating the benefits of environmental improvements: Distance-decay functions for use and non-use values. *Journal of Environmental Management* 3:297–304.

Heller, E., J. M. Rhemtulla, S. Lele, et al. 2012. Global croplands and their water use for food security in the twenty-first century. *Photogrammetric Engineering and Remote Sensing* 78:815–827.

Hicks, C. C., T. R. McClanahan, J. E. Cinner, and J. M. Hills. 2009. Trade-offs in values assigned to ecological goods and services associated with different coral reef management strategies. *Ecology and Society* 14:10.

Higuera, D., B. Martín-López, and A. Sánchez-Jabba. 2013. Social preferences towards ecosystem services provided by cloud forests in the neotropics: Implications for conservation strategies. *Regional Environmental Change*. doi:10.1007/s10113-012-0379-1.

Jackson, B., P. Timothy, F. Sinclair, et al. 2013. Polyscape: A GIS mapping framework providing efficient and spatially explicit landscape-scale valuation of multiple ecosystem services. *Landscape and Urban Planning* 112:74–88.

Jorgensen, B. S., M. A. Wilson, and T. A. Haberlein. 2001. Fairness in the contingent valuation of environmental public goods: Attitude towards paying for environmental improvement at two levels of scope. *Ecological Economics* 36:133–148.

Kenter, J. O., T. Hyde, M. Christie, and I. Fazey. 2011. The importance of deliberation in valuing ecosystem services in developing countries—Evidence from the Solomon Islands. *Global Environmental Change* 21:505–521.

Kheir, R. B., M. H. Greve, P. K. Bocher, et al. 2010. Predictive mapping of soil organic carbon in wet cultivated lands using classification-tree based models: The case study of Denmark. *Journal of Environmental Management* 91:1150–1160.

Konarska, K. M., P. C. Sutton, and M. Castellon. 2002. Evaluating scale dependence of ecosystem service valuation: A comparison of NOAA-AVHRR and Landsat TM datasets. *Ecological Economics* 41:491–507.

Kontogianni, A., G. W. Luck, and M. Skourtos. 2010. Valuing ecosystem services on the basis of service-providing units: A potential approach to address the "endpoint problem" and improve stated preference methods. *Ecological Economics* 7:1479–1487.

Kroll, F., F. Müller, D. Haase, and N. Fohrer. 2012. Rural-urban gradient analysis of ecosystem services supply and demand dynamics. *Land Use Policy* 29:521–535.

Krysanova, V., F. Hattermann, and F. Wechsung. 2007. Implications of complexity and uncertainty for integrated modeling and impact assessment in river basins. *Environmental Modelling and Software* 22:701–709.

Kumar, M., and P. Kumar. 2008. Valuation of the ecosystem services: A psycho-cultural perspective. *Ecological Economics* 64:808–819.

Lamarque, P., F. Quétier, and S. Lavorel. 2011. The diversity of the ecosystem services concept and its implications for their assessment and management. *Comptes Rendus Biologies* 334:441–449.

Lansford, N. H., and L. L. Jones. 1995. Recreational and aesthetic value of water using hedonic price analysis. *Journal of Agriculture Resource Economics* 2:341–355.

Liu, S., C. Robert, F. Stephen, and T. Austin. 2010. Valuing ecosystem services. *Annals of the New York Academy of Sciences* 1185:54–78.

Liu, X., and J. Li. 2008. Application of SCS model in estimation of runoff from small watershed in Loess Plateau of China. *Chinese Geographical Science* 18:235–241.

Lobell, D. B., S. M. Lesch, D. L. Corwin, et al. 2009. Regional-scale assessment of soil salinity in the Red River Valley using multi-year MODIS EVI and NDVI. *Journal of Environmental Quality* 1:35–41.

Lobitz, B., L. Beck, A. Huq, et al. 2000. Climate and infectious disease: Use of remote sensing for detection of *Vibrio cholerae* by indirect measurement. *Proceedings of the National Academy of Sciences of the United States of America* 4:1438–1443.

Loomis, J., P. Kent, L. Strange, L. Fausch, and A. Covich. 2000. Measuring the total economic value of restoring ecosystem services in an impaired river basin: Results from a contingent valuation survey. *Ecological Economics* 33:103–117.

Losey, J. E., and M. Vaughan. 2006. The economic value of ecological services provided by insects. *BioScience* 56:311–323.

Luck, G. W., G. C. Daily, and P. R. Ehrlich. 2003. Population diversity and ecosystem services. *Trends in Ecology and Evolution* 18:331–336.

Luck, G. W., R. Harrington, P. A. Harrison, et al. 2009. Quantifying the contribution of organisms to the provision of ecosystem services. *BioScience* 59:223–235.

MA (Millennium Ecosystem Assessment). 2003. *Ecosystems and human well-being: A Framework for Assessment*. Washington, D. C.: Island Press and World Resources Institute.

MA (Millenium Ecosystem Assessment). 2005. *Ecosystems and human well-being: Current states and trends*. Washington, DC: World Resources Institute.

Maes, J., L. Braat, K. Jax, et al. 2011. *A spatial assessment of ecosystem services in Europe: Methods, case studies and policy analysis-phase 1*. PEER Report No 3. Ispra, Italy: Partnership for European Environmental Research.

Martín-López, B., M. García-Llorente, I. Palomo, and C. Montes. 2011. The conservation against development paradigm in protected areas: Valuation of ecosystem services in the Doñana social–ecological system (southwestern Spain). *Ecological Economics* 70:1481–1491.

Martín-López, B., E. Gómez-Baggethun, P. L. Lomas, and C. Montes. 2009. Effects of spatial and temporal scales on cultural services valuation areas. *Journal of Environmental Management* 2:1050–1059.

Martín-López, B., C. Montes, and J. Benayas. 2007. The non-economic motives behind the willingness to pay for biodiversity conservation. *Biological Conservation* 139:67–82.

McMichael, A. J., D. Campbell-Lendrum, C. F. Corvalan, et al. 2003. *Climate change and human health: Risks and responses*. Geneva: World Health Organization.

Mengzhi, D. 2009. Study on tobacco spatial agglomeration pattern based on remote sensing and GIS methods in Henan province, China. *Geoscience and Remote Sensing Symposium IEEE International, IGARSS 2009* 2:646–649.

Menzel, S., and J. Teng. 2010. Ecosystem services as a stakeholder-driven concept for conservation science. *Conservation Biology* 3:907–909.

Mertes, L. A. K. 2002. Satellite remote sensing for detailed landslide inventories using change detection and image fusion. *Freshwater Biology* 47:799–816.

Minacapilli, M., C. Agnese, F. Blanda, et al. 2009. Estimation of actual evapotranspiration of Mediterranean perennial crops by means of remote-sensing based surface energy balance models. *Hydrology and Earth System Sciences* 13:1061–1074.

Mitchell, R., and R. Carson. 1989. *Using surveys to value public goods: The contingent valuation method*. Washington, DC: Resources for the Future.

Myeong, S., D. J. Nowak, and M. J. Duggin. 2006. A temporal analysis of urban forest carbon storage using remote sensing. *Remote Sensing of Environment* 101:277–282.

Naidoo, R., and T. H. Ricketts. 2006. Mapping the economic costs and benefits of conservation. *PLoS Biology* 11:360.

Nelson, G. C., M. W. Rosegrant, J. Koo, et al. 2009. *Climate change: Impact on agriculture and costs of adaptation*. Washington, DC: IFPRI Food Policy Report.

Nichol, J., and M. S. Wong. 2005. Satellite remote sensing for detailed landslide inventories using change detection and image fusion. *International Journal of Remote Sensing* 26:1913–1926.

Norgaard, R. B. 2010. Ecosystem services: From eye-opening metaphor to complexity blinder. *Ecological Economics* 69:1219–1227.

Olofsson, P., F. Lagergren, A. Lindroth, et al. 2008. Towards operational remote sensing of forest carbon balance across northern Europe. *Biogeosciences* 5:817–832.

Paetzold, A., P. H. Warren, and L. L. Maltby. 2010. A framework for assessing ecological quality based on ecosystem services. *Ecological Complexity* 7:273–281.

Palomo, I., B. Martín-López, C. López-Santiago, and C. Montes. 2011. Participatory scenario planning for natural protected areas management under the ecosystem services framework: The Doñana social–ecological system, SW Spain. *Ecology and Society* 16:23.

Palomo, I., B. Martín-López, M. Potschin, R. Haines-Young, and C. Montes. 2012. National Parks, buffer zones and surrounding landscape: Mapping ecosystem services flows. *Ecosystem Services Journal* 4:104–116. doi:10.1016/j.ecoser.2012.09.001.

Paruelo, J. M., E. G. Jobbágy, and O. E. Sala. 2001. Current distribution of ecosystem functional types in temperate South America. *Ecosystems* 4:683–698.

Pearce, D. W., and R. K. Turner. 1990. *Economics of natural resources and the environment*. Hemel Hempstead and London: Harvester Wheatsheaf.

Perrings, C. 2007. Future challenges. *Proceedings of the National Academy of Sciences of the United States of America* 104:15179–15180.

Perrings, C., A. Duraiappah, A. Larigauderie, and H. Mooney. 2011. The biodiversity and ecosystem services science-policy interface. *Science* 331:17–19.

Polasky, S., E. Nelson, J. Camm, et al. 2008. Where to put things? Spatial land management to sustain biodiversity and economic returns. *Biological Conservation* 141:1505–1524.

Quetier, F., F. Rivoal, P. Marty, J. de Chazal, W. Thuiller, and S. Lavorel. 2010. Social representations of an alpine grassland landscape and socio-political discourses on rural development. *Regional Environmental Change* 10:119–130.

Reed, M. S. 2008. Stakeholder participation for environmental management: A literature review. *Biological Conservation* 141:2417–2431.

Ressurreição, A., J. Gibbons, T. Ponce Dentinho, M. Kaiser, R. S. Santos, and G. Edwards-Jones. 2011. Economic valuation of species loss in the open sea. *Ecological Economics* 4:729–739.

Rodríguez, J. P., T. D. Beard, J. Agard Jr., et al. 2005. Interactions among ecosystem services. In *Ecosystems and human well-being: Scenarios. Volume 2. Working group, Millennium Ecosystem Assessment. Findings of the scenarios*, eds. S. R. Carpenter, P. L. Pingali, E. M. Bennett, and M. B. Zurek, 431–448. Washington, DC: Island Press.

Ropert-Coudert, Y., and R. P. Wilson. 2005. Trends and perspectives in animal-attached remote-sensing. *Frontiers in Ecology and the Environment* 3:437–444.

Sagoff, M. 2011. The quantification and valuation of ecosystem services. *Ecological Economics* 70:497–502.

Sandbrook, C., I. R. Scales, V. Ivan, et al. 2011. Value plurality among conservation professionals. *Conservation Biology* 25:285–294.

Sayadi, S., M. C. G. Roa, and J. C. Requena. 2005. Ranking versus scale rating in conjoint analysis: Evaluating landscapes in mountainous regions in southeastern Spain. *Ecological Economics* 55:539–550.

Schläpfer, F. 2008. Contingent valuation: A new perspective. *Ecological Economics* 64:729–740.

Schulp, C. J. E., and R. Alkemade. 2011. Consequences of uncertainty in global-scale land cover maps for mapping ecosystem functions: An analysis of pollination efficiency. *Remote Sensing* 3:2057–2075.

Seppelt, R., C. F. Dormann, F. V. Eppink, et al. 2011. A quantitative review of ecosystem service studies: Approaches, shortcomings and the road ahead. *Journal of Applied Ecology* 48:630–636.

Seppelt, R., B. Fath, B. Burkhard, et al. 2012. Form follows function? Proposing a blueprint for ecosystem service assessments based on reviews and case studies. *Ecological Indicators* 21:145–154.

Sherrouse, B. C., J. M. Clement, and D. J. Semmens. 2011. A GIS application for assessing, mapping, and quantifying the social values of ecosystem services. *Applied Geography* 31:748–760.

Shrestha, R. K., A. F. Seidl, and A. S. Moraes. 2002. Value of recreational fishing in the Brazilian Pantanal: A travel cost analysis using count data models. *Ecological Economics* 42:289–299.

Spash, C. L., K. Urama, R. Burton, et al. 2009. Motives behind willingness to pay for improving biodiversity in a water ecosystem: Economics, ethics and social psychology. *Ecological Economics* 4:955–964.

Stephen, B., M. Walter, and A. Ivan. 2013. Remote sensing in agricultural livestock welfare monitoring: Practical considerations. In *Wireless sensor networks & ecological monitoring*, eds. S. Mukhopadhyay and J. A. Jiang, 179–193. Heidelberg, Berlin: Springer-Verlag.

Stuart, N., T. Barratt, and C. Place. 2006. Classifying the neotropical savannas of Belize using remote sensing and ground survey. *Journal of Biogeography* 33:476–490.

Syrbe, R. U., and U. Walz. 2012. Spatial indicators for the assessment of ecosystem services: Providing, benefiting and connecting areas and landscape metrics. *Ecological Indicators* 21:80–88.

TEEB (The Economics of Ecosystems and Biodiversity). 2010. *The economics of ecosystems and biodiversity: Ecological and economic foundations*. London: Earthscan.

Troy, A., and M. A. Wilson. 2006. Mapping ecosystem services: Practical challenges and opportunities in linking GIS and value transfer. *Ecological Economics* 60:435–449.

Turner, R. K., S. Morse-Jones, and B. Fisher. 2010. Ecosystem valuation: A sequential decision support system and quality assessment issues. *Annals of the New York Academy of Science* 1185:79–101.

Turpie, J. K., B. J. Heydenrych, and S. J. Lamberth. 2003. Economic value of terrestrial and marine biodiversity in the Cape Floristic region: Implications for defining effective and socially optimal conservation strategies. *Biological Conservation* 112:233–251.

Vejre, H., S. Jensen, F. Jellesmark, et al. 2010. Demonstrating the importance of intangible ecosystem services from peri-urban landscapes. *Ecological Complexity* 7:338–348.

Vrieling, A. 2006. Satellite remote sensing for water erosion assessment: A review. *Catena* 65:2–18.

Webb, T. J., and D. Raffaelli. 2008. Conversations in conservation: Revealing and dealing with language differences in environmental management. *Journal of Applied Ecology* 45:1198–1204.

Wegner, G., and U. Pascual. 2011. Cost-benefit analysis in the context of ecosystem services for human well-being: A multidisciplinary critique. *Global Environmental Change* 21:492–504.

Westerberg, V. H., L. Robert, and S. B. Olsen. 2010. To restore or not? A valuation of social and ecological functions of the Marais des Baux wetland in southern France. *Ecological Economics* 12:2383–2393.

Williams, J. D., S. Dun, D. S. Robertson, et al. 2010. WEPP simulations of dryland cropping systems in small drainages of northeastern Oregon. *Journal of Soil and Water Conservation* 1:22–23.

Wilson, M. A., and R. B. Howarth. 2002. Distributional fairness and ecosystem service valuation. *Ecological Economics* 41:421–429.

Woodward, R. T., and Y. S. Wui. 2001. The economic value of wetland services: A meta-analysis. *Ecological Economics* 37:257–270.

Xenarios, S., and I. Tziritis. 2007. Improving pluralism in multi criteria decision aid approach through focus group technique and content analysis. *Ecological Economics* 62:692–703.

Zander, K. K., and A. Straton. 2010. An economic assessment of the value of tropical river ecosystem services: Heterogeneous preferences among Aboriginal and non-Aboriginal Australians. *Ecological Economics* 69:2417–2426.

Zandersen, M., and R. S. J. Tol. 2009. A meta-analysis of forest recreation values in Europe. *Journal of Forest Economics* 15:109–130.

Zografos, C., and R. B. Howarth. 2008. Towards a deliberative ecological economics. In *Deliberative ecological economics*, eds. C. Zografos and R. B. Howarth, 1–20. Delhi: Oxford University Press.

索　引

A

地上生物量（AGB）
　　估算,27～30
　　测量,19
地上净初级生产力（ANPP）,25,27,88,93,96～97,107,285
地上净初级生产力（ANPP）估算
　　光能利用效率估算,94
　　通过遥感,92～93
　　通过连续生物量收获,89～92
可吸收光,41
光合有效辐射（APAR）,24,26,50
灌溉沟渠,339～340
用于生物量估算的主动传感器,27～30
先进地球观测卫星-2（ADEOS-2）,48
先进扫描微波辐射计-E（AMSR-E）,72,307
高级散射计（ASCAT）,247
先进星载热发射和反射辐射仪（ASTER）,251,424
高级合成孔径雷达（ASAR）,74
高级甚高分辨率辐射计（AVHRR）,64～66,68,70～71,160,336,401,424,470
气溶胶光学厚度（AOD）,128
气溶胶,128,353,380
　　亚马孙热带森林,139～142
农业区地上生物量（AGB）,207
农业扩张,25,127～127,219

基于过程的新型全球动态植被模型,6
机载可见/红外光谱仪（AVIRIS）,336
监测积雪的机载传感器,336～337
反照率,380～382,407～410
反照率估算,389～393,404～407
亚马孙热带森林气溶胶,139～142
　　生物质燃烧排放估计,125～144
　　生物量分配,129
微量气体排放估算,139～142
动物生态学,165～171
动物种类
　　生态学与保护,168～171
　　人口动态,169～170
　　远程监测,157
年度地上净初级生产力（AANPP）,69
ANPP,参见地上净初级生产（ANPP）
AOD,参见气溶胶光学深度（AOD）
可吸收性光合有效辐射估算、光能利用效率估算,94～95
北极苔原
　　碳循环池和过程,63～77
　　绿色化,65,77
　　陆地—大气碳交换,71～73
　　生物量和初级生产力的遥感,65～68
　　土壤碳过程,73～74
阿根廷,大规模毁林,407～410
人工智能服务生态系统（ARIES）,6,268,449～450

ASTER,参见先进星载热发射和反射辐射仪(ASTER)

雅典 Thermopolis 实验活动,423~424

大气和水,235,245

B

裸土,207

BDRoute,207

推移质侵蚀,267

生态系统服务受益者,450451

双向反射分布函数(BRDF)模型,381

生物多样性,8,25,126,202

国家公园网络评估,185

评估和监测,151~172

碳收益,185~186,192~193

经济,180,442,453~458

生态系统功能,165~168

生态系统服务,180

减少,151~152

研究要求的空间分辨率,153~157,472

生物地球化学模型,262

生物量,24

通过收获估算地上净初级生产力,89~92

亚马孙热带森林分布,129

估算(estimation),19

估算(estimations),27~30

收获,106

北极苔原遥感,65~68

北方森林遥感,68~71

存量,23

微量气体排放估算,139~142

生物质燃烧排放估算,125~144

气溶胶,139~142

CCATT-BRAMs 模型,134~135

排放模型评估,142~143

现场数据和清单比较,135~137

火焰辐射能量分布,137~139

火焰辐射功率整合,130~134

材料与方法,129~137

方法,128

结果,137~143

热异常监测,129~130

BiOME-BGC 模型,68

生物物理变量,6~7

在区域气候模型中的变化,363~366

生态系统功能类型,360~363

生态系统服务评估,445,446~448

水文,251

北方生态系统大气研究(BOREAS),69

北方森林

"褐化",65,77

生物量和初级生产力的遥感,68~71

土壤碳过程,73~74

波文比能量平衡(BREB),315

BRDF,参见双向反射分布函数模型

BREB,参见波文比能量平衡

缓冲能力,26~27

C

卡吕普索卫星,235

树冠反射率模拟模型,384~385

碳(C),18

收支,106

野火释放,30~31,75~76

土壤碳过程,73~74

碳循环,7~8

关键流程,18~19

模型,262

监控,23~31

北部高纬度地区观测,63~77

过程,63~65

二氧化碳(CO_2),18,52

同化,50

固定,41,47,181

吸收,49,64
  碳动力学,17~31
  碳收益,25~27
与功能多样性的一致性 185~186,192~193
使用 LUE 模型估算,105~115
国家公园网络,192~193
资格,182~185
  一氧化碳(CO),52,140~141
  评估碳相关生态系统服务的规模问题,22~23
  碳截存,18~19,40,126
  碳汇,64
  碳储存,23,40,63~64
  估算碳吸收,39~55
  类胡萝卜素,45~46
  CASI-1500,参见紧凑型机载光谱成像仪
集水区规模分析,201~221
利比里亚地区土地覆盖分类,208~212
数据集,205~208
讨论,218~220
简介,202~204
模拟土地覆盖和河流生态之间的关系,212~21
结果,216~218
研究区,204~205
  牛,87,91
  旱雀麦,155
  中国的研究,4
  叶片色素中的叶绿素,42~44
  叶绿素荧光,41,49~55
研究进展,51~52
基础、起源和特征,49~51
最终讨论,54
未来,52,54
遥感工具,52~53
  叶绿素指数,43~44

CHRIS(袖珍高分辨率成像光谱仪),48
CHRIS/PROBA,46
CICES,生态系统服务国际共同分类(CICES)
  公民科学,333
  环境信息协调土地分类,208,214
  气候变化,18,127,232,250
  气候调节,40
  气候调节服务,351~374
  干旱时期案例研究,366~372
  讨论与结论,373~374
  生态系统-气候反馈,353~354
  识别生态系统功能类型,357~370
  简介,352~356
  生态系统-大气相互作用建模,354~356
  区域建模,363~372
  气候调节值(CRV)指数,352~353
  云,235,237~38
  CloudSat 卫星,235
  有色溶解有机物(CDOM),274
  燃烧因子,139
  生态系统服务国际共同分类(CICES),127,330
  紧凑型机载光谱成像仪(CASI-1500),51
  保护
生态系统功能,165~171
人类福祉,180
国家公园网络,179~196
野生动物种群,168~171
  保护生物地理学,159~161
  CORINE(环境信息协调),159~161,208,217,251
  化学物质-气溶胶-示踪剂耦合传输模型与巴西区域大气模型系统耦合(CCATT-BRAMs),133~135,139
  生物质燃烧估算,134~135
  覆盖率,265~266

农作物
叶绿素含量,42～44
灌溉,295～298
光使用效率,111
    CRV,参见气候调节值(CRV)指数
    文化服务,5,202,298,304,319,340～341
    低温层,水 245～246

D

    数据
通量,55
点云,29～30
来源,4
光谱,24
    数据同化研究试验(DART),270～271
    最大增强植被指数日期(DMAX),184,187,357～360
        分解,73～74
        毁林,25,127
大规模冲击,407～410
    沿海三角洲地区,267～268
    DEM,参见数字高程模型(DEM)
    DESIREX 实验活动,421～424,430
    数字高程模型(DEM),334
        扰动,21～22
    DMAX,参见最大增强植被指数日期(DMAX)
    双季作物,25
    干旱,247,249,366～372
    动态植被模型,355～356

E

    地球探测器,8,53
    用于物种多样性评估和监测的对地观测(EO),151～172
        对地观测-1号(EO-1),48
        地球辐射收支试验,380

地球辐射收支卫星(ERBS),380
生态变化,232
生态质量定额(EQR),213
生态学,野生动物种群,168～171
生态系统的经济价值评估,453～458
生态系统-大气相互作用建模,354～356
生态系统-气候反馈,353～354
生态系统功能类型(EFTs),181、356、444
国家公园网络评估,185
生物物理特性,360～363
定义,357～359
识别,182～185,357～370
土地服务参数化,360～361
卫星数据记录,357
南美洲南部,359～360
空间格局,186～189
    美国地质调查局与生态系统功能类型衍生的生物物理特性,361～363
    生态系统功能,18
测量,165～171
物种多样性,165～168
    生态系统缓冲能力,26～27
    生态系统服务
受益人,450～451
生物多样性,8,180
生物物理指标,445～448
碳循环,7～8
碳动力学,17～31
级联,126
分类,19～21,127,330
气候调节,351～374
概念,127
文化服务,5,202,298,304,319
定义,3,19
降级,180
需求方,450～458
经济,180,442,453～458

经济估值,453~458

能量平衡,379~394,399~411

能量平衡和陆地光合胁迫的估算,39~55

蒸散,401~407

最终,19~21

水文,229~254,261~276

中间产物,19~21,23~31

制图,6,9,445,449~450

建模,6

监测,23~31

山区,329~344

多维方法评估,441~460

付款,127

供给服务,126,337~338

供给,21~23

量化,262~268,330~332

调节服务,126,202,338~339

遥感,399~411

遥感,3~7,17~18

河流生态系统,201~222

角色,379~380

规模问题,22~23

科研进展,181

社会文化评估,451~453

供应侧,444~450

支持,202

权衡,21~22

用途,17~18

水循环,8

 野火释放的能量,30~31

 能源平衡,9

冠层反射率模拟模型 384~385

生态系统,38~82

生态系统服务相关,379~394

蒸散,399~411

简介,379~383

模拟案例研究,385~387

模拟反射率,387~393

湿地反射能量,383~393

 能源预算,265

 增强型植被指数(EVI),24,25,43,72,106~107,167~168,181,184,289~290,292~295,404

 环境测绘和分析项目(EnMAP),47~48

 环境变异性,170~171

 ENVISAT,参见欧洲环境卫星(ENVISAT)

 EQR,参见生态质量定额(EQR)

 地球辐射预算实验(ERBE)

 ERBS,参见地球辐射收支卫星(ERBS)

 侵蚀控制,23,25

 ESA,参见欧洲航天局(ESA)

 ET,参见蒸散(ET)

 欧洲环境卫星(ENVISAT),53,74

 欧洲航天局(ESA),307

 蒸散量(ET),6,265,379~380

生态系统服务,401~407

能量平衡,399~411

估算,312~317,402~404

使用热卫星影像绘制地图,314~315

遥感,402~404

利用传感器和卫星测量,241~242

 EVI,参见增强植被指数(EVI),292~295

 EVOP,285

F

 fAPAR,植被光合有效辐射吸收比(fAPAR)

 实地观测,4

 最终生态系统服务,19~21

 火焰辐射能(FRE),30,128,137~139,141

 火焰辐射功率(FRP),30~31,128,130~134,140

火焰,24,30~31
生物质燃烧排放估算,125~144
碳排放,75~76
　　火灾热异常(FTA)算法,30
　　洪水,24,247,249
　　荧光,49~54,55
　　荧光探测器(FLEX),52~53
　　荧光成像光谱仪(FIS),52
　　通量数据,55
　　FluxNet 数据库,316~317
　　通量塔,55
　　食物,40
　　粮农组织,251
　　牧草生长率,参见牧草产量
牧草生产监测,87~100
遥感,98~99
　　森林,43
生物量估算,28~30
北方地区,65,68~73
大规模毁林,407~410
　　光合有效辐射吸收的部分
　　(fAPAR),24,43,68,94~95,98~99
　　弗劳恩霍夫线,51~52
　　FRE,参见火焰辐射能(FRE)
　　火焰辐射分布,137~139
　　FRP,参见火焰辐射功率(FRP)
　　FRP 集成,130~134
　　FTA,参见火灾热异常(FTA)
　　功能多样性
评估,185
与碳收益之间的一致性,185~186,192~193

G

　　GCPM,参见静地碳过程测绘仪(GCPM)
　　GDEM,参见全球数字高程模型(GDEM)
　　GEO BON,参见对地观测组生物多样性
观测网(GEO BON)

　　地理信息系统(GIS),203
　　地理信息系统(GIS),311
　　　　GEOSS,参见全球地球观测系统
(GEOSS)
　　静地碳过程测绘仪(GCPM),52~53
　　静地傅里叶变换光谱仪(GeoFTS),53
　　地球静止运行环境卫星(GOES),130,
133,137~139
　　地球静止轨道(GEO),54
　　GIS,见地理信息系统(GIS)
　　全球数字高程模型(GDEM),251
　　全球对地观测系统(GEOSS),4
　　全球成像仪(GLI),48
　　全球土地覆盖2000(GLC2000),161~163
　　全球生态系统服务变化监测系统,3~4
　　全球变暖,参见气候变化全球覆盖,163
　　GOCE,参见重力场和稳态海洋环流探测
器(GOCE)
　　　GOES,参见地球静止业务环境卫星
(GOES)
　　GOSAT(温室气体观测卫星),53
　　重力场恢复与气候实验卫星,248
　　草地,40,220
　　GOCE,重力场和稳态海洋环流探测器
(GOCE),248
　　放牧,26
　　绿色生物量,评估,40
　　绿色经济,绿色经济,232
　　温室气体,30
　　温室气体(GHGs),419~420
　　绿度异常,283~299
气候和卫星数据估算,288~290
确定的,288
　　总初级生产力(GPP),40~44,64,68,76
　　地下水,247~248
依赖性梯度,292~295
检测生态系统依赖,283~299

影响植被动力学,291~292
遥感,247~248
传感器和卫星测量,244
 地下水依赖的生态系统（GDEs）,283~299
 简介,284~286
 研究方法,286~292
 研究结果,292~298
地球观测组,3~4
地球观测小组
对地观测组生物多样性观测网（GEO BON）,4
 古达尔菲奥研究案例,267~268

H

 热损失,41
 草本/灌木植被,207
 HRU,参见水文响应单元（HRU）
 HSI,参见超光谱成像仪（HSI）
 人类活动,21~22,203,379
 人与自然关系,442
 人类福祉,18,40,180
 混合方法,458~459
 水文生态系统服务,339~340
评估,261~276
驱动及压力,249~250
供给,230~231,234~249
遥感,229~254,261~276
社会,230~232
菌株,232
水循环,232~234
 水文生物物理变量,251
 水文循环,263;参见水循环水文建模,262~264
 生态系统服务量化,262~268
 遥感数据整合,250~253,261~276,311~312

遥感,268~272
 水文响应单元（HRU）,264
 水文状态变量,251~252
 Hyperion高光谱成像光谱仪,46,48
 多维空间影像,155~157
 高光谱成像仪（HSI）,48
 高光谱红外成像仪（HyspIRI）,48
 高光谱遥感,44,156~157
 HyspIRI,47

I

冰,240~241,245~246
IGBP-Discover全球土地覆盖数据集,161
伊科诺斯卫星,250
影响函数,21~22
偶然阳光照射,41
指标选择,23~24
INMET,参见国家气象研究所
生态系统服务评估与权衡（InVEST）,6,266~268,449
雷达影像干涉测量（InSAR）,28~29
干涉测量技术,28~29
政府间气候小组（IPCC）,419
生物多样性和生态系统服务政府间平台,442
生物多样性和生态系统服务政府间科学政策平台,3
生态系统中间服务,19~21
中间服务,监测,23~31
国际地圈—生物圈方案数据集,161
国际自然保护联盟,151
入侵物种,153,155
灌溉作物,295~298
IUCN,见国际自然保护联盟（IUCN）
IUCN红色清单,163

K

克鲁斯卡尔-沃利斯检验,112~114

L

陆地-大气层碳交换
北极苔原地区遥感,71~73
北方森林遥感,71~73
   土地清理,25,27
   土地覆盖,127,157~165
生物反应,219~220
普遍,207
生态系统服务制图,449
对河流生态的影响,217~218
比率,265~266
全球土地覆盖项目,161~165
跨类型光利用效率估算,110~112
制图,158~159
   与河流生态完整性的关系建模,212~216
河岸地区土地覆盖(RALC),203~222
空间指标,213~215
   土地覆盖数据库,208
   土地覆盖空间指标(LCSI),212~213
   土地功能动态评估,6
   土地参数反演模型(LPRM),247
   Landsat 卫星 55,72,74~75,155,158,171,265~266,275,335~336,401
   生态系统功能类别土地服务参数化,360~361
   滑坡,249
   地表动力学,249
   地表发射率(LSE),430~432
   城市地区地表发射率,429~433
   陆面温度(LST),168,403,429~430,432~433
   土地利用,127
变化,157,232
生态系统服务制图,449
   土地利用规划,17
   激光成像探测和测距(LIDAR),336

LCSI,参见土地覆盖空间指标(LCSI)
   叶面积指数(LAI),43,264
   叶色素,42~49
叶绿素含量,42~44
循环,44~47
   潜热通量,402,404
   LIDAR,19,28~30,参见激光成像探测和测距(LIDAR)
   光探测和测距(LIDAR),156~235
   光使用效率(LUE),40,43~47,49,55,107
   光使用效率(LUE)估算,105~115
跨组织层面和土地覆盖类型,110~112
材料与方法,107~108
结果,108~114
时间间隔,112~114
   全球标准化生物指数(IBGN),208,212~213
   畜牧系统遥感,87~99
   活植物指数,152
LSE,参见地表发射率(LSE)
LST,404~407,参见陆面温度(LST)
LUE,参见光使用效率(LUE)

M

   马德里 DESIREX 实验活动,421~423
   积雪的维持服务,338~340
   玉米,43
MAP-EVI 区域函数,292
   质量平衡方程,18
   马托格拉索州,141
   最大差异水指数(MDWI),248
   最大 SUHI(SUHIM),426~427
   年降水量期望(MAP),27,288~289
   中分辨率成像光谱仪(MERIS),51,53,247
   MERIS,参见中分辨率成像光谱仪

(MERIS)
    MERIS 陆地叶绿素指数(MTCI),43
    METEOSAT 卫星,250
    甲烷(CH_4),52
生产力,73~74
    微气象塔站点,316~317
    微波辐射(MR),235,245~246,306~307
    微波传感器,337
    中红外辐射,131,246
    千年生态系统评估,3,18~19,232~233,442
    中分辨率成像光谱仪(MODIS),6,43,46,48,66,72,75,129~130,153,168,182,184,235,265,334~335,381,404,424
        FRE 分配,137~139
FRP 整合,130~134
国家公园网络,186~196
    MODIS 土地覆盖产品系列 5(MCD12Q1),163
    蒙特沙漠,286~288
    蒙蒂斯模型,24~25,106~107
    摩洛哥
国家公园网络,183~184
    生态功能类型的空间模式,186~189
    山区生态系统服务,329~344
    文化服务,340~341
    维持服务,338~340
    供给服务,337~338
    调节服务,338~339
    内华达山脉生物圈保护区,341~343
    MR,参见微波辐射(MR)
    多维方法,441~460
需求,442~444
    多传感器技术,235

N

    美国国家航空航天局 QuickScat 卫星,74

    国家环境预测中心/国家大气研究中心(NCEP/NCAR),365
    国家气象研究所(INMET),142
    国家公园网络
碳收益,192~193
功能多样性和碳收益之间的一致性,185~186,192~193
生态系统服务评估,179~196
功能多样性评估,185
摩洛哥,183~184
葡萄牙,183
代表性和稀有性,189~192
西班牙,183
国家极地轨道伙伴关系(NPP),55
    国家统计数据,4
    自然作为系列连接系统,263
    NDII,参见归一化差值红外指数(NDII)
    NDVI,参见归一化差异植被指数(NDVI)
    近红外(NIR),43,64,205,235
    生态系统净交换(NEE),18,71~73,77,106
    生态系统净生产(NEP),18,77
    净初级生产力(NPP),23~24,26,40,64,68,70~71,76,106,165~167,181
估计,24~27
    归一化差红外指数(NDII),248
    归一化差异雪指数(NDSI),334
    归一化差异植被指数(NDVI),24~26,43,64~68,70~72,94~96,106~107,161,166~169,181,211~212,253,265,285,309~310,357~358
    归一化差水指数(NDWI),248
    北部高纬度碳循环过程遥感,63~77
    数值模拟模型,4

O

    基于对象的影像分析(OBIA),209~211,

218~219,220~221
    光学遥感,19
    轨道碳观测卫星-2号(OCO-2),52~53

P

    PAR,参见光合有效辐射(PAR)
    颗粒物,30
    牧场更新,127
    泥炭地,43
    PEM,参见生产效率模型(PEM)
    有害生物控制,127
    光化学反射率指数(PRI),25,44~47,49
    光合作用,18,40,379,401
    光合有效辐射(PAR),24,69,89,97,106~107,167
    光合效率,50~51
    光合胁迫
叶绿素荧光,49~54
估算,39~55
叶色素,42~49
    植物生物量,67,469
    行星边界层(PBL),135
    点云数据,29~30
    污染聚集地,219
    POLYSCAPE,6,450
    葡萄牙
国家公园网络,183,186~196
生态功能类型的空间模式,186~189
    降水,235,238~239,245,248~249,288~289
    使用人工神经网络(PERSIANN-CCS)从遥感信息估算降水,245
    PRI,参见光化学反射指数(PRI)
    初级生产力
地上净初级生产力(ANPP),25,27,88,93,96~97,107
总初级生产力,40~44,64,68,76

净初级生产力,23~24,26,40,64,68,76,106,165~167,181
北极苔原遥感,65~68
高纬度森林遥感,68~71
    星上自主项目卫星(PROBA),48
    过程方程,18
    生产效率模型(PEM),40,68,70
    生产函数,19~21
    巴西亚马孙河流域森林砍伐评估方案(PRODES),142
    供给服务,126
积雪,337~338
    日照计,380

Q

    快鸟卫星数据,155~156,250

R

    雷达,参见无线电探测和测距(RADAR)
    雷达,28~29
    辐射
FAPAR,24,43,68,94~95,98
光合有效辐射吸收比(fAPAR),24,43,68
微波,245~246
光合活性,24,69,89,97
从陆地表面反射,306
太阳,380~382,389~393,403
可见的,235
    辐射利用效率(RUE),24~25,88,99~100
估算,92~98
    辐射传输模型,248
    无线电探测和测距(RADAR),235,251
    辐射指数,24~26
    RALC,见河岸地区土地覆盖(RALC)
    区域建模
生物物理特性变化,363~366

气候调节服务,363~372
    调节服务,126,202
积雪,338~339
    遥感,4
地上净初级生产力估算,92~93
应用SWAT,252~253
碳循环过程,63~77
火灾产生的碳排放,75~76
生态系统服务,3~7,17~18,399~411
估算蒸散,402~404
牧草监测系统,98~99
水文生态系统服务,229~254
水文模拟,250~253,261~276
水文模拟,268~272
高光谱,44,156~157
陆地-大气碳交换,71~73
激光雷达,19,28~30
牲畜系统,87~99
国家公园网络,179~196
光学,19
北极初级生产力,65~68
北部森林初级生产力,68~71
辐射利用效率,24~25
土壤碳过程,73~74
物种多样性,151~172
土壤湿度,303~319
合成孔径雷达(SAR),19,28~29
城市热岛效应,424~429
北极植被生物量,65~68
高纬度森林植被生物量,68~71
    水质监测,272~275
    呼吸,18
    河岸地区土地覆盖(RALC),203~222
分类自动化和准确性评估,211~212
分类,208~209
地图,216~217,475
广义领土制图,218~219

基于对象的影像分析设计,209~211,218~219
    河岸植被对河流生态的影响,201~221
    河岸带,202~203
    河流生态系统
功能过程,203
河岸植被影响,201~221
河岸带,202~203
提供的服务,202
河流生态系统,201~222
均方根误差(RMSE),384~385,429
根区,265
    辐射利用效率估算,93~98
    鲁尔斯大坝,267

S

    SAILHFlood模型,384~385
    盐柏,155
    卫星影像,5~7,24,47~48,181
检测生态系统对地下水的依赖,283~299
多维空间影像,155~157
测量积雪,334~336
测量水循环元素,236~244
卫星传感器监测生态系统服务,3~7
    碳相关生态系统服务评估中的规模问题,22~23
    大气制图法扫描成像吸收光谱仪(SCIAMACHY),53
    扫描多通道微波辐射计(SMMR),73
    Scopus数据库,4
    SDMs,参见物种分布模型(SDM)
    季节性,25,27
    SEBAL,参见地面能量平衡算法(SEBAL)
    沉积物负荷,267
    半自然裸土,207
    哨兵卫星,55

服务提供单元(SPUs),444
SEVIRI,参见旋转增强可见和红外成像仪(SEVIRI)
航天飞机雷达地形任务(SRTM),251
塞拉内华达山脉生物圈保护区,341～343
简化表面能量平衡指数（S-SEBI）,403～404
SMA,参见频谱混合分析(SMA)
SMAP,参见土壤湿度主动和被动卫星(SMAP)
SMMR,参见扫描多通道微波辐射计(SMMR)
SMOS,参见土壤湿度和海洋盐度(SMOS)
积雪遥测,333
雪,240～241,245～246
积雪
 机载传感器,336～337
 监测案例研究,341～343
 文化服务,340～341
 维持服务,338～340
 地图,271
 监控,330～332
 监控服务,332～337
 倾斜照片,333～334
 供给服务,337～338
 调节服务,338～339
 卫星测量,334～336
 现场测量,332～334
  雪盖,329～344
  雪地调查,332～333
  雪地遥测,333
  社会文化评估,451～453
 土壤
光谱混合分析分类,308～309
湿气,243,247
半自然裸土,207

地表水监测,303～319
土壤层,265
 土壤和地下水,247～248
 土壤和水评估工具（SWAT）,252～253,264
 土壤碳过程
北极冻土带遥感,73～74
北方森林遥感,73～74
 土壤湿度主动和被动卫星(SMAP),305
 土壤湿度和海洋盐度(SMOS),305
 土壤呼吸,73～74
 太阳辐射,265,403
效率模型,105～120
估算,389～393,404～407
 太阳辐射反射(反照率),380～382
 南美洲
南部生态服务类型,359～360
气象系统,129
 大豆,43
 西班牙
马德里 DESIREX 实验活动,421～423
国家公园网络,183,186～196
内华达山脉生物圈保护区,341～343
 生态功能类型的空间模式,186～189
 空间尺度,22～23
 特殊微波传感器/成像仪(SSM/I),73
 物种
分布建模,159～161
太空发现,153～157
预测命运,164
全球评估,161～165
侵入性,153,155
损失,151～152
定位,157～165
 物种分布模型(SDM),159～161
 物种多样性,参见生物多样性
评估和监测,151～172

与碳收益之间的一致性,192~193
生态系统功能,165~171
土地覆盖,157~165
    物种能量假说,165~168
    光谱数据,24
    光谱混合分析(SMA),土壤状态分类,308~309
    光谱植被指数,106
    旋转增强可见和红外成像仪(SEVIRI),235
    SPU,参见服务提供单位(SPU)
    SRTM,参见航天飞机雷达地形任务(SRTM)
    SSM/I,参见专用传感器微波/成像仪(SSM/I)
    压力因素,21~22
    国家极地轨道伙伴关系(NPP),55
    支持服务,202
    表面反照率,380~382
    地面能量平衡算法(SEBAL),252
    地表土壤层,265
    地表土壤湿度
估算蒸散以估计,312~317
    简介,304~305
    遥感监测,303~319
    表面温度,168,403,407~410,432~433
    地表城市热岛(SUHI),421~423,426~427
    地表水,242,246~247
    SWAT,参见土壤和水评估工具(SWAT)
    合成孔径雷达(SAR),19,28~29
    环境经济账户体系(SEEA),126

T

    TEEB,参见生态系统和生物多样性经济学(TEEB)
    TEM,参见陆地生态系统模型(TEM)
    温度,41;另见表面温度
    温度和发射率分离(TES)算法,421
    时间动力学,283~299
    时间尺度,22~23
    Terra-AQUA卫星,48
    陆地生态系统模型(TEM),68
    陆地生态系统服务
光合胁迫估算,39~55
    陆地植被模型,262
    TES,参见温度和发射率分离(TES)算法
    TEV,参见总经济价值(TEV)框架
    生态系统与生物多样性经济学(TEEB),180,442
    碳观测用热和近红外传感器—傅里叶变换光谱仪(Tanso-FTS/GOSAT),51
    热异常探测,129~130
    热红外(TIR),421
    热卫星影像,绘制蒸散图,314~315
    碳观测用热和近红外传感器—傅里叶变换光谱仪(Tanso-FTS),53
    thermopolis实验活动,423~424
    木材,126
    总经济价值(TEV)框架,454~455
    微量气体,128
    亚马逊热带森林微量气体排放估算,139~142
    树木植被,207
    热带森林
生物质燃烧排放估算,125~144
    光照利用效率,111
    热带降雨测量卫星(TRMM),235,262
    热带稀树草原,111

U

    UCL,参见城市冠层(UCL)
    UHI,参见城市热岛效应城市地区,207,220

城市冠层(UCL),418
城市热岛效应,417~434
雅典案例研究,423~424
评估,420~424
简介,417~420
地表发射率,429~433
马德里案例研究,421~423
过境时间,427~429
遥感,424~429
空间分辨率,424~427
城市化,219
美国地质调查局,361~363

V

阿克里州环保林资产估值政策(PVAFA),142~143
蒸气压差(VPD),41
植被,24
动态植被模型,355~356
草本/灌木,207
年际变化,363
光使用效率(LUE),105~120
土壤水分胁迫,312~317
树木,207
北极苔原植被生物量遥感,65~68
北方森林遥感,68~71
地下水对植被动力学的影响,291~292
植被指数(VIs),24~26,43
增强植被指数(EVI),24—25,43,72,106~107,167~168,181,184,289,290,292~295,404
归一化差异植被指数(NDVI),24~26,64~68,70~72,94~96,106~107,161,166~169,181,211~212,253,265,285,309~310,357~358
卫星衍生增强,182~185
光谱,106

植被含水量(VWC),248
植被覆盖,265
甚高空间分辨率(VHSR)制图,203~205
VIs,参见植被指数(VIs)
可见光和红外区(VNIR),307
可见辐射(VIS),235
VPD,见蒸气压差(VPD)
VSWIR,48

W

废物处理,126
水,126
大气压,235,245
可用性,230,232
预算,265
分类,207
冰冻圈,245~246
损伤修复,234,248~249
需求,232
检测生态系统对地下水的依赖,283~299
重要作用,230
地下水,244,247~248,283~299
管理,232
质量监控,246~247,272~275
稀缺性,230
土壤和地面,247~248
存储容量,230
表面,242,246~247
植被,248
水波段指数(WBI),248
水循环,8,232~234,261~262
水文服务,232~234
传感器和卫星测量,236~244
水框架指令(WFD),208
遥感水质监测,272~275
水调节,25,126
地中海环境流域综合管理(WiMMed),

266,271
    水供应,235,245~248
    水蒸气,235—237,265
    WBI,可见水波段指数(WBI)
    天气研究与预报,365
    湿地,295~298,382~383
    湿地反射能量,383~393
冠层反射率模拟模型,384~385
模拟案例研究,385~387
模拟反射率,387~393
    WFD,参见水框架指令(WFD)
    野火自动生物质燃烧算法(WFABBA),130
    野火,30~31

生物质燃烧排放估算,125~144
碳排放,75~76
    野生动物种群
评估和监测,151~172
保护,168~171
生态学,168~171
    WiMMED,参见地中海环境流域综合管理(WiMMed)
    Woodlands,林地,126,295~298
    WRF,参见天气研究和预测(WRF)

X

    叶黄素循环,44—46

译 后 记

《生态系统服务中的对地观测》是关于对地观测技术在生态系统服务领域应用的学术论文集。本书作者多明戈·阿尔卡拉斯-塞古拉和卡洛斯·马塞洛·迪贝拉分别是西班牙格拉纳达大学教授和阿根廷国家科学与技术研究理事会研究员。本书英文原版于2014年由美国著名学术出版社CRC出版社出版。

本书系统从生态系统供给和需求角度探讨了利用对地观测技术研究复杂生态系统和自然资源管理实际问题。总体分为导言、碳循环、生物多样性、水循环、地表能量平衡、生态系统服务的其他维度六个部分，概述了对地观测技术在其中的重要作用，指出了利用遥感技术监测生态系统服务关键变量时存在的挑战。

本书第一章、第二章、第五章、第九章、第十一章由自然资源部测绘发展研究中心薛超副研究员翻译；第八章、第十二章、第十四章、第十五章、第十六章、第十八章分别由自然资源部测绘发展研究中心李方舟、陈汉生、常燕卿、张月、周夏、贾宗仁翻译；第三章、第四章、第六章、第七章、第十章由北京建筑大学蒋捷教授及吴巧丽、刘亚男、郭贤、崔慧珍、陈强等人完成翻译；第十三章、第十七章分别由香港中文大学王代堃博士、中国测绘科学研究院杜凯旋完成翻译；第十九、二十章由上海外国语大学王诗雨、王永婷、李鱼儿和香港科技大学李文健共同翻译。北京建筑大学蒋捷教授、国家基础地理信息中心刘若梅副主任和周旭研究员、北京师范大学李京教授及香港中文大学王代堃博士、自然资源部国土卫星遥感应用中心工程师牟兴林、中国科学院地理科学与资源研究所博士后孙凯承担了译校工作；自然资源部测绘发展研究中心李方舟、贾宗仁、周夏、陈汉生、王维、常燕卿、张月、贾丹共同承担了整理和清样工作；薛超承担了统稿工作。

在本书的翻译的过程中，得到了很多同事和专家前辈们的支持与帮助，在此我要表示由衷的感谢！十分感谢自然资源部测绘发展研究中心陈常松主任对我的信任和支持，让我这个年轻人能够有信心担纲如此重要的工作任务。尤其是陈常松主任帮助联系了多位业界专家出色地完成了审稿和校对工作，并为书稿的翻译提出了宝贵的修改建议。感谢北京建筑大学蒋捷教授及团队对本书夜以继日、不辞劳苦地翻译校对勘误，使得本书的专业性得到了保障。我还想特别感谢商务印书馆李娟主任及其各位编辑，正是他们的深入沟通和耐心帮助，才使得本书的出版得以实现。

虽然对这本书的翻译工作投入了很多精力，但是我深知，由于自身能力限制，本书的翻译还有很多不尽如人意的地方，遗漏甚至是错误不可避免，还希望读者能够多多指教。

译　者

2023 年 7 月

图书在版编目(CIP)数据

生态系统服务中的对地观测/(西)多明戈·阿尔卡拉斯-塞古拉,(阿根廷)卡洛斯·马塞洛·迪贝拉,(阿根廷)朱丽叶·维罗妮卡·斯特拉施诺伊编;薛超,蒋捷译.—北京:商务印书馆,2023

("自然资源与生态文明"译丛)
ISBN 978-7-100-22833-6

Ⅰ.①生… Ⅱ.①多…②卡…③朱…④薛…⑤蒋… Ⅲ.①遥感技术—应用—生态系统—环境监测 Ⅳ.①X835

中国国家版本馆 CIP 数据核字(2023)第 156137 号

权利保留,侵权必究。

"自然资源与生态文明"译丛
生态系统服务中的对地观测

〔西〕多明戈·阿尔卡拉斯-塞古拉 〔阿根廷〕卡洛斯·马塞洛·迪贝拉
〔阿根廷〕朱丽叶·维罗妮卡·斯特拉施诺伊 编
薛超 蒋捷 译

商 务 印 书 馆 出 版
(北京王府井大街36号 邮政编码100710)
商 务 印 书 馆 发 行
北 京 冠 中 印 刷 厂 印 刷
ISBN 978-7-100-22833-6
审 图 号:GS (2023) 3064 号

2023年11月第1版　开本710×1000　1/16
2023年11月北京第1次印刷　印张 32¼ 插页 10
定价:168.00元

彩插

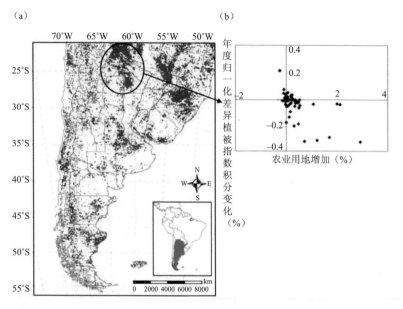

图 2-3

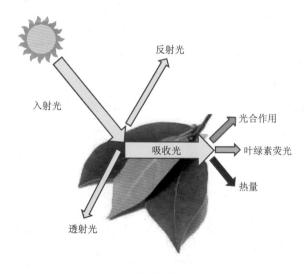

图 3-1

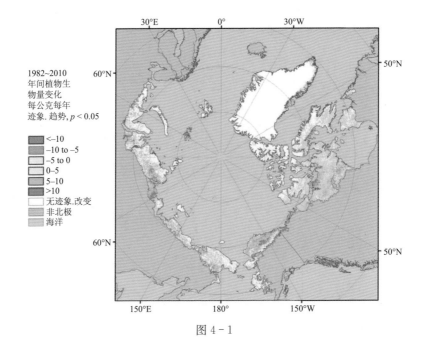

图 4-1

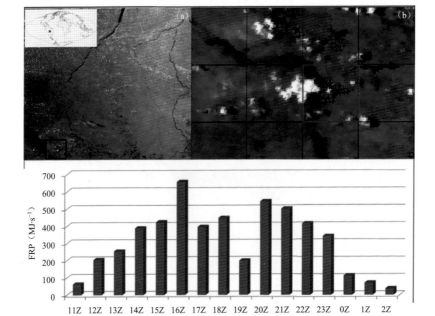

图 7-2

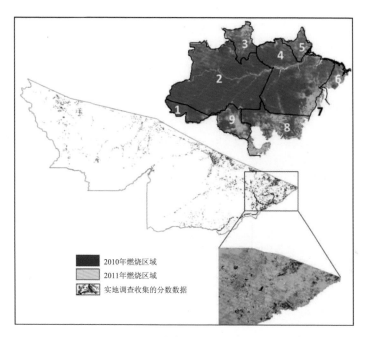

图 7-4

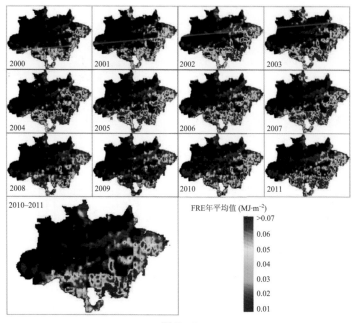

图 7-5

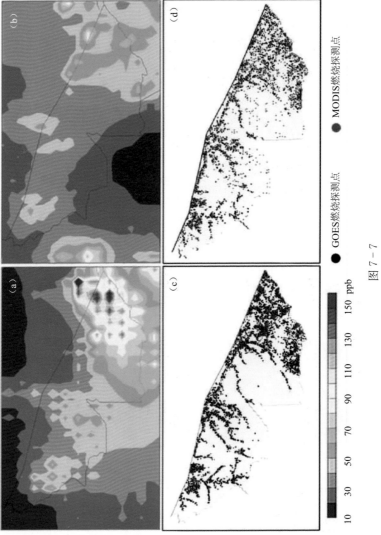

图 7-7

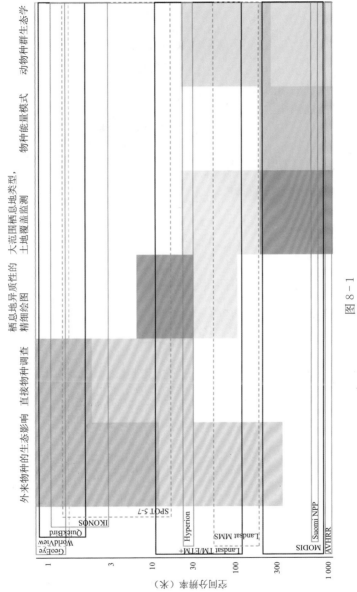

图 8-1

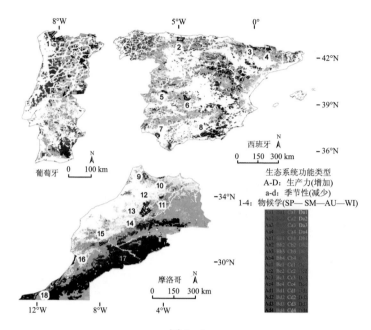

图 9-1

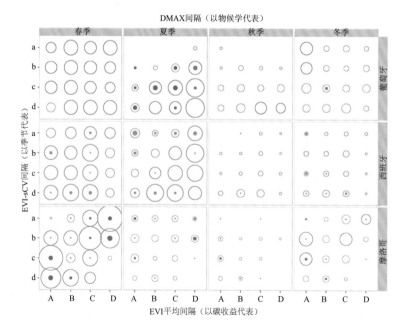

图 9-2

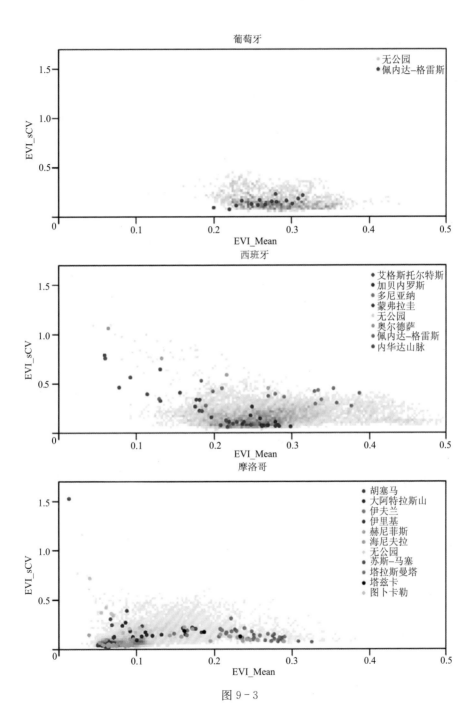

图 9-3

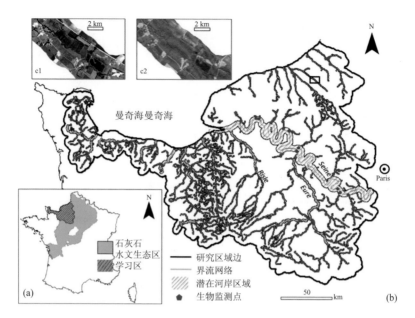

图 10-1

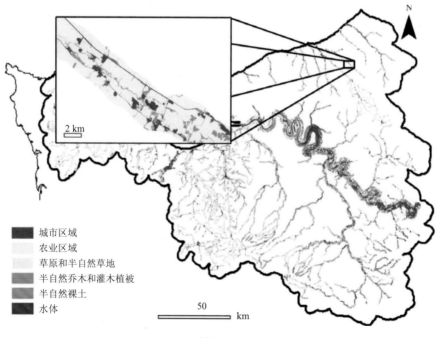

图 10-7

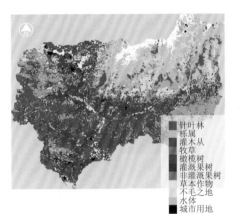

图 12-1

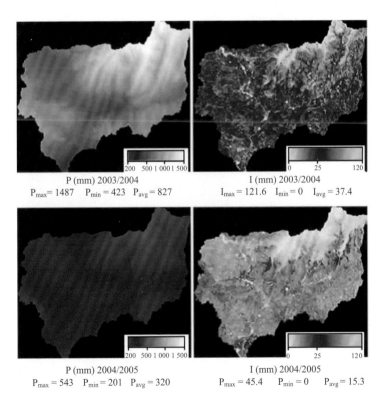

图 12-2

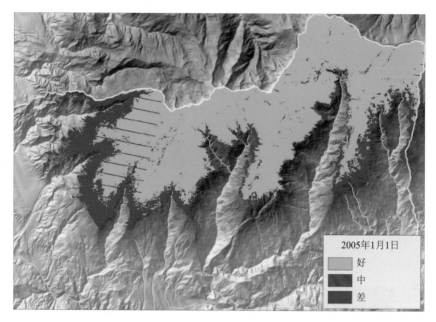

图 12-3

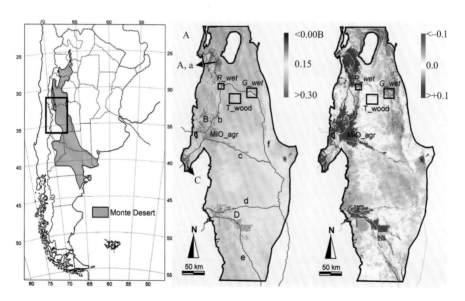

图 13-1

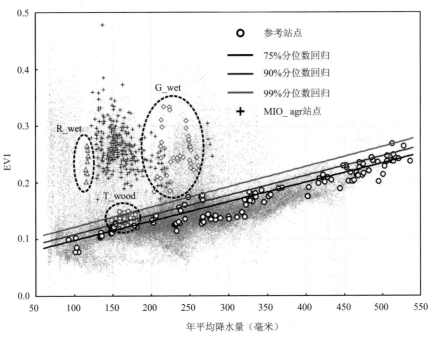

图 13-3

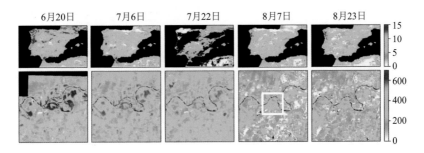

图 14-2

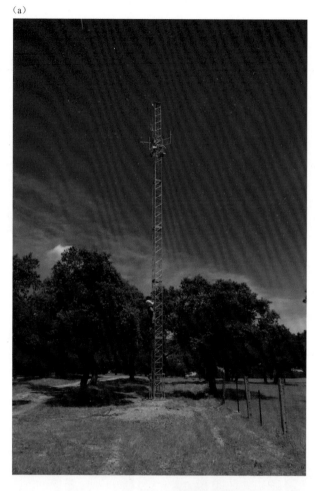

图 14-3

Different types of quantification tools are used to describe the services provided by snow. Quantification of ecosystem services is an attempt to create a "vehicular language" that homogenizes different approaches and ways to describe the natural world. The results obtained by these tools can be used to inform the process of decision making.

Decision layer

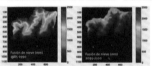

WiMMed is a distributed hydrological model that simulates the snow cover extent, depth, and water content. Maps shows the amount of snow in the present time (left) and future predictions (right).

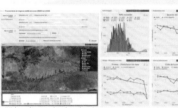

Web application that describes the snow cover status in Sierra Nevada using four indicators: snow cover duration, snow onset date, snow cover melting dates, and number of melting cycles.

Analysis/simulation layer

In situ protocols to obtain information about snow cover extent, water content, depth, density. Meteorological station (left), snow depth transects (right).

Remote sensors to monitor snow cover. Oblique photos (left) are georeferenced to obtain a map of snow cover in spring and early summer. MODIS images (right) are periodically downloaded and processed.

In situ protocols　　　　　*Remote sensors*

Monitoring layer

图 15 - 2

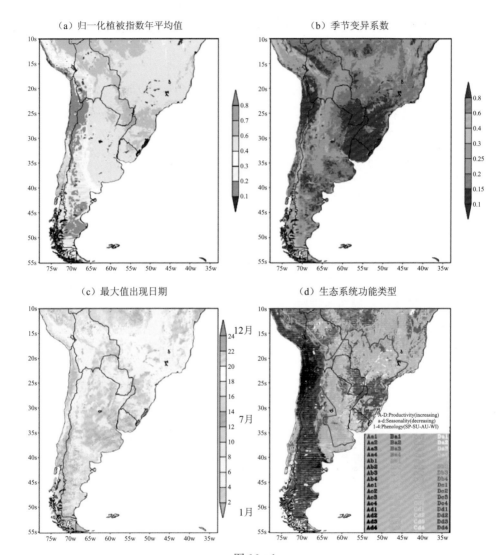

图 16-1

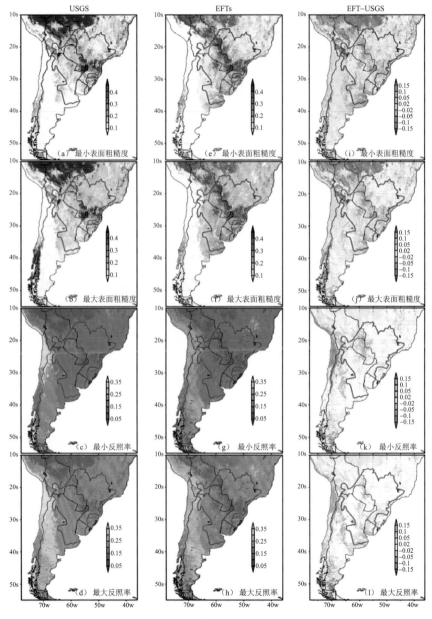

图 16-2

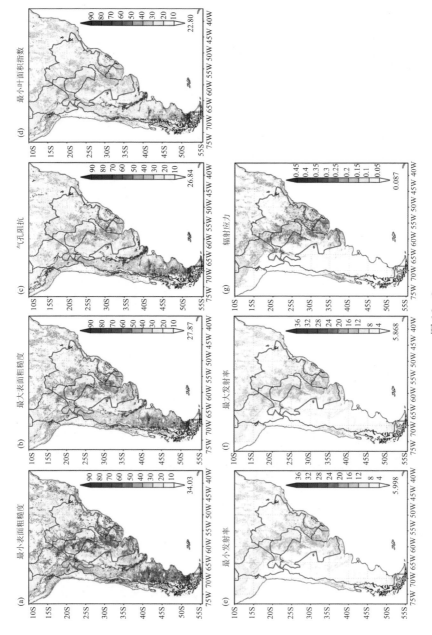

图 16-3

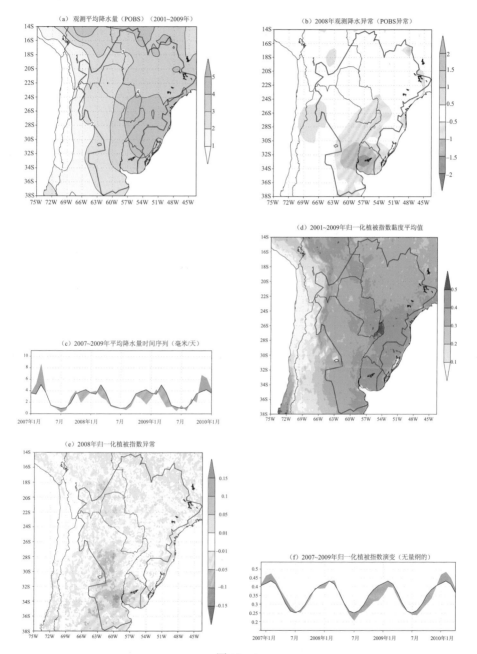

图 16-4

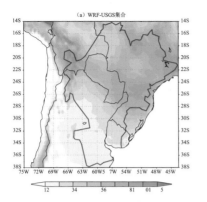
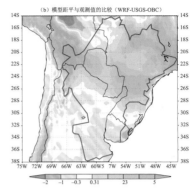
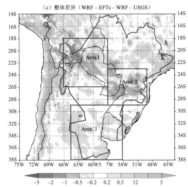

图 16-5

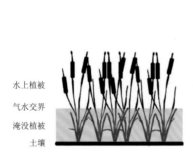
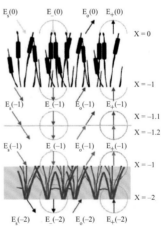

图 17-1

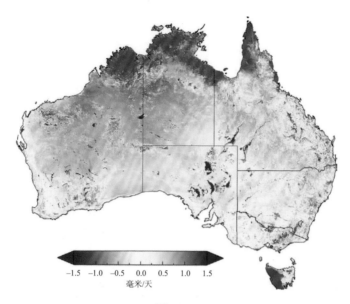

图 18-1

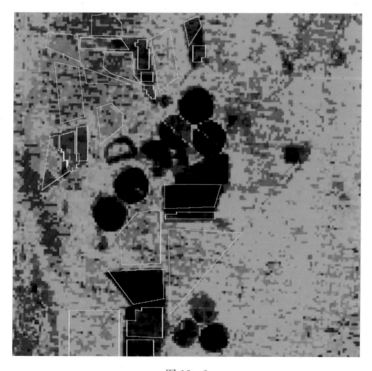

图 18-2

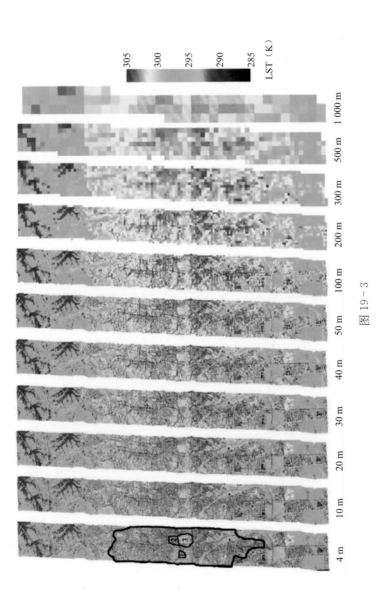

图 19-3